Ergebnisse der Mathematik
und ihrer Grenzgebiete

Volume 53

3. Folge

A Series of Modern Surveys
in Mathematics

Vítor Araújo · Maria José Pacifico

Three-Dimensional Flows

Springer

Vítor Araújo
Maria José Pacifico
Instituto de Matemática
Universidade Federal do Rio de Janeiro
21.945-970 Rio de Janeiro
Brazil
vitor.araujo@im.ufrj.br
pacifico@im.ufrj.br

ISBN 978-3-642-11413-7 e-ISBN 978-3-642-11414-4
DOI 10.1007/978-3-642-11414-4
Springer Heidelberg Dordrecht London New York

Ergebnisse der Mathematik und ihrer Grenzgebiete. 3. Folge / A Series of Modern Surveys in Mathematics ISSN 0071-1136

Library of Congress Control Number: 2010929946

Mathematics Subject Classification (2010): 34Cxx, 34Dxx, 37C10, 37C20, 37C40, 37C70, 37D05, 37D20, 37D25, 37D30, 37D45, 37D50

Cover design: VTEX, Vilnius

Printed on acid-free paper

Springer is part of Springer Science+Business Media (www.springer.com)

Não sou nada.

Nunca serei nada.

Não posso querer ser nada.

À parte isso, tenho em mim todos os sonhos do mundo.

Álvaro de Campos. Tabacaria.

Vítor dedica a Simone, Andreia, Manuela e Domingos.

Maria José Pacifico dedica a Maria, Laura e Ricardo.

Foreword

E.N. Lorenz's twelve pages note *Deterministic nonperiodic flow*, published in 1963 in the Journal of the Atmospheric Sciences [139], surely ranks among the most influential inputs from the experimental sciences in the history of 20th century Mathematics. While investigating the validity of linear regression models for weather forecast, Lorenz hit upon the observation that typical trajectories, even in very simple models, are unstable, or sensitive to small changes in the initial conditions.

This observation had deep philosophical and practical implications. From a mathematician's perspective, Lorenz's paper set the challenge of describing and explaining on rigorous grounds the sensitivity phenomenon, and its stability under small modifications of the dynamical model.

By that time, starting in the early sixties, the mathematical theory of hyperbolic dynamical systems was being developed by Steven Smale and his collaborators and students in the West, and by the Moscow school (Anosov, Arnold, Sinai and coworkers) in the Soviet Union. Hyperbolicity theory does provide a conceptual framework for understanding stable sensitive behavior, and has rapidly become a main paradigm in dynamical systems theory. However, the geometric models proposed by Afraimovich, Bykov, Shil'nikov [2, 4] and Guckenheimer, Williams [98, 274] suggested that the Lorenz system is actually *not* hyperbolic.

This state of perplexity lasted until the renewal of the theory of partial hyperbolicity in the mid nineties. A number of breakthroughs were obtained that led to a very complete theory of stable (or robust) behavior for flows in three-dimensional spaces, including both hyperbolic systems and Lorenz-type systems. It is this theory that the authors present in this book. Their text provides a much needed coherent presentation of one of the fastest developing subjects in recent mathematical research. Theirs was not an easy task: the material is very rich and widespread in a large number of papers; it is also quite recent, so that assessing the relevance of different results may be tricky.

A particularly successful compromise between all these aspects was achieved, and I am sure this book will be a useful reference both for the expert working in the field and the student looking for an introduction to the subject. I also expect it to play a significant role towards the extension of this theory to flows in arbitrary dimensions, which is currently under way.

Enjoy your reading!

Rio de Janeiro

Marcelo Viana

Preface

In this book we present the elements of a general theory for flows on three-dimensional compact boundaryless manifolds, encompassing flows with equilibria accumulated by regular orbits.

The main motivation for the development of this theory was the Lorenz system of equations whose numerical solution suggested the existence of a robust chaotic attractor with a singularity coexisting with regular orbits accumulating on it.

More than three decades passed before the existence of the Lorenz attractor was rigorously established by Warwick Tucker with a computer-assisted proof in the year 2000.

The difficulty in treating this kind of system is both conceptual and numerical. On the one hand, the presence of the singularity accumulated by regular orbits prevents this invariant set from being uniformly hyperbolic. On the other hand, solutions slow down as they pass near the saddle equilibria and so numerical integration errors accumulate without bound.

Trying to address this problem, a successful approach was developed by Afraimovich-Bykov-Shil'nikov and Guckenheimer-Williams independently, leading to the construction of a geometrical model displaying the main features of the behavior of the solutions of the Lorenz system of equations.

In the 1990's a breakthrough was obtained by Carlos Morales, Enrique Pujals and Maria José Pacifico following very original ideas developed by Ricardo Mañé during the proof of the C^1-stability conjecture, providing a characterization of robustly transitive attractors for three-dimensional flows, of which the Lorenz attractor is an example.

This characterization placed this class of attractors within the realm of a weak form of hyperbolicity: they are partially hyperbolic invariant sets with volume expanding central direction (or volume hyperbolic sets). Moreover robustly transitive attractors without singularities were proved to be uniformly hyperbolic. Thus these results extend the classical uniformly hyperbolic theory for flows with isolated singularities.

Once this was established it is natural to try and understand the dynamical consequences of partial hyperbolicity with central volume expansion. It is well known that uniform hyperbolicity has very precise implications for the dynamics, geometry and statistics of the invariant set. It is important to ascertain which properties are implied by this new weak form of hyperbolicity, known today as *singular-hyperbolicity*.

Significant advances at the topological and ergodic level where recently obtained through the work of many authors which deserve a systematic presentation.

This is the main motivation for writing these notes. We hope to provide a global perspective of this theory and make it easier for the reader to approach the growing literature on this subject.

There have been several books and monographs on the subject of Dynamical Systems. But there are many distinct aspects which together make this book unique.

First of all, this book treats mostly continuous time dynamical systems, instead of its discrete counterpart which is exhaustively treated in some of the other texts.

Second, this book treats all the subjects from a mathematical perspective with proofs of most of the results included. Some of the proofs are done in a different way than those in the original papers because, once the theory is organized, it is possible to simplify many of the original proofs. We also extend many of the results about singular-hyperbolicity to higher dimensional flows, adding some new and natural hypotheses on the flow. The proofs about such extensions are also included.

Third, this book is meant to be an advanced graduate textbook and not just a reference book or monograph on the subject. This aspect is reflected in the way the cover material is presented, with careful and complete proofs, and precise references to any topic in the book.

Finally, there is not enough room (or time!) to cover all the topics in an advanced graduate course. This means that the book is not exhaustive: the main topics still constitute a very active area of research, but the book tries to treat the core concepts thoroughly and others enough so the reader will be prepared to read further on the subject and, we hope, also be prepared to contribute with new results on this theory.

It is a pleasure to thank our co-authors Carlos Morales and Enrique Pujals who made definitive contributions and helped build the theory of singular-hyperbolicity. We also thank Ivan Aguilar for providing the figures of his MSc. thesis at UFRJ, and Serafin Bautista and Alfonso Artigue for having communicated to us some arguments which we include in this text. We are indebted to several PhD. students at IM-UFRJ who read previous versions of this text and pointed out to us several places where the presentation should be improved, among them Laura Senos, Regis Castijos and João Reis.

We extend our acknowledgment to Marcelo Viana who, besides being our co-author, encouraged us to write this text.

We are thankful to DMP-FCUP, IM-UFRJ and IMPA for providing us with the necessary time and conditions to write this text. We also profited from partial financial support from CNPq, FAPERJ, PRONEX-Dynamical Systems (Brazil) and CMUP-FCT (Portugal). Finally we thank the scientific committee of the XXVI Brazilian Mathematical Colloquium for selecting our proposal for an earlier version of this text.

Rio de Janeiro Vítor Araújo and Maria José Pacifico

Contents

List of Figures

Chapter 1
Introduction

In this book we present the elements of a general theory for flows on three-dimensional compact manifolds, encompassing flows with equilibria accumulated by regular trajectories.

The main motivation for the development of this theory was the Lorenz system of ordinary differential equations whose numerical solution suggested the existence of a robust chaotic attractor with an equilibrium coexisting with regular trajectories accumulating on it.

An attractor for us is the set of points whose forward trajectories remain inside a bounded region of space forever and such that all nearby trajectories converge to it. To avoid trivial cases, we also assume that an attractor is a closed set with a dense trajectory but which does not coincide with this trajectory. This excludes periodic trajectories from our definition – we regard these as trivial attractors when they arise.

This kind of set plays a central role in the study of the long term behavior of systems of differential equations since attractors are the possible locus of convergence of trajectories when time goes to infinity, at least on compact invariant regions – for example, on three-dimensional compact manifolds. Hence to understand the possible types of asymptotic behavior it is necessary to understand the kinds of attractors a given system of differential equations may have.

A relevant feature of attractors is their robustness, that is, they should persist under small perturbations of the given system of equations. These are the kind of sets one should observe when building mathematical models since there are intrinsic errors in all observed data, so the models are usually at best just an approximation of some relevant phenomenon one wishes to study. The features of our models one can reasonably expect to observe in the real world are those which persist under small errors.

Trivial attractors, like equilibria or closed trajectories or combinations of such, are well understood: the asymptotic behavior of nearby trajectories is either an equilibrium or a periodic trajectory, or a finite number of equilibria and trajectories connecting them. These are in fact typical in two-dimensional systems. These attractors are not chaotic because trajectories starting on nearby points tend to remain nearby in the future.

V. Araújo, M.J. Pacifico, *Three-Dimensional Flows*, Ergebnisse der Mathematik und ihrer Grenzgebiete. 3. Folge / A Series of Modern Surveys in Mathematics 53,
DOI 10.1007/978-3-642-11414-4_1,

For three-dimensional systems the Lorenz example exhibited a new kind of feature: *sensitiveness to initial conditions*, or chaos. Small initial differences are amplified as time passes, causing two trajectories originally coming from practically indistinguishable points to behave in a completely different manner after a short while. Long term predictions based on such models are unfeasible since it is not possible to both specify initial conditions with arbitrary accuracy and numerically calculate with arbitrary precision.

Remarkably, the Lorenz system of equations is *both chaotic and robust*. More than three decades passed before the existence of the Lorenz attractor was rigorously established by Warwick Tucker in the year 2000.

The difficulty in treating this kind of systems is both conceptual and numerical. On the one hand, the presence of an equilibrium accumulated by regular orbits prevents this invariant set to be uniformly hyperbolic. On the other hand, solutions slow down as they pass near the saddle equilibria and so numerical integration errors accumulate without bound.

Hyperbolicity means that at each point there are two complementary directions to the flow direction, one uniformly contracted by the tangent map of the flow generated by the system of ordinary differential equations, and the other uniformly expanded. The invariant foliations associated to these sets are relevant objects of study which we do not deal with in this book. This kind of geometrical notion was introduced in the 1960's by Stephen Smale in a (very successful) attempt to characterize structural stability of a system.

Stability has two main distinct meanings in the theory of dynamical systems. On the one hand, when referring to trajectories, one says that a trajectory is stable if nearby trajectories converge to it as time increases. On the other hand, it can refer to a system as a whole: in this case it means that the global behavior is not affected if the laws of evolution are slightly modified. A system is *structurally stable* if small changes to it leave the whole orbit structure unchanged up to a global continuous change of coordinates. It is a celebrated result that the hyperbolic systems are essentially the structurally stable ones.

An equilibrium accumulated by regular orbits cannot belong to a hyperbolic set, because the dimensions of the expanding and contracting directions have a discontinuous jump at the equilibrium, since the flow direction vanishes at an equilibrium, while the notion of hyperbolicity for smooth systems demands continuity of these directions. Hence there was no mathematical theory able to understand robust objects like the attractor of the Lorenz system until recently.

Trying to address this problem, a very successful approach was developed by Afraimovich-Bykov-Shil'nikov and Guckenheimer-Williams independently, leading to the construction of a geometrical model displaying the main features of the behavior of the solutions of the Lorenz system of equations.

In the 1990's a breakthrough was obtained by Morales, Pujals and Pacifico following very original ideas developed by Mañé during the proof of the C^1-stability conjecture, providing a characterization of robustly transitive attractors for three-dimensional flows, of which the Lorenz attractor is an example.

This characterization placed this class of attractors within the realm of a weak form of hyperbolicity: they are partially hyperbolic invariant sets with volume expanding central direction. Moreover robustly transitive attractors without equilibria were proved to be hyperbolic. Thus these results extend the classical hyperbolic theory for flows with isolated equilibria.

An invariant set for a three-dimensional flow is *partially hyperbolic* if there are two complementary directions at each point. One of them, referred to as *central direction*, is two-dimensional and necessarily contains the flow direction. The other one is one-dimensional and contracted by the tangent map of the flow. Moreover every possible contraction along the central direction is weaker than the contraction along the complementary one-dimensional direction. In this way the presence of equilibria accumulated by regular trajectories is not an obstruction for partial hyperbolicity.

Once this was established it is natural to try and understand the dynamical consequences of partial hyperbolicity with central volume expansion. It is well known that hyperbolicity has very precise implications for the dynamics, geometry and statistics of the invariant set. It is important to ascertain which properties are implied by this new weak form of hyperbolicity, known today as *singular-hyperbolicity*.

Significant advances at the topological and ergodic level where recently obtained through the work of many authors which deserve a systematic presentation. Thus we will focus on flows with equilibria accumulated by regular trajectories.

This is the main motivation for writing these notes. We hope to provide a global perspective of this theory and make it easier for the reader to approach the growing literature on this subject.

1.1 Organization of the Text

We start with an overview of the main results of uniformly hyperbolic dynamical systems to be used throughout the rest of the text, both from the geometrical viewpoint and the measure-theoretical or ergodic point of view, in Chap. 2. We also mention some by now standard generic properties of flows in the C^1 topology, such as the Kupka-Smale vector fields (which are in fact C^r generic for every $r \geq 1$), Pugh's Closing Lemma and Hayashi's Connecting Lemma. We restrict ourselves to the results which will actually be used in the course of the proofs of the main results of the text.

In Chap. 3, we describe the construction of the most simple non-trivial examples of singular-hyperbolic sets: the singular horseshoe of Labarca-Pacifico, and the geometric Lorenz attractor of Afraimovich-Bykov-Shil'nikov and Guckenheimer-Williams.

In Chap. 4, we present the proof that every robustly transitive vector field X in a 3-manifold is an Anosov vector field, i.e., a robustly transitive vector field X on three-dimensional manifold is globally hyperbolic. This is a preliminary result for the main characterization of robustly transitive sets, presented in Chap. 5, and many arguments detailed here are explored in the rest of the book.

Next, in Chap. 5, we characterize robustly transitive sets with singularities as partially hyperbolic attractors with volume expanding central direction, either for the original flow, or for the time reversed flow. This naturally leads to the notion of *singular-hyperbolic set*: a compact partially hyperbolic invariant subset with volume expanding central direction or, more concisely, a volume hyperbolic invariant set. We present also an extension of this notion to higher dimensions, the notion of sectionally expanding attractors.

In Chap. 6, we study some consequences of singular-hyperbolicity, showing that a singular-hyperbolic attractor is a homoclinic class, and provide converse results giving sufficient conditions for a singular-hyperbolic attractor to be robust.

We construct, in Chap. 7, a physical measure with non-zero Lyapunov exponents and positive entropy for singular-hyperbolic attractors. This is another non-trivial consequence of singular-hyperbolicity. Before this construction we dwell on expansiveness and chaotic behavior, showing that every singular-hyperbolic attractor is robustly expansive and robustly chaotic and that, conversely, under mild conditions, a robustly chaotic attractor must be singular-hyperbolic. The existence of a physical measure with non-zero Lyapunov exponents is also another aspect of chaotic behavior.

Chapter 8 is dedicated to relations between singular-hyperbolicity and volume. Here we show, using only the assumptions of singular-hyperbolicity, that singular-hyperbolic attractors either have zero volume or else the flow is globally hyperbolic, that is, an Anosov flow (without singularities). We present a similar result for sectionally expanding attractors. Among incompressible flows on three-dimensional manifolds, a similar result is obtained for invariant sets satisfying the weaker condition of dominated splitting for the linear Poincaré flow.

In Chap. 9 we attempt a description of the Omega-limit set for C^1 generic flows: either the limit set contains an infinite collection of sinks or sources; or is a finite union of basic pieces, either uniformly hyperbolic transitive isolated sets, or singular-hyperbolic attractors or repellers. Then we turn to conservative or incompressible flows on three-dimensional manifolds and present a recent proof of a global dynamical dichotomy: C^1 *generically either the flow is Anosov or else the Lyapunov exponents are zero Lebesgue almost everywhere*.

In an attempt to provide a broader view of the dynamics of flows on three-dimensional manifolds, we close the text briefly mentioning in Chap. 10 many other related results: the *contracting Lorenz attractor* introduced by Rovella, singular cycles exhibiting singular-hyperbolic and/or contracting Lorenz attractors in its unfolding, other attractors resembling the Lorenz attractor, decay of correlations, large deviations, quantitative recurrence results for singular-attractors, and other generic results for conservative flows on three-dimensional manifolds. Among the brief presentation of results we state some conjectures that we feel are achievable goals in the near future.

Chapter 2
Preliminary Definitions and Results

In this book we will consider a boundaryless compact finite dimensional manifold M of dimensions 1 to 3 and study the dynamics of the flow associated to a given smooth vector field X on M from the topological and measure-theoretic or ergodic point of view.

We fix on M some Riemannian metric which induces a distance dist on M and naturally defines an associated Riemannian volume form Leb which we call *Lebesgue measure* or simply *volume*, and always take Leb to be normalized: $\text{Leb}(M) = 1$. For any subset A of M we denote by $\overline{A}$ the (topological) closure of A.

We always assume that a C^r vector field X on M is given, $r \geq 1$, and consider the associated global *flow* $(X^t)_{t\in\mathbb{R}}$ (since X is defined on the whole of M, which is compact, X is bounded and X^t is defined for every $t \in \mathbb{R}$). Recall that the flow $(X^t)_{t\in\mathbb{R}}$ is a family of C^r diffeomorphisms satisfying the following properties:

1. $X^0 = Id : M \to M$ is the identity map of M;
2. $X^{t+s} = X^t \circ X^s$ for all $t, s \in \mathbb{R}$,

and it is *generated by the vector field* X if

3. $\frac{d}{dt} X^t(q)\big|_{t=t_0} = X(X_{t_0}(q))$ for all $q \in M$ and $t_0 \in \mathbb{R}$.

Note that reciprocally a given C^{r+1} flow $(X^t)_{t\in\mathbb{R}}$ determines a unique vector C^r field X whose associated flow is precisely $(X^t)_{t\in\mathbb{R}}$.

In what follows we denote by $\mathfrak{X}^r(M)$ the vector space of all C^r vector fields on M endowed with the C^r topology and by $\mathscr{F}^r(M)$ the space of all flows on M also with the C^r topology. Many times we usually denote the flow $(X^t)_{t\in\mathbb{R}}$ by simply X^t. For details on these topologies the reader is advised to consult standard references on Differential Equations [111] and/or Dynamical Systems [190].

Given $X \in \mathfrak{X}^r(M)$ and $q \in M$, an orbit segment $\{X^t(q); a \leq t \leq b\}$ is denoted by $X^{[a,b]}(q)$. We denote by DX^t *the derivative of* X^t *with respect to the ambient variable* q *and when convenient we set* $D_q X^t = DX^t(q)$. Analogously, DX *is the derivative of the vector field* X with respect to the ambient variable q, and when convenient we write $D_q X$ for the derivative DX at q, also denoted by $DX(q)$.

V. Araújo, M.J. Pacifico, *Three-Dimensional Flows*, Ergebnisse der Mathematik und ihrer Grenzgebiete. 3. Folge / A Series of Modern Surveys in Mathematics 53,
DOI 10.1007/978-3-642-11414-4_2, © Springer-Verlag Berlin Heidelberg 2010

2.1 Fundamental Notions and Definitions

2.1.1 Critical Elements, Non-wandering Points, Stable and Unstable Sets

An *equilibrium* or *singularity* for X is a point $\sigma \in M$ such that $X^t(\sigma) = \sigma$ for all $t \in \mathbb{R}$, i.e. a fixed point of all the flow maps, which corresponds to a zero of the associated vector field X: $X(\sigma) = 0$. We denote by $S(X)$ the set of singularities (zeroes) of the vector field X. Every point $p \in M$ which is not a singularity, that is p satisfies $X(p) \neq 0$, is a *regular* point for X.

An *orbit* of X is a set $\mathcal{O}(q) = \mathcal{O}_X(q) = \{X^t(q) : t \in \mathbb{R}\}$ for some $q \in M$. Hence $\sigma \in M$ is a singularity of X if, and only if, $\mathcal{O}_X(\sigma) = \{\sigma\}$. A *periodic orbit* of X is an orbit $\mathcal{O} = \mathcal{O}_X(p)$ such that $X^T(p) = p$ for some minimal $T > 0$ (equivalently $\mathcal{O}_X(p)$ is compact and $\mathcal{O}_X(p) \neq \{p\}$). We denote by $\mathrm{Per}(X)$ the set of all periodic orbits of X.

A *critical element* of a given vector field X is either a singularity or a periodic orbit. The set $C(X) = S(X) \cup \mathrm{Per}(X)$ is the set of *critical elements* of X.

We say that $p \in M$ is *non-wandering* for X if for every $T > 0$ and every neighborhood U of p there is $t > T$ such that $X^t(U) \cap U \neq \emptyset$. The set of non-wandering points of X is denoted by $\Omega(X)$. If $q \in M$, we define $\omega_X(q)$ as the set of accumulation points of the positive orbit $\{X^t(q) : t \geq 0\}$ of q. We also define $\alpha_X(q) = \omega_{-X}$, where $-X$ is the time reversed vector field X, corresponding to the set of accumulation points of the negative orbit of q. It is immediate that $\omega_X(q) \cup \alpha_X(q) \subset \Omega(X)$ for every $q \in M$.

A subset Λ of M is *invariant* for X (or X-invariant) if $X^t(\Lambda) = \Lambda$, $\forall t \in \mathbb{R}$. We note that $\omega_X(q)$, $\alpha_X(q)$ and $\Omega(X)$ are X-invariant. For every compact invariant set Λ of X we define the *stable set* of Λ

$$W^s_X(\Lambda) = \{q \in M : \omega_X(q) \subset \Lambda\},$$

and also its *unstable set*

$$W^u_X(\Lambda) = \{q \in M : \alpha_X(q) \subset \Lambda\}.$$

2.1.2 Limit Sets, Transitivity, Attractors and Repellers

We say that a compact X^t-invariant set Λ is *isolated* (or *maximal*) if there exists a neighborhood U of Λ such that $\Lambda = \cap_{t\in\mathbb{R}} X^t(U) = \cap_{t\in\mathbb{R}} \overline{X^t(U)}$. A compact invariant set Λ is *transitive* if $\Lambda = \omega_X(q)$ for some $q \in \Lambda$, and *attracting* if $\Lambda_X(U) = \cap_{t\geq 0} X^t(U)$ equals Λ for some neighborhood U of Λ satisfying $\overline{X^t(U)} \subset U$, for all $t > 0$. In this case the neighborhood U is called an *isolating* neighborhood of Λ. Note that $\Lambda_X(U)$ is in general different from $\cap_{t\in\mathbb{R}} X^t(U)$, but for an attracting set the extra condition $\overline{X^t(U)} \subset U$ for all $t > 0$ ensures that $X^{-t}(U) \supset U$ and so

$$\Lambda \supset \bigcap_{t\in\mathbb{R}} X^t(U) = \bigcap_{t\leq 0} X^t(U) \cap \bigcap_{t>0} X^t(U) \supset U \cap \Lambda = \Lambda$$

thus every attracting set is isolated. An *attractor* of X is a transitive attracting set of X and a *repeller* is an attractor for $-X$. We say that Λ is a *proper* attractor or repeller if $\emptyset \neq \Lambda \neq M$.

The *limit set* $L(X)$ is the closure of $\cup_{x\in M}\alpha_X(x)\cup\omega_X(x)$. Clearly $L(X)\subset\Omega(X)$. Using these notions we have the following simple and basic

Lemma 2.1 *For any flow X the limit set $L(X)$ can neither be a proper attractor nor a proper repeller.*

Proof Suppose $L(X)$ is a proper attractor with isolating open neighborhood U (and $U \neq M$). Let $z \in U$. Then $\alpha(z) \in L(X) \subset U$ and so $X^{-t_n}(z) \in U$ for a sequence $t_n \to +\infty$, that is $z \in X^{t_n}(U)$ for all n. But since $X^{t_n-t}(U) \subset U$ for $0 < t < t_n$ by definition of U, we have that $z \in X^{t_n}(U) \subset X^t(U)$ (recall that each X^t is an invertible map) for all $0 < t < t_n$, and so $z \in X^t(U)$ for all $t > 0$. We conclude that $z \in L(X)$. Thus $L(X) \supset U$ and $L(X)$ is simultaneously open and closed, hence it cannot be a proper subset of the connected manifold M. The proper repeller case is similar. □

The following relation between transitivity as we have defined it, and topological transitivity is very useful and we include the proof here for completeness.

Lemma 2.2 *For a homeomorphism h of a compact metric space Λ, h has a dense forward orbit, that is*

$$\bigcap_{\ell\geq 1}\overline{\{h^n(x_0) : n \geq \ell\}} = \Lambda \quad \textit{for some } x_0 \in \Lambda$$

if, and only if, $\Omega(h) = \Lambda$ and h admits a dense full orbit, i.e. for some $z \in \Lambda$

$$\bigcap_{\ell\geq 1}\overline{\{h^n(z) : n \geq \ell\}} = \Lambda = \bigcap_{\ell\geq 1}\overline{\{h^n(z) : n \leq -\ell\}}.$$

In particular, a transitive compact invariant set Λ for a flow X^t admits a point w whose *full orbit* is dense: $\omega_X(w) = \Lambda = \alpha_X(w)$.

Proof If h has a dense forward orbit, we check first that every point of Λ is non-wandering. Arguing by contradiction, assume there exists some non-empty open subset U such that the family $\{h^k(U) : k \in \mathbb{Z}\}$ is pairwise disjoint. For some n_0 we have $h^{n_0}(x_0) \in U$, thus $h^{n_0+n}(x_0) \in h^n(U)$ for each $n \geq 0$ and so only the orbit segment $\{x_0, h(x_0), \ldots, h^{n_0-1}(x_0)\}$ can belong to $\cup_{i\geq 1}h^{-i}(U)$. But this contradicts the assumption that the forward orbit of x_0 is dense. To obtain a dense full orbit, we note that given a pair U, V of open sets in Λ we can always find $n > m$ such that $h^n(x_0) \in U, h^m(x_0) \in V$, so that $h^{n-m}(V) \cap U \neq \emptyset$, and likewise exchanging U and V. Hence if $(U_i)_{i\in\mathbb{N}}$ is a countable basis for the topology of Λ, then we have that both $\cup_{j\geq\ell}h^j(U_i)$ and $\cup_{j\leq-\ell}h^j(U_i)$ are open and dense for each $i, \ell \geq 1$. Hence we have that $\cap_{i\geq 1}\cap_{\ell\geq 1}\cup_{j\geq\ell}h^j(U_i)$ is both non-empty (it is a countable intersection of open and dense subsets and Λ is a Baire space, see e.g. [126]) and formed by points whose omega-limit is dense. Analogously $\cap_{i\ge 1}\cap_{\ell\ge 1}\cup_{j\le -\ell} h^j(U_i)$ is non-empty and formed by points whose alpha-limit is dense. The point z is any point in the (dense) intersection of these last two dense sets. The reciprocal is trivial. □

We state and prove here some simple but very useful results on attracting set which are well known, see for instance Conley [73].

Lemma 2.3 *Let Λ be an isolated set of $X \in \mathfrak{X}^r(M)$, $r \ge 0$. Then for every isolating block U of Λ and every $\varepsilon > 0$ there is a neighborhood $\mathscr{U}$ of X in $\mathfrak{X}^r(M)$ such that $\Lambda_Y(U) \subset B(\Lambda,\varepsilon)$ and $\Lambda \subset B(\Lambda_Y(U),\varepsilon)$ for all $Y \in \mathscr{U}$.*

We recall that the Hausdorff distance between compact subsets $K, L \in \mathscr{K}$ of a metric space is given by (see e.g. [85])

$$d_H(K,L) = \inf\{r>0 : K \subset B(L,r) \quad \text{and} \quad L \subset B(K,r)\}.$$

So the previous lemma ensures that the map $(\mathfrak{X}^1(M), C^1) \to (\mathscr{K}, d_H)$, $Y \mapsto \Lambda_Y(U)$ is continuous.

Proof We have by assumption $\Lambda = \Lambda_X(U) = \cap_{t\in\mathbb{R}} X^t(U) = \cap_{t\in\mathbb{R}} X^t(\overline{U})$. Given $\varepsilon > 0$ there is a big enough $T > 0$ such that

$$\Lambda_X^T(U) := \bigcap_{-T\le t\le T} X^t(\overline{U}) \subset B\left(\Lambda, \frac{\varepsilon}{4}\right).$$

Then using the continuous dependence of the flow with the vector field and the compactness of $\overline{U}$, there exists a neighborhood $\mathscr{U}$ of X in $\mathfrak{X}^r(M)$ such that

$$\Lambda_X^T(U) \subset B\left(\Lambda_Y^T(U), \frac{\varepsilon}{4}\right) \quad \text{and} \quad \Lambda_Y^T(U) \subset B\left(\Lambda_X^T(U), \frac{\varepsilon}{4}\right)$$

for each $Y \in \mathscr{U}$. Hence we deduce that, on the one hand

$$\Lambda_Y(U) \subset \Lambda_Y^T(U) \subset B\left(\Lambda_X^T(U), \frac{\varepsilon}{4}\right) \subset B\left(\Lambda, \frac{\varepsilon}{4} + \frac{\varepsilon}{4}\right) \subset B(\Lambda,\varepsilon);$$

and, on the other hand

$$\Lambda \subset \Lambda_X^T(U) \subset B\left(\Lambda_Y^T(U), \frac{\varepsilon}{4}\right) \subset B\left(\Lambda_Y(U), \frac{\varepsilon}{4} + 3\frac{\varepsilon}{4}\right) = B(\Lambda_Y(U),\varepsilon);$$

because

$$\Lambda_Y(U) \subset B\left(\Lambda, 2\frac{\varepsilon}{4}\right) \subset B\left(\Lambda_X^T(U), 2\frac{\varepsilon}{4}\right) \subset B\left(\Lambda_Y^T(U), 2\frac{\varepsilon}{4} + \frac{\varepsilon}{4}\right),$$

for all $Y \in \mathscr{U}$. □

We remark that if $X^t(U) \subset U$ for all $t > 0$, since the flow X^t is a family of diffeomorphism we deduce that $X^{t-s}(U) \subset U$ for all $s < t$, or $X^t(U) \subset X^s(U)$. Thus the family $(X^t(U))_{t>0}$ is a nested family of sets and so

$$\Lambda_X(U) = \bigcap_{t>0} X^t(U) = \lim_{t\to+\infty} X^t(U)$$

that is, given any neighborhood V of Λ there exists $T = T(V)$ such that $X^t(U) \subset V$ for all $t > T$.

Lemma 2.4 *If Λ is an attracting set and a repelling set of $X \in \mathfrak{X}^1(M)$, then $\Lambda = M$.*

Proof Suppose that Λ is an attracting set and a repelling set of X. Then there are neighborhoods V_1 and V_2 of Λ satisfying $X^t(V_1) \subset V_1$, $X^{-t}(V_2) \subset V_2$ for every $t \geq 0$,

$$\Lambda = \bigcap_{t\geq 0} X^t(V_1) \quad \text{and} \quad \Lambda = \bigcap_{t\geq 0} X^{-t}(V_2).$$

Define $U_1 = \text{int}(V_1)$ and $U_2 = \text{int}(V_2)$. Clearly $X^t(U_1) \subset U_1$ and $X^{-t}(U_2) \subset U_2$ for all $t \geq 0$, since X^t is a diffeomorphism. As U_2 is open and contains Λ, the first equality implies that there is $t_2 > 0$ such that $X^{t_2}(V_1) \subset U_2$. As $X^{t_2}(U_1) \subset X^{t_2}(V_1)$ it follows that $U_1 \subset X^{-t_2}(U_2) \subset U_2$ proving

$$U_1 \subset U_2.$$

Similarly, as U_2 is open and contains Λ, the second equality implies that there is $t_1 > 0$ such that $X^{-t_1}(V_2) \subset U_1$. As $X^{-t_1}(U_2) \subset X^{-t_1}(V_2)$ it follows that $U_2 \subset X^{t_1}(U_1) \subset U_1$ proving

$$U_2 \subset U_1.$$

Thus, $U_1 = U_2$. From this we obtain $X^t(U_1) = U_1$ for all $t \geq 0$ proving $\Lambda = U_1$. As Λ is compact by assumption we conclude that Λ is open and closed. As M is connected and Λ is not empty we obtain that $\Lambda = M$ as desired. □

The lemma below gives a sufficient condition for an isolated set to be attracting.

Lemma 2.5 *Let Λ be an isolated set of $X \in \mathfrak{X}^1(M)$. If there are an isolating block U of Λ and an open set W containing Λ such that $X^t(W) \subset U$ for every $t \geq 0$, then Λ is an attracting set of X.*

Proof Let Λ and X be as in the statement. To prove that Λ is attracting we have to find a neighborhood V of Λ such that $X^t(V) \subset V$ for all $t > 0$ and

$$\Lambda = \cap_{t\geq 0} X^t(V). \tag{2.1}$$

To construct V we let W be the open set in the statement of the lemma and define $V = \cup_{t>0} X^t(W)$. Clearly V is a neighborhood of Λ satisfying $X^t(V) \subset V$ for each $t > 0$.

We claim that V satisfies (2.1). Indeed, as $X^t(W) \subset U$ for every $t > 0$ we have that $V \subset U$ and so $\cap_{t \in IR} X^t(V) \subset \Lambda$ because U is an isolating block of Λ. But $V \subset X^{-t}(V)$ for every $t \geq 0$ since V is forward invariant. So $V \subset \cap_{t \leq 0} X^t(V)$ and from this we have

$$\bigcap_{t \geq 0} X^t(V) \subset V \cap \bigcap_{t>0} X^t(V)$$
$$= \bigcap_{t \leq 0} X^t(V) \cap \bigcap_{t>0} X^t(V) = \bigcap_{t \in \mathbb{R}} X^t(V).$$

Thus, $\cap_{t \geq 0} X^t(V) \subset \Lambda$. Now, as $\Lambda \subset V$ and Λ is invariant, we have $\Lambda \subset X^t(V)$ for every $t \geq 0$. Then $\Lambda \subset \cap_{t \geq 0} X^t(V)$, proving (2.1). □

2.1.3 Hyperbolic Critical Elements

A *sink* of X is a singularity of X which is also an attractor of X, it is a trivial attractor of X. A *source* of X is a trivial repeller of X, i.e. a singularity which is a attractor for $-X$.

A *singularity* σ *is hyperbolic* if the eigenvalues of $DX(\sigma)$, the derivative of the vector field at σ, have a real part different from zero. In particular sinks and sources are hyperbolic singularities, where all the eigenvalues of the former have negative real part and those of the latter have positive real part.

A *periodic orbit* $\mathcal{O}_X(p)$ *of* X *is hyperbolic* if the eigenvalues of $DX^T(p) : T_pM \to T_pM$, the derivative of the diffeomorphism X^T, where $T > 0$ is the period of p, are all different from 1. In Sect. 2.3 we will define hyperbolicity in a geometric way.

When a critical element is hyperbolic, then its stable and unstable sets have the structure of an embedded manifold (a consequence of the Stable Manifold Theorem, see Sect. 2.3), and are called *stable* and *unstable manifolds*.

2.1.4 Topological Equivalence, Structural Stability

Given two vector fields $X, Y \in \mathfrak{X}^r(M), r \geq 1$, we say that X and Y are *topologically equivalent* if there exists a homeomorphism $h : M \to M$ taking orbits to orbits and preserving the time orientation, that is

- $h(\mathcal{O}_X(p)) = \mathcal{O}_Y(h(p))$ for all $p \in M$, and
- for all $p \in M$ and $\varepsilon > 0$ there exists $\delta > 0$ such that for $t \in (0, \delta)$ there is $s \in (0, \varepsilon)$ satisfying $h(X^t(p)) = Y^s(h(p))$.

The map h is then said a *topological equivalence* between X and Y. This is an equivalence relation in $\mathfrak{X}^r(M)$.

We say that $X, Y \in \mathfrak{X}^r(M)$ are *conjugate* if there exists a topological equivalence h between X and Y which preserves the time, i.e. $X^t(h(p)) = h(Y^t(p))$ for all $p \in M$ and $t \in \mathbb{R}$. This is also an equivalence relation on $\mathfrak{X}^r(M)$.

In many respects the behavior of two topologically equivalent vector fields are the same, as the following result shows.

Proposition 2.6 *Let h be a topological equivalence between $X, Y \in \mathfrak{X}^r(M)$. Then*

1. $p \in S(X)$ *if, and only if,* $h(p) \in S(Y)$;
2. $\mathcal{O}_X(p)$ *is closed if, and only if,* $\mathcal{O}_Y(h(p))$ *is closed*;
3. $h(\omega_X(p)) = \omega_Y(h(p))$ *and* $h(\alpha_X(p)) = \alpha_Y(h(p))$.

We say that a vector field $X \in \mathfrak{X}^r(M), r \geq 1$, is C^r*-structurally stable* if there exists a neighborhood $\mathcal{V}$ of X in $\mathfrak{X}^r(M)$ such that every $Y \in \mathcal{V}$ is topologically equivalent to X.

Roughly speaking, a vector field is structurally stable if its qualitative features are robust under small perturbations.

2.2 Low Dimensional Flow Versus Chaotic Behavior

2.2.1 One-Dimensional Flows

The only connected one-dimensional compact boundaryless manifold M is the circle $\mathbb{S}^1$, which we represent by $\mathbb{R}/\mathbb{Z}$ or by the unit interval $I = [0, 1]$ with its endpoints identified $0 \sim 1$.

Let X_0 be one of the two unit vector fields on $\mathbb{S}^1$, i.e., either $X_0 \equiv 1$ or $X_0 \equiv -1$. Then every $X \in \mathfrak{X}^r(\mathbb{S}^1)$ can be written in a unique way as $X(p) = f(p) \cdot X_0(p)$ for $p \in S_1$, where $f : \mathbb{S}^1 \to \mathbb{R}$ is a C^r-function.

It is well known (see for example [136, 177]) that given any compact set $K \subset \mathbb{S}^1$ and $r \geq 1$ there exists $f : \mathbb{S}^1 \to \mathbb{R}$ of class C^r with $f^{-1}(\{0\}) = K$. Thus K is the set of singularities of $X = f \cdot X_0$. Since topological equivalence preserves singularities, we see that there exist at least as many topological equivalence classes of vector fields in $\mathbb{S}^1$ as there are homeomorphism classes of compact subsets of $\mathbb{S}^1$. Hence *the problem of classifying smooth vector fields on $\mathbb{S}^1$ up to topological equivalence is hopeless*, and we need to restrict our attention to a subset of $\mathfrak{X}^r(M)$ which is open and dense, or residual or, at least, dense.

Here by a *residual* subset of the space $\mathfrak{X}^r(M)$ we mean a set $\mathcal{R}$ which contains a countable intersection of open and dense subsets of $\mathfrak{X}^r(M)$: $\mathcal{R} \supset \cap_{n\geq 1}\mathcal{R}_n$ where each $\mathcal{R}_n$ is an open and dense subset of $\mathfrak{X}^r(M)$.

We say that *a generic vector field in $\mathfrak{X}^r(M)$ satisfies a property* (P) if there is a residual subset $\mathcal{R}$ of $\mathfrak{X}^r(M)$ such that (P) holds for every $X \in \mathcal{R}$.

A singularity $\sigma \in S(X)$ is *non-degenerate* if $DX(\sigma) \neq 0$ or $Df(\sigma) \neq 0$ where $X = f \cdot X_0$. It can be a sink $(Df(\sigma) < 0)$ or a source $(Df(\sigma) > 0)$ and in either case a non-degenerate singularity is *isolated*: there exists a neighborhood U of σ in M such that σ is the only zero of $f \mid U$.

Let $\mathcal{G} \subset \mathfrak{X}^r(\mathbb{S}^1)$ be the subset consisting of vector fields whose singularities are all non-degenerate. Since these are isolated there are only finitely many of them. It is not difficult to show that $\mathcal{G}$ is open and dense, that the number of singularities is even and that $X, Y \in \mathcal{G}$ are topologically conjugate if, and only if, the number of singularities is the same (see e.g. [190, 258]). Moreover the elements of $\mathcal{G}$ are precisely the structurally stable vector fields of $\mathbb{S}^1$, that is *generically a smooth vector field on the circle is structurally stable.*

2.2.2 *Two-Dimensional Flows*

Surfaces have a simple enough topology (albeit much more complex that the topology of the circle) to enable one to characterize the non-wandering set of the flow of a vector field. The most representative result in this respect is the Poincaré-Bendixson's Theorem on planar flows or flows on the two-dimensional sphere (essentially the result depends on the Jordan Curve Theorem: any closed simple curve splits the manifold in two connected components, see e.g. [99, 157, 177]).

Theorem 2.7 (Poincaré-Bendixson) *Let $X \in \mathfrak{X}^r(\mathbb{S}^2)$, $r \geq 1$, be a smooth vector field with a finite number of singularities. Let $p \in \mathbb{S}^2$ be given. Then the omega-limit set $\omega_X(p)$ satisfies one of the following*:

1. *$\omega_X(p)$ is a singularity*;
2. *$\omega_X(p)$ is a periodic orbit*;
3. *$\omega_X(p)$ consists of singularities $\sigma_1, \ldots, \sigma_n$ and regular orbits $\gamma \in \omega_X(p)$ such that $\alpha_X(\gamma) = \sigma_i$ and $\omega_X(\gamma) = \sigma_j$ for some $i, j = 1, \ldots, n$.*

The proof of this basic result may be found e.g. in [111, 190]. This answers essentially all the questions concerning the asymptotic dynamics of the solutions of autonomous ordinary differential equations on the plane or on the sphere.

Observe that now hyperbolic singularities σ can be of three types: sink ($DX(\sigma)$ with two eigenvalues with negative real part), source ($DX(\sigma)$ whose eigenvalues have positive real part, see Fig. 2.1) or a saddle ($DX(\sigma)$ with eigenvalues having negative and positive real parts, see Fig. 2.3).

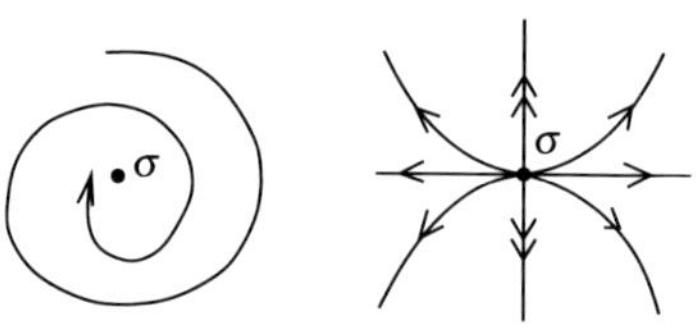

Fig. 2.1 A sink and a source

Historically the characterization of structurally stable vector fields on compact surfaces by Peixoto, based on previous work of Poincaré [206–208] and Andronov and Pontryagin [14], was the origin of the notion of structural stability for Dynamical Systems. In this setting structural stability is still synonymous with a hyperbolic non-wandering set containing finitely many orbits. We now write S for any compact connected two-manifold without boundary.

Theorem 2.8 (Peixoto) *A C^r vector field on a compact surface S is structurally stable if, and only if:*

1. *the number of critical elements is finite and each is hyperbolic;*
2. *there are no orbits connecting saddle points;*
3. *the non-wandering set consists of critical elements alone.*

Moreover if S is orientable, then the set of structurally stable vector fields is open and dense in $\mathfrak{X}^r(S)$.

The proof of this celebrated result can be found in [194, 195] and for a more detailed exposition of this results and sketch of the proof see [97]. The last part of the statement uses a version of Pugh's C^1-Closing Lemma [212, 213], which is a fundamental tool to be used repeatedly in many proofs in this book, see Sect. 2.5.6 for the statement of this result.

The extension of Peixoto's characterization of structural stability for C^r flows, $r \geq 1$, on non-orientable surfaces is known as *Peixoto's Conjecture*, and up until now it has been proved for the projective plane $\mathbb{P}^2$ [190], the Klein bottle $\mathbb{K}^2$ [150] and $\mathbb{L}^2$, the torus with one cross-cap [101].

In an attempt to extend Peixoto's result to higher dimensions, Steve Smale considered in [252] the following type of vector field which preserves the main features of the structurally stable vector fields on surfaces.

We say that a vector field $X \in \mathfrak{X}^r(M)$, $r \geq 1$, is *Morse-Smale* (where now M is a compact manifold of any dimension) if

1. the number of critical elements of X is finite and each one of them is hyperbolic;
2. every stable and unstable manifold of each critical element intersects transversely the unstable or stable manifold of any other critical element;
3. the non-wandering set consists only of the critical elements of X: $\Omega(X) = C(X)$.

Hence *structurally stable vector fields in two-dimensions are Morse-Smale and they are open and dense on the set of all smooth vector fields of an orientable surface.*

There exists a similar notion of Morse-Smale diffeomorphisms on any compact manifold. *Smale's Horseshoe*, presented in [252], showed that Morse-Smale diffeomorphisms are neither dense on the space of all diffeomorphisms, nor the only structurally stable type of diffeomorphisms.

Moreover the singular horseshoe, which we present in Sect. 3.1, is a compact invariant set for a flow similar to a Smale Horseshoe which is structurally stable but non-hyperbolic, defined on manifolds with boundary.

It is well known that Morse-Smale vector fields are structurally stable in any dimension, see e.g. [190, 191]. However early hopes that they might form an open and dense subset of the space of all smooth vector fields or that they are the representatives of structurally stable vectors fields were shattered in higher dimensions, as the following section explains.

2.2.3 Three Dimensional Chaotic Attractors

In 1963 the meteorologist Lorenz published in the Journal of Atmospheric Sciences [139] an example of a parametrized polynomial system of differential equations

$$\begin{aligned} \dot{x} &= a(y-x), & a &= 10 \\ \dot{y} &= rx - y - xz, & r &= 28 \\ \dot{z} &= xy - bz, & b &= 8/3 \end{aligned} \tag{2.2}$$

as a very simplified model for thermal fluid convection, motivated by an attempt to understand the foundations of weather forecast. Later W. Markus [257, Chap. 9] and Lorenz [140] together with other experimental researches showed that the equations of motions of a certain laboratory waterwheel are given by (2.2) with $b = 1$ (see [257, Chap. 9] for the details of the derivation of the equations for this mechanical system) but this restriction does not change the qualitative behavior. Hence equations (2.2) can be deduced directly as a model of a mechanical system, instead of as an approximation to a partial differential equation. This waterwheel was built at MIT (Massachusetts Institute of Technology) in the 1970s and helped to convince the skeptical physicists and engineers of the reality of chaos in concrete systems.

Numerical simulations for an open neighborhood of the chosen parameters suggested that almost all points in phase space tend to a stranger attractor, called the *Lorenz attractor*; see Fig. 2.2. However Lorenz's equations proved to be very resistant to rigorous mathematical analysis, and also presented very serious difficulties to rigorous numerical study.

A very successful approach was taken by Afraimovich, Bykov and Shil'nikov [2–4], and Guckenheimer, Williams [98], independently: they constructed the so-called *geometric Lorenz models* (see Sect. 3.3) for the behavior observed by Lorenz. These models are flows in 3-dimensions for which one can rigorously prove the existence of an attractor that contains an equilibrium point of the flow, together with regular solutions. The accumulation of regular orbits near a singularity prevents such sets to be hyperbolic (see Sect. 2.3). Moreover, for almost every pair of nearby initial conditions, the corresponding solutions move away from each other exponentially fast as they converge to the attractor, that is, the attractor is *sensitive to initial conditions*: this unpredictability is a characteristic of *chaos*. Most remarkably, this attractor is robust: it can not be destroyed by any small perturbation of the original flow.

Another approach was through rigorous numerical analysis. In this way, it could be proved, by [105, 106, 158, 159], that the equations (2.2) exhibit a suspended Smale Horseshoe. In particular, they have infinitely many closed solutions, that is, the attractor contains infinitely many periodic orbits. However, proving the existence

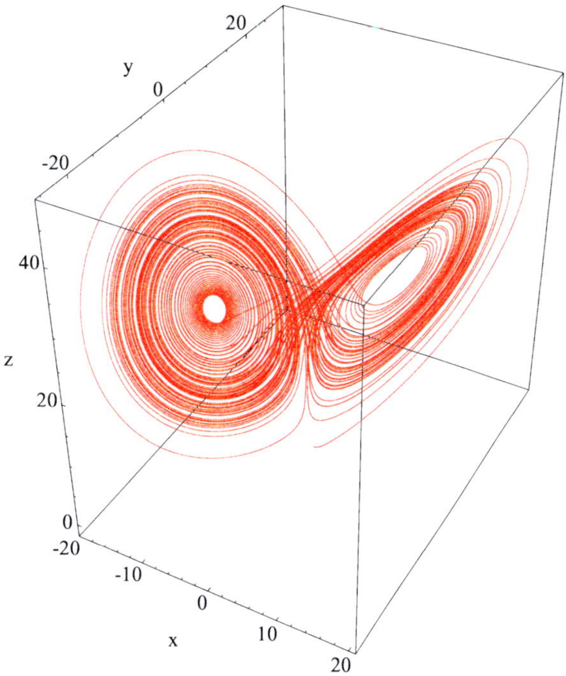

Fig. 2.2 Lorenz strange attractor

of a strange attractor as in the geometric models is an even harder task, because one cannot avoid the main numerical difficulty posed by Lorenz's equations, which arises from the very presence of an equilibrium point: solutions slow down as they pass near the origin, which means unbounded return times and, thus, unbounded integration errors.

As a matter-of-fact, proving that equations (2.2) support a strange attractor was listed by Steve Smale in [253] as one of the several challenging problems for the twenty-first century. In the year 2000 this was finally settled by Tucker who gave a mathematical proof of the existence of the Lorenz attractor, see [260–262].

The algorithm developed by Tucker incorporates two kinds of ingredients: a numerical integrator, based on the theory of interval arithmetic, used to compute good approximations of trajectories of the flow far from the equilibrium point sitting at the origin, together with quantitative results from normal form theory, that make it possible to handle trajectories close to the origin. Moreover in this work it was also proved that the Lorenz attractor is indeed robust: for an open set of parameters (a, r, b) in (2.2) there is an invariant set containing a dense non-periodic orbit and a equilibrium, which is the ω-limit set of the orbits of all nearby points. This shows that the Lorenz attractor is *singular-hyperbolic*, as we show in Chap. 5.

The consequences of the sensitiveness to initial conditions on a (albeit simplified) model of the atmosphere were far-reaching: assuming that the weather behaves according to this model, then long-range weather forecasting is impossible. Indeed the unavoidable errors in determining the present state of the weather system are magnified as time goes by casting off any reliability of the values obtained by numerical integration within a small time period.

This observation was certainly not new. Since the development of the kinetic theory of gases and thermodynamics in the end of the nineteenth century it was known that gas environments, specifically the Earth atmosphere, are very complex systems whose dynamics involves the interaction of a huge number of particles, so it is not surprising that the evolution of such systems is hard to predict. What bewildered mathematicians was the simplicity of the Lorenz system, the fact that it arises naturally as a model of a physical phenomenon (convection) and that its solutions exhibit sensitiveness with respect to the initial conditions. This suggests that sensitiveness is the rule rather than the exception in the natural sciences.

For an historical account of the impact of the Lorenz paper [139] on Dynamical Systems and an overview of the proof by Tucker see [267].

The robustness of this example provides an open set of flows which are not Morse-Smale, nor hyperbolic, and also non-structurally stable, as we will see in Sect. 3.3.

2.3 Hyperbolic Flows

In an attempt to identify what properties were common among stable systems, Stephen Smale introduced in [252] the notion of *Hyperbolic Dynamical System*. Remarkably it turned out that stable systems are essentially the hyperbolic ones, plus certain transversality conditions. In the decades of 1960 and 1970 an elegant and rather complete mathematical theory of hyperbolic systems was developed, culminating with the proof of the Stability Conjecture, by Mañé in the 1990's in the setting of C^1 diffeomorphisms, followed by Hayashi for C^1 flows.

In what follows we present some results of this theory which will be used throughout the text.

Let $X \in \mathfrak{X}^r(M)$ be a flow on a compact manifold M. Denote by $m(T) = \inf_{\|v\|=1} \|T(v)\|$ the *minimum norm* of a linear operator T. A compact invariant set $\Lambda \subset M$ of X is *hyperbolic* if

1. admits a continuous DX-invariant tangent bundle decomposition $T_\Lambda M = E^s_\Lambda \oplus E^X_\Lambda \oplus E^u_\Lambda$, that is we can write the tangent space TxM as a direct sum $E^s_x \oplus E^X_x \oplus E^u_x$, where E^X_x is the subspace in $T_x M$ generated by $X(x)$, satisfying
   - $DX^t(x) \cdot E^i_x = E^i_{X^t(x)}$ for all $t \in \mathbb{R}$, $x \in \Lambda$ and $i = s, X, u$;
2. there are constants $\lambda, K > 0$ such that
   - E^s_Λ *is* (K, λ)*-contracting*, i.e. for all $x \in \Lambda$ and every $t \geq 0$
   $$\|DX^t(x) \mid E^s_x\| \leq K^{-1} e^{-\lambda t},$$
   - E^u_Λ *is* (K, λ)*-expanding*, i.e. for all $x \in \Lambda$ and every $t \geq 0$
   $$m(DX^t \mid E^u) \geq K e^{\lambda t}.$$

Fig. 2.3 A saddle singularity σ for bi-dimensional flow

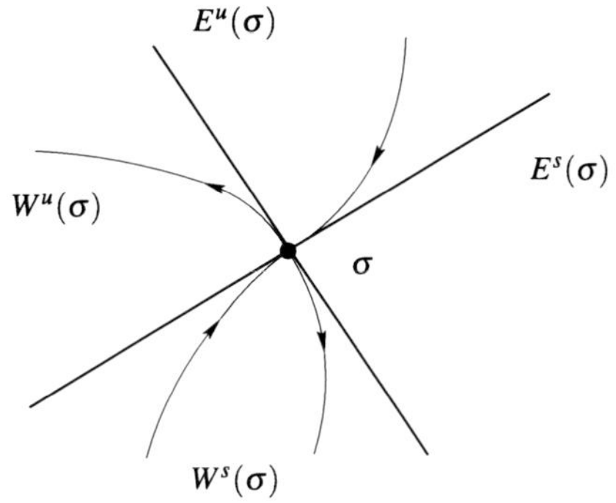

By the Invariant Manifold Theory [110] it follows that for every $p \in \Lambda$ the sets

$$W_X^{ss}(p) = \{q \in M : \text{dist}(X^t(q), X^t(p)) \xrightarrow[t\to\infty]{} 0\}$$

and

$$W_X^{uu}(p) = \{q \in M : \text{dist}(X^t(q), X^t(p)) \xrightarrow[t\to-\infty]{} 0\}$$

are invariant C^r immersed manifolds tangent to E_p^s and E_p^u respectively at p. Here dist is the *distance on M induced by some Riemannian norm.*

If $\mathcal{O} = \mathcal{O}_X(p) \subset \Lambda$ is an orbit of X one has that

$$W_X^s(\mathcal{O}) = \cup_{t\in\mathbb{R}} W_X^{ss}(X^t(p)) \quad \text{and} \quad W_X^u(\mathcal{O}) = \cup_{t\in\mathbb{R}} W_X^{uu}(X^t(p))$$

are invariant C^r-manifolds tangent to $E_p^s \oplus E_p^X$ and $E_p^X \oplus E_p^u$ at p, respectively. We shall denote $W_X^s(p) = W_X^s(\mathcal{O}_X(p))$ and $W_X^u(p) = W_X^u(\mathcal{O}_X(p))$ for the sake of simplicity; see Fig. 2.4.

A *singularity* or a *periodic orbit* of X is *hyperbolic* if its orbit is a hyperbolic set of X. Note that $W_X^{ss}(\sigma) = W_X^s(\sigma)$ and $W_X^{uu}(\sigma) = W_X^u(\sigma)$ for every hyperbolic singularity σ of X. A sink and a source are both hyperbolic singularities. A *hyperbolic* singularity which is *neither* a sink *nor* a source is called a *saddle*.

For every hyperbolic critical element, the dimension of its contracting direction of the tangent bundle decomposition is the *index* of that element. Hence a sink has maximal index, equal to the dimension of the ambient space, and a source has zero index.

A hyperbolic set Λ of X is called *basic* if it is transitive and *isolated*, that is $\Lambda = \cap_{t\in\mathbb{R}} \overline{X^t(U)}$ for some neighborhood U of Λ. It follows from the Shadowing Lemma [181] that every hyperbolic basic set of X either reduces to a singularity or else has no singularities and it is the closure of its periodic orbits.

We say that X is *Axiom A* if the non-wandering set $\Omega(X)$ is both hyperbolic and the closure of its periodic orbits and singularities. The *Spectral Decomposition Theorem* asserts that if X is Axiom A, then there is a disjoint decomposition $\Omega(X) = \Lambda_1 \cup \cdots \cup \Lambda_k$, where each Λ_i is a hyperbolic basic set of X, $i = 1, \ldots, k$.

A *cycle* of a Axiom A vector field X is a sub-collection $\{\Lambda_{i_0}, \ldots, \Lambda_{i_k}\}$ of $\{\Lambda_1, \ldots, \Lambda_n\}$ such that $i_0 = i_k$ and $W_X^u(\Lambda_{i_j}) \cap W_X^s(\Lambda_{i_{j+1}}) \neq \emptyset$, $\forall 0 \leq j \leq k-1$.

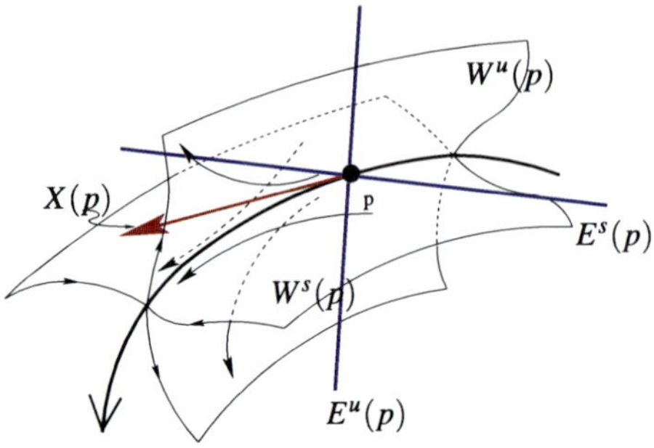

Fig. 2.4 The flow near a hyperbolic saddle periodic orbit through p

2.3.1 Hyperbolic Sets and Singularities

The continuity of the DX-invariant splitting on the tangent space of a uniformly hyperbolic set Λ is a consequence of the uniform expansion and contraction estimates (see e.g. [190]). This means that if $x_n \in \Lambda$ is a sequence of points converging to $x \in \Lambda$, and we consider orthonormal basis $\{e_i^n\}_{i=1,\dots,\dim E^s(x_n)}$ of $E^s(x_n)$, $\{f_i^n\}_{i=1,\dots,\dim E^u(x_n)}$ of $E^u(x_n)$ and $X(x_n)$ of $E^X(x_n)$, then these vectors converge to a basis of $E^s(x)$, $E^u(x)$ and $E^X(x)$ respectively. In particular the dimension of the subspaces in the hyperbolic splitting is constant if Λ is transitive.

This shows that a uniformly hyperbolic basic set Λ cannot contain singularities, except if Λ is itself a singularity. Indeed, if $\sigma \in \Lambda$ is a singularity then it is hyperbolic but the dimension of the central sub-bundle is zero since the flow is zero at σ. Therefore the dimensions of either the stable or the unstable direction at σ and those of a transitive regular orbit in Λ do not match.

In other words *an invariant subset Λ containing a singularity accumulated by regular orbits cannot be uniformly hyperbolic.*

2.3.2 Examples of Hyperbolic Sets and Axiom A Flows

Any hyperbolic singularity or hyperbolic periodic orbit is a hyperbolic invariant set. Also any finite collection of hyperbolic critical elements is a hyperbolic set. We refer to these sets as *trivial hyperbolic sets.*

The first examples of a non-trivial (different from a singularity or a periodic orbit) hyperbolic basic set (on the whole manifold) was the *geodesic flow on any Riemannian manifold with negative curvature*, studied by Anosov [15], whose name is attached to this type of systems today, and the *Smale Horseshoe*, presented in [252] in the setting of diffeomorphisms.

We use a global construction of a (linear) Anosov diffeomorphism (hyperbolic with dense orbit) on the 2-torus and then consider its suspension on the solid (3-)torus to obtain an example of a transitive Axiom A flow.

2.3.2.1 A Linear Anosov Diffeomorphism on the 2-Torus

Consider the linear transformation $A : \mathbb{R}^2 \to \mathbb{R}^2$ with the following matrix in the canonical base

$$\begin{pmatrix} 1 & 1 \\ 2 & 1 \end{pmatrix}.$$

Consider the 2-torus $\mathbb{T}^2$ as the quotient $\mathbb{R}^2/\mathbb{Z}^2 = [0,1]^2/\sim$, where $(x,0) \sim (x,1)$ and $(y,0) \sim (y,1)$ for all $x, y \in [0,1]$, that is the square $[0,1]^2$ whose parallel sides are identified. We denote by $\pi : \mathbb{R}^2 \to \mathbb{T}^2$ the quotient map or projection from $\mathbb{R}^2$ to $\mathbb{T}^2$. Since A preserves $\mathbb{Z}^2$, i.e. $A(\mathbb{Z}^2) \subset \mathbb{Z}^2$, then there exists a well defined quotient map $F_A : \mathbb{T}^2 \to \mathbb{T}^2$. This is a linear automorphism of $\mathbb{T}^2$, see e.g. [147, 269].

The matrix A is hyperbolic: its eigenvalues are $\lambda_1, \lambda_2 = 1 \pm \sqrt{2}$ and the corresponding eigenvectors $v_1, v_2 = (\pm\sqrt{2}/2, 1)$, with irrational slope. Given any point $p \in \mathbb{T}^2$, if we take the projection $W_i(p)$ of the line L_i through p parallel to v_i, $W_i(p) = \pi(L_i)$, then distances along $W_i(p)$ are multiplied by λ_i under the action of F_A, for $i = 1, 2$. These are the stable and unstable manifolds of p. Due to the irrationality of the slope every such "line" is dense in the torus. Moreover there is a transitive orbit and a dense set of periodic orbits for the map F_A (see e.g. [78]). The entire torus is then a uniformly hyperbolic set.

2.3.2.2 Definition of Suspension Semiflow over a Roof Function

Let (X, d) be a metric space with distance d and $r : X \to \mathbb{R}$ be a strictly positive function. The *phase space* X_r of the suspension flow is defined as

$$X_r = \{(x, y) \in X \times [0, +\infty) : 0 \le y < r(x)\}.$$

Let $f : X \to X$ be a map on X. The *suspension semi-flow over f with roof r* is the following family of maps $X_f^t : X_r \to X_r$ for $t \ge 0$: X^0 is the identity and for each $x = x_0 \in X$ denote by x_n the nth iterate $f^n(x_0)$ for $n \ge 0$. Denote also $S_n r(x_0) = \sum_{j=0}^{n-1} r(x_j)$ for $n \ge 1$. Then for each pair $(x_0, y_0) \in X_r$ and $t > 0$ there exists a unique $n \ge 1$ such that $S_n r(x_0) \le y_0 + t < S_{n+1} r(x_0)$ and we define (see Fig. 2.5)

$$X_f^t(x_0, y_0) = \big(x_n, y_0 + t - S_n r(x_0)\big).$$

This construction is the basis of many examples and also of many techniques to pass from a flow with a transverse section to a suspension flow and viceversa, enabling us to transfer results which are easy to prove for suspension flows, due to their "almost product structure", to more general flows. In this text we will see several examples of this.

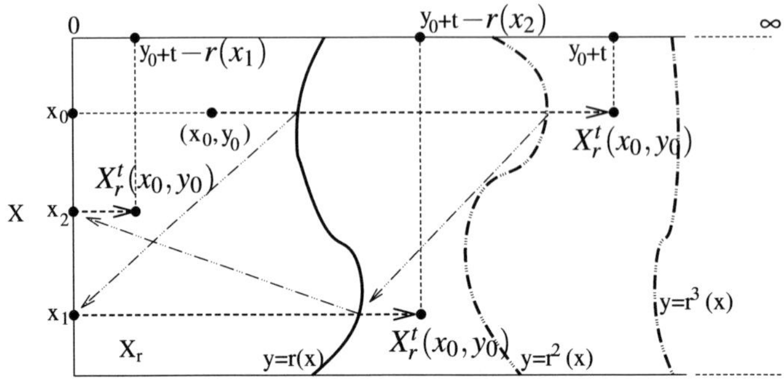

Fig. 2.5 The equivalence relation defining the suspension flow of f over the roof function r

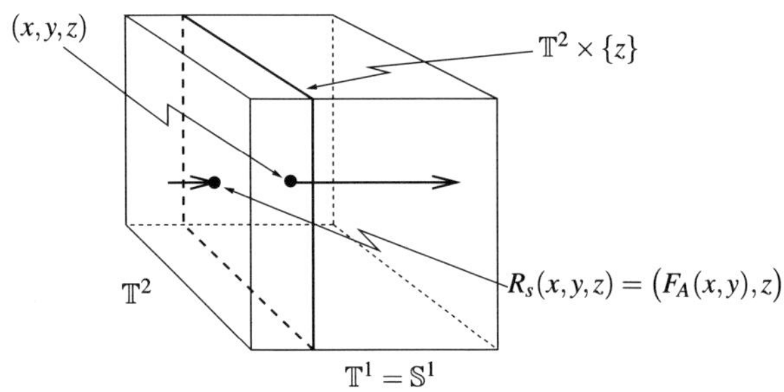

Fig. 2.6 Suspension flow over Anosov diffeomorphism with constant roof

2.3.2.3 An Anosov Flow on a Three-Dimensional Manifold Though the Suspension of an Anosov Diffeomorphism

Consider the suspended flow X_r over $F_A : \mathbb{T}^2 \to \mathbb{T}^2$ defined in Sect. 2.3.2.1 with a constant roof function $r \equiv 1$. Then X_r is the 3-cube $[0, 1]^3$ with parallel sides identified, that is, we obtain a flow on the 3-torus such that the *first return map* R_z from any section $\mathbb{T}^2 \times \{z\}$ to itself can be naturally identified with F_A, see Fig. 2.6.

This flow $X^t_{F_A}$ is uniformly hyperbolic since the hyperbolic structure exhibited by the map F_A is naturally carried by the flow to $\mathbb{T}^3$, e.g. it has a dense orbits and a dense set of periodic orbits, each of which are the suspension of the corresponding dense orbit and periodic orbits for F_A. The invariant manifolds of a point (x, y, s) are simply the translate of the corresponding invariant manifolds of (x, y) for F_A: $W^k_{X_r}(x, y, z) = W^k(x, y) \times \{z\}$ for $k = uu, ss$ and any $z \in [0, 1]$.

We will see in Sect. 10.2.3 that this Anosov flow *is not topologically mixing*. We note that since the definition of suspension flow involves the identification of points through the base map, the resulting three-dimensional manifold of the suspension presented above is *not* the 3-torus.

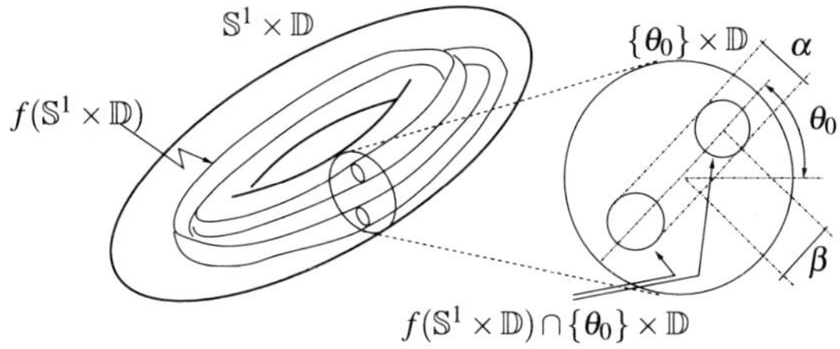

Fig. 2.7 The solenoid attractor

2.3.2.4 The Solenoid Attractor

Consider now the solid 2-torus $\mathbb{S}^1 \times \mathbb{D}$ where $\mathbb{D} = \{z \in \mathbb{C} : |z| < 1\}$ is the unit disk in $\mathbb{C}$, together with the map $f : \mathbb{S}^1 \times \mathbb{D} \to \mathbb{S}^1 \times \mathbb{D}$ given by

$$(\theta, z) \mapsto (2\theta, \alpha z + \beta e^{i\theta/2}),$$

$\theta \in \mathbb{R}/\mathbb{Z}$ and $\alpha, \beta \in \mathbb{R}$ with $\alpha + \beta < 1$. This transformation maps $\mathbb{S}^1 \times \mathbb{D}$ strictly inside itself, that is $\overline{f(\mathbb{S}^1 \times \mathbb{D})} \subset \mathbb{S}^1 \times \mathbb{D}$. The maximal positively invariant set $\Lambda = \bigcap_{n \geq 0} f^n(\mathbb{S}^1 \times \mathbb{D})$ is a uniformly hyperbolic basic set: the $\mathbb{S}^1$ direction is uniformly expanding and the $\mathbb{D}$ direction is uniformly contracting, see Fig. 2.7. This set is transitive and has a dense subset of periodic orbits [78, 230].

2.3.2.5 Uniformly Hyperbolic Basic Set for a Flow

Consider the suspension of the solenoid map f of the previous subsection over the constant roof function $r \equiv 1$ to get a flow with an attractor $\Lambda_f = \bigcap_{t \geq 0} X^t_f((\mathbb{S}^1 \times \mathbb{D})_r)$ which is a uniformly hyperbolic basic set for the flow X_f.

This is an example of an Axiom A attractor for a flow. As before X^t_f on Λ is *not* topologically mixing.

2.4 Expansiveness and Sensitive Dependence on Initial Conditions

The development of the theory of dynamical systems has shown that models involving expressions as simple as quadratic polynomials (as the *logistic family* or *Hénon attractor*), or autonomous ordinary differential equations with a hyperbolic singularity of saddle-type accumulated by regular orbits, as the *Lorenz flow*, exhibit *sensitive dependence on initial conditions*, a common feature of *chaotic dynamics*: small initial differences are rapidly augmented as time passes, causing two trajectories originally coming from practically indistinguishable points to behave in a completely different manner after a short while. Long term predictions based on such models are unfeasible since it is not possible to both specify initial conditions with arbitrary accuracy and numerically calculate with arbitrary precision.

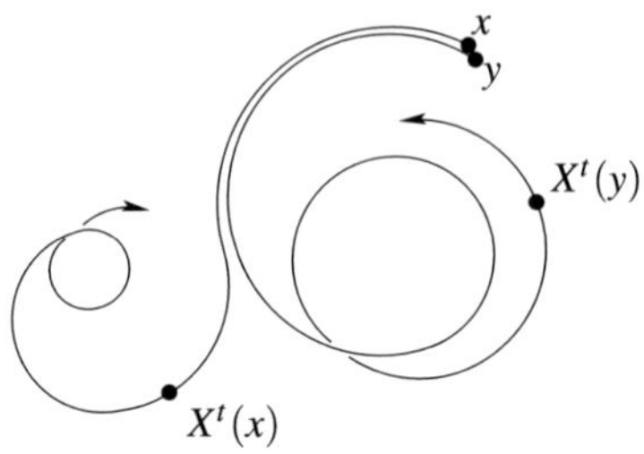

Fig. 2.8 Sensitive dependence on initial conditions

Formally the definition of sensitivity is as follows for a flow X^t: a X^t-invariant subset Λ is *sensitive to initial conditions* or has *sensitive dependence on initial conditions*, or simply *chaotic* if, for every small enough $r > 0$ and $x \in \Lambda$, and for any neighborhood U of x, there exists $y \in U$ and $t \neq 0$ such that $X^t(y)$ and $X^t(x)$ are r-apart from each other: $\operatorname{dist}(X^t(y), X^t(x)) \geq r$. See Fig. 2.8.

2.4.1 Chaotic Systems

We distinguish between forward and backward sensitive dependence on initial conditions. We say that an invariant subset Λ for a flow X^t is *future chaotic with constant* $r > 0$ if, for every $x \in \Lambda$ and each neighborhood U of x in the ambient manifold, there exists $y \in U$ and $t > 0$ such that $\operatorname{dist}(X^t(y), X^t(x)) \geq r$. Analogously we say that Λ is *past chaotic with constant* r if Λ is future chaotic with constant r for the flow generated by $-X$. If we have such sensitive dependence both for the past and for the future, we say that Λ is *chaotic*. Note that in this language sensitive dependence on initial conditions is weaker than chaotic, future chaotic or past chaotic conditions.

An easy consequence of chaotic behavior is that it prevents the existence of sources or sinks, either attracting or repelling singularities or periodic orbits, inside the invariant set Λ. Indeed, if Λ is future chaotic (for some constant $r > 0$) then, were it to contain some attracting periodic orbit or singularity, any point of such orbit (or singularity) would admit no point in a neighborhood whose orbit will move away in the future. Likewise, reversing the time direction, a past chaotic invariant set cannot contain repelling periodic orbits or singularities. As an almost reciprocal we have the following.

Lemma 2.9 *If* $\Lambda = \overline{\cap_{t\in\mathbb{R}} X^t(U)}$ *is a isolated proper subset for* $X \in \mathfrak{X}^1(M)$ *with isolating neighborhood* U *and* Λ *is* not future chaotic (*respective not past chaotic*), *then* $\Lambda_X^-(U) = \overline{\cap_{t>0} X^{-t}(U)}$ (*respective* $\Lambda_X^+(U) = \overline{\cap_{t>0} X^t(U)}$) *has non-empty interior.*

Proof If Λ is not future chaotic, then for every $r > 0$ there exists some point $x \in \Lambda$ and a neighborhood V of x such that $\operatorname{dist}(X^t(y), X^t(x)) < r$ for all $t > 0$ and each $y \in V$. If we choose $0 < r < \operatorname{dist}(M \setminus U, \Lambda)$ (we note that if $\Lambda = U$ then

Λ would be open and closed, and so, by connectedness of M, Λ would not be a proper subset), then we deduce that $X^t(y) \in U$, that is, $y \in X^{-t}(U)$ for all $t > 0$, hence $V \subset \Lambda_X^-(U)$. Analogously if Λ is not past chaotic, just by reversing the time direction. □

In particular *if an invariant and isolated set Λ with isolating neighborhood U is given such that the volume of both $\Lambda_X^+(U)$ and $\Lambda_X^-(U)$ is zero, then Λ is chaotic.*

Sensitive dependence on initial conditions is part of the definition of *chaotic system* in the literature, see e.g. [78]. It is an interesting fact that sensitive dependence is a consequence of another two common features of most systems considered to be chaotic: existence of a dense orbit and existence of a dense subset of periodic orbits. We present a short proof of this based on [32]. An extensive discussion of this and related topics can be found in [94].

Proposition 2.10 *A compact invariant subset Λ for a flow X^t with a dense subset of periodic orbits and a dense (regular and non-periodic) orbit is chaotic.*

This result provides sensitive dependence on initial conditions for singular-hyperbolic attractors, and shows that this class of attractors are both past and future chaotic, once we have shown that these attractors have a dense subset of periodic orbits, in Chap. 6.

Proof Let Λ have a dense regular orbit and a dense subset of periodic orbits.

Claim There exists $\delta_0 > 0$ such that for each $x \in \Lambda$ we can find $p \in \mathrm{Per}(X) \cap \Lambda$ such that $\mathrm{dist}(\mathcal{O}_X(p), x) > \delta_0/2$.

Indeed there are at least two distinct periodic orbits $\mathcal{O}_X(q_1), \mathcal{O}_X(q_2)$ in Λ and we can set $\delta_0 = \mathrm{dist}(\mathcal{O}_X(q_1), \mathcal{O}_X(q_2)) > 0$. Then we have

$$\mathrm{dist}\left(\mathcal{O}_X(q_1), \mathcal{O}_X(q_2)\right) \le \mathrm{dist}\left(\mathcal{O}_X(q_1), x\right) + \mathrm{dist}\left(x, \mathcal{O}_X(q_2)\right)$$

and so either $\mathrm{dist}(\mathcal{O}_X(q_1), x) > \delta_0/2$ or $\mathrm{dist}(x, \mathcal{O}_X(q_2)) > \delta_0/2$.

Now we show that X on Λ is future chaotic with constant $r = \delta_0/8$. Let $x \in \Lambda$ and a neighborhood U of x be given.

The denseness of periodic orbits in Λ ensures we can find $p \in \mathrm{Per}(X)\Lambda$ such that $p \in U \cap B(x, r)$. Let $\tau > 0$ be the period of $\mathcal{O}_X(p)$. By the previous claim we can also find a periodic orbit $\mathcal{O}_X(q) \subset \mathrm{Per}(X) \cap \Lambda$, which we can assume is distinct from $\mathcal{O}_X(p)$ and with smaller period, such that $\mathrm{dist}(\mathcal{O}_X(q), x) > 4r$. We now define

$$V := \bigcap_{s=0}^{\tau} X^{-s}\left(B(X^s(q), r)\right).$$

Since $\mathcal{O}_X(q) \subset V$ this set is a non-empty neighborhood of $\mathcal{O}_X(q)$: we just have to use the continuity of the flow X^t.

Now using the assumption of the existence of a dense orbit, we can find $s > 0$ and $y \in \Lambda$ such that $X^s(y) \in V$. We take now $s_0 := [s/\tau + 1]$ the integer part of $s/\tau + 1$. By construction $s_0\tau - s \le \tau$ and

$$X^{s_0\tau}(y) = X^{s_0\tau - s}\big(X^s(y)\big) \in X^{s_0\tau - s}(V) \subset B\big(X^{s_0\tau - s}(q), r\big)$$

The triangle inequality implies

$$\begin{aligned} \operatorname{dist}\big(p, X^{s_0\tau}(y)\big) &= \operatorname{dist}\big(X^{s_0\tau}(p), X^{s_0\tau}(y)\big) \\ &\ge \operatorname{dist}\big(x, X^{s_0\tau - s}(q)\big) - \operatorname{dist}(x, p) - \operatorname{dist}\big(X^{s_0\tau}(y), X^{s_0\tau - s}(q)\big) \\ &> 4r - r - r = 2r. \end{aligned}$$

Again by the triangle inequality we get

$$\operatorname{dist}\big(p, X^{s_0\tau}(x)\big) + \operatorname{dist}\big(X^{s_0\tau}(x), X^{s_0\tau}(y)\big) > 2r$$

and so either $\operatorname{dist}(p, X^{s_0\tau}(x)) > r$ or $\operatorname{dist}(X^{s_0\tau}(x), X^{s_0\tau}(y)) > r$ and in each case we have found a point in the neighborhood U whose orbit is at a distance bigger than r from the orbit of x. Since $r = \delta_0/8$ is fixed, $s_0\tau > 0$ and x and $U \ni x$ are an arbitrary point in Λ with some neighborhood of x, the proof that Λ is future chaotic is complete.

Finally the assumption of existence of a dense subset of periodic orbits and of a dense orbit are invariant under time reversal (see Lemma 2.2), so the same proof can be repeated with negative time, showing that Λ is also past chaotic. The proof is complete. □

2.4.2 *Expansive Systems*

A related concept is that of expansiveness, which roughly means that points whose orbits are always close for all time must coincide. The concept of expansiveness for homeomorphisms plays an important role in the study of transformations. Bowen and Walters [63] gave a definition of expansiveness for flows which is now called *C-expansiveness* [121]. The basic idea of their definition is that two points which are not close in the orbit topology induced by $\mathbb{R}$ can be separated at the same time even if one allows a continuous time lag—see below for the technical definitions. The equilibria of C-expansive flows must be isolated [63, Proposition 1] which implies that the Lorenz attractors and geometric Lorenz models are not C-expansive.

Keynes and Sears introduced [121] the idea of restriction of the time lag and gave several definitions of expansiveness weaker than C-expansiveness. The notion of *K-expansiveness* is defined allowing only the time lag given by an increasing surjective homeomorphism of $\mathbb{R}$. Komuro [124] showed that the Lorenz attractor (presented in Sect. 2.2.3) and the geometric Lorenz models (to be presented in Sect. 3.3) are not K-expansive. The reason for this is not that the restriction of the time lag is insufficient but that the topology induced by $\mathbb{R}$ is unsuited to measure the closeness of two points in the same orbit.

Taking this fact into consideration, Komuro [124] gave a definition of *expansiveness* suitable for flows presenting equilibria accumulated by regular orbits. This concept is enough to show that two points which do not lie on a same orbit can be separated.

Let $C(\mathbb{R},\mathbb{R})$ be the set of all continuous functions $h:\mathbb{R}\to\mathbb{R}$ and define

$$\mathscr{K}_0=\{h\in C(\mathbb{R},\mathbb{R}):h(0)=0\},$$

and

$$\mathscr{K}=\{h\in C\big(\mathbb{R},\mathbb{R}\big):h(\mathbb{R})=\mathbb{R},\ h(0)=0 \quad\text{and}\quad h(s)>h(t) \text{ for every } s>t\}.$$

A flow X is *C-expansive* (*K-expansive* respectively) on an invariant subset $\Lambda\subset M$ if for every $\varepsilon>0$ there exists $\delta>0$ such that if $x,y\in\Lambda$ and for some $h\in\mathscr{K}_0$ (respectively $h\in\mathscr{K}$) we have

$$\operatorname{dist}\big(X^t(x),X^{h(t)}(y)\big)\le\delta\quad\text{for all } t\in\mathbb{R}, \tag{2.3}$$

then $y\in X^{[-\varepsilon,\varepsilon]}(x)=\{X^t(x):-\varepsilon\le t\le\varepsilon\}$.

We say that the flow X is *expansive* on Λ if for every $\varepsilon>0$ there is $\delta>0$ such that for $x,y\in\Lambda$ and some $h\in\mathscr{K}$ (note that now we do not demand that 0 be fixed by h) satisfying (2.3), then we can find $t_0\in\mathbb{R}$ such that $X^{h(t_0)}(y)\in X^{[t_0-\varepsilon,t_0+\varepsilon]}(x)$.

Observe that expansiveness on M implies sensitive dependence on initial conditions for any flow on a manifold with dimension at least 2. Indeed if ε,δ satisfy the expansiveness condition above with h equal to the identity and we are given a point $x\in M$ and a neighborhood U of x, then taking $y\in U\setminus X^{[-\varepsilon,\varepsilon]}(x)$ (which always exists since we assume that M is not one-dimensional) there exists $t\in\mathbb{R}$ such that $\operatorname{dist}(X^t(y),X^t(x))\ge\delta$. The same argument applies whenever we have expansiveness on an X-invariant subset Λ of M containing a dense regular orbit of the flow.

Clearly by definition we have

$$\text{C-expansive} \Longrightarrow \text{K-expansive} \Longrightarrow \text{expansive}.$$

When a flow has no fixed point then the notion of C-expansiveness is equivalent to K-expansiveness [183, Theorem A]. In [63] Bowen and Walters proved the following.

Lemma 2.11 *If X^t is a C-expansive flow on M, then each fixed point of X^t is an isolated point of M.*

Hence, on a connected manifold, a C-expansive flow has no fixed points.

Proof Let us fix $x\in M$ such that $X^t(x)=x$ for all $t\in\mathbb{R}$. Let $\varepsilon>0$ be given and $\delta>0$ be the corresponding number from the definition of C-expansiveness.

If $\operatorname{dist}(x, y) < \delta$, then for $h(t) \equiv 0$ for each $t \in \mathbb{R}$ we get $\operatorname{dist}(X^t(x), X^t(y)) \leq \delta$ for all $t \in \mathbb{R}$. Hence $y = X^t(x) = x$. Thus x is an isolated point of M. □

In fact, K-expansiveness and C-expansiveness are equivalent in full generality.

Proposition 2.12 *A flow is C-expansive on a manifold M if, and only if, it is K-expansive.*

Proof From Lemma 2.11, a C-expansive flow admits only finitely many isolated fixed points on a compact space M. We assume now that X^t has non-isolated fixed points in M, that is, there exists at least a singularity σ which is accumulated by other points of M (this always holds on a connected manifold). Then X is not C-expansive by Lemma 2.11. We now show that it is not K-expansive either, proving the proposition.

Using the continuity of X^t we have that for all $R > 0$ and $T > 0$ there exists $x \in M \setminus \{\sigma\}$ such that $\operatorname{dist}(X^t(x), \sigma) \leq R$ whenever $|t| < T$.

Let $\varepsilon, \delta > 0$ be given and let us set $T = 3\varepsilon$ and $R = \delta/2$. Define $y = X^\varepsilon(x)$ and

$$h(t) = \begin{cases} t+\varepsilon & \text{if } t \notin (-2\varepsilon, \varepsilon) \\ 2t & \text{if } 0 \leq t < \varepsilon \\ t/2 & \text{if } t \in (-2\varepsilon, 0) \end{cases},$$

which is a monotonously increasing homeomorphism of $\mathbb{R}$ with $h(0) = 0$.

Next we verify that $\operatorname{dist}(X^t(y), X^{h(t)}(x)) \leq \delta/2$ for all $t \in \mathbb{R}$:

- if $t \notin (-2\varepsilon, \varepsilon)$ then $X^{h(t)}(x) = X^{t+\varepsilon}(x) = X^t(y)$ and so we are done;
- if $t \in (0, \varepsilon)$ then $h(t) = 2t < T$ and so $\operatorname{dist}(X^{h(t)}(x), \sigma) \leq \delta/2$ which implies $\operatorname{dist}(X^t(y), X^{h(t)}(x)) \leq \operatorname{dist}(X^t(y), \sigma) + \operatorname{dist}(X^{h(t)}(x), \sigma)$. As $X^t(y) = X^{t+\varepsilon}(x)$ and for $t < \varepsilon$ we have $t + \varepsilon < 3\varepsilon = T$ we obtain $\operatorname{dist}(X^t(y), \sigma) < \delta/2$. Hence $\operatorname{dist}(X^t(y), X^{h(t)}(x)) < \delta$;
- if $t \in (-2\varepsilon, 0)$ then $|h(t)| = |t/2| < 3\varepsilon$ and so $\operatorname{dist}(X^{h(t)}(x), \sigma) \leq \delta/2$.

  We have now that

$$\operatorname{dist}(X^t(y), X^{h(t)}(x)) \leq \operatorname{dist}(X^t(y), \sigma) + \operatorname{dist}(X^{h(t)}(x), \sigma) \leq \operatorname{dist}(X^t(y), \sigma) + \delta/2.$$

  But $t \in (-2\varepsilon, 0)$ and $t + \varepsilon \in (-\varepsilon, \varepsilon)$, hence $|t + \varepsilon| < \varepsilon$ implying that

$$\operatorname{dist}(X^{t+\varepsilon}(x), \sigma) < \delta/2.$$

  Thus, as $X^\varepsilon(x) = y$, we get $\operatorname{dist}(X^t(y), \sigma) < \delta/2$ and replacing this in the inequality above we obtain $\operatorname{dist}(X^t(y), X^{h(t)}(x)) < \delta$.

All together we have proved $\operatorname{dist}(X^t(y), X^{h(t)}(x)) \leq \delta/2$ for all $t \in \mathbb{R}$. Now there are two possibilities:

1. either $X^t(x) \neq y$ for all $|t| < \varepsilon$, and we are done, or
2. or there exists $s \in \mathbb{R}$ such that $X^s(x) = y$, and in this case x is a periodic orbit with period $\tau \leq s - \varepsilon < 2\varepsilon$. Thus $\operatorname{dist}(X^t(x), X^{h(t)}(\sigma)) < \delta$.

Either way we found a pair of points (x and y in case (1), x and σ in case (2)) which remain δ-close even when time is reparametrized through h in one of the orbits, and both points are not connected through any X-orbit in a time less than ε. In this construction we may take $\delta > 0$ arbitrarily close to zero for a fixed $\varepsilon > 0$, so we have shown that X is not K-expansive. □

We will prove in Sect. 7.2 that singular-hyperbolic attractors are expansive. In particular, the Lorenz attractor and the geometric Lorenz examples are all expansive and sensitive to initial conditions. Since these families of flows exhibit equilibria accumulated by regular orbits, we see that expansiveness is compatible with the existence of fixed points by the flow.

We observe that Peixoto's characterization of structurally stable vector fields on surfaces and its genericity implies that, for such vector field X on S, there is an open and dense subset B of S such that the positive orbit $X^t(p), t \geq 0$ of $p \in B$ converges to one of finitely many attracting equilibria. Therefore *no sensitive dependence on initial conditions arises for an open and dense subset of all vector fields in orientable surfaces*. The same lack of sensitiveness holds generically for all non-orientable surfaces where Peixoto's conjecture has been successfully proved.

This explains in part the great interest attached to the Lorenz attractor, as one of the first examples of sensitive dependence on initial conditions.

2.5 Basic Tools

Here we state two basic classical results which enable us to understand in many cases the local dynamics near many flow orbits. Then we state the powerful *closing* and *connecting lemmas* which will be used in a fundamental way in several key points in the following chapters.

2.5.1 The Tubular Flow Theorem

The following result shows that the local behavior of orbits near a regular point of any flow is very simple.

Theorem 2.13 (Tubular flow) *Let $X \in \mathfrak{X}^r(M)$ and let $p \in M^n$ be a regular point of X where $n \geq 1$ is the dimension of M. Let $V = \{(x_1, \ldots, x_n) \in \mathbb{R}^n : \|x_i\| < 1\}$ and Y be the vector field on V given by $Y = (1, 0, \ldots, 0)$. Then there is a C^r diffeomorphism $h : U \to V$ for some neighborhood U of p in M, which takes trajectories of X to trajectories of Y, that is $X \mid U$ is topologically equivalent to $Y \mid V$.*

This shows that near a regular point p every smooth flow can be smoothly linearized: under a change of coordinates orbits near p look like the orbits of a constant flow, see Fig. 2.9.

Fig. 2.9 Linearization of orbits near a regular point of a flow

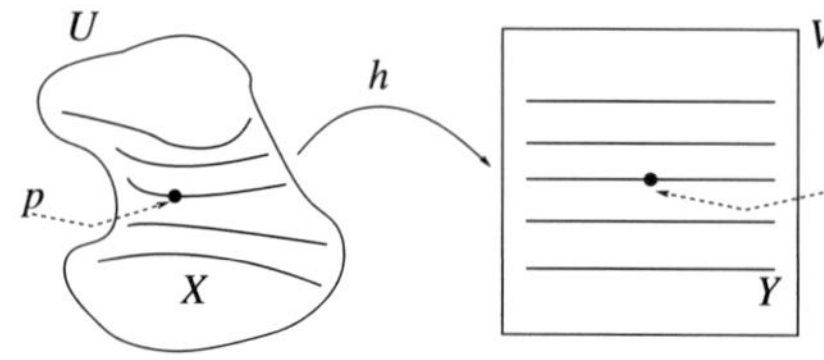

2.5.2 *Transverse Sections and the Poincaré Return Map*

Now we describe a standard and extremely useful consequence of the tubular flow theorem, which provides a converse to the construction of suspensions semiflows (presented in Sect. 2.3.2).

Let $X \in \mathfrak{X}^1(M^3)$ be a flow on a three-dimensional manifold and let S be an embedded surface in M which is transverse to the vector field X at all points, i.e. for every $x \in S$ we have $T_x S + E_x^X = T_x M$ or equivalently $X(x) \notin T_x S$. We say in what follows that such S is a *cross-section* to the flow X^t or to the vector field X.

Let S_0 and S_1 be a pair of cross-sections to X and $x_0 \in S_0$ be a regular point of X and suppose that there exists $T > 0$ such that $x_1 = X^T(x_0) \in S_1$. Applying the Tubular Flow Theorem 2.13 to a finite open covering of the compact arc $\gamma = X^{[0,T]}(x_0)$ we obtain a tubular flow in a neighborhood of γ. This shows that there exists a smooth map R from a neighborhood V_0 of x_0 in S_0 to a neighborhood V_1 of x_1 in S_1, with the same degree of smoothness of the flow, such that $R(x) = X^{T(x)}(x)$ for all $x \in V_0$ with $R(x_0) = x_1$ and $T : V_0 \to \mathbb{R}$ also smooth with $T(x_0) = T$. Moreover R is a bijection and thus a diffeomorphism.

We can reapply the Tubular Flow Theorem and extend the domain of definition of R to its maximal domain relative to S_0 and S_1 and to the connection time T. Notice that x_1 need not be the first entry to S_1, that is T might be bigger than $\inf\{t > 0 : X^t(x_0) \in S_1\}$.

Note that if x_0 is a periodic orbit of X then taking $S_1 = S_0$ we see that x_0 is a fixed point of R and the local behavior of the flow near x_0 can be studied through the map R acting on a space with less dimension than M. This is an important example where we can reduce the study of a flow to a lower dimensional transformation. The power and applicability of this method should be clear after Chaps. 3 and 7.

2.5.3 *The Hartman-Grobman Theorem on Local Linearization*

The following result due to Hartman and Grobman [96, 104] shows that a flow of a vector field X is locally equivalent to its linear part at a hyperbolic singularity. Since linear hyperbolic flows can be completely classified by topological equivalence, this result enables us to classify the local behavior of the flow of any smooth vector field near a hyperbolic singularity. See [192] for generalizations and more references on this subject.

Theorem 2.14 (Hartman-Grobman) *Let $X \in \mathfrak{X}^r(M)$ and let $p \in M$ be a hyperbolic singularity of X. Let $Y = DX^0 : T_pM \to T_pM$ be the linear vector field on T_pM given by the linear transformation DX^0. Then there exists a neighborhood U of p in M, a neighborhood V of 0 in T_pM and a homeomorphism $h : U \to V$ which takes trajectories of X to trajectories of Y, that is $X \mid U$ is topologically equivalent to $Y \mid V$.*

2.5.4 The (Strong) Inclination Lemma (or λ-Lemma)

These are basic results of dynamics near a hyperbolic singularity which are extremely useful to obtain intersections between stable and unstable manifolds through simple geometric arguments.

2.5.4.1 The Inclination Lemma

Let $\sigma \in M$ be a hyperbolic singularity of $X \in \mathfrak{X}^r(M)$ for some $r \geq 1$, with its local stable and unstable manifolds $W^s_{\text{loc}}(\sigma)$, $W^u_{\text{loc}}(\sigma)$. Fix an embedded disk B in $W^u_{\text{loc}}(\sigma)$ which is a neighborhood of σ in $W^u_{\text{loc}}(\sigma)$, and a neighborhood V of this disk in M. Then let D be a transverse disk to $W^s_{\text{loc}}(\sigma)$ at z with the same dimension as B, and write D^t for the connected component of $X^t(D) \cap V$ which contains $X^t(z)$, for $t \geq 0$, see Fig. 2.10.

Lemma 2.15 (Inclination lemma [190]) *Given $\varepsilon > 0$ there exists $T > 0$ such that for all $t > T$ the disk D^t is ε close to B in the C^r-topology.*

This means that the embeddings whose images are the disks B and D^t are close in the C^r topology.

2.5.4.2 The Strong Inclination Lemma

In the same setting as above but imposing that the eigenvalues of $DX(\sigma)$ closest to the imaginary axis be real and simple it is possible to improve the convergence

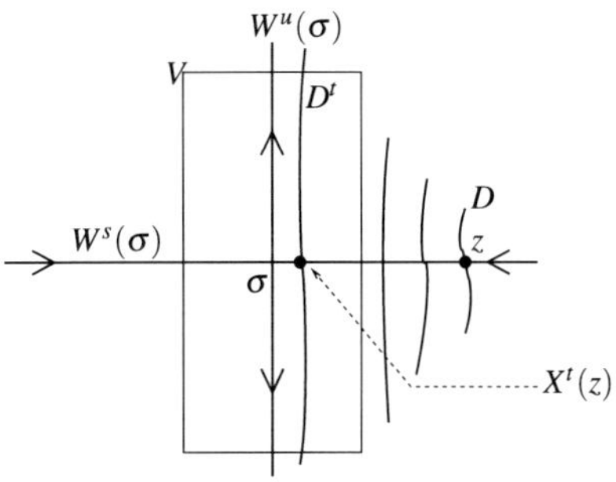

Fig. 2.10 The inclination lemma

estimates. This condition on $DX(\sigma)$ is satisfied in particular by all hyperbolic singularities with distinct real eigenvalues, and so also by the so-called Lorenz-like singularities, see Definition 3.2. These are the only kind of singularities allowed on singular-hyperbolic sets, see Chap. 5.

Lemma 2.16 (Strong inclination lemma [77]) *There are $c, \lambda, T > 0$ such that for all $t > T$ the C^r distance between the embeddings of B and of D^t is bounded by $c \cdot e^{-\lambda t}$.*

2.5.5 *Homoclinic Classes, Transitiveness and Denseness of Periodic Orbits*

Given a hyperbolic periodic orbit p of saddle-type for a flow $X \in \mathfrak{X}^1$ we can define its associated *homoclinic class* $H_X(p)$ by the *closure of the set of transverse intersections between the stable and unstable manifolds of* p

$$H_X(p) = \overline{W_X^u(p) \pitchfork W_X^s(p)}.$$

Note that there are cases where $W_X^u(p)$ coincides with $W_X^s(p)$, a saddle-connection, and then $H_X(p) = \emptyset$. Observe that a nonempty homoclinic class is always an invariant subset of the flow.

Otherwise we have the following important classical result from the early works of Poincaré [205] (who showed that transverse homoclinic orbits are accumulation points of other homoclinic orbits) and developed by Birkhoff [49] (transverse homoclinic orbits are accumulation points of periodic orbits) and by Smale [251].

Theorem 2.17 (Birkhoff-Smale) *Any non-empty homoclinic class has a dense orbit and contains a dense set of periodic orbits.*

See [193] for a general modern presentation of this result including motivation, proofs and other non-trivial dynamical consequences.

The transitiveness part of this theorem is a consequence of the inclination lemma and we present a short proof here.

Lemma 2.18 *Every homoclinic class H of a flow X is topologically transitive.*

Proof Let $q, r \in H = \text{closure}[W_X^s(p) \pitchfork W_X^u(p)]$ be distinct points and U, V two disjoint neighborhoods of q, r in H, respectively. Let q_1, r_1 be points of intersection between the stable and unstable manifolds of p in U and V, respectively. Then for some future time $t > 0$ very large and some $s > 0$ close to the period of p we have that $X^{t+s}(q_1)$ is on $W^s(p)$ very close to p and $X^{-t}(r_1)$ is on $W^u(p)$ very close to p also.

The invariance of the stable and unstable manifolds and the inclination lemma imply that there exists a point w in the intersection between $W^{uu}(X^{t_1}(q_1))$ and $W^{ss}(X^{-t_2}(r_1))$ for some $t_1, t_2 > t$. Hence $X^{-t_1}(w)$ is inside U near q_1 and $X^{t_2}(w)$ is inside V near r_1. Then $X^{t_1+t_2}(U) \cap V \neq \emptyset$. □

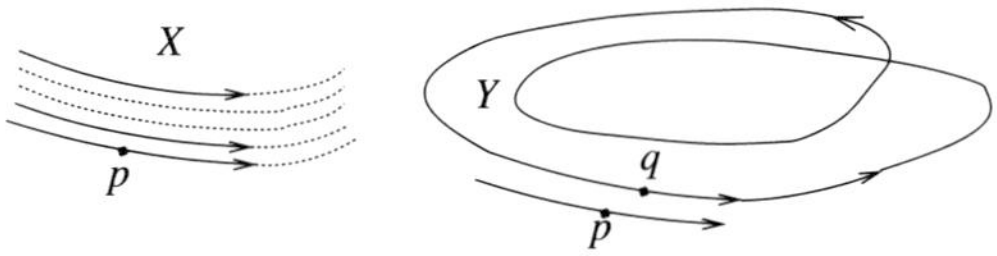

Fig. 2.11 Closing a recurrent orbit

2.5.6 The Closing Lemma

This celebrated result, proved by Charles Pugh [212, 213, 215], says that every regular orbit which accumulates on itself can be closed by an arbitrarily small C^1 perturbation of the vector field, as sketched in Fig. 2.11. The question whether a vector field with a recurrent trajectory through a point p can be perturbed so that the solution through p for the new vector field is closed, albeit trivial in class C^0, is a deep problem in class C^r for $r \geq 1$, as first remarked by Peixoto [195].

In [212, 213] Pugh proved the C^1 Closing Lemma for compact manifolds of dimensions two and three and generalized the result for arbitrary dimensions and to the case of closing a non-wandering trajectory, rather than a recurrent one. In [214] he proved that for a weaker type of recurrent point, for which $\alpha_X(p) \cap \omega_X(p) \neq \emptyset$, the C^2 double-closing is not always possible on the 2-torus $\mathbb{T}^2$. Later Pugh and Robinson [215] established the Closing Lemma when M is non-compact, provided the point q to be closed satisfies $\alpha_X(q) \cap \omega_X(q) \neq \emptyset$.

We remark that the C^r Closing Lemma, for $r \geq 1$, in the case of M being the 2-torus and the vector field has no equilibria, was proved earlier by Peixoto [195] and later by Gutierrez [102] for the "constant type" vector fields on the 2-torus with finitely many equilibria. In [103] Gutierrez gave a counter-example to the C^2 Closing Lemma for the punctured torus. A closely related and also extremely useful result is the "Ergodic Closing Lemma" proved by Ricardo Mañé, see Sect. 2.5.8.

Theorem 2.19 (C^1-closing lemma) *Let $X \in \mathfrak{X}^1(M)$ be a C^1-flow on a compact boundaryless finite dimensional manifold M and $p \in M$ be a non-wandering point of X. Given a C^1-neighborhood $\mathcal{U}$ of X and a neighborhood V of p, then there exists $Y \in \mathcal{U}$ and $q \in V$ such that q belongs to a periodic orbit of Y.*

Observe that in the Closing Lemma above the point whose orbit is closed is not necessarily the initial non-wandering point, but only a point arbitrarily close to it. The same situation appears in the "Ergodic Closing Lemma" of Mañé, see Sect. 2.5.8. Later this was improved in the Connecting Lemma by Hayashi, see the next subsection. Moreover the Ergodic Closing Lemma does provides a bound for the distance between the original and the periodic approximating orbit, which is not given by the Closing Lemma.

2.5.7 The Connecting Lemma

The connecting lemma is motivated by the following situation often faced when studying dynamical systems. Suppose the unstable manifold of a hyperbolic peri-

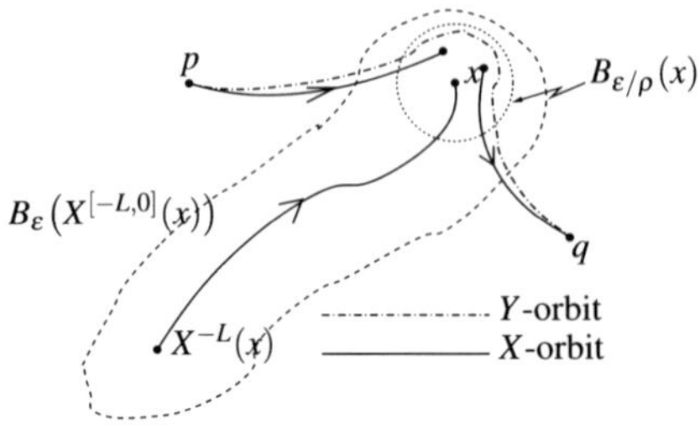

Fig. 2.12 The connecting lemma for C^1 flows

odic orbit accumulates on the stable manifold of another hyperbolic periodic orbit. We would like to find a vector field close to the given one such that the continuation of the invariant manifolds of the periodic orbits above really intersect.

Observe that although very similar to the closing lemma, now we are demanding that the orbits whose manifolds intersect are continuations of the original ones, so by a change of coordinates we can assume they are the same! The closing lemma only provides a point arbitrarily close to the initially given recurrent point. The Ergodic Closing Lemma of Mañé and Wen, presented in Sect. 2.5.8, does provides a bound on the distance between the original orbit and the approximating periodic orbit.

The result below is the flow version of [271, Theorem E, p. 5214] first proved by Hayashi [108, 109] (see also [23]). This shows that if two distinct points p, q have orbits which visit a given neighborhood of a point x and the points p, q are far way from a piece of the negative orbit of x, then we can find a C^1-close vector field such that p, q are in the same orbit, see Fig. 2.12.

Theorem 2.20 (Connecting lemma (Hayashi)) *Let $X \in \mathfrak{X}^1(M)$ and $x \notin S(X)$. For any C^1 neighborhood $\mathscr{U}$ of X there are $\rho > 1$, $L > 0$ and $\varepsilon_0 > 0$ such that for every $0 < \varepsilon \le \varepsilon_0$ and any two points $p, q \in M$ satisfying*

1. $p, q \notin B_\varepsilon(X^{[-L,0]}(x))$;
2. $\mathscr{O}^+_X(p) \cap B_{\varepsilon/\rho}(x) \neq \emptyset$;
3. $\mathscr{O}^-_X(q) \cap B_{\varepsilon/\rho}(x) \neq \emptyset$,

there is $Y \in \mathscr{U}$ such that $Y = X$ outside of $B_\varepsilon(X^{[-L,0]}(x))$ and such that $q \in \mathscr{O}^+_Y(p)$.

There is an extension of this result [55] showing that it is possible to connect pseudo-orbits in the C^1 setting.

Theorem 2.20 above gives a solution to the problem of connecting stable and unstable manifolds of periodic orbits. In fact this result can be stated in a slightly different way, more adapted to our needs in Chap. 5.

Theorem 2.21 *Let $X \in \mathfrak{X}^1(M)$ and $\sigma \in S(X)$ be hyperbolic. Suppose that there are $p \in W^u_X(\sigma) \setminus \{\sigma\}$ and $q \in M \setminus C(X)$ such that:*

(H1) *For all neighborhoods U, V of p, q (respectively) there is $x \in U$ such that $X^t(x) \in V$ for some $t \ge 0$.*

Then there are Y arbitrarily C^1 close to X and $T > 0$ such that $p \in W^u_Y(\sigma(Y))$ and $Y^T(p) = q$. If in addition $q \in W^s_X(x) \setminus \mathcal{O}_X(x)$ for some $x \in C(X)$ hyperbolic, then Y can be chosen so that $q \in W^s_Y(x(Y)) \setminus \mathcal{O}_Y(x(Y))$.

Moreover we can use it to connect orbits of two distinct points which accumulate a third point, but with one of the points in the unstable manifold of a hyperbolic singularity. This singularity persists under perturbation and the connecting orbits will still be in its unstable manifold.

Theorem 2.22 *Let $X \in \mathfrak{X}^1(M)$ and $\sigma \in S(X)$ be hyperbolic. Suppose that there are $p \in W^u_X(\sigma) \setminus \{\sigma\}$ and $q, x \in M \setminus C(X)$ such that:*

(H2) *For all neighborhoods U, V, W of p, q, x (respectively) there are $x_p \in U$ and $x_q \in V$ such that $X^{t_p}(x_p) \in W$ and $X^{t_q}(x_q) \in W$ for some $t_p > 0$, $t_q < 0$.*

Then there are Y arbitrarily C^1 close to X and $T > 0$ such that $p \in W^u_Y(\sigma(Y))$ and $Y^T(p) = q$.

2.5.8 The Ergodic Closing Lemma

In several proofs in this text we shall use the Ergodic Closing Lemma for flows in a fundamental way. This result shows that any given invariant measure gives full weight to the set of points whose orbits can be well approximated by periodic orbits of C^1 nearby flows. The main feature of this approximation result is that it provides a bound for the distance between the original orbit and the approximating orbit. Such amount of control is not allowed to by either the Closing Lemma or the Connecting Lemma.

The Ergodic Closing Lemma was first proved by Mañé [145] for diffeomorphisms and for flows by Wen [270]. It has become an extremely useful tool in dynamics, as the Closing and Connecting Lemmas.

We need the following definition. A point $x \in M \setminus S(X)$ is δ-*strongly closed* if for any C^1 neighborhood $\mathcal{U} \subset \mathfrak{X}^1(M)$ of X, there are $Z \in \mathcal{U}$, $z \in M$ and $T > 0$ such that $Z^T(z) = z$, $X = Z$ on $M \setminus B_\delta(X^{[0,T]}(x))$ and $\operatorname{dist}(Z^t(z), X^t(x)) < \delta$, for all $0 \le t \le T$.

Denote by $\Sigma(X)$ the set of points of M which are δ-strongly closed for any δ sufficiently small.

Theorem 2.23 (Ergodic closing lemma, flow version) *Let μ be any X-invariant Borel probability measure. Then $\mu(S(X) \cup \Sigma(X)) = 1$.*

We remark that although much stronger that the Closing Lemma, since it provides a bound on the distance of the approximating periodic orbit to the original orbit, the Ergodic Closing Lemma *cannot be applied to every orbit*, just to a full probability subset for any invariant probability measure (a *total probability set*).

2.5.9 A Perturbation Lemma for Flows

A very useful result of Franks [89, Lemma 1.1] shows that it is possible to modify a diffeomorphism to achieve a desired derivative at a finite number of points, as long as the modification is made in the C^1 topology. Here we state a version for vector fields of this result: under some mild conditions, any C^2 perturbation of the derivative of the vector field along a compact orbit segment is realized by the derivative of a C^1 nearby vector field. Hence this result allows one to locally change the derivative of the flow along a compact trajectory, while the original result of Franks allows only perturbations on a finite number of points of the orbit of a diffeomorphism.

The version we present here is an unpublished work by Pacifico and E. R. Pujals. It is very useful and it was already used in several published works [79, 172], and [173] but a proof was never provided.

To simplify notations we shall state it for flows defined on compact sets of $\mathbb{R}^n$. Using local charts it is straightforward to obtain the result for flows on compact boundaryless n-manifolds. Let M be an open subset of $\mathbb{R}^n$.

Theorem 2.24 *Let us fix $Y \in \mathfrak{X}^2(M)$, $p \in M$ and $\varepsilon > 0$. Given an orbit segment $Y^{[a,b]}(p)$, a neighborhood U of $Y^{[a,b]}(p)$, and a C^2 parametrized family of invertible linear maps $A_t : \mathbb{R}^n \longrightarrow \mathbb{R}^n$, $t \in [a,b]$ (i.e. the coefficients of the matrices A_t with respect to a fixed basis are C^2 functions of t), such that for all s, t with $t+s \le b$ we have*

1. $A_0 = Id : \mathbb{R}^n \to \mathbb{R}^n$ *and* $A_t(Y(Y^s(p))) = Y(Y^{t+s}(p))$,
2. $\|\partial_s A_{t+s} A_t^{-1}|_{s=0} - DY(Y^t(p))\| < \varepsilon$,

then there is $Z \in \mathfrak{X}^1(M)$ such that $\|Y - Z\|_1 \le \varepsilon$ and Z coincides with Y in $M \setminus U$. Moreover $Z^s(p) = Y^s(p)$ for every $a \le s \le b$ and $DZ^t(p) = A_t$ for every $t \in [a,b]$.

A proof of this result is presented in Appendix B.

Assume that there is such Z as in Theorem 2.24. On one hand A_t must preserve the direction of the vector field along the orbit segment $Y^{[a,b]}(p)$ for all $t \in [a,b]$ by item 1 above. On the other hand since

$$\begin{aligned}\partial_s A_{t+s} A_t^{-1}|_{s=0} &= \frac{\partial}{\partial s} DZ^{t+s}(p)\big(DZ^t(p)\big)^{-1}\Big|_{s=0} = \frac{\partial}{\partial s} DZ^{t+s}(p) DZ^{-t}(Z^t(p))\Big|_{s=0}\\ &= \frac{\partial}{\partial s} DZ^s(Z^t(p))\Big|_{s=0} = DZ(Z^t(p))\end{aligned}$$

we see that item 2 above ensures that Z is C^1 near Y along the orbit segment $Y^{[a,b]}(p)$.

We observe that although we start with a C^2 vector field we obtain at the end a C^1 vector field nearby the original one. If we increase the class of differentiability of the initial vector field Y and of A_t with respect to the parameter t, then we obtain Z of higher order of differentiability. But even in this setting we can only control the distance between Y and the final vector field in the C^1 topology, by results of Pujals and Sambarino in [218] which we now explain.

There is an example of a homoclinic class H (recall Sect. 2.5.4 for the definition of homoclinic class) of a C^2 diffeomorphism f on a compact surface with a unique fixed point which is a saddle-node, i.e. one of its eigenvalues is equal to one, corresponding to an indifferent direction, and the other is smaller than one in modulus, corresponding to a contracting direction. Hence there are periodic orbits x_n with arbitrarily large period p_n whose normalized Lyapunov exponent λ_n^{1/p_n} tends to 1 when $n \to +\infty$, where λ_n is an eigenvalue of $Df^{p_n}(x_n)$.

Therefore if it were possible to have a C^2 perturbation lemma analogous to Theorem 2.24, then we would obtain a C^2 diffeomorphisms arbitrarily close to f in the C^2 topology exhibiting a non-hyperbolic periodic orbit.

However in [218] Pujals and Sambarino show that for homoclinic classes H of C^2 diffeomorphisms, if k is the maximum period of non-hyperbolic periodic orbits in H, then every periodic point with period $2k$ must be hyperbolic for every C^2 close diffeomorphisms (a kind of C^2 rigidity result). This shows that a straightforward extension of Theorem 2.24 for C^2 diffeomorphisms is impossible.

2.5.10 Generic Vector Fields and Lyapunov Stability

Recall that a compact set $L \subset M$ is called *Lyapunov stable* for $X \in \mathfrak{X}^1(M)$ if for every neighborhood U of L there is a neighborhood $V \subset U$ of L such that $X^t(V) \subset U$, for all $t \geq 0$. Every attractor is a transitive Lyapunov stable set but not conversely.

The following lemmas summarize some classical properties of Lyapunov stable sets, see Chap. V in [46] for proofs.

Lemma 2.25 *Let Λ be a Lyapunov stable set of X. Then,*

1. *If $x_n \in M$ and $t_n \geq 0$ satisfy $x_n \to x \in \Lambda$ and $X^{t_n}(x_n) \to y$, then $y \in \Lambda$; and conversely, that is, this property is equivalent to Lyapunov stability;*
2. *$W^u_X(\Lambda) \subset \Lambda$;*
3. *if Γ is a transitive invariant set of X and $\Gamma \cap \Lambda \neq \emptyset$, then $\Gamma \subset \Lambda$.*

We remark that the last item ensures that if a Lyapunov stable set is transitive, then it is *maximally transitive*: it cannot be properly contained in another transitive invariant set.

Proof The first item follows easily from the definition of Lyapunov stability. For the second item take z such that $X^{-t}(z)$ approaches Λ when $t \to +\infty$. Then for any given neighborhood U we can find V as in the definition of Lyapunov stability and $T > 0$ such that $X^{-t}(z) \in V$ for all $t > T$. Thus $z \in X^t(V) \subset U$. Since U was an arbitrary neighborhood of Λ and Λ is closed, we get $z \in \Lambda$.

Finally let Γ be a transitive invariant subset such that $\Gamma \cap \Lambda \neq \emptyset$ and let U, V be neighborhoods of Λ according to the definition of Lyapunov stability. Then $V \cap$

$\Gamma \neq \emptyset$. Thus there exists in $V \cap \Gamma$ a point w whose positive orbit is dense in Γ. Hence $\Gamma = \omega_X(w) \subset U$. Since U was an arbitrary neighborhood of Λ, we deduce $\Gamma \subset \Lambda$. □

The following provides a necessary and sufficient conditions for a Lyapunov stable set to be an attractor.

Lemma 2.26 *A Lyapunov stable set Λ of a vector field X is an attracting set of X if, and only if, there exists a neighborhood U of Λ such that $\omega_X(x) \subset \Lambda$, for all $x \in U$.*

Proof We assume that such a neighborhood U exists and show that Λ must be an attracting set. In fact, we prove that $\Lambda = \cap_{t>0} \overline{X^t(V)}$ for some neighborhood V of Λ *without using that Λ is Lyapunov stable*.

Fix $\varepsilon > 0$ and write B_ε for the ε-neighborhood Λ. Let V be an open neighborhood of Λ such that its closure $\overline{V}$ is contained in U. We show that for every big enough $t > 0$ we have $X^t(V) \subset B_\varepsilon$.

Arguing by contradiction, assume that for some sequence $t_n > 0$ with $t_n \to +\infty$ there exists $x_n \in V$ such that $X^{t_n}(x_n) \notin B_\varepsilon$. Hence we can find $x \in \overline{V}$ and a subsequence n_k such that $x_{n_k} \to x$ when $k \to +\infty$ and since $M \setminus B_\varepsilon$ is closed and compact, we have $X^{t_i}(x_{n_k}) \xrightarrow[k\to+\infty]{} X^{t_i}(x) \notin B_\varepsilon$ for all i. This implies that $\omega(x) \not\subset \Lambda$, which contradicts the assumption on the neighborhood U. Therefore no such sequences exist and we have $X^t(V) \subset B_\varepsilon$ for all big enough $t > 0$. Since $\varepsilon > 0$ was arbitrarily chosen, this proves that

$$\bigcap_{t>0} X^t(V) \subset \bigcap_{\varepsilon>0} B_\varepsilon = \overline{\Lambda} = \Lambda.$$

Now assume that Λ is Lyapunov stable and that there exists a neighborhood U of Λ such that $\Lambda = \cap_{t>0} X^t(U)$. Take another neighborhood $V \subset U$ of Λ such that $X^t(V) \subset U$ for all $t \geq 0$. Then for all $s \geq t \geq 0$ we have

$$X^s(V) = X^{s-t}\big(X^t(V)\big) \subset X^{s-t}(U)$$

so in fact $X^s(V) \subset \cap_{t=0}^{s} X^t(U)$ for all $s > 0$. This implies that every accumulation point of the positive orbit of each point of V lies in $\cap_{t\geq 0} X^t(U) = \Lambda$. □

Let us collect some properties for generic vector fields $X \in \mathfrak{X}^1(M)$ for future reference.

L1. X is *Kupka-Smale*, i.e. every periodic orbit and singularity of X is hyperbolic and the corresponding invariant manifolds intersect transversely.

Among conservative or incompressible vector fields $\mathfrak{X}^1_\mu(M)$ with respect to a volume form μ induced by the Riemannian metric on M, there exists also a residual subset of vector fields such that every singularity and periodic orbit is of saddle-type or elliptic, and the corresponding invariant manifolds intersect transversely.

For the proof of this classic result see [190]. In particular, $S(X)$ is a finite set.

L2. $\Omega(X) = \overline{\text{Per}(X) \cup S(X)}$, as a consequence of the *Closing Lemma*, see [212] and Sect. 2.5.6.
L3. $\overline{W^u_X(\sigma)}$ is Lyapunov stable for X for each $\sigma \in S(X)$ with one-dimensional unstable manifold.
L4. $\overline{W^s_X(\sigma)}$ is Lyapunov stable for $-X$, for every $\sigma \in S(X)$ with one-dimensional stable manifold.
L5. If $\sigma \in S(X)$ and $\dim(W^u_X(\sigma)) = 1$ then $\omega_X(q)$ is Lyapunov stable for X, for every $q \in W^u_X(\sigma) \setminus \{\sigma\}$.
L6. If $\sigma \in S(X)$ and $\dim(W^s_X(\sigma)) = 1$ then $\alpha_X(q)$ is Lyapunov stable for $-X$, for all $q \in W^s_X(\sigma) \setminus \{\sigma\}$.

The proofs of items L3 to L6 following [66] are presented in Appendix A.

2.6 The Linear Poincaré Flow

The following notion can be defined for a flow on any finite dimensional Riemannian manifold. If x is a regular point of X (i.e. $X(x) \neq \mathbf{0}$), denote by

$$N_x = \{v \in T_x M : v \cdot X(x) = 0\}$$

the orthogonal complement of $X(x)$ in $T_x M$. Denote by $O_x : T_x M \to N_x$ the orthogonal projection of $T_x M$ onto N_x. For every $t \in \mathbb{R}$ define

$$P^t_x : N_x \to N_{X^t(x)} \quad \text{by} \quad P^t_x = O_{X^t(x)} \circ DX^t(x).$$

It is easy to see that $P = \{P^t_x : t \in \mathbb{R}, X(x) \neq 0\}$ satisfies the cocycle relation

$$P^{s+t}_x = P^t_{X^s(x)} \circ P^s_x \quad \text{for every } t, s \in \mathbb{R}.$$

The family P is called the *Linear Poincaré Flow* of X.

2.6.1 Hyperbolic Splitting for the Linear Poincaré Flow

Let a non trivial (i.e., containing a regular orbit) compact subset Λ invariant under the flow of $X \in \mathfrak{X}^1(M)$ be given, where M is a finite dimensional Riemannian manifold. Assume that Λ is *hyperbolic and transitive*, as defined in Sect. 2.3. Then the normal space N_x is defined for all $x \in \Lambda$, since Λ does not contain singularities. Hence the Linear Poincaré Flow is defined everywhere on the family of normal spaces $N_\Lambda = \{N_x\}_{x \in \Lambda}$. Compactness and absence of singularities enable us to obtain the following characterization of transitive hyperbolic subsets for flows.

Theorem 2.27 *Let Λ be a transitive compact invariant subset for $X \in \mathfrak{X}^1(M)$. Then Λ is (uniformly) hyperbolic if, and only if, the Linear Poincaré Flow is everywhere defined over Λ and P_Λ admits a (uniformly) hyperbolic splitting of N_Λ.*

Proof If Λ is a transitive compact invariant hyperbolic set for X, then the corresponding DX^t-invariant tangent bundle decomposition $T_\Lambda M = E^s \oplus E^X \oplus E^u$ projects into a normal bundle decomposition $N_\Lambda = N^s \oplus N^u$ through the orthogonal projection as: $N_x^s = O_x(E_x^s)$ and $N_x^u = O_x(E_x^u)$ for all $x \in \Lambda$. Since Λ does not contain singularities, this is well defined and the splitting of the normal bundle is hyperbolic since the orthogonal projection does not increase norms:

- for $v \in E_x^s$ we have that $\|P^t v\| = \|O_{X^t(x)} DX_x^t \cdot v\| \le \|DX_x^t \cdot v\|$ is uniformly contracted for $t > 0$;
- for $u \in E_x^u$ we have that $\|P^{-t} u\| = \|O_{X^{-t}(x)} DX_x^{-t} \cdot u\| \le \|DX_x^{-t} \cdot u\|$ is also uniformly contracted for $t > 0$. Hence for some $K, \lambda > 0$ we get $\|u\| = \|P^{-t} P^t u\| \le K e^{-\lambda t} \|P^t u\|$ and so $\|P^t u\| \ge K e^{\lambda t} \|u\|$ as needed for hyperbolicity of P^t.

For the converse, assume that there exists a P^t-invariant hyperbolic decomposition $N_\Lambda = N^s \oplus N^u$ of the normal bundle over Λ. In particular, this ensures that Λ has no singularities.

Set $E^{cu} = E^X \oplus N^u$ over Λ. Each vector $v \in E_x^{cu}$ can be written as $v = \alpha X(x) + u$ for $u \in N_x^u$ and $\alpha \in \mathbb{R}$. This is clearly a DX^t-invariant bundle since $DX_x^t \cdot v = DX_x^t \cdot u + \alpha \cdot X(X^t(x))$ and $O_{X^t(x)} DX_x^t \cdot u \in N_{X^t(x)}^u$, thus this vector belongs to $N_{X^t(x)}^u \oplus E_{X^t(x)}^X$, for all $x \in \Lambda$ and $t \in \mathbb{R}$.

The uniform expansion of vectors along the N_x^u direction under P^t ensures that, for some $K, \lambda > 0$ not depending on $x \in \Lambda$ and $t > 0$

$$\|DX_x^t \cdot v\| \ge \|O_x DX_x^t \cdot u\| \ge K e^{\lambda t} \|u\|, \tag{2.4}$$

where $v \in E_x^{cu}$ decomposes as explained above and

$$\|DX_x^t \cdot \alpha X(x)\| = |\alpha| \cdot \|X(X^t(x))\| \le |\alpha| \cdot \max_{x \in \Lambda} \|X(x)\|.$$

Hence for all big enough $s \in \mathbb{R}$ we have for some $C > 0$

$$m(DX^s \mid E_x^{cu}) > C e^{\lambda s} \|DX^s \mid E_x^X\|, \quad x \in \Lambda. \tag{2.5}$$

Consider now the space $\mathscr{L} = \mathscr{L}([x]_\Lambda, E_\Lambda^{cu})$ of families of maps

$$\Big\{\ell_x : N_x^u \to E_x^X \text{ is linear for all } x \in \Lambda \text{ and } \|(\ell_x)\| := \sup_{x \in \Lambda} \|\ell_x\| < \infty\Big\}.$$

The norm of each linear map is well defined since the spaces have finite dimension and are endowed with the induced norm from the Riemannian structure of the manifold. This is a Banach space since Λ is compact.

Now we show that the action of DX^s on $\mathscr{L}$ given by the graph transform is a contraction for s as in (2.5). Indeed the image of an element $(\ell_x) \in \mathscr{L}$ is

$$\mathscr{D}^s\big((\ell_x)\big) = \big\{(O_{X^s(x)} + DX_x^s \mid E_x^X \circ \ell_x) \circ (O_{X^s(x)} \circ DX_x^s)^{-1} : x \in \Lambda\big\}$$

and $\|\mathscr{D}^s((\ell_x))\| \le \sup_{x\in\Lambda}[\|DX^s \mid E^X_x\|/m(DX^s \mid E^{cu}_x)] \le e^{-\lambda s}/C < 1$. We have then a fixed element ℓ for $\mathscr{D}^s$. This corresponds to a sub-bundle E^u given at each $x \in \Lambda$ by the graph of ℓ_x which is DX^s-invariant for this particular value of s.

Now we show that ℓ not only does not depend on s but also that it is $\mathscr{D}^t$-invariant for all t. Indeed, it is easy to see that the actions commute

$$\mathscr{D}^s\mathscr{D}^r = \mathscr{D}^{(s+r)} = \mathscr{D}^r\mathscr{D}^s \quad s, r \in \mathbb{R}$$

even for negative s, r since the flow is complete, and for all big enough $s, r > 0$ both $\mathscr{D}^r$ and $\mathscr{D}^s$ are contractions. Thus the fixed point is the same: $\mathscr{D}^s\mathscr{D}^r(\ell) = \mathscr{D}^r\mathscr{D}^s(\ell) = \mathscr{D}^r(\ell)$ hence $\mathscr{D}^r(\ell) = \ell$. Therefore given any $t \in \mathbb{R}$ we can find $s, r > 0$ very big with $s - r = t$ and obtain

$$\mathscr{D}^t(\ell) = \mathscr{D}^{(s-r)}(\ell) = \mathscr{D}^s\mathscr{D}^{-r}(\ell) = \mathscr{D}^s(\ell) = \ell.$$

Finally note that the same argument as in (2.4) shows that the sub-bundle E^u is uniformly expanded by DX^t.

Analogously we obtain the contracting direction E^s reasoning with $E^{cs} = N^s \oplus E^X$ and the action $\mathscr{D}^{-s}$ for some big enough $s > 0$. □

2.6.2 Dominated Splitting for the Linear Poincaré Flow

We say that an invariant subset Λ without singularities (not necessarily compact) has a (C, λ)-*dominated splitting for the Linear Poincaré Flow* if $N_x = N^{cs}_x \oplus N^{cu}_x$ is a P^t-invariant splitting defined for all $x \in \Lambda$ and there are constants $\lambda, C > 0$ such that for all $t > 0$

$$\|P^t \mid N^{cs}_x\| \le Ce^{-\lambda t} \cdot m(P^t \mid N^{cu}_x), \tag{2.6}$$

where $m(\cdot)$ denotes the minimum norm of a linear map L on a normed space, that is $m(L) := \min\{\|Lv\| : \|v\| = 1\}$. We observe that clearly a hyperbolic splitting is a dominated splitting. We show in Sect. 2.6.3 that for an invariant subset, without singularities on its closure, with respect to a volume preserving flow on a 3-manifold, hyperbolicity is equivalent to the existence of a dominated splitting for the Linear Poincaré Flow.

We can also define a (C, λ)-*dominated splitting for the flow* on an invariant set Λ: it is a splitting $E^{cs} \oplus E^{cu}$ of the tangent bundle $T_\Lambda M$ over Λ, invariant under DX^t and satisfying

$$\|DX^t \mid E^{cs}_x\| \le Ce^{-\lambda t} \cdot m(DX^t \mid E^{cu}_x). \tag{2.7}$$

We remark that the same proof of Theorem 2.27 shows that *a dominated splitting $E^{cs} \oplus E^{cu}$ for the flow over Λ induces a dominated splitting for the Linear Poincaré Flow P^t over Λ* setting $N^{cs}_x := \mathscr{O}_x \cdot E^{cs}_x$ and $N^{cu}_x := \mathscr{O}_x \cdot E^{cu}_x$ for $x \in \Lambda$.

A dominated splitting must be a continuous splitting in the Whitney sense, as follows.

Lemma 2.28 *Let a sequence of points $x_k \in \Lambda$ and a orthonormal basis $\mathcal{B}_k = (v_{k,1}, \dots, v_{k,l})$ of N^{cs} at x_k be given. If $x_n \to x \in \Lambda$ and $\mathcal{B}_k \to (v_1, \dots, v_l)$ as $k \to \infty$, then $(v_1, \dots, v_l)$ is also a basis for N^{cs}_x. The analogous statement is also true for the sub-bundle N^{cu}. In particular the dimensions of the sub-bundles of a dominated decomposition are continuous functions of the base point.*

We remark that a dominated splitting induces a kind of hyperbolic dynamics on the projective bundle of $\dim N^{cu}$-planes orthogonal to the flow direction. Indeed in this bundle every element converges to N^{cu} under the action of the Linear Poincaré Flow. This is because domination means that the component on the direction of N^{cs} of any vector orthogonal to the flow direction on $T_\Lambda M$ vanishes when compared to the component along the N^{cu} direction. Analogously P^{-t} acts on $\dim N^{cs}$-planes orthogonal to the flow direction in such a way that they converge to N^{cs}. This behavior ensures that *a dominated splitting for the linear Poincaré flow with given bundle dimensions is unique*. It is easy to see that there can be distinct dominated splittings for the Linear Poincaré Flow on invariant sets in dimensions bigger than 3, just consider a hyperbolic periodic orbit $\mathcal{O}_X(p)$ with period τ whose "period map" $X^\tau \mid T_p M$ has 4 distinct eigenvalues $\lambda < 1 < \sigma_1 < \sigma_2$, where $\dim(M) = 4$.

Proof of Lemma 2.28 In the setting of the statement, assume by contradiction that there exists some vector v in the span of $(v_1, \dots, v_l)$ which is not in E_x. Since $(v_1, \dots, v_l)$ is still orthonormal, this means that some element, say v_1, of this linearly independent family is not in E_x. Hence we may consider the projections $v_1 = \pi_s v_1 + \pi_u v_1 = v_{1,s} + v_{1,u}$ of v_1 on $N^{cs}_x \oplus N^{cu}_x$ and of its iterates under P^t: $v^t_{1,s} + v^t_{1,u} = v^t_1$, $t > 0$ and we know that $v^t_1 \neq \mathbf{0}$.

The domination implies now that $\alpha^t := \|v^t_{1,u}\| / \|v^t_{1,s}\| \geq K e^{\lambda t}$ for some constant $K > 0$ depending only on C and $\|v_{1,u}\| / \|v_{1,s}\|$. Fix $s > 0$ big enough so that $\alpha^s > 1$. The continuity of DX^s_x as a function of x ensures that

$$\beta^s_k := \frac{\|\pi_u P^s v_{k,1}\|}{\|\pi_s P^s v_{k,1}\|} \xrightarrow[k\to\infty]{} \alpha^s > 1$$

but the invariance of the splitting under P^t implies that $P^s v_{k,1} \in E^{cs}_{x_k}$ for all $k \geq 1$, thus $\beta^s_k = 0$. This contradiction shows that all v_i are in N^{cs}_x, $i = 1, \dots, l$, so

$$\dim(N^{cs}_x) \geq \limsup_{k\to\infty} \dim(N^{cs}_{x_k}). \tag{2.8}$$

Clearly the same argument is true for N^{cu}. Therefore we can write

$$\begin{aligned}
\limsup_{k\to\infty} \dim(N^{cs}_{x_k}) &= \dim(M) - 1 - \liminf_{n\to\infty} \dim(N^{cu}_{x_k}) \\
&\leq \dim(N^{cs}_x) = \dim(M) - 1 - \dim(N^{cu}_x) \\
&\leq \dim(M) - 1 - \limsup_{k\to\infty} \dim(N^{cu}_{x_k})
\end{aligned}$$

that is

$$\limsup_{k\to\infty} \dim(N_{x_k}^{cu}) \le \dim(N_x^{cu}) \le \liminf_{n\to\infty} \dim(N_{x_k}^{cu}).$$

Hence the dimensions of N_x^{cu} and N_x^{cs} are in fact continuous functions of the base point. In particular the family $(v_1, \ldots, v_l)$ is a base for N_x^{cs}. □

We also have that a dominated splitting is *robust* in the following sense.

Lemma 2.29 *Given a connected invariant subset Λ with a (C,λ)-dominated splitting for the linear Poincaré flow with respect to a vector field X, then there exists a neighborhood U of Λ and $\delta > 0$ such that the set $\Lambda_Y(U)^* := \cap_{t\in\mathbb{R}} Y^t(U \setminus S(Y))$ has a (C',λ')-dominated splitting for the linear Poincaré flow with respect to any vector field $Y \in \mathscr{U}$ which is δ-C^1-close to X, where C', λ' are positive constants depending only on δ, U and (C,λ), and satisfy $(C',\lambda') \to (C,\lambda)$ as $\delta \to 0$ and $U \to \Lambda$.*

We present a proof of this lemma in Appendix C. This means that perturbing the original flow X to Y around an invariant dominated set, we can get (2.6) for every regular orbit of Y which remains nearby Λ with domination constants C', λ' close to the original ones for Λ.

Assume that a C^1 flow X admits a compact attracting set with isolating neighborhood U, that is $\Lambda = \Lambda_X(U) = \cap_{t\in\mathbb{R}} \overline{X^t(U)}$ and $\overline{X^t(U)} \subset U$ for every $t > 0$. Hence there exists a C^1-neighborhood $\mathscr{U}$ of X such that if $Y \in \mathscr{U}$, $x \in \mathrm{Per}(Y)$ and $\mathscr{O}_Y(x) \cap U \ne \emptyset$, then

$$\mathscr{O}_Y(x) \subset \Lambda_Y(U). \tag{2.9}$$

Given $Y \in \mathscr{U}$ define $\Lambda_Y^*(U) = \Lambda_Y(U) \setminus S(Y)$. In what follows P^t stands for the linear Poincaré flow of X over $\Lambda_X^*(U)$.

Definition 2.30 We say that a singularity σ of a flow X^t is *(generalized) Lorenz-like* if there are two real eigenvalues $\lambda < 0 < \mu$ of $DX(\sigma)$ such that every other eigenvalue is contained in the union of the sets

$$\lambda^-(\sigma) = \{z \in \mathbb{C} : z \ne \lambda \text{ is an eigenvalue of } DX(\sigma) \text{ with } \Re(z) < 0\}, \text{ and}$$

$$\mu^+(\sigma) = \{z \in \mathbb{C} : z \ne \mu \text{ is an eigenvalue of } DX(\sigma) \text{ with } \Re(z) > 0\};$$

and satisfy $\sup\{\Re(z) : z \in \lambda^-(\sigma)\} < \lambda < 0 < -\lambda < \mu \le \inf\{\Re(z) : z \in \mu^+(\sigma)\}$ and also $\lambda^-(\sigma) \ne \emptyset$.

We remark that for a vector field in a three-dimensional manifold such singularities have only real eigenvalues $\lambda_2 \le \lambda_3 \le \lambda_1$ such that $\lambda_2 < \lambda_3 < 0 < -\lambda_3 < \lambda_1$. We will see that these singularities are the only ones allowed on robustly transitive sets in three-dimensional manifolds.

Remark 2.31 Metzger and Morales in [156] present a more general definition of generalized Lorenz-like equilibrium. However, we can be more specific and show that equilibria, properly accumulated by regular orbits in attracting sets for higher dimensional flows, must satisfy this more restrictive structure, see Lemma 5.22 and Remark 5.26.

Definition 2.32 We say that a vector field $X \in \mathfrak{X}^1(M)$ is *homogeneous* on a set $U \subset M$ if there exists a C^1 open set $\mathcal{U}$ in $\mathfrak{X}^1(M)$ containing X such that

- for all $X \in \mathcal{U}$ there are no sinks nor sources in U, and
- every critical element of X in $\Lambda_X(U)$ is hyperbolic.

In addition, if the dimension of M is greater than three, we impose also the condition

- the index of the continuation on $\mathcal{U}$ of every critical element does not change (i.e. there are no bifurcations of critical elements).

Using (2.9) and the same arguments as in [79, Theorem 3.2] (see also [270, Theorem 3.8] and [132]) we obtain

Theorem 2.33 (Homogeneous flows and dominated splitting) *Let $X \in \mathfrak{X}^1(M)$ be a homogeneous vector field in an open subset U of M and let $\Lambda_Y(U)$ be connected and a subset of $\Omega(Y)$ for all $Y \in \mathfrak{X}^1(M)$ in a C^1 neighborhood of X. Then there exists an invariant, continuous and dominated splitting $N_{\Lambda_X^*(U)} = N^{cs,X} \oplus N^{cu,X}$ for the Linear Poincaré Flow P^t on $\Lambda_X^*(U)$. Moreover*

1. *for all hyperbolic sets $\Gamma \subset \Lambda_X^*(U)$ with splitting $E^{s,X} \oplus E^X \oplus E^{u,X}$*
$$E_x^{s,X} \subset N_x^{cs,X} \oplus E_x^X \quad \text{and} \quad E_x^{u,X} \subset N_x^{cu,X} \oplus E_x^X, \quad \text{for every } x \in \Gamma.$$
2. *If $Y_n \to X$ in $\mathfrak{X}^1(M)$ and $x_n \to x$ in M, with $x_n \in \Lambda_{Y_n}^*(U), x \in \Lambda_X^*(U)$, then $N_{x_n}^{cs,Y_n} \xrightarrow[n\to\infty]{} N_x^{cs,X}$ and $N_{x_n}^{cu,Y_n} \xrightarrow[n\to\infty]{} N_x^{cu,X}$.*
3. *If $\sigma \in S(X) \cap \Lambda_X(U)$ is a (generalized) Lorenz-like singularity and $x \in W^s(\sigma) \setminus \{\sigma\}$, then on N_x one invariant and dominated splitting for the Linear Poincaré Flow is given by $N_x^{cs} = N_x \cap T_x W^s(x)$ and $N_x^{cu} = N_x \cap T_x W^u(x)$.*

We present a proof of all items above, plus a sketch of the argument proving the existence of the dominated splitting on regular orbits.

The proof of existence is detailed in Sect. 4.2 of Chap. 4.

Proof Given a hyperbolic compact invariant subset Γ inside $\Lambda_X(U)^*$, the projections of the contracting E^s and expanding E^u subbundles of $T\Gamma$ on the normal subbundle N_Γ over Γ must coincide with N^{cs} and N^{cu}, respectively. Indeed, these projections are P^t-invariant and form a dominated decomposition of N_Γ (recall the proof of Theorem 2.27), so the equality follows from the uniqueness of such dominated decomposition after the dimensions have been fixed.

We observe that if the dominated splitting for the Linear Poincaré Flow is constructed for $\Lambda_Y(U)^*$ for all Y sufficiently C^1-close to X, then the continuity statements in item 2 follow from the same argument in the proof of Lemma 2.28.

The uniqueness statement of item 3 is a consequence of uniqueness of dominated splittings when the dimensions of the sub-bundles are fixed, since along orbits of a point $x \in U$ converging to σ the linearization of the flow of X near σ defines a dominated splitting with the dimensions of $N_x \cap W^u(x)$ and $N_x \cap W^s(x)$ uniquely. In the three-dimensional case the sub-bundles N^{cs} and N^{cu} are both one-dimensional, and thus there exists a unique invariant and dominated splitting for the Linear Poincaré Flow.

To construct such splitting we use the Closing Lemma of Pugh, Theorem 2.19, since we are assuming that $\Lambda_Y(U) \subset \Omega(Y)$. For every orbit of $x \in \Lambda_Y(U)$ we find a periodic orbit $\mathcal{O}_Z(p) \subset U$ for a C^1 nearby flow Z which is close to a long piece of the orbit of x. The homogeneity condition on the vector field X ensures that this is a hyperbolic periodic orbit and that the dimensions of the strong-stable and strong-unstable manifolds of p do not depend either on the periodic orbit $\mathcal{O}(p)$ inside U or on the vector field Z. The connectedness assumption on $\Lambda_Y(U)$ enables us to show that it is enough to carry this argument over to periodic orbits whose periods are bounded from below away from zero, since the complement in $\Lambda_Y(U)$ of the periodic orbits with small period is dense in $\Lambda_Y(U)$.

The hyperbolicity ensures that the splitting $E^s \oplus E^Z \oplus E^u$ along $\mathcal{O}_Z(p)$ induces a dominated splitting $N^s \oplus N^u$ on $N_{\mathcal{O}(p)}$ invariant under P_Z^t. If we can show that this is a (C, λ)-dominated splitting with (C, λ) not depending either on the periodic orbit or on the flow, then we would be able to define a dominated splitting for the flow along the Y orbit of x as the limit of the sub-bundles $N_{p_n}^s$, $N_{p_n}^u$ over the periodic orbits $\mathcal{O}_{Z_n}(p_n)$ of flows Z_n^t converging to Y^t.

The uniform (C, λ)-dominated splitting for Z along its periodic orbits inside U is provided by the arguments of Linear Systems of Matrixes in [145] and in [79]. The basic idea is to show that if there exist periodic orbits with a (C, λ) dominated splitting either with λ very close to 1 or C very big, then it is possible to find a C^1 flow Z_0 close to Z exhibiting a periodic orbit with a different index from the other periodic orbits in U, contradicting the homogeneous condition on X. □

2.6.3 *Incompressible Flows, Hyperbolicity and Dominated Splitting*

In the particular case of *incompressible* or *volume preserving* flows on three-dimensional manifolds, we show that in the absence of singularities, a dominated splitting of the linear Poincaré Flow is equivalent to hyperbolicity.

A C^1 flow X in an n-manifold M endowed with a Riemannian metric is *incompressible* or *conservative* if the volume form $\omega = \text{Leb}$ is invariant under DX^t, that is $(DX^t)^*\omega = \omega$. This means that for any given basis $v_1, \dots, v_n$ of T_xM we have that $\omega(DX_x^t \cdot v_1, \dots, DX_x^t \cdot v_n) = \omega(v_1, \dots, v_n)$ for all $x \in M$ and $t \in \mathbb{R}$.

Proposition 2.34 *Let $X \in \mathfrak{X}^1_\omega(M^3)$ have an invariant set $\Lambda \neq \emptyset$ (not necessarily compact) such that its closure $\overline{\Lambda}$ has no singularities of X and admits a dominated splitting for the Linear Poincaré Flow. Then $\overline{\Lambda}$ is an hyperbolic set.*

Before the proof we need the following definition, which will be used throughout the rest of the book: given two subspaces $A \subset T_x M$ and $B \subset T_x M$ such that $A \cap B = \{\mathbf{0}\}$, the angle $\measuredangle(A, B)$ or $\angle(A, B)$ between A and B is defined as

$$\angle(A, B) = \inf\{\measuredangle(\mathbf{v}, \mathbf{w}) : \mathbf{v} \in A \setminus \{\mathbf{0}\}, \mathbf{w} \in B \setminus \{\mathbf{0}\}\},$$

where $\measuredangle(\mathbf{v}, \mathbf{w}) = \arccos\left\langle \frac{\mathbf{v}}{\|\mathbf{v}\|}, \frac{\mathbf{w}}{\|\mathbf{w}\|} \right\rangle$ is well defined for non-zero vectors in a given tangent space by the Riemannian inner product.

Proof On a 3-manifold the domination of the Linear Poincaré Flow can only occur if the dimension of the sub-bundles in the splitting of the normal bundle is constant an equal to 1. Since there are no singularities on $\overline{\Lambda}$ the dominated splitting extends to the closure. Then there exists $K > 1$ and $\beta > 0$ such that for all $x \in \overline{\Lambda}$ the flow size is bounded above and below and the angles between the two sub-bundles of the dominated splitting are bounded away from zero, as follows

$$K^{-1} \le \|X(x)\| \le K \quad \text{and} \quad \theta_t := \measuredangle(N^u_{X^t(x)}, N^s_{X^t(x)}) \ge \beta.$$

Since the flow is incompressible we must have

$$\left|\det(DP^t_x)\right| \cdot \frac{\|X(X^t(x))\|}{\|X(x)\|} = 1 \quad \text{and we set} \quad x(t) = \frac{\|X(X^t(x))\|}{\|X(x)\|}.$$

We can write this in a convenient way

$$\sin(\theta_0) = \|P^t_x \mid N^u_x\| \cdot \|P^t_x \mid N^s_x\| \cdot \sin(\theta_t) \cdot x(t)$$

and so we can relate with the domination as follows

$$\|P^t_x \mid N^s_x\|^2 = \frac{\sin(\theta_0)}{x(t)\sin(\theta_t)} \cdot \frac{\|P^t_x \mid N^s_x\|}{\|P^t_x \mid N^u_x\|} \le \frac{K^2 C}{\sin(\beta)} e^{-\lambda t}.$$

Hence N^s is uniformly contracted by the Linear Poincaré Flow. We proceed analogously for the unstable direction and obtain that the dominated splitting is in fact hyperbolic. Then by Theorem 2.27 we have that $\overline{\Lambda}$ is hyperbolic.

The reciprocal is easier: by Theorem 2.27 the Linear Poincaré Flow over $\overline{\Lambda}$ is hyperbolic and hyperbolic splittings are always dominated. □

2.7 Ergodic Theory

The ergodic theory of uniformly hyperbolic systems was initiated by Sinai's Theory of Gibbs States for Anosov flows [16, 250] and was extended to Axiom A flows and diffeomorphisms by Bowen and Ruelle [60, 62]. The special measures studied by these authors are commonly referred to by their combined name *Sinai-Ruelle-Bowen* or just *SRB* in the literature since.

Recall that an *invariant probability measure* μ for a flow $X \in \mathfrak{X}^r(M)$ is a probability measure such that $\mu((X^t)^{-1}A) = \mu(A)$ for all measurable subsets A and any $t > 0$ or, equivalently, $\int \varphi \circ X^t d\mu = \int \varphi \, d\mu$ for all continuous functions $\varphi : M \to \mathbb{R}$ and any $t > 0$.

A simple and useful property of invariant probability measures is given by the Recurrence Theorem of Poincaré: the set of points x such that $x \in \omega_X(x)$, that is, the set of *recurrent points*, has full μ measure on the ambient space.

Recall also that an invariant measure μ is *ergodic* if any X-invariant subset have either measure 0 or 1 with respect to μ. Equivalently, any X-invariant function $\varphi \in L^1(\mu)$, i.e., $\varphi \circ X^t = \varphi$ almost everywhere for all $t > 0$ is constant almost everywhere with respect to μ. The cornerstone of Ergodic Theory is the following celebrated result of George David Birkhoff (see [48] or for a recent presentation [269]).

Theorem 2.35 (Ergodic theorem) *Let $f : M \to M$ be a measurable transformation, μ an f-invariant probability measure and $\varphi : M \to \mathbb{R}$ a bounded measurable function. Then the time average $\tilde{\varphi}(x) = \lim \frac{1}{n} \sum_{j=0}^{n-1} \varphi(f^n(x))$ exists for μ-almost every point $x \in M$. Moreover $\tilde{\varphi}$ is f-invariant and $\int \tilde{\varphi}\, d\mu = \int \varphi \, d\mu$. In addition, if μ is ergodic, then $\tilde{\varphi} = \int \varphi \, d\mu$ almost everywhere with respect to μ.*

For a flow X^t just replace in the statement of Theorem 2.35 above the discrete time average with $\lim_{T \to +\infty} \frac{1}{T} \int_0^T \varphi(X^t(z))\, dt$ and f-invariance by X-invariance. For invertible transformations or flows forward and backward (i.e. with $T \to -\infty$) time averages are equal μ almost everywhere.

Every invariant probability measure μ is a generalized convex linear combination of ergodic measures in the following sense: for μ-a.e. x there exists an ergodic measure μ_x satisfying for every continuous function φ

$$\int \varphi \, d\mu_x = \lim_{T \to \pm\infty} \frac{1}{T} \int_0^T \varphi\big(X^t(z)\big)\, dt$$

and for every bounded measurable function ψ we have

$$\int \psi \, d\mu = \int \left(\int \psi \, d\mu_x \right) d\mu(x).$$

2.7.1 Physical or SRB Measures

The chaotic nature of hyperbolic phenomena prevents accurate long term predictions for many models of physical, biological or economic origin. Inspired by an analogous situation of unpredictability faced in the field of Statistical Mechanics/Thermodynamics—although due to the large number of particles involved, whereas dynamical systems exhibit unpredictability even for models expressed with few variables and simple mathematical formulas, e.g. the Lorenz flow in Sect. 2.2.3—researchers focused on the statistics of the data provided by the time averages of some observable (a continuous function on the manifold) of the system. Time averages are guaranteed to exist for a positive volume subset of initial states (also called an *observable subset*) on the mathematical model if the transformation, or the flow associated to the ordinary differential equation, admits a smooth invariant measure (a density) or a *physical* measure.

Indeed, if μ_0 is an ergodic invariant measure for the transformation T_0, then the Ergodic Theorem ensures that for every μ-integrable function $\varphi : M \to \mathbb{R}$ and for μ-almost every point x in the manifold M the time average

$$\tilde{\varphi}(x) = \lim_{n\to+\infty} n^{-1} \sum_{j=0}^{n-1} \varphi(T_0^j(x))$$

exists and equals the space average $\int \varphi \, d\mu_0$. A *physical measure* μ is an invariant probability measure for which it is *required* that *time averages of every continuous function* φ *exist for a positive Lebesgue measure (volume) subset of the space and be equal to the space average* $\mu(\varphi)$.

We note that if μ is a density, that is, is absolutely continuous with respect to the volume measure, then the Ergodic Theorem ensures that μ is physical. However not every physical measure is absolutely continuous. To see why in a simple example we just have to consider a singularity p of a vector field which is an attracting fixed point (a sink), then the Dirac mass δ_p concentrated on p is a physical probability measure, since every orbit in the basin of attraction of p will have asymptotic time averages for any continuous observable φ given by $\varphi(p) = \delta_p(\varphi) = \int \varphi \, d\delta_p$.

Physical measures need not to be unique or even exist in general, but when they do exist it is desirable that *the set of points whose asymptotic time averages are described by physical measures* (such set is called the *basin* of the physical measures) *be of full Lebesgue measure*—only an exceptional set of points with zero volume would not have a well defined asymptotic behavior. This is yet far from being proved for most dynamical systems, in spite of much recent progress in this direction.

There are robust examples of systems admitting several physical measures whose basins together are of full Lebesgue measure, where *robust* means that there are whole open sets of maps of a manifold in the C^2 topology exhibiting these features. For typical parametrized families of one-dimensional unimodal maps (maps of the circle or of the interval with a unique critical point) it is known that the above scenario holds true for Lebesgue almost every parameter [143]. It is known that there are systems admitting no physical measure [118], but the only known cases are not robust, i.e. there are systems arbitrarily close which admit physical measures.

2.7.1.1 Physical Probability Measures for a Flow

Given an invariant probability measure μ for a flow X^t, let $B(\mu)$ be the *(ergodic) basin* of μ, i.e., the set of points $z \in M$ satisfying for all continuous functions $\varphi : M \to \mathbb{R}$

$$\lim_{T \to +\infty} \frac{1}{T} \int_0^T \varphi\big(X^t(z)\big)\, dt = \int \varphi \, d\mu.$$

We say that μ is a *physical* (or *SRB*) measure for X if $B(\mu)$ has positive Lebesgue measure: $\mathrm{Leb}(B(\mu)) > 0$.

The notion of *SRB* measure captures the intuitive idea that the natural measure for a dynamical system should be one which gives probabilistic information on the asymptotic behavior of trajectories starting from a "big" set of initial states. Here the notion of "big" can arguably be taken to mean "positive volume". In this sense an *SRB* measure provides information on the behavior of trajectories starting from a set of initial states which is in principle "physically observable" in practice, say when modeling some physical experiment. That is why the name *physical measure* is also attached to them

This kind of measures was first constructed for C^2 Anosov flows by Anosov and Sinai [16] and later for every Axiom A attractor for C^2 flows and for C^2 diffeomorphisms by Bowen and Ruelle [60, 62]. Moreover if the attractor is transitive (i.e. a basic piece in the spectral decomposition of an Axiom A flow), then there is a unique such measure supported in the attractor whose basin covers a full neighborhood of the attractor except for a volume zero subset. In addition, in the setting of diffeomorphisms these measures are ergodic and mixing (see Sect. 10.2.3 for the definition of mixing for an ergodic probability measure).

The existence of physical measures shows that uniformly hyperbolic attractors have well defined asymptotic behavior in a probabilistic sense for Lebesgue almost all points in a neighborhood.

2.7.2 Gibbs Measures Versus SRB Measures

The concept of SRB measure is closely related to the concept of Gibbs measure introduced in the setting of uniformly hyperbolic flows and transformation by Sinai [16, 250] and by Bowen and Ruelle [60, 62]. To explain these notions we need to recall the definition and basic properties of Lyapunov exponents for flows.

2.7.2.1 Multiplicative Ergodic Theorem for a Smooth Flow

We recall that for a given flow X the *Lyapunov exponent of x in the direction of $v \in T_x M \setminus \{0\}$* is the number

$$L(x, v) = \liminf_{t \to +\infty} \frac{1}{t} \log \left\| DX^t(x) v \right\|. \tag{2.10}$$

Given an invariant ergodic probability measure μ for the flow X^t, the multiplicative ergodic theorem of Oseledets [185, 269] ensures that μ almost every x there exists a DX^t-invariant splitting (for all $t > 0$) $T_xM = E_1 \oplus \cdots \oplus E_k$ and numbers $\lambda_1 < \cdots < \lambda_k$ such that for all $i = 1, \dots, k$ and $v_i \in E_i \setminus \{0\}$

$$\lambda_i = \lim_{t\to+\infty} \frac{1}{t} \log \| DX^t(x)v_i \|. \tag{2.11}$$

Moreover the angles between these invariant subspaces decrease along orbits of the flow at most subexponentially, so that the rates of growth expressed by the exponents dominate the possibly small angles along most orbits. More precisely, we have that for any $I \subset \{1,\dots,k\}$ and μ-almost every point

$$\lim_{t\to\pm\infty} \frac{1}{t} \log \sin \measuredangle \left(\bigoplus_{i\in I} E^i_{X^t(x)}, \bigoplus_{j\notin I} E^j_{X^t(x)} \right) = 0.$$

These facts imply the following regularity condition:

$$\lim_{t\to\pm\infty} \frac{1}{t} \log \left| \det\left(DX^t \mid \bigoplus_{i\in I} E^i_x \right) \right| = \sum_{i\in I} \lambda_i(x) \dim E^i_x$$

for μ-almost every point and, in particular

$$\lim_{t\to\pm\infty} \frac{1}{t} \log |\det DX^t(x)| = \sum_{i=1}^{k} \lambda_i(x) \dim E^i_x.$$

We note that since M is compact and X is smooth, then we have that the invariant direction given by $E^X_z (:= \{\alpha X(z) : \alpha \in \mathbb{R}\})$ cannot have positive Lyapunov exponent, since for all $t > 0$ and $z \in M$

$$\frac{1}{t} \log \| DX^t(z) \cdot X(z) \| = \frac{1}{t} \log \| X(X^t(z)) \| \le \frac{1}{t} \log \| X \|_0, \tag{2.12}$$

where $\| X \|_0 = \sup\{\| X(z) \| : z \in M\}$ is a constant. Analogously this direction cannot have positive exponent for negative values of time, thus the Lyapunov exponent along the flow direction must be zero.

Consequently *the flow direction is never tangent to a direction along which all Lyapunov exponents are non-zero.* In particular, E^X *is never tangent either to a strong-stable or strong-unstable direction.*

Additionally, the subexponential control of the angles between Oseledets subspaces ensures that, along regular orbits of a flow, *Lyapunov exponents off the direction of the flow coincide with the Lyapunov exponents of the Linear Poincaré Flow* (see Sect. 2.6). Indeed, since $\| P^t_X \mathbf{v} \| = \| (\mathcal{O}_{X^t(x)} \circ DX^t(x))\mathbf{v} \|$ for all vectors $\mathbf{v}$ in the normal direction N_x to E^X_x at T_xM, the normal projections $\widetilde{E}^i_x := \mathcal{O}_x(E^i_x)$ of the Oseledets subspaces on the normal direction to the flow have the same exponential growth rates, because the norm of the projection $\mathcal{O}_{X^t(x)}$ depends on the angles between $X(X^t(x))$ and the other Oseledets subspaces, which decreases at most subexponentially along μ-almost every orbit.

2.7.2.2 Non-uniform Hyperbolicity

In the uniformly hyperbolic setting, it is well known that physical measures for hyperbolic attractors admit a disintegration into conditional measures along the unstable manifolds of almost every point which are absolutely continuous with respect to the induced Lebesgue measure on these sub-manifolds, see [60, 62, 201, 266]. We explain the meaning of this technical notion in what follows.

Assume that an ergodic invariant probability measure μ for the flow X has a positive Lyapunov exponent. In this setting the existence of unstable manifolds through μ-almost every point x and tangent at x to $E^u_x := \bigoplus_{\lambda_i>0} E_i(x)$ is guaranteed by the non-uniform hyperbolic theory of Pesin [197]: the strong-unstable manifolds $W^{uu}(x)$ are the "integral manifolds" in the direction of the (measurable) sub-bundle E^u, tangent to E^u_x at almost every x. The sets $W^{uu}_{loc}(x)$ are embedded sub-manifolds in a neighborhood of x which, in general, depend only measurably (including its size) on the base point x. Let $W^u_{loc}(x)$ be the unstable manifold through x whenever the strong-unstable manifold $W^{uu}_{loc}(x)$ is defined (see Sect. 2.3). These manifolds are tangent at x to the center-unstable direction $E^X_x \oplus E^u_x$.

Analogously we have the strong-stable manifold tangent at almost every x to $E^s_x := \bigoplus_{\lambda_i<0} E_i(x)$ and the center-stable manifold.

Hyperbolic Blocks and Bounded Distortion Along Invariant Manifolds The measurable dependence of the invariant manifolds on the base point means that for each $\kappa \in \mathbb{N}$ we can find a compact *hyperbolic block* (or *Pesin set*) $\mathcal{H}(\kappa)$ and positive numbers τ_x, C_x satisfying

- $\operatorname{dist}(X^t(y), X^t(x)) \le C_x e^{-t\tau_x} \cdot \operatorname{dist}(y, x)$ for all $t > 0$ and $y \in W^{ss}_{loc}(x)$, and analogously for $y \in W^{uu}_{loc}(x)$ exchanging the sign of t;
- $C_x \le \kappa$ and $\tau_x \ge \kappa^{-1}$ for every $x \in \mathcal{H}(\kappa)$;
- $\mathcal{H}(\kappa) \subset \mathcal{H}(\kappa+1)$ for all $k \ge 1$ and $\mu(\mathcal{H}(\kappa)) \to 1$ as $\kappa \to +\infty$;
- the C^1 strong-stable and strong-unstable disks $W^{ss}_{loc}(x)$ and $W^{uu}_{loc}(x)$ vary continuously with $x \in \mathcal{H}(\kappa)$ (in particular the sizes of these disks and the angle between them are uniformly bounded from zero for x in $\mathcal{H}(\kappa)$).

Now we have the following *bounded distortion property*. Here the Hölder condition on the derivative of X^t is crucial.

Theorem 2.36 ([34, Theorems 11.1 & 11.2]) *Fix $\kappa \in \mathbb{N}$ such that $\mu(\mathcal{H}(\kappa)) > 0$. Then the function*

$$h^s(x, y) := \prod_{i\ge 0} \frac{|\det Df \mid E^s_{f^i(x)}|}{|\det Df \mid E^s_{f^i(y)}|}$$

is Hölder-continuous for every $x \in \mathcal{H}(\kappa)$ and $y \in W^{ss}_{loc}(x)$, where $f := X^1$ is the time-1 map of the flow X^t and E^s is the direction c corresponding to negative Lyapunov exponents.

An analogous statement is true for a function h^u on the unstable disks in $\mathcal{H}(\kappa)$ exchanging E^s with the direction E^u corresponding to positive Lyapunov exponents and reversing the sign of i in the product h^s above.

We remark that, because $\mathcal{H}(\kappa)$ is compact, we can find $0 < h_\kappa < \infty$ such that $\max\{h^u, h^s\} \le h_\kappa$ on $\mathcal{H}(\kappa)$.

2.7.2.3 Absolutely Continuous Disintegration

Here the Hölder control on the derivatives is also crucial. Assume that $n = \dim(M)$ and $l = \dim(F)$.

Given $x \in M$ let S be a co-dimension one submanifold of M everywhere transverse to the vector field X and $x \in S$, which we call a *cross-section of the flow at x*. Let ξ_0 be the connected component of $W^u(x) \cap S$ containing x. Then ξ_0 is a smooth submanifold of S and we take a parametrization $\psi : [-\varepsilon, \varepsilon]^l \times [-\varepsilon, \varepsilon]^{n-l-1} \to S$ of a compact neighborhood S_0 of x in S, for some $\varepsilon > 0$, such that

- $\psi(0,0) = x$ and $\psi((-\varepsilon, \varepsilon)^l \times \{0^{n-l-1}\}) \subset \xi_0$;
- $\xi_1 = \psi(\{0^l\} \times (-\varepsilon, \varepsilon)^{n-l-1})$ is transverse to ξ_0 at x: $\xi_0 \pitchfork \xi_1 = \{x\}$.

Consider the family $\Pi(S_0)$ of connected components ζ of $W^u(z) \cap S_0$ which *cross* S_0. We say that a *submanifold ζ crosses S_0* if it can be written as the graph of a map $\xi_0 \to \xi_1$.

Given $\delta > 0$ we let $\Pi_\delta(x) = \{X_{(\delta,\delta)}(\zeta) : \zeta \in \Pi(S_0)\}$ be a family of co-dimension one submanifolds inside unstable leaves in a neighborhood of x crossing S_0, see Fig. 2.13. The volume form Leb induces a volume form Leb_γ on each $\gamma \in \Pi_\delta(x)$ naturally. Moreover, since $\gamma \in \Pi_\delta(x)$ is a measurable family of submanifolds (S_0 is compact and each curve is tangent to a measurable sub-bundle E^{cu}), it forms a measurable partition of $\hat{\Pi}_\delta(x) = \cup\{\gamma : \gamma \in \Pi_\delta(x)\}$. We say that $\Pi_\delta(x)$ is a *δ-adapted foliated neighborhood of x*.

Hence (see [231]) $\mu \mid \hat{\Pi}_\delta(x)$ can be disintegrated along the partition $\Pi_\delta(x)$ into a family of *conditional measures* $\{\mu_\gamma\}_{\gamma \in \Pi_\delta(x)}$ such that

$$\mu \mid \hat{\Pi}_\delta(x) = \int \mu_\gamma \, d\hat{\mu}(\gamma),$$

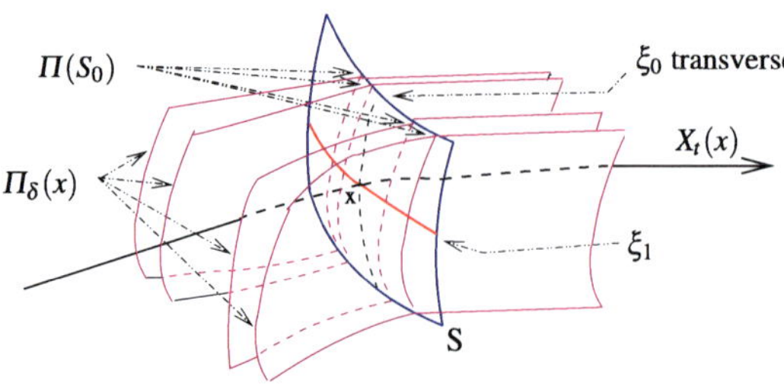

Fig. 2.13 Disintegration along centre-unstable leaves

where $\hat{\mu}$ is a measure on $\Pi_\delta(x)$ defined by

$$\hat{\mu}(A) = \mu(\cup_{\gamma \in A} \gamma) \quad \text{for all Borel sets } A \subset \Pi_\delta(x).$$

In this setting we say that μ *has an absolutely continuous disintegration along the center-unstable direction* or a *Gibbs state* if *for μ-almost every $x \subset M$, each δ-adapted foliated neighborhood $\Pi_\delta(x)$ of x induces a disintegration $\{\mu_\gamma\}_{\gamma \in \Pi_\delta(x)}$ of $\mu \mid \hat{\Pi}_\delta(x)$, for all small enough $\delta > 0$, such that $\mu_\gamma \ll \mathrm{Leb}_\gamma$ for $\hat{\mu}$-almost all $\gamma \in \Pi_\delta(x)$*. In this setting we also say that μ *is a Gibbs measure* for the flow X.

Note that completely dual properties and definitions can be stated for the strong-stable $W^{ss}(x)$ and stable leaves $W^s(x)$ of μ-almost every point x for a system with an invariant probability measure μ having a negative Lyapunov exponent.

2.7.2.4 Absolute Continuity of Foliations

In the same setting above, assume that x has a unstable leaf $W^u(x)$ and let D_1, D_2 be embedded disks in M transverse to $W^u(x)$ at x_1, x_2, that is $T_{x_i} D_i \oplus T_{x_i} W^u(x) = T_{x_i} M$, $i = 1, 2$. Then the strong-unstable leaves through the points of D_1 which cross D_2 define a map h between a subset of D_1 to D_2: $h(y_1) = y_2 = W^{uu}(y_1) \cap D_2$, called the *holonomy* map of the strong-unstable foliation between the transverse disks D_1, D_2. The holonomy is injective if D_1, D_2 are close enough due to uniqueness of the strong-unstable leaves through μ-a.e. point.

We say that h is *absolutely continuous* if there is a measurable map $J_h : D_1 \to [0, +\infty]$, called the *Jacobian of h*, such that

$$\mathrm{Leb}_2\big(h(A)\big) = \int_A J_h \, d\,\mathrm{Leb}_1 \quad \text{for all Borel sets } A \subset D_1,$$

and J_h is integrable with respect to Leb_1 on D_1, where Leb_i denotes the Lebesgue measure induced on D_i by the Riemannian metric, $i = 1, 2$.

The foliation $\{W^{uu}(x)\}$ is *absolutely continuous* (Hölder continuous) if every holonomy map is absolutely continuous (or J_h is Hölder continuous, respectively).

Since the pioneering work of Anosov and Sinai [15, 16] it became clear that for C^2 transformations or flows (in fact it is enough to have transformations or flows which are C^1 with α-Hölder derivative for some $0 < \alpha < 1$) the strong-unstable foliation is absolutely continuous and Hölder continuous. See also [147]. *When the leaves are of co-dimension one, then the Jacobian J_h of the holonomy map h coincides with the derivative h' since h is a map between curves in M*. In this case the holonomy map can be seen as a $C^{1+\alpha}$ transformation between subsets of the real line.

Going back to the case of the unstable foliation for a flow, see Fig. 2.14, we have that for any pair of disks γ_1, γ_2 inside S_0 transverse to $W^u(x) \cap S_0$ at distinct points y_1, y_2, the holonomy H between γ_1 and γ_2 along the leaves $W^u(z) \cap S_0$ crossing S_0 is also Hölder continuous *if the flow is C^2*.

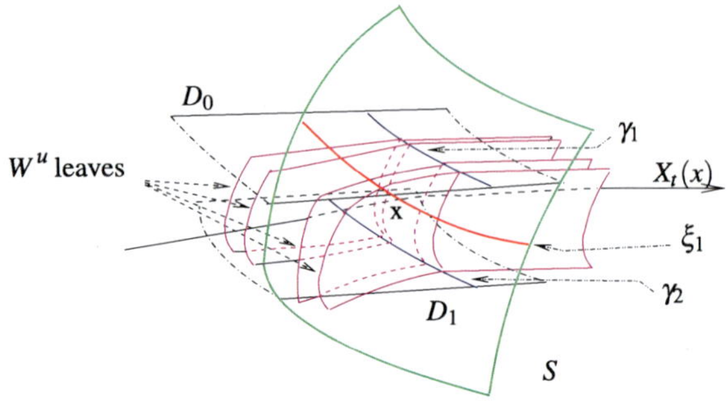

Fig. 2.14 The holonomy maps

Indeed note that this holonomy map H can be obtained as a composition of the holonomy map h between two disks D_1, D_2 transverse to the strong-unstable leaves which cross S_0, and the "projection along the flow" sending $w \in X^{(-\delta,\delta)}(S_0)$ to a point $X^t(w) \in S_0$ uniquely defined, with $t \in (-\delta, \delta)$. The disks are defined simply as $D_i = X^{(-\varepsilon,\varepsilon)}(\gamma_i)$ for $0 < \varepsilon < \delta$ and satisfy $D_i \cap S_0 = \gamma_i$, $i = 1, 2$. Since the holonomy h is Hölder continuous and the projection along the flow has the same differentiability class of the flow (due to the Tubular Flow Theorem 2.13), we see that the holonomy H is also Hölder continuous.

A very important consequence of absolute continuity is the following.

Lemma 2.37 *Assume that for some given submanifold W of M one knows that through* Leb_W*-almost every point $x \in W$ (*Leb_W *is the induced volume form on W by the volume form* Leb *of M) there passes a strong-stable manifold $W^{ss}(x)$ transverse to W. Then the union of the points of all these strong-stable manifolds has positive volume in M.*

Proof In a neighborhood of one of its points W can be written as $\mathbb{R}^k \times \{0^{n-k}\}$ and by the transversely assumption on $W^{ss}(x)$ these submanifolds can be written as graphs $\mathbb{R}^{n-k} \to \mathbb{R}^k$ on a neighborhood of 0^{n-k} which depends measurably on $x \in \mathbb{R}^k$. This change of coordinates through some local chart of M affects the derivatives of maps and holonomies at most by multiplication by bounded smooth functions.

The measurability ensures that given $\varepsilon > 0$ we can find $\delta, \alpha > 0$ small enough such that there exists $G \subset \mathbb{R}^k$ satisfying:

1. for every $x \in G$:
   a. $W^{ss}(x)$ is the graph of a map $\gamma_x : B_\delta(0^{n-k}) \to \mathbb{R}^k$ defined on a δ-ball around the origin;
   b. the slope of the tangent space to $W^{ss}(x)$ at every point is smaller than α (meaning that $\|D\gamma_x(w)\| \le \alpha$ for all w);
2. $\mathrm{Leb}_W(G) > 1 - \varepsilon$.

Then the submanifold $W_t = \mathbb{R}^k \times \{t\}$ for $t \in \mathbb{R}^{n-k}$ near 0^{n-k} is transverse to $W^{ss}(x)$ for all $x \in G$. Thus the holonomy map h_t from a subset of W_t to $W = W_0$ contains G in its image, which has positive volume in W_0. By absolute continuity of h_t, the intersection $W_t \cap \cup_x W^s(x)$ has positive volume in W_t. Hence $\mathrm{Leb}(\cup_x W^s(x)) = \int \mathrm{Leb}_{W_t}(W_t \cap \cup_x W^s(x))\, d\,\mathrm{Leb}^{n-k}(t) > 0$, and this concludes the proof. $\square$

2.7.2.5 Hyperbolic Measures, Gibbs Property and Construction of Physical Measures

These technical notions have crucial applications in the construction of physical measures for a dynamical system. Indeed, if the measure μ is ergodic and *hyperbolic*, meaning that all Lyapunov exponents are non-zero except the one corresponding to the flow direction, and also a Gibbs measure, then transverse to a center-unstable manifold $W^u(z)$ there exist strong-stable manifolds through μ_γ-almost every point and also through Leb_γ-almost all points $w \in W^u(z)$. Along strong-stable manifolds forward time averages of continuous functions are constant and along center-unstable manifolds backward time averages of continuous functions are constant. Moreover forward and backward time averages are equal μ-almost everywhere and through disintegration and ergodic decomposition, we deduce that μ-almost every z has a strong-unstable manifold $W^{uu}(z)$ where Leb-a.e. point has the same forward and backward time averages.

We are in the setting of Lemma 2.37 thus the absolute continuity of the strong-stable foliation implies that the family of all the strong-stable manifolds through $W^u(z)$ covers a positive Lebesgue measure subset of M *if the flow is of class* C^2. By the previous observations this set is inside the (ergodic) basin of μ. Hence *a hyperbolic ergodic invariant probability measure for a* C^2 *flow which is a Gibbs measure is also a physical measure.*

2.8 Stability Conjectures

The search for a characterization of stable systems, from Smale's seminal work in the sixties [252], led to several conjectures some of which are still open.

The famous stability conjecture, by Palis and Smale [191], states that a vector field X is structurally stable if, and only if, the non-wandering set is hyperbolic, coincides with the closure of the set of critical elements, there are no cycles between the stable and unstable manifolds of the critical elements and the intersection between the stable and unstable manifolds of points at the non-wandering set is transverse. In short terms, this conjecture states that *a system is structurally stable if, and only if, its non-wandering set is uniformly hyperbolic, the periodic orbits are dense and it satisfies the strong transversality condition.*

This conjecture was proved in the setting of C^1 diffeomorphisms by the combined work of several people along the years. First Robbin [222] showed that if a diffeomorphisms f is C^2, $\Omega(f)$ is Axiom A and satisfies the strong transversality condition, then f is C^1-structurally stable. Then Wellington de Melo [76] obtained the same result for C^1 diffeomorphisms on surfaces and finally Robinson [226] showed that for C^1 diffeomorphisms on any compact manifold the strong transversality condition plus Axiom A is sufficient for C^1 structural stability. The proof of this conjecture, in the C^1 topology, was completed by Mañé [144, 145, 148] (see also Liao [133] for a proof for surface diffeomorphisms) who showed that

C^1-structural stability implies that the non-wandering set is uniformly hyperbolic and satisfies the strong transversality condition.

For flows the proof that uniform hyperbolicity together with strong transversality is sufficient for C^1 structural stability was given by Robinson [224, 225]. Finally, that these conditions are also necessary for structural stability was proved much later by Hayashi [108] using the Connecting Lemma.

Developments in the last decades led Palis to conjecture [189] that the set of dynamical systems exhibiting finitely many attractors is dense in the set of all dynamical systems (in a suitable topology) and, moreover, each attractor supports a physical/SRB measure and the union of the (ergodic) basin of all physical measures covers Lebesgue almost every point of the ambient manifold. This conjecture admits a version for parametrized families where denseness is to be taken in the set of parameters corresponding to finitely many attractors whose basins cover the ambient manifold Lebesgue almost everywhere.

In the context of three-dimensional flows, one has to consider another homoclinic phenomenon involving singularities of the vector field: the situation in which the stable and unstable manifolds of a singularity have intersections other than the singularity itself. In this case, it is said that the vector field has a singular cycle.

In the setting of C^1 surface diffeomorphisms this conjecture was proved true by E. Pujals and M. Sambarino [217]. In the setting of real analytic families of unimodal maps of the interval or the circle, this was obtained by M. Lyubich [143]. In higher dimensions this conjecture is still wide open in spite of much recent progress, see e.g [56] and references therein.

Chapter 3
Singular Cycles and Robust Singular Attractors

A cycle Γ for a flow X^t is a finite sequence $\{\sigma_i, 0 \le i \le n\} \subset C(X)$ of hyperbolic critical elements of X^t, with $\sigma_0 = \sigma_n$, such that $W^u(\sigma_j) \cap W^s(\sigma_{j+1}) \neq \emptyset$ for $0 \le j < n$, that is the unstable manifold of one element intersects the stable manifold of the next element. A cycle is *singular* if at least one of its critical elements is a fixed point of X^t.

Cycles play an important role in the bifurcation theory of Dynamical Systems. A singular cycle is one of the mechanisms to go from a Morse-Smale flow (whose non-wandering set is a finite collection of hyperbolic critical elements) to a hyperbolic flow (whose non-wandering set is a finite collection of basic sets) through a one parameter family of flows.

In this chapter we shall describe three types of singular cycles, that will be used in the sequel. Nowadays the first one, presented in Sect. 3.1, is denominated *singular-horseshoe*. It was introduced by Labarca and Pacifico in [128] as a model for stable non hyperbolic flows in the context of boundary manifolds. We show that this set satisfies some properties which, in Chap. 5, will be defined as singular-hyperbolicity. This generalization of (uniform) hyperbolicity will characterize a much broader class of invariant sets for flows.

The second cycle is a homoclinic connection associated to a hyperbolic singularity of saddle-type. There are several possibilities for these cycles which are used in the proofs presented in the following chapters. We provide a brief description of the dynamics of perturbations of these cycles here. One of them is an inclination flip cycle. This was studied by many authors, see e.g. [69, 115] among others. The study of this type of cycle is crucial for the proof, in Chap. 5, that a robust transitive set with singularities for a 3-flow is either an attractor or a repeller, together with the Shil'nikov bifurcation, first considered in [243]. These are presented in Sect. 3.2.

Finally the third one is the Lorenz geometrical model, introduced by Guckenheimer and Williams [98] and presented in Sect. 3.3. This is a model for a robust attractor with singularities for a 3-flow, as we will see in Chap. 5.

V. Araújo, M.J. Pacifico, *Three-Dimensional Flows*, Ergebnisse der Mathematik und ihrer Grenzgebiete. 3. Folge / A Series of Modern Surveys in Mathematics 53,
DOI 10.1007/978-3-642-11414-4_3,

3.1 Singular Horseshoe

We start in Sect. 3.1.1 with the description of a map defined on a rectangle into itself which resembles the Smale horseshoe map [252]. For this reason this type of map is nowadays denominated singular horseshoe.

Afterward, in Sect. 3.1.2, we exhibit a singular cycle presenting a singular horseshoe map as a first return map. Then, in Sect. 3.1.3, we show in several stages that the singular horseshoe is a transitive partially hyperbolic set with volume expanding central direction.

3.1.1 A Singular Horseshoe Map

Given $\delta > 0$ small enough, $\lambda < 1/2$ and $\mu > 1$, let $Q = [-1, 0] \times [0, 1 + \delta]$ and define

$$R_\delta = Q \setminus \big((\mu^{-1}(1+\delta), 1/2 - \delta) \times (1/2, 1)\big).$$

Let $F : R_\delta \to Q$, $(x, y) \mapsto (g(x, y), f(y))$ be a smooth map satisfying:

(a) $|\partial_x g(x, y)| < 1/2$ for all $(x, y) \in R_\delta$ and

$$g(x, y) = \lambda \cdot x \quad \text{for } 0 \le y \le \mu^{-1}(1 + 2\delta).$$

(b) $f : I \setminus (J \cup K) \to I$ where $I = [0, 1]$, $J = (\mu^{-1}(1 + 2\delta), 1/2 - \delta)$ and $K = (1/2, 1)$ satisfying
 (i) $f(y) = \mu \cdot y$ for $0 \le y \le \mu^{-1}(1 + 2\delta)$,
 (ii) $f'(y) \gg \mu$ for $y \in [1/2 - \delta, 1/2] \cup [1, 1 + \delta]$.
(c) $F(x, 1) = F(x, 1/2) = (\alpha, 0)$ for $-1 \le x \le 0$ with a fixed $-1 < \alpha < \lambda$.
(d) The following sets

$$\gamma_{-1} = F(\{-1\} \times (1, 1 + \delta]), \qquad \gamma_0 = F(\{0\} \times (1, 1 + \delta]),$$

$$\beta_0 = F(\{0\} \times [1/2 - \delta, 1/2)), \qquad \beta_{-1} = F(\{-1\} \times [1/2 - \delta, 1/2))$$

are disjoint C^1 curves, except for the point $(\alpha, 0)$ where all are tangent. These curves are contained in $(-1, -\lambda) \times [0, 1+\delta]$ and are transverse to the horizontal lines. Moreover, if $d(A, B)$ denotes the distance between the sets A and B, and $L = \{-1\} \times [0, 1 + \delta]$ then

$$d(\gamma_{-1}, L) < d(\gamma_0, L) < d(\beta_0, L) < d(\beta_{-1}, L).$$

Figure 3.1 displays the main features of the map F.

Observe that, by construction, the horizontal lines $\{x\} \times [0, 1 + \delta]$, for $x \in [-1, 0]$, are invariants by F. They are also uniformly contracted by a factor $0 < c_0 < 1/2$. This guaranties that Q has a uniformly contracted (strong-)stable foliation invariant by F that we denote by $\mathscr{F}^{ss}(Q)$.

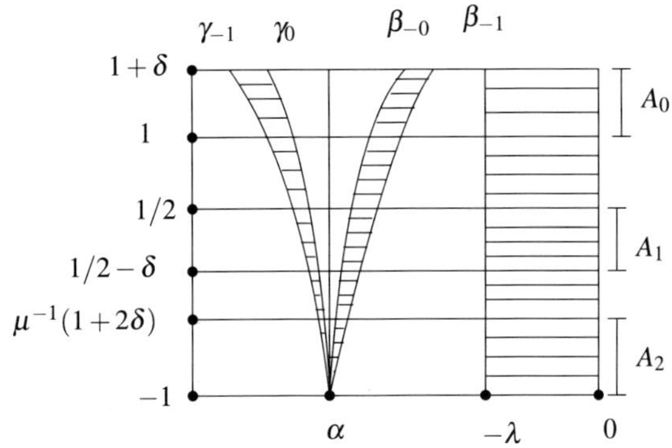

Fig. 3.1 A singular-horseshoe map

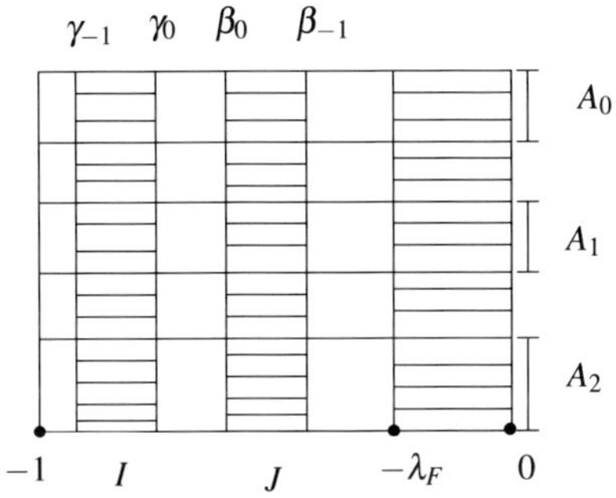

Fig. 3.2 A Smale horseshoe map

Define the following rectangles

$$A_0 = [-1, 0] \times [1, 1 + \delta], \qquad A_1 = [-1, 0] \times [1/2 - \delta, 1/2],$$

$$A_2 = [-1, 0] \times [0, \mu^{-1}(1 + 2\delta)].$$

Note that

$$R_\delta = \bigcup_{i=0}^{i=1} A_i \quad \text{and define} \quad \Omega_F = \bigcap_{n\in\mathbb{Z}} F^n(R_\delta).$$

It is clear that $F^{-1}(\Omega_F) = \Omega_F$.

3.1.1.1 Singular Symbolic Dynamics

We now associate a symbolic dynamics to the restriction $F \mid \Omega_F$. For this, consider a map $\widetilde{F} : R_\delta \to Q$ such that $\widetilde{F}$ has the same properties described for F, except that $\widetilde{F}([-1, 1] \times \{1\})$ and $\widetilde{F}([-1, 1] \times \{1/2\})$ are disjoint intervals I and J contained in the interior of $[-1, \lambda] \times \{0\}$ as in Fig. 3.2. Define $\Omega = \cap_{n\in\mathbb{Z}} \widetilde{F}^n(R_\delta)$.

Clearly $\widetilde{F}$ is a Smale horseshoe map. Roughly speaking, F is obtained from $\widetilde{F}$ pinching the intervals I and J into a unique point in such a way that the resulting boundary lines $\widetilde{\gamma}_{-1}$, $\widetilde{\gamma}_0$, $\widetilde{\beta}_0$, and $\widetilde{\beta}_{-1}$ are tangent at this point.

Let Σ^3 be the set of doubly infinite sequences of symbols in $\{0, 1, 2\}$ endowed with the topology given by the distance

$$d(x, y) = \sum_{i \in \mathbb{Z}} \frac{|x_i - y_i|}{3^{|i|}}$$

and $\sigma : \Sigma^3 \to \Sigma^3$ be the left shift map $\sigma(x)_i = x_{i+1}$.

It is well known (see e.g. [252] but also the textbooks of e.g. Devaney [78] or Robinson [230]) that there exists a homeomorphism $\widetilde{H} : \Omega \to \Sigma^3$ which conjugates $\widetilde{F}$ and σ, i.e. $\widetilde{H} \circ \widetilde{F} = \sigma \circ \widetilde{H}$. The image $\widetilde{H}(x)$ of $x \in \Omega$ is the sequence $(\widetilde{H}(x)_i) \in \Sigma^3$ defined by

$$\widetilde{H}(x)_i = j \in \{0, 1, 2\} \iff \widetilde{F}^i(x) \in A_j, \quad i \in \mathbb{Z}. \tag{3.1}$$

Recall that the set of periodic orbits for σ is dense in Σ^3 and that there exists a dense orbit.

We now describe the sequences associated, in a similar way, to points Ω_F.

Observe that the tangency point $(\alpha, 0)$ is the unique point of Ω_F outside of $[-\lambda, 0] \times [0, 1 + \delta]$ which remains forever in the bottom boundary of Q. This line corresponds to the local stable manifold of the fixed point $(0, 0)$ of F.

- Since $[-1, 0] \times \{0\} = \cap_{n \le 0} \widetilde{F}^n(A_2)$ we have $z \in [-1, 0] \times \{0\} \cap \Omega$ if, and only if, $\theta_i(z) = 2$ for all $i \ge 0$, i.e. $\widetilde{H}(z) = (\dots, x_{-1}, 2, 2, 2, \dots)$.

The points belonging to this line which are outside of $[-\lambda, 0] \times [0, 1 + \delta]$ are the points of the local stable manifold of $(0, 0)$ which are different from $(0, 0)$, i.e. their corresponding codes differ from the constant sequence $x_i \equiv 2$ at some coordinate with negative index. Defining Σ^3_* the subset of Σ^3 of those sequences $(x_i)_{i \in \mathbb{Z}}$ with $x_0 \in \{0, 1\}$ and $x_i = 2$ for all $i \ge 1$, then

$$W^s_{\widetilde{F}}(\widetilde{H}(0, 0)) \setminus \widetilde{H}([-\lambda, 0] \times [0, 1 + \delta]) = \bigcup_{k \ge 1} \sigma^k \Sigma^3_* = \widetilde{\Sigma}^3_*.$$

Note that $\sigma^{-1} \widetilde{\Sigma}^3_* \subseteq \widetilde{\Sigma}^3_*$. Defining an equivalence relation on Σ^3 by $\theta \sim \widetilde{\theta}$ if and only if $\theta, \widetilde{\theta} \in \widetilde{\Sigma}^3_*$, then this relation is preserved by the shift.

Let $\widetilde{\Sigma}^3$ be the corresponding quotient space and $\widetilde{\sigma}$ the associated quotient shift map. This map can be seen as the original full shift map on three symbols after identifying the sequences on $\widetilde{\Sigma}^3_*$, which correspond to the points which are taken to $(\alpha, 0)$ by F.

By the above considerations and the dynamics of F we get

Lemma 3.1 *There exists a homeomorphism $H_F : \Omega_F \to \widetilde{\Sigma}^3$ which conjugates $F \mid \Omega_F$ and H_F, that is $H_F \circ (F \mid \Omega_F) = \widetilde{\sigma} \circ H_F$.*

The homeomorphism H_F is defined just as in (3.1) replacing $\widetilde{F}$ by F.

Observe that the set of periodic orbits for $\widetilde{\sigma}$ is the same set of periodic orbits for σ. Note also that the dense orbit for σ is not contained in $\widetilde{\Sigma}^3_*$. Therefore the set of periodic orbits for $\widetilde{\sigma}$ is dense in $\widetilde{\Sigma}^3$ and this space contains a dense orbit. The existence of the conjugation above ensures that Ω_F has a dense subset of periodic orbits and a dense orbit for the dynamics of F.

3.1.2 A Singular Cycle with a Singular Horseshoe First Return Map

We start by recalling the definition of a special type of singularity of a vector field X in a 3-manifold.

Definition 3.2 We say that a singularity σ of a 3-flow X^t is Lorenz-like if the eigenvalues λ_i, $1 \le i \le 3$ are real and satisfy

$$\lambda_2 < \lambda_3 < 0 < -\lambda_3 < \lambda_1. \tag{3.2}$$

Next we shall exhibit a singular cycle $\mathscr{C}$ having a Lorenz-like singularity p and a hyperbolic saddle-type closed orbit σ, connected through a branch of the unstable manifold associated to p: this branch is contained in the stable manifold associated to σ. Moreover there are two orbits of transverse intersection between $W^s(p)$ and $W^u(\sigma)$. The cycle will be constructed in such away that it is contained in the maximal invariant set $\Lambda(X)$ of a vector field X in a neighborhood U of $\mathscr{C}$, and the first return map associated to $\mathscr{C}$ is a singular horseshoe map, see Fig. 3.3.

We start with a vector field $X_0 \in \mathfrak{X}^r(\mathbb{D}^3)$ on the 3-disk $\mathbb{D}^3$ in $\mathbb{R}^3$. This vector field has one repeller singularity r_1 at the north pole. Outside a neighborhood of r_1, X_0 has four singularities which we denote by p, p_1, p_2, r_2, plus a hyperbolic closed orbit σ. These satisfy the following:

1. p is a Lorenz-like singularity.
2. (p,σ) is a saddle connection with a branch $\gamma^u(p)$ of $W^u(p) \setminus \{p\}$ whose ω-limit set is σ. By the Hartman-Grobman Theorem there exists a neighborhood $p \in \mathscr{N} \subset \mathbb{R}$ such that the restriction of X_0 to $\mathscr{N}$ is equivalent to the linear vector field $L(x_1,x_2,x_3) = (\lambda_2 x_1, \lambda_1 x_2, \lambda_3 x_3)$.

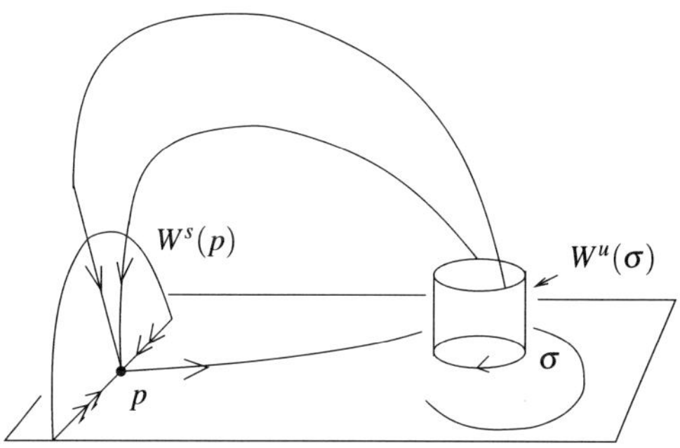

Fig. 3.3 A singular cycle

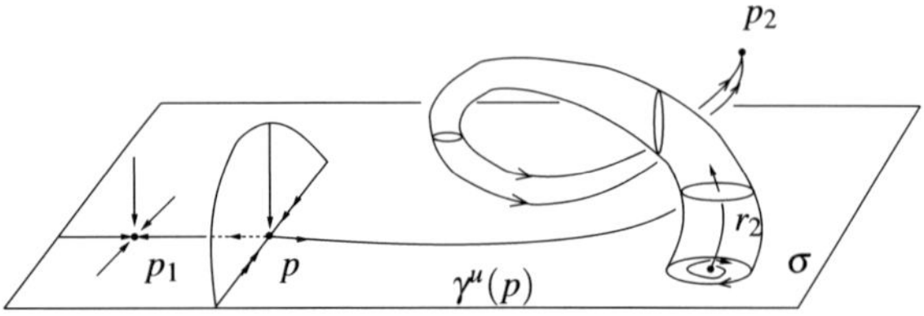

Fig. 3.4 The vector field X_0

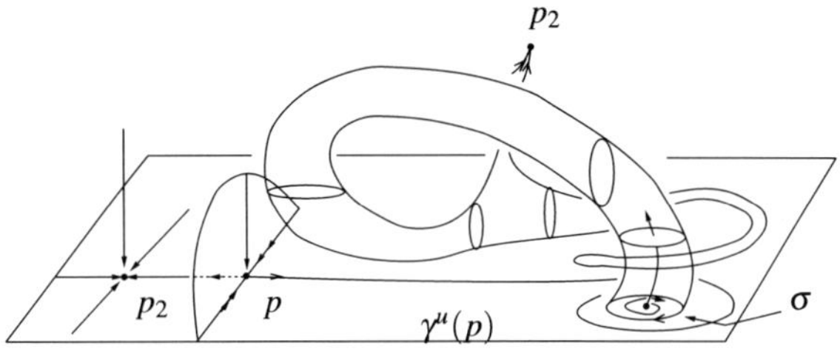

Fig. 3.5 Producing a unique tangency

3. p_1 is an attractor and is also the ω-limit set of the other branch of $W^u(p) \setminus \{p\}$.
4. p_2 is an attractor and is the ω-limit of $W^u(\sigma) \setminus \{\sigma\}$.
5. r_2 is a repeller contained in the interior of the 2-disk $\mathbb{D}^2$ bounded by σ in $\mathbb{S}^2$.
6. We assume that
   a. $p_1, p, \gamma^u(p), \sigma$ and $\mathbb{D}^2$ are contained in the boundary $\partial(\mathbb{D}^3) = \mathbb{S}^2$ of the 3-disk;
   b. the eigenvalues of $DX_0(r_2)$ corresponding to eigenvectors in $T\mathbb{S}^2$ are complex conjugates. Therefore the part of $W^u(r_2) \setminus \{r_2\}$ in $\mathbb{S}^2$ is a spiral whose ω-limit set is σ;
   c. the strong unstable manifold $W^{uu}(r_2) \setminus \{r_2\}$ is contained in the interior of $\mathbb{D}^3$ and its ω-limit set is the attractor p_2.
7. The α-limit set of $W^s(p) \setminus \{p\}$ is the repeller r_1 and $W^s(p)$ separates the two attractors.

Figure 3.4 shows the essential features of the vector field X_0 outside a neighborhood of r_1. Observe that X_0 constructed in this way is a Morse-Smale vector field.

Now we can modify the vector field X_0 away from its critical elements, in particular away from the neighborhood $\mathscr{N}$ of p, in order to produce a unique tangency between $W^s(p)$ and $W^u(\sigma)$, see Figs. 3.5 and 3.6.

By another slight perturbation of the above vector field we get a vector field X such that $W^u(\sigma)$ is transverse to $W^s(p)$ at two orbits, see Fig. 3.7.

3.1.2.1 The First Return Map Associated to $\mathscr{C}$ is a Singular Horseshoe Map

Now we study the first return map associated to $\mathscr{C}$ and show that it is a singular horseshoe map.

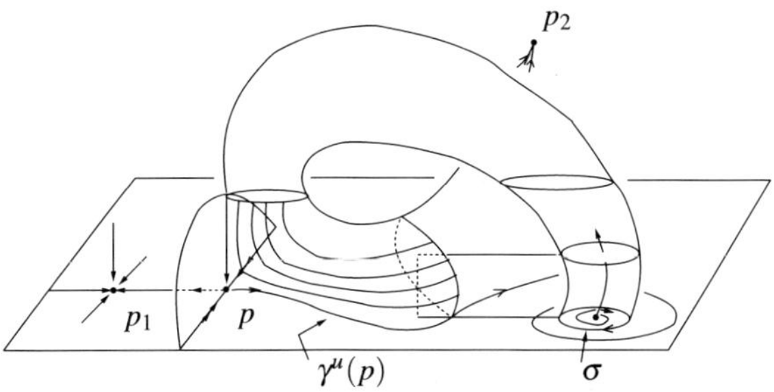

Fig. 3.6 One point of tangency

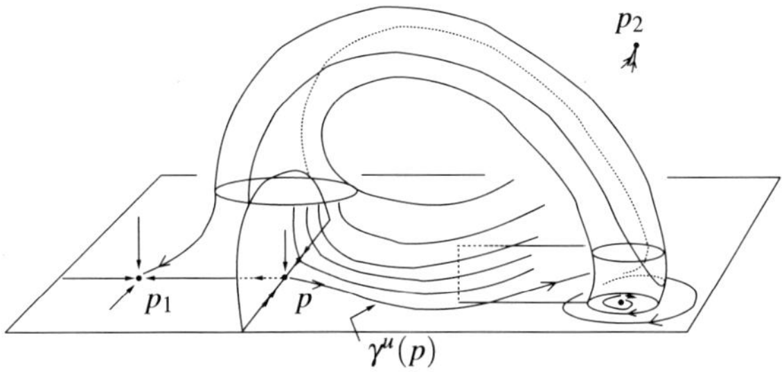

Fig. 3.7 The final vector field

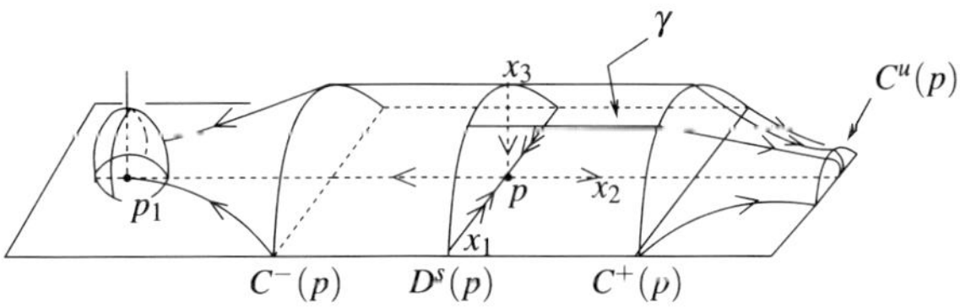

Fig. 3.8 The cross-section C^s at p

Let S be a cross-section to X at $q \in \sigma$. Reparametrizing X, if necessary, we can assume that the period of σ is equal to one and that S is invariant by X^1: there exists a small neighborhood $U \subset S$ of q such that $X^1(S \cap U) \subset S$.

Since there are two orbits of transverse intersection of $W^u(\sigma)$ with $W^s(p)$ and the branch $\gamma^u(p)$ has σ as ω-limit set, there exists a first return map F defined on subsets of S, taking points of S back to S under the action of the flow. The goal now is to describe F.

From now we assume mild non-resonant conditions on the eigenvalues of p to ensure that there are C^1 linearizing coordinates (x_1, x_2, x_3) in a neighborhood U_0 containing p.

Let $D^s(p) \subset U_0$ and $D^u(p) \subset U_0$ be fundamental domains for the action of the flow inside $W^s(p)$ and $W^u(p)$ respectively. That is $D^s(p)$ is a circle in $W^s(p) \setminus \{p\}$ containing p in its interior and transverse to X, and $D^u(p)$ is a pair of points, one in each branch of $W^u(p) \setminus \{p\}$.

Let $C^s(p) \subset U_0$ be a cross-section to X, as in Fig. 3.8, with several components: $C^s = C^s(p) = C^+(p) \cup D^s(p) \cup C^-(p)$. We assume that $C^-(p)$ is contained in the stable manifold of the attractor p_1. We also assume that the plane $\{x_1 = 0\}$ is a center-unstable manifold for p and we denote it by $W^{cu}(p)$. Let $C^u(p)$ be a cross section to X formed by a 2-disk through the point of $\gamma^(p) \cap D^u(p)$.

Observe that if γ is a C^1 curve transverse to $D^s(p)$ and $\gamma \cap W^{ss}(p) = \emptyset$, then

$$C^u(p) \cap \left(\bigcup_{t \geq 0} X^t(\gamma) \right)$$

is a C^1 curve tangent to $W^{cu}(p) \cap C^u(p)$ at $D^u(p) \cap \gamma^u(p)$.

Let $D^s(p_2) \subset \mathbb{D}^3$ be a fundamental domain for the dynamics on $W^s(p_2)$, i.e. the boundary of a 3-ball containing p_2. Let $V \subset S$ be a small neighborhood of $q \in \sigma$, where we have C^1 linearizing coordinates (x, y) for the Poincaré first return map F associated to σ. The eigenvalues of $DF(q)$ are λ, μ both bigger than 1.

Let $Q = [-1, 1] \times [0, 1]$ be a rectangle contained in the interior of V. Assume that

$$[-1, 1] \times \left\{ \frac{1}{2}, 1 \right\} \subset W^s(p) \quad \text{and} \quad [-1, 1] \times \{0\} \subset \mathbb{S}^2.$$

There are only two orbits of transverse intersection between $W^u(\sigma)$ and $W^s(p)$, and the points in $\{1\} \times (1/2, 1)$ will fall in the stable set of p_1, by construction of the vector field X. Since $W^s(p_1)$ is open we can assume that $[-1, 1] \times (1/2, 1) \subset W^s(p_1)$ (taking V small enough) and also

$$X^1\big([-1, 1] \times (1/2, 1)\big) \subset C^-(p)$$

through a reparametrization of time if necessary. Assume further that there exists $\delta > 0$ such that $(1 + 2\delta)\mu^{-1} < 1/2 - \delta$ and

(a) for $A_0 = [-1, 1] \times (1, 1 + \delta]$ we have $X^1(A_0) \subset C^+(p)$;
(b) for $A_1 = [-1, 1] \times [1/2 - \delta, 1]$ we have $X^1(A_1) \subset C^+(p)$;
(c) $X^2([-1, 1] \times [1 + \delta, 1 + 2\delta]) \subset D^s(p_2)$;
(d) $X^2([-1, 1] \times [1/2 - 2\delta, 1/2 - \delta]) \subset D^s(p_2)$;
(e) for $A_2 = [-1, 1] \times [0, (1 + 2\delta)\mu^{-1}]$ we have

$$X^1(A_2) = [-\lambda, 0] \times [0, 1 + 2\delta] \subset Q.$$

Now define

$$H_1(X) = \bigcup_{t \geq 0} X^t(X^1(A_0)) \cap C^u(p), \qquad H_2(X) = \bigcup_{t \geq 0} X^t(X^1(A_1)) \cap C^u(p).$$

Clearly $H_i(X)$ are cones tangent to $W^{cu}(p) \cap C^u(p)$ at $D^u(p)$ for $i = 1, 2$, see Figs. 3.9 and 3.10.

Let α be the first intersection point between $W^u(p)$ and Q. We can assume that $\tau_i(X) = X^1(H_i(X))$ is contained in Q and that these sets are cones tangent to $W^{cu}(p) \cap Q$ at α, for $i = 1, 2$.

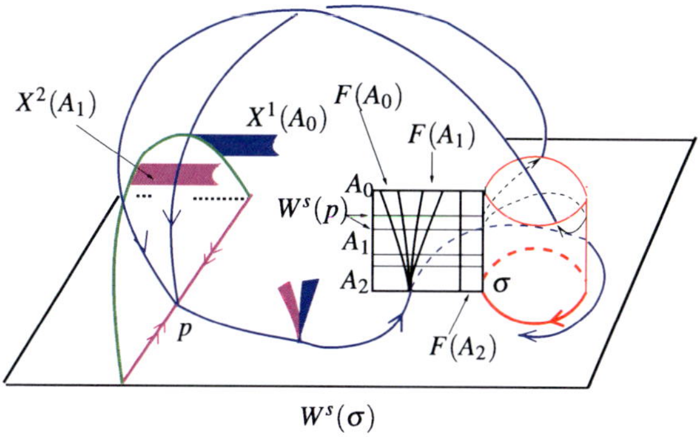

Fig. 3.9 The first return map to Q

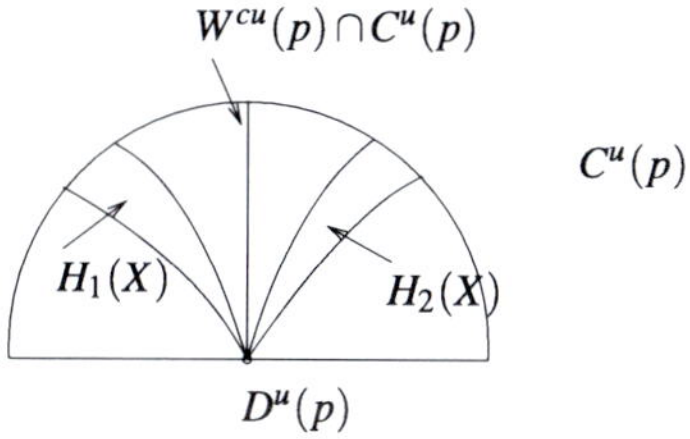

Fig. 3.10 The first return map at $D^u(p)$

Clearly we can also assume that

$$X^3\big([-1,1] \times \{1+\delta\}\big) \subset \tau_1(X) \quad \text{and} \quad X^3\left([-1,1] \times \left\{\frac{1}{2}-\delta\right\}\right) \subset \tau_2(X).$$

If necessary, we modify the vector field X in order to have (see Fig. 3.11):

(a) horizontal lines $\{y = constant\}$ going to horizontal lines in $\tau_i(X)$;
(b) writing π_y for the projection on the y-axis in V

$$\pi_y\Big(X^3\big([-1,1] \times \{1+\delta\}\big)\Big) = \{1+2\delta\} \quad \text{and}$$

$$\pi_y\left(X^3\left([-1,1] \times \left\{\frac{1}{2}-\delta\right\}\right)\right) = \{1+2\delta\};$$

(c) for $D^s_\sigma = [-1,\lambda] \times [0, 1+\delta]$ we have $\tau_i(X) \subset \operatorname{int} D^s_\sigma$ for $i = 1, 2$.

Now we describe the first return map F.

- If we take a point (x, y) with $1+\delta < y \le 1+2\delta$, then (x, y) is contained in the stable manifold of the attractor p_2 and F is not defined at these points.
- For either a point $(x, 1) \in Q$ or $(x, 1/2) \in Q$ we define $F(x, 1) = \alpha = F(x, 1/2)$.
- For points $(x, y) \in Q$ with $0 \le y \le \mu^{-1}(1+2\delta)$ we define $F(x, y) = (\lambda x, \mu y)$.
- For points (x, y) such that either $1 < y \le 1+\delta$ or $1/2-\delta \le y < 1/2$, we define $F(x, y)$ as the first intersection of the positive orbit through (x, y) with the rectangle $Q_\delta = [-1,1] \times [0, 1+2\delta]$.

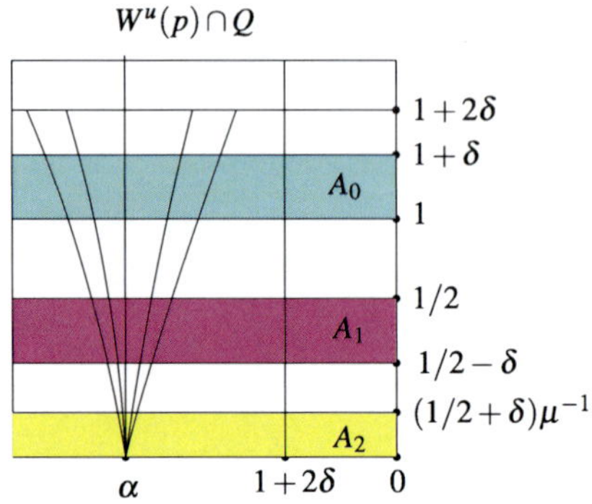

Fig. 3.11 The singular horseshoe return map

- For points (x, y) with $1/2 < y < 1$ the first return F is not defined, since these points are in the stable manifold of the attractor p_1.
- F is also not defined for points (x, y) with $\mu^{-1}(1+2\delta) < y < 1/2 - \delta$. Indeed, these points are such that the projection on the y-axis of their first return to S is larger than $1 + 2\delta$. So these points return once to S and then they are taken to the attractor p_2.

Then the first return map F has the expression:

$$F(x, y) = \begin{cases} (\lambda x, \mu y) & \text{if } 0 \le y \le \mu^{-1}(1+2\delta) \\ (g_1(x, y), f_1(y)) & \text{if } \quad 1 \le y \le 1+\delta \\ (g_2(x, y), f_2(y)) & \text{if } \ 1/2-\delta \le y \le 1/2 \end{cases}$$

with

- $g_i(x, y)$ is some smooth function with $|\partial_x g_i| < c < \frac{1}{2}$, and
- f_i is a smooth function satisfying $f_i'(y) > \mu$ and $0 \le f_i(y) \le 1+2\delta$, for $i = 1, 2$.

We assume that the image $F(\{0\} \times [0, 1+\delta])$ is transverse to the horizontal lines in Q_δ.

The non-trivial dynamics of F is concentrated in the square Q_δ.

Let $\Omega_F = \overline{\cap_{n\ge 0} F^n(Q_\delta)}$. Observe that the non-wandering set $\Omega(X)$ is the disjoint union of the critical elements $\{r_1, r_2, p_1, p_2\}$ and Λ, where Λ is the closure of the saturation by the flow X^t of the non-wandering set of the first return map F described above, i.e., $\Lambda = \overline{\cup_{t\ge 0} X^t(\Omega_F)}$.

The set Λ is the maximal invariant set containing the singular cycle $\mathscr{C}$ in the neighborhood U chosen at the beginning of the construction. This invariant set is the so-called *singular horseshoe*.

Remark 3.3 On the boundary of the manifold $\mathbb{D}^3$, which is preserved by the flow, we have a Morse-Smale system. Hence any vector field Y close to X preserving the boundary will have the same features as X on the boundary.

Moreover the features of X depend on the transverse intersection of certain invariant manifolds of the hyperbolic critical elements, all of which lie on the boundary of the ambient manifold. Hence every vector field close to X preserving the boundary will exhibit the same critical elements and the same transversality relations between them, and so *the singular-horseshoe is robust among the vector fields which preserve the boundary of* $\mathbb{D}^3$.

3.1.3 The Singular Horseshoe Is a Partially Hyperbolic Set with Volume Expanding Central Direction

We start by constructing local stable and unstable manifolds through points of Ω_F with respect to F. The stable and unstable foliation of the singular horseshoe Λ is than obtained as the saturation by the flow of these manifolds. Then we explain how to obtain the strong-stable foliation. Having these foliations we can define a splitting of the tangent space at Λ which will behave much like a hyperbolic splitting.

3.1.3.1 Stable Manifold for Points in Ω_F

Let $F : Q_\delta \to Q$ be the singular horseshoe map defined in the previous subsection.

It is easy to see that any horizontal line crossing Q is uniformly contracted by a factor of $c \in (0, 1/2)$ by the definition of F. Then, given any pair of points x, y of Ω_F in the same horizontal line, one has

$$\operatorname{dist}\left(F^k(x), F^k(y)\right) \le c^k \xrightarrow[k\to+\infty]{} 0.$$

Hence these curves are the local stable manifolds through points of Ω_F with respect to F. Saturating these curves by the flow we obtain the foliation of stable manifolds $\mathscr{F}^s$ through the points of the singular horseshoe.

For the particular case of the saddle singularity p and the periodic orbits σ the stable leaves are given by the stable manifolds of these hyperbolic critical points.

3.1.3.2 Unstable Manifolds for Points of Ω_F

Define $R_0 = Q \cap F(A_0)$, $R_1 = Q \cap F(A_1)$ and $R_2 = Q \cap F(A_2)$. Then R_0 and R_1 are, except for their vertices, disjoint cones. R_2 is a rectangle, crossing Q from bottom to top.

For each $i, j \in \{0, 1, 2\}$, let $R_{ij} = R_i \cap F(R_j)$. Then $F(R_j) = \cup_{i=0}^2 R_{ij}$. Since $F(x, y) = (g(x, y), f(y))$ with $|g_x(x, y)| < c < 1/2$, we have that the horizontal lines are contracted by a factor of c when iterated by F. Thus, except for R_{22} (which is a rectangle strictly contained in R_2), R_{ij} is a cone strictly contained in R_i.

Inductively, given any sequence of n-symbols $x_1, x_2, \ldots, x_n$ with $x_i \in \{0, 1, 2\}$ and $n \ge 2$, define $R_{ix_1x_2\cdots x_n} = R_i \cap F(R_{x_1\cdots x_n})$ for $i = 0, 1, 2$. Then

$$F(R_{x_1\cdots x_n}) = \bigcup_{i=0}^{2} R_{ix_1\cdots x_n}.$$

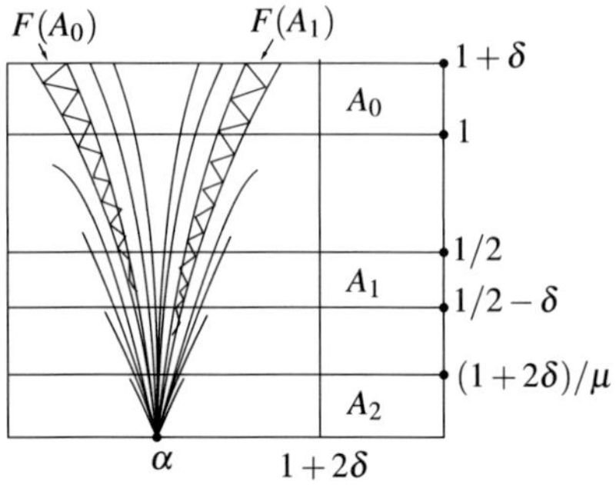

Fig. 3.12 The unstable curves of Ω_F tangent at $(\alpha, 0)$

Note that

- If all the x_i are equal to 2, then

$$R_2, \quad R_2 \cap F(R_2), \quad \dots, \quad R_2 \cap F(R_2) \cap \dots \cap F^n(R_2)$$

  is a strictly decreasing sequence of rectangles converging, in the C^1 topology, to the vertical line $\{0\} \times [0, 1+\delta]$.
- If there is any $x_{i_0} \in \{0, 1\}$, then the sequence

$$R_{x_0}, \quad R_{x_0} \cap F(R_{x_1}), \quad \dots, \quad R_{x_0} \cap F(R_{x_1}) \cap F^2(R_{x_2}) \cap \dots \cap F^n(R_{x_n})$$

  is a strictly decreasing sequence of C^1-cones. Hence this sequence converges to a C^1 curve, denoted by $\gamma(x_0, x_1, \dots)$, which crosses Q from top to bottom, that is, γ intersects each horizontal line of Q in a unique point, see Fig. 3.12.

Note that every point $x \in \Omega_F \setminus \{(\alpha, 0)\}$ has a corresponding code $H_F(x)$ in $\tilde{\Sigma}^3$ whose coordinates with positive index define a unique regular curve $\gamma = \gamma(x_1, x_2, \dots)$ as above. This curve γ is the same for every $y \in \Omega_F$ having a code $H_F(y)$ with the same coordinates as $H_F(x)$ at positive indices. Such points y form the unstable manifold of x with respect to F, since $d(\sigma^{-k} H_F(x), \sigma^{-k} H_F(y)) \xrightarrow[k\to+\infty]{} 0$.

Indeed, from the description of the map F, it is clear that γ is expanded by all iterates of F whenever its image is defined. Or, reversing time, by the construction of γ, the pre-image of any pair of points $y, z \in \gamma$ by F^k is well defined for all $k \ge 1$ and, moreover, for any pair y_{-k}, z_{-k} of such pre-images under the same sequence of inverse branches of F, we have

$$\operatorname{dist}\left(y_{-k}, z_{-k}\right) \le c^k \xrightarrow[k\to+\infty]{} 0.$$

Saturating these curves by the flow we obtain the central-unstable foliation $\mathscr{F}^u$ through the points of Λ.

The point $(\alpha, 0)$ has already a well defined unstable manifold: the vertical line crossing Q through $(\alpha, 0)$, corresponding to the intersection of the unstable manifold of the orbit of $W^u(p)$ connecting the saddle singularity p to the periodic orbit σ, see Fig. 3.9.

In addition, the saddle singularity p and the periodic orbit σ also have a well defined unstable foliation compatible with the leaves defined above.

3.1.3.3 Strong-Stable Foliation for the Singular-Horseshoe

The previous observations show that *every periodic orbit of F on Ω_F is hyperbolic of saddle-type*. Since F is the Poincaré first return map to Q of the flow X, we deduce that *every periodic orbit of X in Λ is hyperbolic of saddle-type*. Moreover the density of periodic orbits for $F \mid \Omega_F$ implies that *the family of periodic orbits of X in Λ is dense in Λ*.

In addition the stable foliation of the periodic orbits coincides with the stable foliation defined above for all points, including the singularity p and the periodic orbit σ. Hence the strong-stable leaves $\widetilde{\mathscr{F}}^{ss}$ defined on the periodic orbits extend (by continuity and coherence) to a strong-stable foliation $\mathscr{F}^{ss}$ defined throughout Λ. Notice that at the singularity p the strong-stable foliation coincides with its strong-stable manifold corresponding to the weakest contracting eigenvalue.

3.1.3.4 Partial Hyperbolicity

The flow invariance of the stable $\mathscr{F}^s$, strong-stable $\mathscr{F}^{ss}$ and unstable $\mathscr{F}^u$ foliations through points of Λ and the smoothness of their leaves enables us to define the following DX invariant sub-bundles: for every point $z \in \Lambda$

$$E_z = T_z \mathscr{F}^{ss}(z) \quad \text{and} \quad F_z = T_z \mathscr{F}^u(z)$$

satisfy $DX^t \cdot E_z = E_{X^t(z)}$ and $DX^t \cdot F_z = F_{X^t(z)}$, for all $t \in \mathbb{R}$.

Now we show that the flow X contracts E uniformly, and contracts more strongly than any contraction along F. Then we conclude by showing that X expands volume along F.

Let V be a neighborhood of p where linearizing coordinates are defined. Assume without loss of generality that $X^1(Q) \subset V$. In V the solutions of the linear flow can be given explicitly as in (3.3).

Write $J_t^c(z)$ for the absolute value of the determinant of the linear map $DX^t \mid F_z : F_z \to F_{X^t(z)}$ where z is any point of Λ and $t \in \mathbb{R}$.

For points z in $X^1(Q)$ and for $s > 0$ such that $X^t(z)$ remains in V for $0 \le t \le s$ we have

- $\|DX^t \mid E_z\| = e^{\lambda_2 t}$;
- $\|DX^t \mid E_z\| = e^{(\lambda_2 - \lambda_3)t} \cdot m(DX^t \mid F_z)$;
- $|\det DX^t \mid F_z| = e^{(\lambda_1 + \lambda_3)t}$,

where $m(\cdot)$ denotes the minimum norm of the linear map. Note that because $\lambda_1 + \lambda_3 > 0$ the flow in V expands volume along the F direction. Moreover since $\lambda_2 < \lambda_3$ the flow contracts along the E direction more strongly than it expands along the F direction, by the second item above. We say that F *dominates* E, see Chap. 5 for more on dominated splitting. Observe that the above properties are also valid for the singularity p and the periodic orbit σ.

In what follows we extend these properties for the action of X on points of Λ for all times.

Notice that the flow takes a finite amount of time, bounded from above and from below, to take points in Q to $X^1(Q)$, and from $D^u(p)$ to Q (these times are constant and equal to 1 in our construction).

Hence if we are given a point $z \in \Lambda \setminus \{p, \sigma\}$, then its negative orbit $X^{-t}(z)$ for $t > 0$ will have consecutive and alternate hits on $D^u(p)$ and Q, at times $t_1 < s_1 < t_2 < s_2 < \cdots < t_n < s_n < \cdots$ respectively, with $t_0 = s_0 = 0$ and $r_n = |t_{n+1} - s_n|$ bounded from below by T_0 independently of $n \geq 1$.

Let $B > 0$ be a upper bound on $\|DX^{-t}(z)\|$ from 0 to T_1 and for all $z \in \Lambda$. Then from the volume expansion on V we have for $t_n < t \leq s_n$

$$
\begin{aligned}
|\det DX^{-t} \mid F_z| &\leq \exp\left(B \cdot n - (\lambda_1 + \lambda_3) \cdot \left(t - \sum_{i=1}^{n-1} r_i\right)\right) \\
&= \exp\left(t \cdot (\lambda_1 + \lambda_3) \cdot \left(\frac{Bn}{t(\lambda_1 + \lambda_3)} - 1 + \frac{\sum_{i=1}^{n-1} r_i}{t}\right)\right).
\end{aligned}
$$

Since $t > T_0 n$ and $\sum_{i=1}^{n-1} r_i < t$ we see that there exists $K > 0$ such that $|\det DX^{-t} \mid F_z| \leq K^{-1} \cdot e^{(\lambda_1+\lambda_3)t}$, which is equivalent to volume expansion.

The uniform contraction along E and the domination of F over E are obtained by similar arguments, see also Sect. 3.3.3.

3.2 Bifurcations of Saddle-Connections

A homoclinic orbit associated to a singularity σ of $X \in \mathfrak{X}^1(M)$ is a regular orbit $\mathcal{O}(q)$ satisfying $\lim_{t\to+\infty} X^t(q) = \sigma$ and $\lim_{t\to-\infty} X^t(q) = \sigma$. Here we focus on the dynamics close to $\mathcal{O}(q)$ for small perturbations of the flow.

3.2.1 Saddle-Connection with Real Eigenvalues

Consider the following one-parameter system of ordinary differential equations in $\mathbb{R}^3$:

$$
\begin{cases}
\dot{x} = \lambda_1 x + f_1(x, y, z; \mu) \\
\dot{y} = \lambda_2 y + f_2(x, y, z; \mu) \\
\dot{z} = \lambda_3 z + f_3(x, y, z; \mu)
\end{cases}
\qquad (x, y, z, \mu) \in \mathbb{R}^4
$$

where f_i are C^2 functions which vanish together with Df_i at the origin of $\mathbb{R}^4$. So $\sigma = (0, 0, 0)$ is a singularity. We assume that the eigenvalues λ_i, $i = 1, 2, 3$, of σ are real and $\lambda_2 \leq \lambda_3 < 0 < \lambda_1$.

Note that any other case of a hyperbolic saddle singularity *with only real eigenvalues* for a three-dimensional flow can be reduced to the present case by considering the time reversed flow.

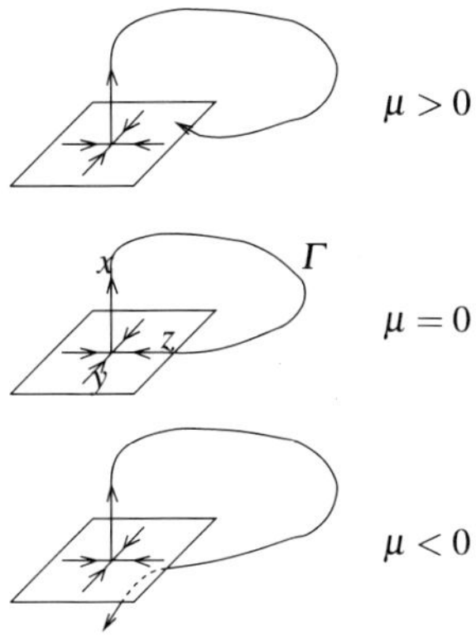

Fig. 3.13 Breaking the saddle-connection

The hyperbolicity of σ ensures the existence of C^1 stable $W^s(\sigma)$ and unstable manifolds $W^u(\sigma)$. The manifold $W^s(\sigma)$ is tangent at σ to the eigenspace $\{0\} \times \mathbb{R}^2$ associated to the eigenvalues λ_2, λ_3, and $W^u(\sigma)$ is tangent at σ to the eigenspace $\{(0,0)\} \times \mathbb{R}$ associated to λ_1.

In this setting a homoclinic orbit associated to σ is any orbit $\Gamma = \mathcal{O}_X(q)$ of a point $q \in W^s(\sigma) \cap W^u(\sigma) \setminus \{\sigma\}$. We assume that there exists such a homoclinic orbit. Moreover we make the supposition that the saddle-connection breaks as in Fig. 3.13.

Using linearizing coordinates and an analysis of the return maps to convenient cross-sections near σ one can prove the following.

Theorem 3.4 *For $\mu \neq 0$ small enough a periodic orbit bifurcates from Γ. This periodic orbit is*

1. *a sink for $\lambda_1 < -\lambda_3 \leq -\lambda_2$;*
2. *a saddle for $\lambda_1 < -(\lambda_2 + \lambda_3)$, $-\lambda_2 < \lambda_1$ and/or $-\lambda_3 < \lambda_1$;*
3. *a source for $-(\lambda_2 + \lambda_3) < \lambda_1$.*

A proof of this result can be found in [272, pp. 207–219].

Observe that if σ is Lorenz-like (recall Definition 3.2), then only item 2 above is possible. That is, *a Lorenz-like saddle equilibrium is the only possibility, among hyperbolic saddle equilibria in three-dimensional flows, which generates saddle periodic orbits after the unfolding of an associated saddle-connection with real eigenvalues*, all the other possibilities generate either attracting or repelling periodic orbits (periodic sinks or sources) after the unfolding. It is natural that these are the only allowed singularities for robustly transitive attractors, see Sect. 5.

3.2.2 Inclination Flip and Orbit Flip

Here we consider degenerate homoclinic orbits. We assume that σ satisfies some generic conditions: the eigenvalues $\lambda_i, i = 1, 2, 3$, of σ are real and distinct and

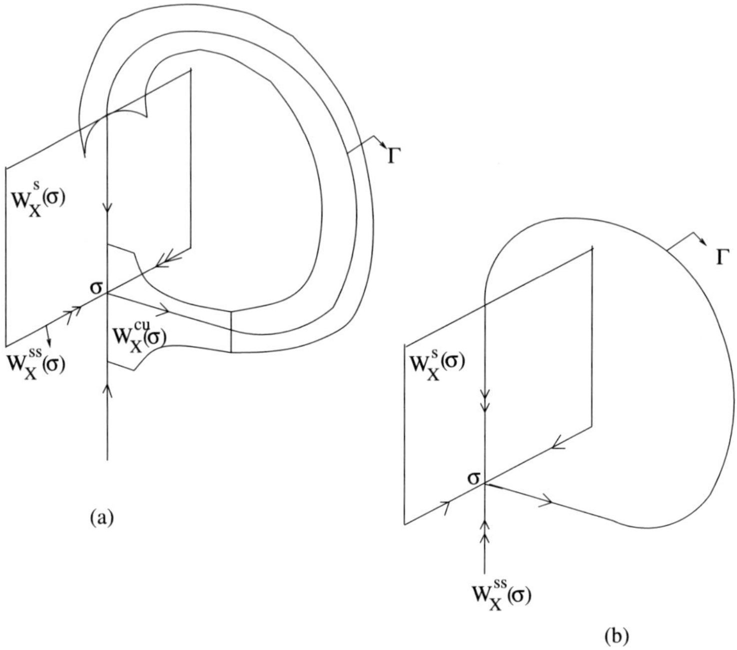

Fig. 3.14 **a** Inclination-flip. **b** Orbit-flip

satisfy $\lambda_2 < \lambda_3 < 0 < -\lambda_3 < \lambda_1$, that is, σ is a Lorenz-like singularity, as in Definition 3.2.

The condition $\lambda_2 < \lambda_3 < 0$ ensures that there is an invariant C^1 manifold $W^{ss}(\sigma)$, the *strong-stable manifold*, tangent at σ to the eigendirection of the eigenvalue λ_2. There are also invariant manifolds $W^{cu}(\sigma)$ containing σ, called *center-unstable* manifolds, tangent at σ to the eigendirection generated by the eigenvectors associated to λ_3, λ_1. There are several of these center-unstable manifolds but all of them are tangent along $W^u(\sigma)$ at σ (the reader should consult Hirsch, Pugh and Shub [110] for a proof of these facts).

Let Γ be a homoclinic orbit associated to σ. The following conditions are generic, that is, both are true for a residual subset of flows in $\mathfrak{X}^1(M)$ exhibiting a homoclinic orbit associated to a Lorenz-like singularity:

(G1) $W^{cu}(\sigma)$ intersects $W^s(\sigma)$ transversely along Γ, i.e.,

$$\Gamma = W^{cu}(\sigma) \pitchfork W^s(\sigma); \quad \text{and}$$

(G2) $\Gamma \cap W^{ss}(\sigma) = \emptyset$.

We are going to study what happens when such generic conditions fail.

Definition 3.5 Let $X \in \mathfrak{X}^r(M)$, $r \geq 1$, be a smooth vector field and let Γ be a homoclinic orbit associated to a Lorenz-like singularity $\sigma \in S(X)$. We say that Γ is of *inclination-flip* type if (G1) fails and of *orbit-flip* type if (G2) fails; see Fig. 3.14.

Generically inclination-flip homoclinic orbits are not orbit-flip and conversely.

Every C^r vector field ($r \geq 1$) exhibiting orbit-flip homoclinic orbits can be C^r approximated by a smooth vector field exhibiting an inclination-flip homoclinic orbits, as stated in the following

Theorem 3.6 *Let X be a C^1 vector field in M exhibiting an orbit-flip homoclinic orbit associated to a singularity σ of X. Suppose that σ has real eigenvalues $\lambda_2 < \lambda_3 < 0 < \lambda_1$ satisfying $-\lambda_3 < \lambda_1$. Then X can be C^1 approximated by C^1 vector fields exhibiting an inclination-flip homoclinic orbit.*

The proof of this theorem can be found in [167] and follows from standard perturbation techniques (see e.g. [190]). Observe that a vector field exhibiting a inclination-flip type homoclinic orbit cannot have a dominated splitting for the linear Poincaré flow. Indeed, the definition of inclination-flip implies the existence of a tangency between the strong-stable and center-unstable manifolds along a regular orbit of the flow.

As a consequence, by Theorem 3.6, for vector fields having every critical element hyperbolic and no sinks or sources inside an isolating neighborhood U in a robust way, there cannot be either orbit-flip or inclination-flip type homoclinic orbits because of Theorem 2.33, since this would contradict the existence of a dominated splitting for the linear Poincaré flow.

3.2.3 Saddle-Focus Connection and Shil'nikov Bifurcations

Consider the following one-parameter system of ordinary differential equations in $\mathbb{R}^3$:

$$\begin{cases} \dot{x} = -\rho x + \omega y + f_1(x, y, z; \mu) \\ \dot{y} = \omega x - \rho y + f_2(x, y, z; \mu) \qquad (x, y, z, \mu) \in \mathbb{R}^4 \\ \dot{z} = \lambda z + f_3(x, y, z; \mu) \end{cases}$$

where $\lambda, \omega, \rho > 0$ and f_i are C^2 functions which vanish together with Df_i at the origin of $\mathbb{R}^4$. Then $\sigma = (0, 0, 0)$ is a saddle-focus with eigenvalues λ and $-\rho \pm \omega i$.

These families exhibit very interesting dynamics when there exists a homoclinic orbit Γ associated to σ, see Fig. 3.15.

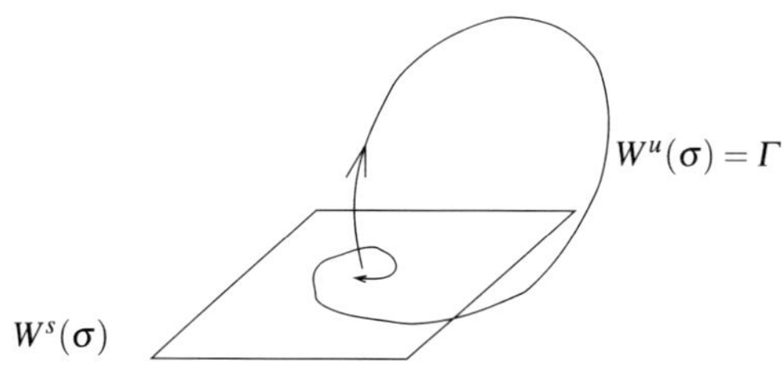

Fig. 3.15 Saddle-focus connection

Again by the use of linearizing coordinates and an analysis of the return maps to convenient cross-sections near σ one can prove the following.

Theorem 3.7 *For $\mu \neq 0$ small enough, we can find near Γ:*

1. *an attracting periodic orbit (a periodic sink), for $\rho > \lambda$;*
2. *infinitely many generic unfoldings of homoclinic tangencies when $\mu \to 0$, inducing in particular the appearance of attracting or repelling periodic orbits near Γ, for $\rho < \lambda$.*

The setting of the second item above is known as *Shil'nikov bifurcation*. The proof of these results can be found in Shil'nikov's work [243] and also in [25, 219, 259, 272].

3.2.3.1 The Shil'nikov Bifurcation for Incompressible Vector Fields

The same configuration as above of a saddle-focus connection for a conservative vector field was also studied by Biragov and Shil'nikov in [47]. In this case the eigenvalues satisfy $\lambda = 2\rho$.

Theorem 3.8 *For an incompressible family of C^7 vector fields X_μ exhibiting a saddle-focus connection as above and for $\mu \neq 0$ small enough, we can find near Γ periodic orbits whose eigenvalues are all zero, that is, elliptic closed trajectories of the vector field X_μ.*

This result implies, in particular, that invariant sets for incompressible flows, having a dominated splitting for the Linear Poincaré Flow, cannot accumulate singularities having a saddle-focus connection. This will be used in Chap. 8.

3.2.3.2 Higher Dimensional Saddle-Focus Connections

For higher dimensional saddle-focus connections there are works of Shil'nikov [244–246] and Fowler and Sparrow [88] which describe the local behavior near the saddle-connection for small perturbations of the vector field. The situation is essentially reduced to either the previous situations or a four-dimensional model case, as we now explain.

The understanding of the behavior of the solutions near the saddle-focus connecting orbit $\mathcal{O}$, for nearby vector fields, is obtained by studying the return map to a small neighborhood U of the saddle singularity σ. This return map is highly nonlinear since the distortion due to the passage near a hyperbolic saddle is very big.

To be able to analyze the return map, we use the fact that the time it takes a solution to go out and return to U near the saddle connecting orbit $\mathcal{O}$ is essentially constant, compared with the much larger time during which points stay close to the hyperbolic saddle. The effect near the saddle is well understood by classical linearization results (like the Grobman-Hartman Theorem). We therefore take U very small so that we consider only trajectories which spend a long time in U.

From this procedure it follows that, in the passage through U, points in the transversal directions to the connecting orbit $\mathcal{O}$ having the largest eigenvalues in absolute value, will exit from *U sufficiently far from $\mathcal{O}$ that they will not return to U* either in the future (for positive eigenvalues) or in the past (for negative eigenvalues). So only those points along the weakest eigenvalues (those with smaller absolute value) will stay close enough to $\mathcal{O}$ in order to return to U many times. This implies that the Poincaré return map on a cross-section in U can be well approximated taking in consideration only those eigendirections of $\mathcal{O}$ whose eigenvalues have real parts closest to 0. These ideas were used by Neimark and Shil'nikov [179, 245] to study near-homoclinic behavior in higher-dimensions, see also Afraimovich and Hsu [5]. This general argument is usually referred to as "central manifold argument". A standard reference for the existence and properties of central manifolds are the lecture notes of Hirsch, Pugh and Shub [110].

Consequently, the only essentially different cases are the following. The stable eigenvalue λ with the smallest negative real part and the unstable eigenvalue μ with the smallest positive real part are

A: both real, i.e., $\lambda < 0 < \mu$. This does not correspond to a saddle-focus connection, but to a saddle-connection with real eigenvalues, and so the results of Sect. 3.2.1 apply to the eigenspace corresponding to the two eigenvalues plus one of the other least stable eigenvalues.

B: one real and the other a complex pair. This case can be seen as a saddle-focus connection in the three-dimensional eigenspace associated to the two eigenvalues. In particular, *homogeneous vector fields do not have this type of saddle-connection*.

C: both are complex pairs (the so-called *double-focus structure*). This case is different and more difficult to analyze, but Fowler and Sparrow [88] have proved, in particular, the following.

Theorem 3.9 *Vector fields arbitrarily near a double-focus connection $\mathcal{O}$ have periodic orbits close to $\mathcal{O}$ with different indices, and periodic orbits undergoing a saddle-node or a period-doubling bifurcation.*

Hence *homogeneous vector fields*, that is, vector fields X whose periodic orbits do not change index under C^1 perturbations Y in a small neighborhood of X, *do not have such double-focus connections*.

3.3 Lorenz Attractor and Geometric Models

Here we present a study of the Lorenz system of equations (2.2) and then explain the construction of the geometric Lorenz models, which initially where intended to mimic the behavior of the solutions of the system (2.2), but actually give an accurate description of this flow. Recall the relation between the Lorenz flow and the associated geometrical model, with sensitive dependence on initial conditions and its historical impact, briefly touched upon in Sect. 2.2.3.

3.3.1 Properties of the Lorenz System of Equations

Here we list analytical properties directly obtained from the Lorenz equations, which can be found with much more detail in the books of Sparrow [95] and Guckenheimer-Holmes [97].

Let $X : \mathbb{R}^3 \to \mathbb{R}^3$ be the flow defined by the equations (2.2).

1. *Singularities of* X. The origin $\sigma_0 = (0, 0, 0)$ is a singularity of the field X which does not depend on the parameters of X. The others are

$$\sigma_1 = (-\sqrt{b(r-1)}, -\sqrt{b(r-1)}, r-1) \quad \text{and}$$
$$\sigma_2 = (\sqrt{b(r-1)}, \sqrt{b(r-1)}, r-1).$$

2. *Symmetry of* X. The map $(x, y, z) \mapsto (-x, -y, z)$ preserves the Lorenz system of equations, that is, if $(x(t), y(t), z(t))$ is a solution of the system of equations, then $(-x(t), -y(t), z(t))$ will also be a solution.
3. *Divergence of* X. We have

$$DX(x, y, z) = \begin{pmatrix} \partial_x(\dot{x}) & \partial_y(\dot{x}) & \partial_z(\dot{x}) \\ \partial_x(\dot{y}) & \partial_y(\dot{y}) & \partial_z(\dot{y}) \\ \partial_x(\dot{z}) & \partial_y(\dot{z}) & \partial_z(\dot{z}) \end{pmatrix} = \begin{pmatrix} -a & a & 0 \\ r-z & -1 & -x \\ y & x & -b \end{pmatrix}$$

and hence

$$\operatorname{div} X(x, y, z) = \nabla \cdot X = \operatorname{trace}\big(DX(x, y, z)\big) = -(a + 1 + b) < 0.$$

This shows the strongly dissipative character of this flow and implies that the flow contracts volume: if V_0 is the initial volume of a subset B of $\mathbb{R}^3$ we have by Liouville's Formula that the volume $V(t)$ of the image $X^t(B)$ is $V(t) = V_0 e^{-(a+b+1)t}$. For the parameters in (2.2) we have $V(t) = V_0 e^{-\frac{41}{3}t}$.

In particular any maximally positively invariant subset under X^t has zero volume: $\operatorname{Leb}(\cap_{t>0} X^t(U)) = 0$ for any open subset U of $\mathbb{R}^3$.
4. *Eigenvalues of the singularities.* For the parameters in (2.2) the singularities are, besides σ_0,

$$\sigma_1 = (-6\sqrt{2}, -6\sqrt{2}, 27) \quad \text{and} \quad \sigma_2 = (6\sqrt{2}, 6\sqrt{2}, 27).$$

For $DX(\sigma_0)$ we have the eigenvalues

$$\lambda_1 = -11/2 + \sqrt{1201}/2 \approx 11.83;$$
$$\lambda_2 = -11/2 - \sqrt{1201}/2 \approx -22.83;$$
$$\lambda_3 = -8/3 \approx -2.67.$$

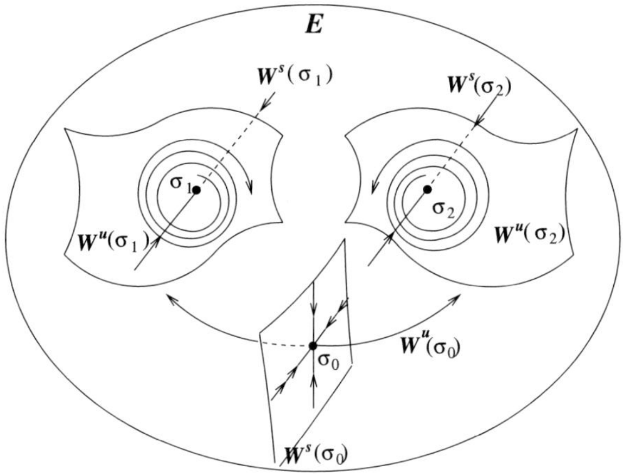

Fig. 3.16 Local stable and unstable manifolds near σ_0, σ_1 and σ_2, and the ellipsoid E

Note that $-\lambda_2 > \lambda_1 > -\lambda_3 > 0$ which corresponds to a *Lorenz-like* singularity (Definition 3.2).

For σ_1 the characteristic polynomial of $DX(\sigma_1)$ is of odd degree $p(x) = x^3 + \frac{41}{3}x^2 + \frac{304}{3}x + 1440$ and its derivative $p'(x) = 3x^2 + \frac{82}{4}x + \frac{304}{3}$ is strictly positive for all $x \in \mathbb{R}$, and hence there exists a single real root λ of p. Since $p(0) > 0 > p(-15)$ the root is negative and simple numerical calculations show that $\lambda \approx -13.85457791$. Factoring p we get

$$\begin{aligned} p(x) &= (x-\lambda)(x^2 - 0.187911244x + 103.9367643) \\ &= (x-\lambda)(x-z)(x-\overline{z}) \end{aligned}$$

and thus $z \approx 0.093955622 + 10.19450522i$.

For σ_2 the eigenvalues are the same by the symmetry of X.

Using this we obtain the following Fig. 3.16 of the local invariant manifolds and thus the local dynamics near the singularities.

5. *The trapping ellipsoid.* There exists an ellipsoid E into which every positive orbit of the flow enters eventually. Moreover E is transverse to the flow X. Therefore the open region V bounded by E is a *trapping region* for X, that is, $\overline{X^t(V)} \subset V$ for all $t > 0$.

   This is obtained by finding an appropriate Lyapunov function. We follow Sparrow [254, Appendix C] (see also the original work of Lorenz [139]). Consider

$$L(x, y, z) = rx^2 + ay^2 + a(z - 2r)^2.$$

   Then along solutions of the system (2.2) we have

$$\frac{dL}{dt} = -2a(rx^2 + y^2 + bz^2 - 2brz).$$

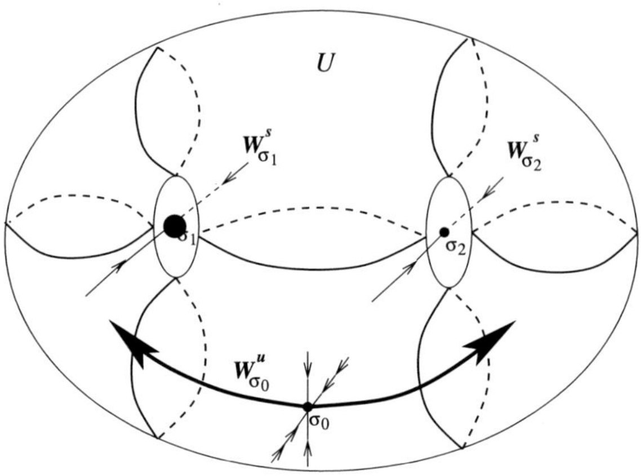

Fig. 3.17 The trapping bi-torus

Let D be the domain where $dL/dt \geq 0$ and let M be the maximum of L in D. Now define V to be the set of points such that $L \leq M + \varepsilon$ for some $\varepsilon > 0$ small. Since $D \subset V$, for x outside V we have $dL/dt = \nabla L \cdot X < -\delta < 0$ where $\delta = \delta(\varepsilon) > 0$ and X is the vector field defined by the equations (2.2). Then after a finite time the solution of the Lorenz system through x will enter the set V. Moreover for the values $(a, r, b) = (10, 28, 8/3)$ it is routine to check that ∇L points to the exterior of V over $\partial V = E$ and so all trajectories through E move towards the interior of V. Once in V any trajectory will remain there forever in the future.

Since $\overline{V}$ is compact the maximal positively invariant set $A = \cap_{t>0}\overline{X^t(V)}$ is an attracting set where trajectories of the flow accumulate when t grows without limit.

In fact numerical simulations show that there exists a subset B homeomorphic to a bi-torus such that every positive trajectory crosses B transversely and never leaves it. Hence the open set U bounded by B (see Fig. 3.17) is a better candidate for the trapping region of the set with interesting limit dynamics for X, since σ_1 and σ_2 are isolated points in the ω-limit set of X. Hence $\Lambda = \cap_{t>0}\overline{X^t(U)}$ is also an attracting set and the origin is the only singularity contained in U.

3.3.1.1 The Evolution of a Regular Orbit Inside the Attracting Basin

Lorenz observed numerically what today is known as *sensitive dependence on initial conditions*, see Sect. 2.4. Due to this the actual path of any given orbit is impossible to calculate for all large values of integration time.

The "butterfly" which appears on computer screens can be explained heuristically through the analytical properties already determined and by some numerical results. In fact the set of points whose orbits will draw the butterfly is the complement $\mathbb{R}^3 \setminus N$ of the union $N = W^s(\sigma_0) \cup W^s(\sigma_1) \cup W^s(\sigma_2)$ of the stable manifolds of the three singularities. Note that N is a bi-dimensional immersed surface in $\mathbb{R}^2$ and so has zero volume.

Figure 3.18 provides a very general view of how the orbit of a generic point in the trapping region U evolves in time. The trajectory starts spiraling around one of the

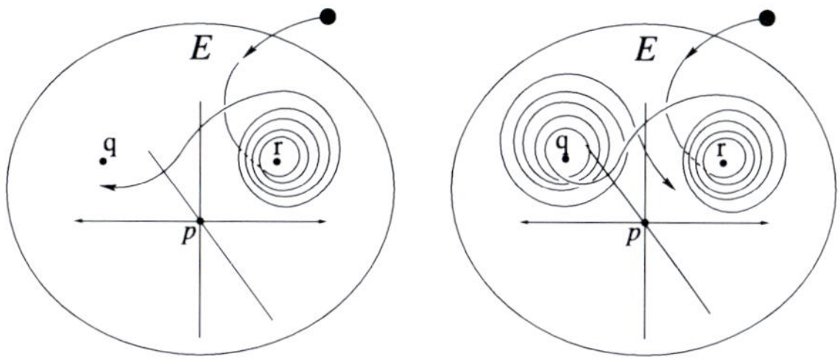

Fig. 3.18 The evolution of a generic orbit inside U

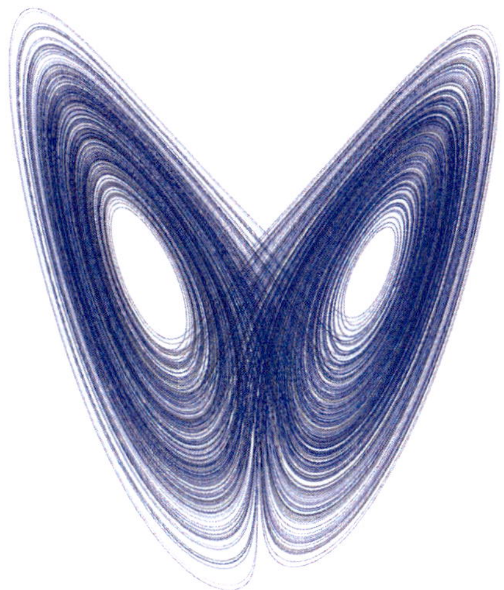

Fig. 3.19 Another view of the Lorenz attractor

singularities, σ_2 say, and suddenly "jumps" to the other singularity and then starts spiraling around σ_2. This process repeats endlessly. The number of turns around each singularity is essentially random. The ω-limit of a generic orbit is the following "butterfly" in Fig. 3.19.

3.3.2 *The Geometric Model*

The work of Lorenz on the famous flow was published in 1963 [139] but more than 10 years passed before new works on the subject appeared. Williams [273] wrote (in 1977):

> *... Several years ago Jim Yorke figured out some things about the Lorenz equation and got other mathematicians interested. He gave some talks on the subject, including one here at Berkeley. Ruelle, Lanford and Guckenheimer became interested and did some work on these equations. Unfortunately, except for the preprint of Ruelle, Guckhenheimer's paper, is the only thing these four people ever wrote on the subject as far as I know.*

Lorenz had already conjectured the existence of a strange attractor according to the available numerical simulations. The rigorous proof of this fact took many years due to the presence of a singularity accumulated by regular orbits of the flow, which prevents this set from being uniformly hyperbolic—see e.g. Sect. 2.3.

An important breakthrough in the understanding of the dynamics of the solutions of the Lorenz system of equations was achieved through the introduction of geometric models independently by Afraimovich, Bykov, Shil'nikov [2–4] in 1977 and by Guckenheimer, Williams [98] in 1979. These models were based on the properties suggested by the numerical simulations. In fact they were able to show the existence of a strange attractor for the geometric model.

This model inspired many others. Today there are different extensions and there are singular attractors which are not of the "Lorenz type": neither conjugated nor equivalent to the Lorenz geometrical model, see e.g. [173].

As explained in Sect. 2.2.3, in 1998 a positive answer to the existence of a strange attractor in the original Lorenz system of equations was given by Tucker [260] in his PhD thesis, through the theory of normal forms together with rigorous numerical algorithms.

3.3.2.1 Construction of the Geometric Model

To present the detailed construction of the geometric Lorenz model we first analyze the dynamics in a neighborhood of the singularity at the origin, and then we imitate the effect of the pair of saddle singularities in the original Lorenz flow, as in Fig. 3.16.

Near the Singularity By the Hartman-Grobman Theorem or by the results of Sternberg [256], in a neighborhood of the origin the Lorenz equations are equivalent to the linear system $(\dot{x}, \dot{y}, \dot{z}) = (\lambda_1 x, \lambda_2 y, \lambda_3 z)$ through conjugation, thus

$$X^t(x_0, y_0, z_0) = (x_0 e^{\lambda_1 t}, y_0 e^{\lambda_2 t}, z_0 e^{\lambda_3 t}), \tag{3.3}$$

where $\lambda_1 \approx 11.83$, $\lambda_2 \approx -22.83$, $\lambda_3 = -8/3$ and $(x_0, y_0, z_0) \in \mathbb{R}^3$ is an arbitrary initial point near $(0, 0, 0)$.

Consider $S = \{(x, y, 1) : |x| \le 1/2, |y| \le 1/2\}$ and

$$S^- = \big\{(x, y, 1) \in S : x < 0\big\}, \qquad S^+ = \big\{(x, y, 1) \in S : x > 0\big\} \quad \text{and}$$
$$S^* = S^- \cup S^+ = S \setminus \Gamma, \quad \text{where} \qquad \Gamma = \big\{(x, y, 1) \in S : x = 0\big\}.$$

Assume that S a transverse section to the flow so that every trajectory eventually crosses S in the direction of the negative z axis as in Fig. 3.20. Consider also $\Sigma = \{(x, y, z) : |x| = 1\} = \Sigma^- \cup \Sigma^+$ with $\Sigma^\pm = \{(x, y, z) : x = \pm 1\}$. For each $(x_0, y_0, 1) \in S^*$ the time τ such that $X^\tau(x_0, y_0, 1) \in \Sigma$ is given by $\tau(x_0) = -\frac{1}{\lambda_1} \log |x_0|$, which depends on $x_0 \in S^*$ only and is such that $\tau(x_0) \to +\infty$ when $x_0 \to 0$. This is one of the reasons many standard numerical algorithms were unsuited to tackle the Lorenz system of equations. Hence we get (where $\operatorname{sgn}(x) =$

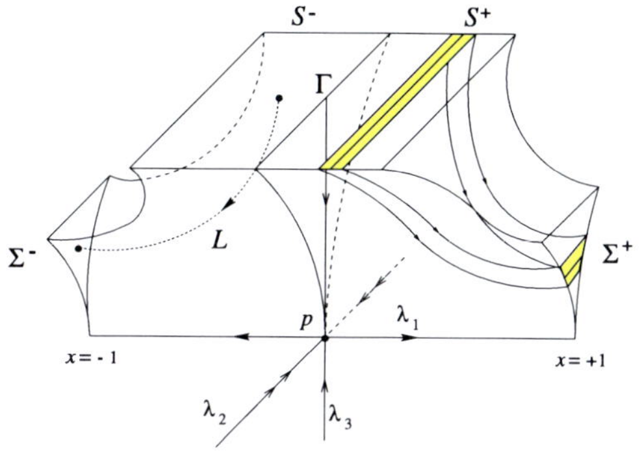

Fig. 3.20 Behavior near the origin

$x/|x|$ for $x \neq 0$ as usual)

$$X^{\tau}(x_0, y_0, 1) = \left(\operatorname{sgn}(x_0), y_0 e^{\lambda_2 \tau(x_0)}, e^{\lambda_3 \tau(x_0)}\right) = \left(\operatorname{sgn}(x_0), y_0 |x_0|^{-\frac{\lambda_2}{\lambda_1}}, |x_0|^{-\frac{\lambda_3}{\lambda_1}}\right).$$

Since $0 < -\lambda_3 < \lambda_1 < -\lambda_2$, we have $0 < \alpha = -\frac{\lambda_3}{\lambda_1} < 1 < \beta = -\frac{\lambda_2}{\lambda_1}$. Let $L : S^* \to \Sigma$ be given by

$$L(x, y, 1) = \left(\operatorname{sgn}(x), y|x|^{\beta}, |x|^{\alpha}\right). \tag{3.4}$$

It is easy to see that $L(S^{\pm})$ has the shape of a triangle without the vertex $(\pm 1, 0, 0)$. In fact $(\pm 1, 0, 0)$ are cusp points of the boundary of each of these sets.

From now on we denote by $\Sigma^{\pm}$ the closure of $L(S^{\pm})$. Clearly each line segment $S^* \cap \{x = x_0\}$ is taken to another line segment $\Sigma \cap \{z = z_0\}$ as sketched in Fig. 3.20.

The Effect of the Saddles To imitate the random turns of a regular orbit around the origin and obtain a butterfly shape for our flow, as in the original Lorenz flow (see Figs. 3.19 and 2.2), we proceed as follows.

The sets $\Sigma^{\pm}$ should return to the cross-section S through a flow described by a suitable composition of a rotation $R_{\pm}$, an expansion $E_{\pm\theta}$ and a translation $T_{\pm}$. The rotation is roughly around $W^s(\sigma_1)$ and $W^s(\sigma_2)$. We are assuming that the "triangles" $L(S^{\pm})$ are compressed in the y-direction and stretched on the other transverse direction. This is related to the eigenvalues of σ_1, σ_2 of the original Lorenz flow as explained below. We assume that this return map takes line segments $\Sigma \cap \{z = z_0\}$ into line segments $S \cap \{x = x_1\}$, as sketched in Fig. 3.21.

We recall that the equilibrium at the origin is hyperbolic and so its stable $W^s(0)$ and unstable $W^u(0)$ manifolds are well defined (see Chap. 2). We also note that $W^u(0)$ has dimension one, and hence $W^u(0) \setminus \{0\}$ has two connected components $W^{u,\pm}(p)$, and $W^u(0) = W^{u,+}(0) \cup \{0\} \cup W^{u,-}(0)$.

The rotation $R_{\pm}$ has axis parallel to the y-direction, which is orthogonal to the x-direction (the x-direction is parallel to the local branches $W^{u,\pm}(p)$). More precisely,

Fig. 3.21 *R* takes $\Sigma^{\pm}$ to *S*

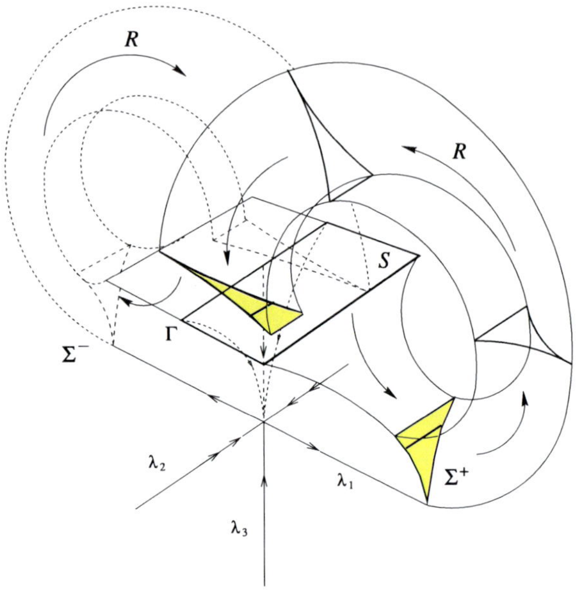

if $(x, y, z) \in \Sigma^{\pm}$, then

$$R_{\pm}(x, y, z) = \begin{pmatrix} 0 & 0 & \pm 1 \\ 0 & 1 & 0 \\ \pm 1 & 0 & 0 \end{pmatrix}.$$

The expansion occurs only along the x-direction, and so E_θ is given by

$$E_{\pm\theta}(x, y, z) = \begin{pmatrix} \theta & 0 & 0 \\ 0 & 1 & 0 \\ 0 & 0 & 1 \end{pmatrix}$$

with $\theta 2^{-\alpha} < 1$ and $\theta\alpha 2^{1-\alpha} > 1$. The first condition ensures that the image of the resulting map is contained in S. The second condition makes a certain one dimensional induced map to be piecewise expanding. This point will be discussed below.

The translations $T_{\pm} : \mathbb{R}^3 \to \mathbb{R}^3$ are chosen such that the unstable direction starting from the origin is sent to the boundary of S and the images of both $\Sigma^{\pm}$ are disjoint. These transformations $R_{\pm}, E_{\pm\theta}, T_{\pm}$ take line segments $\Sigma^{\pm} \cap \{z = z_0\}$ into line segments $S \cap \{x = x_1\}$ as shown in Fig. 3.21, and so does the composition $T_{\pm} \circ E_{\pm\theta} \circ R_{\pm}$.

This composition of linear maps describes a vector field in a region outside $[-1, 1]^3$, in the sense that one can use the above linear maps to define a vector field X such that the time one map of the associated flow realizes $T_{\pm} \circ E_{\pm\theta} \circ R_{\pm}$ as a map $\Sigma^{\pm} \to S$. Since the explicit choice of the vector field is not important for our purposes, we will not construct such vector field. We observe that the flow on the attractor we are constructing will pass through the region between $\Sigma^{\pm}$ and S in a

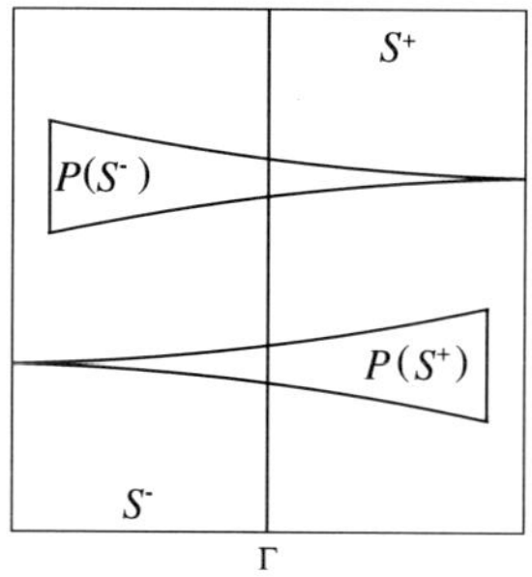

Fig. 3.22 The return map image $P(S^*)$

relatively small time with respect the linearized region. The linearized regions will then dominate all estimates of expansion/contraction.

The above construction enables us to describe, for $t \in \mathbb{R}^+$, the orbit $X^t(x)$ of each point $x \in S$: the orbit will start following the linear field until $\tilde{\Sigma}^{\pm}$ and then it will follow X coming back to S and so on. Let us denote by $W = \{X^t(x), x \in \Sigma, t \in \mathbb{R}^+\}$ the set where this flow acts. The geometric Lorenz flow is then the couple (W, X^t).

A consequence of all this is that every $x \in S$ has a positive orbit disjoint from $W^{ss}(\sigma)$. Since every point $x \in W \setminus \{\sigma\}$ has a positive orbit that will eventually cross S by construction, we see that

$$W^{ss}(\sigma) \cap \Lambda = \{\sigma\}. \tag{3.5}$$

The Poincaré first return map $P : S^* \to S$ can be defined as

$$P(x, y) = \begin{cases} T_+ \circ E_{+\theta} \circ R_+ \circ L(x, y, 1) & \text{for } x > 0 \\ T_- \circ E_{-\theta} \circ R_- \circ L(x, y, 1) & \text{for } x < 0 \end{cases}.$$

The above combined effects imply that the foliation of S given by the lines $S \cap \{x = x_0\}$ is invariant under the return map, meaning that, for any given leaf γ of this foliation, its image $P(\gamma)$ is contained in a leaf of the same foliation. Hence P must have the form $P(x, y) = (f(x), g(x, y))$ for some functions $f : I \setminus \{0\} \to I$ and $g : (I \setminus \{0\}) \times I \to I$, where $I = [-1/2, 1/2]$.

Taking into account the definition of L from the linearized region we see that

$$f(x) = \begin{cases} f_1(x^\alpha), & \text{if } x < 0 \\ f_0(x^\alpha), & \text{if } x > 0 \end{cases}; \quad \text{with } f_i = (-1)^i \theta \cdot x + b_i, \quad i = 0, 1;$$

and

$$g(x, y) = \begin{cases} g_1(x^\alpha, y \cdot x^\beta), & \text{if } x < 0 \\ g_0(x^\alpha, y \cdot x^\beta), & \text{if } x > 0 \end{cases},$$

where $g_1 | I^- \times I \to I$ and $g_0 | I^+ \times I \to I$ are suitable affine maps, with $I^- = [-1/2, 0)$, $I^+ = (0, 1/2]$.

Properties of the One-Dimensional Map f Now we specify the properties of the one-dimensional map f that follow from the previous construction:

(f1) the symmetry of the Lorenz equations implies $f(-x) = -f(x)$;
(f2) f is discontinuous at $x = 0$ with lateral limits $f(0^-) = +\frac{1}{2}$ and $f(0^+) = -\frac{1}{2}$, since P is not defined at Γ because $\Gamma \subset W^s(0,0,0)$;
(f3) f is differentiable on $I \setminus \{0\}$ and $f'(x) > \sqrt{2}$;
(f4) the lateral limits of f' at $x = 0$ are $f'(0^-) = +\infty$ and $f'(0^+) = -\infty$.

On the other hand $g : S^* \to I$ is defined in such a way that it contracts the second coordinate: $g'_y(w) \le \mu < 1$ for all $w \in S^*$. This is suggested by the eigenvalues $\lambda_2 \approx -22.83$ of σ_0 and $\lambda \approx -13.8545$ of the saddles σ_1, σ_2 (see Sect. 3.3.1). Moreover the rate of contraction of g on the second coordinate should be much higher than the expansion rate of f. Figure 3.22 sketches $P(S^*)$. In addition the expansion rate is big enough to obtain a strong mixing property for f (it is locally eventually onto, see Sect. 3.3.5).

Expression for the Derivative *DP* We recall that $P = T_\pm \circ E_{\pm\theta} \circ R_\pm \circ L$. From the definition of each of the maps in the composition above we see that: given $q = (x, y) \in S^*$ with $x > 0$

$$DL(x, y, 1) = \begin{pmatrix} \beta y x^{\beta-1} & x^\beta \\ \alpha x^{\alpha-1} & 0 \end{pmatrix}.$$

Restricting the rotation and the other linear maps to $\Sigma^\pm$ and composing the resulting matrices we get

$$DP(x, y) = \begin{pmatrix} \theta \alpha x^{(\alpha-1)} & 0 \\ \beta y x^{(\beta-\alpha)} & x^\beta \end{pmatrix}. \tag{3.6}$$

The expression for DP at $q = (x, y)$ with $x < 0$ is similar.

Properties of the Map *g* We note that by its definition the map g is piecewise C^2. The above expression (3.6) provides the following bounds on its partial derivatives:

1. For all $(x, y) \in S^*$ with $x > 0$, we have $\partial_y g(x, y) = x^\beta$. As $\beta > 1$, $|x| \le 1/2$, there is $0 < \lambda < 1$ such that
$$|\partial_y g| < \lambda. \tag{3.7}$$
The same bound also holds for $x < 0$.
2. For $(x, y) \in S^*$ with $x \neq 0$, we have $\partial_x g(x, y) = \beta x^{\beta-\alpha}$. As $\beta - \alpha > 0$ and $|x| \le 1/2$, we get $|\partial_x g| < \infty$.

We note that from the first item above there follows the uniform contraction of the foliation given by the lines $S \cap \{x = constant\}$. The foliation is contracting in the following sense: there exists a constant $C > 0$ such that, for any given leaf γ of the foliation and for $y_1, y_2 \in \gamma$, then

$$\operatorname{dist}\big(P^n(y_1), P^n(y_2)\big) \le C\lambda^n \operatorname{dist}(y_1, y_2) \quad \text{when } n \to \infty.$$

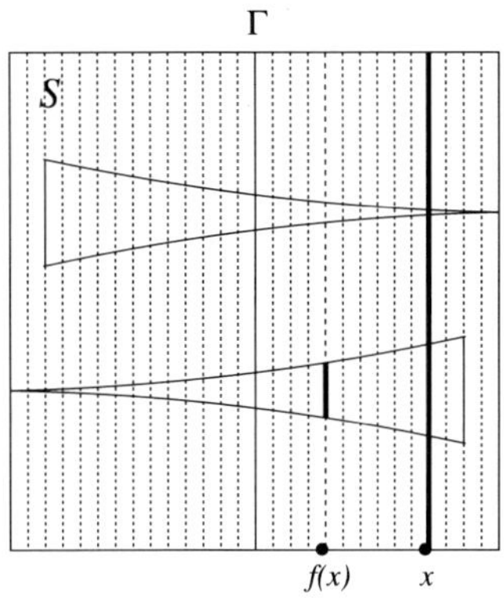

Fig. 3.23 Projection on I through the stable leaves and a sketch of the image of one leaf under the return map

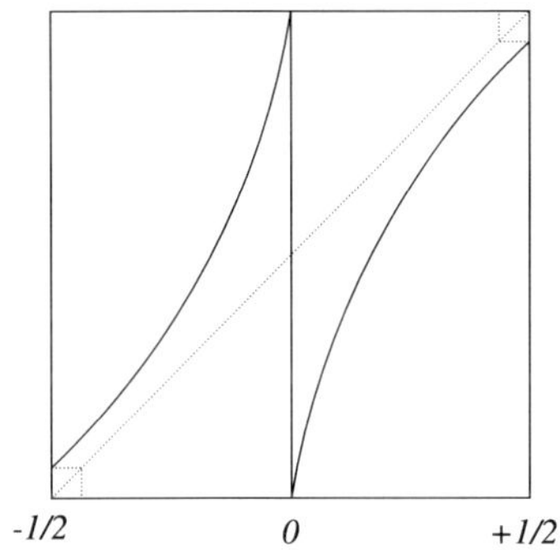

Fig. 3.24 The Lorenz map f

Thus the study of the 3-flow can be reduced to the study of a bi-dimensional map and, moreover, the dynamics of this map can be further reduced to a one-dimensional map, since the invariant contracting foliation enables us to identify two points on the same leaf, since their orbits remain forever on the same leaf and the distance of their images tends to zero under iteration. See Fig. 3.23 for a sketch of this identification.

The quotient map obtained through this identification will be called *the Lorenz map*. Figure 3.24 shows the graph of this one-dimensional transformation.

3.3.3 The Geometric Lorenz Attractor Is a Partially Hyperbolic Set with Volume Expanding Central Direction

Observe that the time $t(w)$ it takes a point $w \in \Sigma$ to go to S, that is, $X^{t(w)}(w) \in S$ and $X^t(w) \in \mathbb{R}^3 \setminus (S \cup \Sigma)$ for $0 < t < t(w)$, is bounded by some constant independently of the point: $t(w) \le t_0$. This ensures that the behavior of the flow on the maximal positively invariant subset of the trapping region is prescribed by the behavior from the cross-section S to the cross-sections Σ^+, Σ^-, as we now explain.

Figure 3.20 makes it clear that the linear flow (3.3) preserves lines in the direction of the y-axis when taking points from S to Σ. Moreover it is not difficult to check that *its derivative* DX^t also preserves planes orthogonal to the y-axis.

In addition, by the choice of the flow from Σ to S and as Fig. 3.21 suggests, horizontal lines at Σ, i.e., parallel to the y-axis, are taken to lines parallel to same axis in S, that is, *the flow preserves lines parallel to the y-axis from Σ to S. Since the flow from Σ to S performs rotation and a translation, we can assume that its derivative also preserves planes orthogonal to the y-axis.*

From this we deduce that the following splitting of $\mathbb{R}^3$: $E = \{0\} \times \mathbb{R} \times \{(0\}$ and $F = \mathbb{R} \times \{0\} \times \mathbb{R}$, is preserved by the flows, that is, $DX^t_w \cdot E = E$ and $DX^t_w \cdot F = F$ for all t and every point w in an orbit inside the trapping ellipsoid.

Moreover we can check that for w on the linearized part of the flow, from S to Σ, we have for $t > 0$ such that $X^{[0,t]}(w)$ is contained in the domain of the linearized coordinates:

- $\|DX^t_w \mid E\| = e^{\lambda_2 t}$;
- $\|DX^t_w \mid E\| = e^{(\lambda_2 - \lambda_3)t} \cdot m\big(DX^t \mid F\big)$,

where $m(DX^t \mid F)$ is the minimum norm of the linear map. Since $\lambda_2 < 0$ we see that E is uniformly contracting, this being a stable direction. But $\lambda_2 - \lambda_3 < 0$ and so the contraction along the direction of F is weaker than the contraction along E. This kind of splitting $E \oplus F$ of $\mathbb{R}^3$ is called a *partially hyperbolic splitting*.

Observe also that since $\lambda_1 + \lambda_3 > 0$ we have $|\det DX^t \mid F| = e^{(\lambda_1+\lambda_3)t}$ and so the flow *expands volume* along the F direction.

We will see in Chap. 5 that these properties characterize compact invariant sets which are robustly transitive.

However we have only checked these properties in the linearized region. But if the orbit of a point w passes outside the linear region k times from Σ to S lasting $s_1 + \cdots + s_k$ from time 0 to time t, then $t > s_1 + \cdots + s_k$ and for some constant $b > 0$ bounding the derivatives of DX^t from 0 to t_0, we have

$$\|DX^t_w \mid E\| \le e^{bk+\lambda_2(t-s_1-\cdots-s_k)} = \exp\left\{\lambda_2 t\left(1 - \frac{bk}{\lambda_2 t} - \frac{s_1 + \cdots + s_k}{t}\right)\right\},$$

and so the last expression in brackets is bounded. We see that E is (K, λ_2)-contracting for some $K > 0$.

An entirely analogous reasoning shows that the direction E dominates F uniformly for all t and that DX^t expands volume along F also uniformly.

Thus the maximal positively invariant set in the trapping ellipsoid is partially hyperbolic and the flow expands volume along a bi-dimensional invariant direction.

3.3.4 Existence and Robustness of Invariant Stable Foliation

Now we prove, partially following the work of [98], that the geometric Lorenz attractor constructed in the previous subsection is *robust*, that is, it persists for all nearby vector fields.

More precisely: there exists a neighborhood U in $\mathbb{R}^3$ containing the attracting set Λ such that, for all vector fields Y which are C^1-close to X, the maximal invariant subset $\Lambda_Y = \cap_{t\geq 0} Y^t(U)$ in U is still a transitive Y-invariant set.

This is a striking property of these flows since the Lorenz flow exhibits sensitive dependence on initial conditions. The robustness will be a consequence of the persistence of the invariant contracting foliation on the cross-section S to the flow.

Numerically this is expected since, in spite of the huge integration errors involved and the various integration algorithms used, the solutions obtained always have a shape similar to the one in Fig. 3.19, independently of the initial point chosen to start the integration.

We start by obtaining the persistence of the stable foliation for points in the attractor, then explain why these attractors, although robust, are *not structurally stable* in Sect. 3.3.5.3.

We note that C^1-robustly transitive attractors in 3-manifolds were completely described from the geometrical point of view in [174] and the proof of this result is presented in Chap. 5.

3.3.4.1 Geometric Idea of the Proof

Theorem 3.10 (Persistence of contracting foliation) *Let X be the vector field obtained in the construction of the geometric Lorenz model and $\mathscr{F}_X$ the invariant contracting foliation of the cross-section S. Then any vector field Y which is sufficiently C^1-close to X admits an invariant contracting foliation $\mathscr{F}_Y$ on the cross-section S.*

We note that $\mathscr{F}_Y$ is a continuous foliation with C^1 leaves. It can be shown that the holonomies along the leaves are in fact Hölder-C^1 (see Sect. 2.7.2.4). Moreover, if we have a *strong dissipative* condition on the equilibrium σ, that is, if $\beta > \alpha + k$ for some $k \in \mathbb{Z}^+$ (see the definitions of α, β as functions of the eigenvalues of σ in (3.4)), it can be shown that $\mathscr{F}_Y$ is then a C^k smooth foliation, and so the holonomies along its leaves are C^k maps. See Remark 3.15 at the end of the proof of Theorem 3.10. In particular, *for strongly dissipative Lorenz attractors with $\beta > \alpha + k$ the one-dimensional quotient map is C^k smooth away from the singularity.*

We first present a geometric idea of the proof and then proceed to the details in the following Sect. 3.3.4.2.

Observe first that the cross-section S remains transverse to any flow C^1-close to X and that the singularities $\sigma_0, \sigma_1, \sigma_2$ persist with eigenvalues satisfying the same relations as before since they are hyperbolic. In addition, since $W^u_X(\sigma_0)$ intersects S transversely, then just by the C^1 continuous variation of compact parts of the unstable manifolds of a hyperbolic singularity we have that $W^u_Y(\sigma_0(Y))$ still intersects S transversely for all Y close to X in the C^1 norm.

Without loss we can assume, after a C^1 change of coordinates, that the Lorenz-like singularity $\sigma_0(Y)$ remains at the origin and that the eigenvectors of $DY(\sigma_0(Y))$ have the directions of the coordinate axis as before, with the plane $x = 0$ containing the stable manifold of $\sigma_0(Y)$.

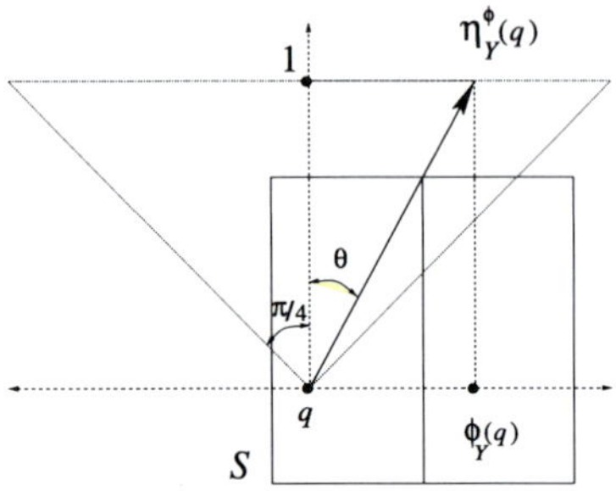

Fig. 3.25 The field η_Y^ϕ

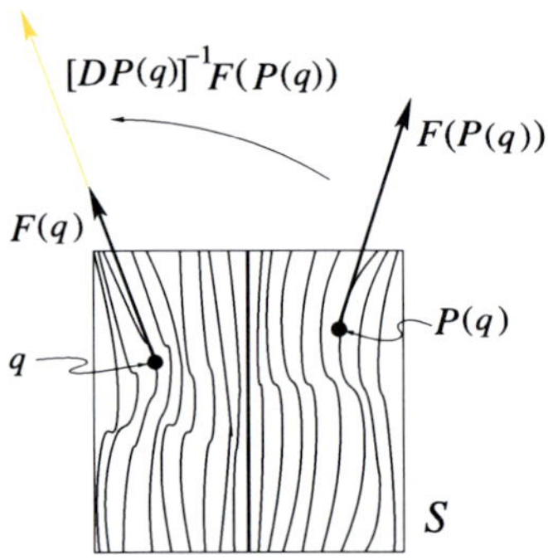

Fig. 3.26 The field F and the parallel condition

Thus for a neighborhood $\mathcal{U}$ of X in the C^1 topology and for each $Y \in \mathcal{U}$ we can define the Poincaré first return map $P_Y : S^* \to S$ as $P_Y = R_Y \circ L_Y$ where $L_Y : S^* \to \Sigma$ is such that $L_Y(x, y) = (y|x|^\beta, |x|^\alpha)$ with $\alpha = -\frac{\lambda_3(Y)}{\lambda_1(Y)}$ and $\beta = -\frac{\lambda_2(Y)}{\lambda_1(Y)}$ (note that $\beta - \alpha > 1$).

On the other hand $R_Y : \Sigma \to S$ is a C^1 diffeomorphism which can be expressed by the composition $R_Y = J_Y \circ R_0$, where J_Y is a C^1 perturbation of the identity and R_0 is the diffeomorphism associated to X_0.

Now let $\mathcal{A}$ be the space of continuous maps $\phi : \mathcal{U} \times S \to [-1, +1]$. For each $Y \in \mathcal{U}$ and $\phi \in \mathcal{A}$ we define $\phi_Y : S \to [-1, 1]$ by $\phi_Y(q) = \phi(Y, q)$ for all $q \in S$. We associate to ϕ_Y a vector field $\eta_Y^\phi : S \to [-1, 1] \times \{1\}$ given by $\eta_Y^\phi(q) = (\phi_Y(q), 1)$ which we view as a vector on $T_q S = \mathbb{R}^2$; see Fig. 3.25. Integrating the field η_Y^ϕ we get a family of curves which induces a foliation on S; see Fig. 3.26. We must show that there exists $\phi \in \mathcal{A}$ such that η_Y^ϕ induces an invariant foliation under P_Y. Before explaining the proof of this fact we state a necessary and sufficient condition for the invariance of this foliation.

Let F be a continuous vector field defined on S and P the map defined above. Integrating F we get a foliation of S. Let $q \in S^*$ have image $P(q)$ and consider the vectors $F(q)$ and $F(P(q))$. Observe that the foliation induced by F is invariant under P if

$$DP(q)\big(F(q)\big) \quad \text{and} \quad F\big(P(q)\big) \quad \text{are parallel,} \quad \text{or}$$

$$F(q) \quad \text{and} \quad \big[DP(q)\big]^{-1} F\big(P(q)\big) \quad \text{are parallel.}$$

On the other hand if we consider the slope of vectors with respect to the vertical direction (0, 1), two vectors are parallel if, and only if, their slope is the same. For $(a, b) \in \mathbb{R}^2$ we set slope $(a, b) = a/b$, and hence to check that the foliation defined by F is invariant under P amounts to showing that

$$\text{slope}\left(F(q)\right) = \text{slope}\left(\left[DP(q)\right]^{-1} F\left(P(q)\right)\right).$$

Translating this for η_Y^ϕ we obtain the condition

$$\phi_Y(q) = \text{slope}\left(\eta_Y^\phi(q)\right) = \text{slope}\left(\left[DP_Y(q)\right]^{-1} \eta_Y^\phi\left(P_Y(q)\right)\right).$$

The last term above depends on ϕ, Y and q and, if we define $T : \mathscr{A} \to \mathscr{A}$ as

$$T(\phi_Y)(q) = \text{slope}\left(\left[DP_Y(q)\right]^{-1} \eta_Y^\phi\left(P_Y(q)\right)\right),$$

then the condition of invariance becomes $T(\phi_Y)(q) = \phi_Y(q)$, that is, $T(\phi) = \phi$. Hence the element $\phi \in \mathscr{A}$ for which η_Y^ϕ induces an invariant foliation on S is a fixed point of T. Thus we are left to prove that the operator T has a fixed point.

For this, we first show that T is well defined and then that T is a contraction in an appropriate space, which concludes the proof of Theorem 3.10.

3.3.4.2 Proof of Existence of Invariant Stable Foliation

The Poincaré map P_Y associated to $Y \in \mathscr{U}$ can be written

$$P_Y(q) = \left(f_Y(q), g_Y(q)\right)$$

for $q \in S^*$. We rewrite T as a function of f and g. First we calculate

$$\left(DP_Y(q)\right)^{-1} = \frac{1}{\Delta} \begin{pmatrix} \partial_y g_Y(q) & -\partial_y f_Y(q) \\ -\partial_x g_Y(q) & \partial_x f_Y(q) \end{pmatrix} \quad \text{with } \Delta = \det DP_Y(q).$$

Then it is not difficult to see that the slope of

$$\left(DP_Y(q)\right)^{-1} \eta_Y^\phi\left(P_Y(q)\right) = \frac{1}{\Delta} \begin{pmatrix} \partial_y g_Y(q) & -\partial_y f_Y(q) \\ -\partial_x g_Y(q) & \partial_x f_Y(q) \end{pmatrix} \begin{pmatrix} \phi_Y\left(P_Y(q)\right) \\ 1 \end{pmatrix}$$

is

$$\text{slope}\left(\left(DP_Y(q)\right)^{-1} \eta_Y^\phi\left(P_Y(q)\right)\right) = \frac{[\phi_Y(P_Y(q))]\partial_y g_Y(q) - \partial_y f_Y(q)}{-[\phi_Y(P_Y(q))]\partial_x g_Y(q) + \partial_x f_Y(q)}.$$

Writing $\widehat{P}(Y, q) = (Y, P_Y(q))$we get

$$T(\phi_Y)(q) = \frac{(\phi \circ \widehat{P})\partial_y g - \partial_y f}{\partial_x f - (\phi \circ \widehat{P})\partial_x g}(Y, q).$$

Lemma 3.11 *Let $a_0 \in (0, 1/5)$ and let Y be a vector field C^1-close to X. If $P_Y(q) = P(Y,q) = (f(Y,q), g(Y,q))$, then there are positive constants k_i, $i = 1, 2, 3$, such that for all $q \in S^*$*

1. $\left|\frac{\partial_x g(Y,q)}{\partial_x f(Y,q)}\right| \le a_0$; $\frac{|\partial_y g(Y,q)|}{|\partial_x f(Y,q)|} \le k_1|x|^{(\beta-\alpha+1)}$, *and* $\frac{|\partial_y f(Y,q)|}{|\partial_x f(Y,q)|} \le k_2|x|^{(\beta-\alpha+1)}$;
2. $\|D_q P(Y,q)\| \le k_3|x|^{(\alpha-1)}$, *and* $|\det D_q P(Y,q)| \le a_0|x|^{(\beta+\alpha-1)}$;
3. $\sup_{S^*}\left\{\frac{|\partial_y g|}{|\partial_x f|}, \frac{|\partial_y f|}{|\partial_x f|}, \frac{|\partial_x g|}{|\partial_x f|}, |\det DP|\right\} < a_0$.

Proof We provide the calculations for $x > 0$, the other case being analogous. Since $P_Y = R_Y \circ L_Y$ we have $DP_Y(q) = DR_Y(L_Y(q))DL_Y(q)$ and

$$DL_Y(x,y) = \begin{pmatrix} \beta y x^{\beta-1} & x^\beta \\ \alpha x^{\alpha-1} & 0 \end{pmatrix}.$$

Recall that $R_Y = J_Y \circ R_0$ and so we have that $DR_Y(L_Y(x,y))$ can be written as $DJ_Y(R_0(L_Y(x,y)))\cdot DR_0(L_Y(x,y))$. Since J_Y is close to the identity we may write $J_Y(x,y) = (x+\varepsilon_1, y+\varepsilon_2)$ with $(\varepsilon_1,\varepsilon_2) = \varepsilon(X,y,z)$ small in the C^1-norm. Thus we have

$$DJ_Y = \begin{pmatrix} 1+\partial_x\varepsilon_1 & \partial_y\varepsilon_1 \\ \partial_x\varepsilon_1 & 1+\partial_y\varepsilon_2 \end{pmatrix} \quad \text{and} \quad DR_0 = \begin{pmatrix} 0 & M \\ \sigma & 0 \end{pmatrix}$$

and so

$$DR_Y = \begin{pmatrix} \sigma\cdot\partial_y\varepsilon_1 & M + M\cdot\partial_x\varepsilon_1 \\ \sigma+\sigma\cdot\partial_y\varepsilon_2 & M\cdot\partial_x\varepsilon_1 \end{pmatrix} = \begin{pmatrix} \varepsilon_1 & M+\varepsilon_2 \\ \sigma+\varepsilon_3 & \varepsilon_4 \end{pmatrix}.$$

Finally we multiply the last pair of matrices to get

$$\begin{aligned} DP_Y(x,y) &= \begin{pmatrix} \varepsilon_1\beta y x^{\beta-1} + M\alpha x^{\alpha-1} + \varepsilon_2\alpha x^{\alpha-1} & \varepsilon_1 x^\beta \\ \sigma\beta y x^{\beta-1} + \varepsilon_3\beta y x^{\beta-1} + \varepsilon_4\alpha x^{\alpha-1} & \sigma x^\beta + \varepsilon_3 x^\beta \end{pmatrix} \\ &= \begin{pmatrix} \partial_x f & \partial_y f \\ \partial_x g & \partial_y g \end{pmatrix}. \end{aligned}$$

We may now assume without loss in what follows that $\varepsilon = \varepsilon_i$ since $\varepsilon_i \to 0$ when $Y \to X$. Hence

$$\begin{pmatrix} \partial_x f & \partial_y f \\ \partial_x g & \partial_y g \end{pmatrix} = \begin{pmatrix} \left[\varepsilon\beta y x^{(\beta-\alpha)} + (M+\varepsilon)\alpha\right]x^{(\alpha-1)} & \varepsilon x^\beta \\ \left[(\sigma+\varepsilon)\beta y x^{(\beta-\alpha)} + \varepsilon\alpha\right]x^{(\alpha-1)} & (\sigma+\varepsilon)x^\beta \end{pmatrix}.$$

Now we may find the stated bounds as follows.

1. When $\varepsilon \to 0$ we have both

$$\frac{\sigma+\varepsilon}{\varepsilon\beta y x^{(\beta-\alpha)} + (M+\varepsilon)\alpha} \to \frac{\sigma}{M\alpha} \quad \text{and} \quad \frac{\varepsilon}{\varepsilon\beta y x^{(\beta-\alpha)} + (M+\varepsilon)\alpha} \to 0,$$

and hence these quotients are bounded: there are k_1 and k_2 so that $\frac{|\partial_y g|}{|\partial_x f|} \le k_1|x|^{(\beta-\alpha+1)}$ and $\frac{|\partial_y f|}{|\partial_x f|} \le k_2|x|^{(\beta-\alpha+1)}$.

On the other hand, again when $\varepsilon \to 0$ we get

$$\frac{|\partial_x g|}{|\partial_x f|} = \frac{(\sigma+\varepsilon)\beta y x^{\beta-\alpha} + \varepsilon\alpha}{\varepsilon\beta y x^{\beta-\alpha} + (M+\varepsilon)\alpha} \to \frac{\sigma\beta y x^{\beta-\alpha}}{M\alpha} \le \frac{\sigma\beta}{2\cdot 2^{\beta-\alpha} M\alpha}$$

were the bound above follows because $\beta - \alpha > 0$. Since $0 < \sigma < 1$ and $M > 1$ where chosen arbitrarily in the construction, we may assume that, σ is very small and M big enough so that $\frac{|\partial_x g|}{|\partial_x f|} \le a_0$.

2. It is clear that since $0 < \alpha < 1 < \beta$ and

$$\begin{pmatrix} \partial_x f & \partial_y f \\ \partial_x g & \partial_y g \end{pmatrix} \xrightarrow[\varepsilon\to 0]{} \begin{pmatrix} M\alpha x^{(\alpha-1)} & 0 \\ \sigma\beta y x^{(\alpha-1)} & \sigma x^{\beta} \end{pmatrix},$$

the norm of the matrix is dominated by the value of $|x|^{\alpha-1}$ for $x \approx 0$, and thus there exists k_3 such that $\|D_q P(Y,q)\| \le k_3|x|^{\alpha-1}$. On the other hand

$$\begin{aligned} |\det D_q P(Y,q)| &= |\partial_x f \partial_y g - \partial_y f \partial_x g| \le |\partial_x f||\partial_y g| + |\partial_y f||\partial_x g| \\ &\le r_1|x|^{\beta+\alpha-1} + r_2|x|^{\beta+\alpha-1} \le K|x|^{\beta+\alpha-1} \end{aligned}$$

where the existence of $r_1, r_2 > 0$ as above is a consequence of both $|\partial_x f| \cdot |\partial_y g| \xrightarrow[\varepsilon\to 0]{} M\alpha\sigma x^{(\beta+\alpha-1)}$ and $|\partial_y f| \cdot |\partial_x g| \xrightarrow[\varepsilon\to 0]{} 0$ and also of $\beta + \alpha > 1$. Note that we may assume $K \le a_0$ by setting $M\sigma$ small (that is, we assume that the volume is contracted).

3. Finally for the quotients of the entries of DP_Y note that we can again use the bounds already obtained and then get smaller than a_0 by taking σ close to 0 and M big enough.

The proof is complete. □

Proposition 3.12 *Let T be defined as before depending on f and g. Then*

1. *$T(\mathscr{A}) \subset \mathscr{A}$, that is, $T : \mathscr{A} \to \mathscr{A}$ is well defined;*
2. *$T : \mathscr{A} \to \mathscr{A}$ is a contraction.*

Proof First we show that $T(\phi)$ is continuous for $\phi \in \mathscr{A}$, and $|T(\phi)| \le 1$, which would prove the first item of the statement. According to the definition of T we have

$$\begin{aligned} |T(\phi)(Y,q)| &= \frac{|(\phi\circ\widehat{P})\partial_y g - \partial_y f|}{|\partial_x f - (\phi\circ\widehat{P})\partial_x g|}(Y,q) = \frac{|(\phi\circ\widehat{P})\frac{\partial_y g}{\partial_x f} - \frac{\partial_y f}{\partial_x f}|}{|1 - (\phi\circ\widehat{P})\frac{\partial_x g}{\partial_x f}|}(Y,q) \\ &\le \frac{|\frac{\partial_y g}{\partial_x f}| + |\frac{\partial_y f}{\partial_x f}|}{1 - |\frac{\partial_x g}{\partial_x f}|}(Y,q) \le \frac{k_1|x|^{\beta-\alpha+1} + k_2|x|^{\beta-\alpha+1}}{1-a_0} \le K|x|^{\beta-\alpha+1}. \end{aligned}$$

Thus $|T(\phi)(Y,q)| \to 0$ if $|x| \to 0$, which shows that $T(\phi)$ is continuous at 0. Then $T(\phi)$ is continuous since the expression is continuous away from $x = 0$. Moreover

$$|T(\phi)(Y,q)| \le \frac{|\frac{\partial_y g}{\partial_x f}| + |\frac{\partial_y f}{\partial_x f}|}{1 - |\frac{\partial_x g}{\partial_x f}|}(Y,q) \le \frac{2a_0}{1-a_0} < 1,$$

by Lemma 3.11(3) and because $a_0 \in (0, 1/5)$ implies $\frac{2a_0}{1-a_0} < 1/2$.

Now the contraction is easy, since for $\phi_1, \phi_2 \in \mathscr{A}$ and for fixed $(Y,q) \in \mathscr{U} \times S$

$$\begin{aligned} |T(\phi_1) - T(\phi_2)| &= \frac{|\det DP(q)| \cdot |\phi_1 \circ \widehat{P} - \phi_2 \circ \widehat{P}|}{|\partial_x f - (\phi_1 \circ \widehat{P})\partial_x g| \cdot |\partial_x f - (\phi_2 \circ \widehat{P})\partial_x g|} \\ &\le \frac{a_0}{(1-a_0)^2}|\phi_1 - \phi_2| \end{aligned}$$

and again $\frac{a_0}{(1-a_0)^2} < 1/2$, as long as $\mathscr{U}$ is taken small enough around X so that Lemma 3.11 remains valid. □

We have shown that there exists a unique fixed point for T on $\mathscr{A}$ as we wanted and so we have an invariant foliation on S.

3.3.4.3 Differentiability of the Foliation

Now we prove that the fixed point $\phi(Y,q)$ depends on Y, q continuously on the C^1 topology. We do this by showing that $D\phi_Y$ depends continuously on (Y,q) and that the operator T is also a contraction on the C^1 norm.

Again using the definition of T at a point (Y,q) we obtain the following expression

$$\begin{aligned} DT(\phi) &= \frac{D\big[(\phi \circ \widehat{P})\partial_y g - \partial_y f\big]}{\partial_x f - (\phi \circ \widehat{P})\partial_x g} - \frac{(\phi \circ \widehat{P})\partial_y g - \partial_y f}{\big(\partial_x f - (\phi \circ \widehat{P})\partial_x g\big)^2} \cdot D\big[\partial_x f - (\phi \circ \widehat{P})\partial_x g\big] \\ &= V_1(\phi) + T(\phi)V_2(\phi) + N(\phi)D\phi\big(\widehat{P}(X,q)\big), \end{aligned}$$

where we have used

$$\begin{aligned} V_1(\phi) &= \frac{\phi \circ \widehat{P}}{\partial_x f - (\phi \circ \widehat{P})\partial_x g} \cdot D\partial_y g - \frac{1}{\partial_x f - (\phi \circ \widehat{P})\partial_x g} D\partial_y f; \\ V_2(\phi) &= \frac{\phi \circ \widehat{P}}{\partial_x f - (\phi \circ \widehat{P})\partial_x g} \cdot D\partial_x f - \frac{1}{\partial_x f - (\phi \circ \widehat{P})\partial_x g} D\partial_x g; \\ N(\phi) &= \frac{\det DP(X,q)}{\big(\partial_x f - (\phi \circ \widehat{P})\partial_x g\big)^2}. \end{aligned}$$

Now define the space $\mathscr{A}_1$ of continuous maps $A : \mathscr{U} \times S \to \mathscr{L}(\mathscr{X} \times \mathbb{R}^2, \mathbb{R})$ such that

$$\sup_{(X,q)} |A(X,q)| < 1 \quad \text{and} \quad A\big(X, (0,y)\big) = 0 \quad \text{for all } y \in \left[-\frac{1}{2}, \frac{1}{2}\right].$$

Consider the operator $\widetilde{T} : \mathscr{A} \times \mathscr{A}_1 \to \mathscr{A} \times \mathscr{A}_1$ such that for $\phi \in C^1$ we have $\widetilde{T}(\phi, D\phi) = (T(\phi), DT(\phi))$, defined as $\widetilde{T}(\phi, A) = (T\phi, S(\phi, A))$ where $S(\phi, A)$ is given by

$$S(\phi, A)(Y,q) = \big[V_1(\phi) - T(\phi)V_2(\phi) + N(\phi)(A \circ \widehat{P}) \cdot D\widehat{P}\big](Y,q).$$

Here $V_1(\phi)$, $V_2(\phi)$ and $N(\phi)$ were defined previously during the calculation of $DT(\phi)$.

Again we need to show that $\widetilde{T}$ is well defined and a contraction.

Lemma 3.13 *Take Y C^1-close to X such that the estimates of Lemma* 3.11 *are valid. If $P_Y(q) = P(Y,q) = (f(Y,q), g(Y,q))$ then there are positive constants $k_i = 4, \ldots, 8$ such that for all $q \in S^*$*

1. $\frac{|D\partial_y g|}{|\partial_x f|} \le k_4 |x|^{\beta-\alpha}$, $\frac{|D\partial_x g|}{|\partial_x f|} \le k_5 |x|^{-1}$;
2. $\frac{|D\partial_y f|}{|\partial_x f|} \le k_6 |x|^{\beta-\alpha}$, $\frac{|D\partial_x f|}{|\partial_x f|} \le k_7 |x|^{-1}$;
3. $\frac{|\det D_q P|}{|\partial_x f|^2} |D\widehat{P}| \le k_8 |x|^{\beta}$, $|N(\phi)| \cdot |D\widehat{P}| < 1/2$;
4. $\sup_{S^*} \left\{ \frac{|D\partial_y g|}{|\partial_x f|}, \frac{|D\partial_y f|}{|\partial_x f|} \right\} < a_0$.

Proof Using Lemma 3.11, since $\partial_x f \underset{\varepsilon \to 0}{\longrightarrow} M\alpha x^{\alpha-1}$ we see there are K_1, K_2 satisfying

$$K_1 |x|^{\alpha-1} \le |\partial_x f| \le K_2 |x|^{\alpha-1}. \tag{3.8}$$

On the other hand, taking derivatives we see that

$$\begin{aligned}
\partial_X(\partial_y g) &= x\beta \partial_X \varepsilon + (\sigma + \varepsilon)x^{\beta} \log(\beta) \partial_X \beta \\
\partial_x(\partial_y g) &= \partial_x \varepsilon x^{\beta} + (\sigma + \varepsilon)\beta x^{\beta-1} \\
\partial_y(\partial_y g) &= \partial \varepsilon x^{\beta}.
\end{aligned}$$

Then $|D\partial_y g| \le K_3 |x|^{\beta-1}$ and by (3.8) we see there exists k_4 such that $\frac{|D\partial_y g|}{|\partial_x f|} \le k_4 |x|^{\beta-\alpha}$. Analogously we may estimate the derivatives $\partial_X(\partial_x g)$, ∂_x^2, $\partial_y(\partial_x g)$ obtaining

$$|\partial_X(\partial_x g)| \le K|x|^{\alpha-1}, \quad |\partial_x^2 g| \le K|x|^{\alpha-2}, \quad |\partial_y(\partial_x g)| \le K|x|^{\beta-1}$$

and thus $|D\partial_x g| \le K|x|^{\alpha-2}$. Then by (3.8) we get k_5 so that $\frac{|D\partial_x g|}{|\partial_x f|} \le k_5 |x|^{-1}$. This proves the first item of the statement.

Again analogously we obtain $|D\partial_y f| \le K|x|^{\beta-1}$ and by (3.8) also $\frac{|D\partial_y f|}{|\partial_x f|} \le k_6|x|^{\beta-\alpha}$ for a constant k_6.

From the explicit expression of $\partial_x f$ we get $|\partial_X(\partial_x f)| \le K|x|^{\alpha-1}$ and also

$$\partial_x(\partial_x f) = \beta y x^{\beta-1}\partial_x\varepsilon + \varepsilon\beta(\beta-1)yx^{\beta-2} + \alpha x^{\alpha-1}\partial_x\varepsilon + \alpha(M+\varepsilon)(\alpha-1)x^{\alpha-2}$$

implying that $|\partial_x(\partial_x f)| \le K|x|^{\alpha-2}$. We also have

$$\partial_y(\partial_x f) = \beta y x^{\beta-1}\partial_y\varepsilon + \varepsilon\beta x^{\beta-1} + \alpha x^{\alpha-1}\partial_y\varepsilon$$

which implies that $\partial_y(\partial_x f) \le K|x|^{\beta-1}$, and so $|D\partial_x f| \le K|x|^{\alpha-2}$ showing the existence of k_7 such that $\frac{|D\partial_x f|}{\partial_x f} \le k_7|x|^{-1}$, and proving the second item of the statement.

Now recall the definition of $N(\phi)$ and use Lemma 3.11 to deduce that

$$\begin{aligned} |N(\phi)||D\widehat{P}| &= \frac{|\det D_q P|}{(\partial_x f - (\phi\circ\widehat{P})\partial_x g)^2}|D\widehat{P}| \\ &\le \frac{a_0|x|^{\beta+\alpha-1}}{|\partial_x f|^2(1-\frac{\partial_x g}{\partial_x f})^2}|D\widehat{P}| \le \frac{a_0}{(1-a_0)^2}|x|^\beta \le \frac{a_0}{2^\beta(1-a_0)^2} \end{aligned}$$

which concludes the proof of the third item since $\beta > 1$. □

Now using the estimates of Lemmas 3.11 and 3.13 we can prove the following.

Proposition 3.14 *The map $S : \mathscr{A}\times\mathscr{A}_1 \to \mathscr{A}_1$ is well defined, continuous and $S(\phi,\cdot) : \mathscr{A}_1 \to \mathscr{A}_1$ is a contraction whose contraction rate is independent of ϕ.*

Note that this shows that for every derivable $\phi \in \mathscr{A}$ there exits $A \in \mathscr{A}_1$ such that $S(\phi, A) = A$.

Proof We can estimate

$$\begin{aligned} |V_1(\phi)| &\le \frac{|\phi\circ\widehat{P}|}{|\partial_x f - (\phi\circ\widehat{P})\partial_x g|}\cdot|D\partial_y g| + \frac{1}{|\partial_x f - (\phi\circ\widehat{P})\partial_x g|}|D\partial_y f| \\ &\le \left(\frac{|\partial_x f|^{-1}}{1-|\frac{\partial_x g}{\partial_x f}|}\right)\cdot|D\partial_y g| + \left(\frac{|\partial_x f|^{-1}}{1-|\frac{\partial_x g}{\partial_x f}|}\right)\cdot|D\partial_y f| \\ &\le \frac{1}{|1-a_0|}\cdot\left(|\partial_x f|^{-1}|D\partial_y g| + |\partial_x f|^{-1}|D\partial_y f|\right) \le K|x|^{\beta-\alpha} \end{aligned}$$

and

$$|T(\phi)V_2(\phi)| \le K|x|^{\beta-\alpha+1}\frac{1}{1-a_0}\left\{\frac{|D\partial_x g|}{|\partial_x f|} + \frac{|D\partial_x f|}{|\partial_x f|}\right\}$$

$$\leq K|x|^{\beta-\alpha+1}|x|^{-1} \leq K|x|^{\beta-\alpha}$$

and also

$$|N(\phi)| \cdot |(A \circ \widehat{P}) \cdot D\widehat{P}| \leq \frac{1}{(1-a_0)^2} |\det D_q P| \cdot |A| \cdot |D\widehat{P}| \cdot |\partial_x f|^{-2} \leq K|x|^{\beta}.$$

Therefore we arrive at

$$|S(\phi, A)| \leq |V_1(\phi)| + |T(\phi)| \cdot |V_2(\phi)| + |N(\phi)| \cdot |(A \circ \widehat{P}) D\widehat{P}| \leq K|x|^{\beta-\alpha}.$$

Since $\beta - \alpha > 0$ we see that S is continuous at $x = 0$. Moreover

$$|V_1(\phi)| < \frac{a_0}{1-a_0} \quad \text{and} \quad |T(\phi)V_2(\phi)| < \frac{a_0^2}{(1-a_0)^2}$$

and for $a_0 \in (0, 1/5)$ we get $\frac{a_0^2}{(1-a_0)^2} < 1/2$, and so $|S(\phi, A)| \leq 1$ and thus S is well defined.

Finally taking $A_1, A_2 \in \mathscr{A}_1$ and fixing $\phi \in \mathscr{A}$ we get

$$S(\phi, A_1) - S(\phi, A_2) = N(\phi) \cdot \left[A_1 \circ \widehat{P} - A_2 \circ \widehat{P}\right] \cdot D\widehat{P}$$

and hence

$$|S(\phi, A_1) - S(\phi, A_2)| \leq |N(\phi)| \cdot |A_1 - A_2| \cdot |D\widehat{P}| < \frac{1}{2}|A_1 - A_2|$$

and we conclude that $S(\phi, \cdot)$ is a contraction as stated. □

Remark 3.15 If we know that $\beta > \alpha + \gamma$ for some $\gamma \geq 1$, it can be shown (following the above techniques adapted from [110]) that S is not only continuous but differentiable, and so the contracting foliation $\mathscr{F}_Y$ is in fact smooth, i.e., it can be linearized by C^1 charts. In fact, if $\gamma \geq k$ for some $k \in \mathbb{Z}^+$, then $\mathscr{F}_Y$ is a C^k foliation.

This shows that $\widehat{T}$ has a fixed point (ϕ_0, A_0) where ϕ_0 is a fixed point of T. Clearly (ϕ_0, A_0) is a global attractor inside $\mathscr{A} \times \mathscr{A}_1$. In particular, by taking ϕ of class C^1 we obtain

$$\widehat{T}^n(\phi, D\phi) = \left(T^n(\phi), D(T^n(\phi))\right) \xrightarrow[n\to+\infty]{} (\phi_0, D\phi_0).$$

Then $A_0 = D\phi_0$ and hence ϕ_0 is continuously differentiable.

3.3.5 Robustness of the Geometric Lorenz Attractors

Here we conclude the proof that the geometric Lorenz attractor is a robustly transitive attractor and show that it is not structurally stable. Here we drop condition (f1) on the symmetry of the one-dimensional map f.

3.3.5.1 Robust Properties of the One-Dimensional Map f

We start by showing that the properties of the one-dimensional map f are robust for small C^1 perturbations of X.

Indeed, note that since the stable foliation is robust, we can define the one-dimensional map f_Y as the quotient map of the corresponding Poincaré map P_Y over the leaves of the foliation $\mathscr{F}_Y$, for all flows Y close to X in the C^1 topology.

Moreover since the leaves of $\mathscr{F}_Y$ are C^1 close to those of $\mathscr{F}$ it follows that f_Y is C^1 close to f and thus there exists $c \in [-1/2, 1/2]$ which play for f_Y the same role of 0, so that properties (f2)–(f4) from Sect. 3.3.2.1 are still valid for f_Y on a subinterval $[-b, b]$ for some $0 < b < 1/2$ close to $1/2$.

This implies that every f_Y is *locally eventually onto* for all Y close to X, that is, for any interval $J \subset (-b, b)$ there exists an iterate $n \geq 1$ such that $f_Y^n(J) = (-b, b)$.

Lemma 3.16 *Let* $f : [-1/2, 1/2] \setminus \{0\} \to [-1/2, 1/2]$ *be given satisfying the properties (f2)–(f4) in Sect.* 3.3.2.1. *Then* f *is locally eventually onto: for any open interval* J *not containing* 0 *there exists* n *such that* $f^n \mid J$ *is a diffeomorphism between* J *and one of the intervals* $(-1/2, 0)$ *or* $(0, 1/2)$ *(and the next iterate covers the interval* $(f(-1/2), f(1/2))$*).*

This implies in particular that the maps f_Y are (robustly) transitive and periodic points are dense. Moreover this also implies that the pre-orbit set $\cup_{n\geq 0} f^{-n}\{x\}$ is dense for every $x \neq 0$.

Proof Let $J_0 \subset (-1/2, 1/2)$ be an open interval with $0 \notin J_0$ and let $\eta = \inf |f'| > \sqrt{2}$.

Since $0 \notin J_0$ then $f(J_0)$ is such that $\ell(f(J_0)) \geq \eta \ell(J_0)$, where $\ell(\cdot)$ denotes length, and $f(J_0)$ is connected.

1. If $0 \notin f(J_0)$, set $J_1 = f^2(J_0)$ and then $\ell(J_1) \geq \eta^2 \ell(J_0)$.
2. If $0 \in f(J_0)$, then $f^2(J_0) = I^- \cup I^+$, where I^+ is the biggest connected component. Thus

$$\ell(I^+) \geq \frac{\ell(f^2(J_0))}{2} \geq \frac{\eta^2}{2} \ell(J_0).$$

Now replace J_0 by I^+ in case (2) or by J_1 in case (1). Since $\min\{\eta, \eta^2/2\} > 1$ we obtain after finitely many steps one of the intervals $(-1/2, 0)$ or $(0, 1/2)$. One more iterate then covers the interval $(f(-1/2), f(1/2))$.

This concludes the proof. □

3.3.5.2 Transitivity and Denseness of Periodic Orbits

We deduce these features from a stronger property: we show that the geometric Lorenz attractor is a homoclinic class (see Sect. 2.5.4).

Proposition 3.17 *There exists a periodic orbit $\mathcal{O}_X(p)$ in the geometric Lorenz attractor Λ such that $\Lambda = H_X(p) = \overline{W^s_X(p) \pitchfork W^u_X(p)}$.*

We prove this in Sect. 3.3.6. Observe that *every periodic orbit $\mathcal{O}(p)$ in Λ must be hyperbolic* since

- the uniformly contracting foliation obtained in Sect. 3.3.4 provides a uniformly contracting direction and a stable manifold for $\mathcal{O}(p)$: if $\mathcal{F}(p)$ is the leaf of $\mathcal{F}$ through $p = \mathcal{O}(p) \cap S$, then

$$W^s\big(\mathcal{O}(p)\big) = \bigcup_{t \geq 0} X^{-t}\big(\mathcal{F}(p)\big);$$

- the expansion of the one-dimensional projection map f (property (f3) from Sect. 3.3.2.1) ensures that there exists a forward DP-invariant expanding cone field around the horizontal direction, which in turn ensures the existence of a DP-invariant expanding direction at p.

Following the Birkhoff-Smale Theorem 2.17, Proposition 3.17 implies that the geometric Lorenz attractor Λ has a dense orbit and a dense subset of periodic orbits.

Since the arguments we use to prove Proposition 3.17 depend only on the properties of f and these properties are robust, we conclude that the geometric Lorenz attractors are robustly transitive.

3.3.5.3 The Geometric Lorenz Models are not Structurally Stable

The dynamics of two nearby geometric Lorenz models are in general not topologically equivalent. In fact Guckenheimer and Williams [98, 274] show that the conjugacy classes are completely described by two parameters: the *kneading sequences* of the two singular values

$$f(0^+) = \lim_{x \to 0^+} f(x) \quad \text{and} \quad f(0^-) = \lim_{x \to 0^-} f(x)$$

with respect to the singular point 0—*a pair of one-dimensional Lorenz-like maps are conjugate if, and only if, they have the same pair of kneading sequences* and, moreover, the corresponding flows are topologically equivalent if, and only if, the one-dimensional maps are conjugated (recall that we have dropped condition (f1)).

The kneading sequence of $x^+ = f(0^+)$ with respect to 0 is a sequence defined by

$$a_n = \begin{cases} 0 \text{ if } f^n(x^+) < 0 \\ 1 \text{ otherwise} \end{cases}; \quad \text{for } n \geq 0,$$

and analogously we define the kneading sequence $(b_n)_{n \geq 0}$ for x^-.

It is easy to see that *if two nearby geometric Lorenz flows are topologically conjugated* (see Sect. 2 for definitions and basic properties) *then the kneading sequences must be equal*, since the equivalence relation preserves the orbit structure and in particular preserves also the first return iterates to the cross-section S.

Now given a geometric Lorenz flow X with corresponding kneading sequences $(a_n)_{n\geq 0}$ and $(b_n)_{n\geq 0}$, we can through a small perturbation find a C^1 close vector field Y whose corresponding one-dimensional map has kneading sequences $(a'_n)_{n\geq 0}$ and $(b'_n)_{n\geq 0}$ distinct from the pair $(a_n)_{n\geq 0}$ and $(b_n)_{n\geq 0}$.

Indeed, if one of the orbits of $x^{\pm}$ is dense in $(-1,1)$, then one of its iterates is arbitrarily close to 0. Thus a small perturbation of the map will flip one of the elements of the kneading sequence from 0 to 1 or vice versa. Otherwise there exists $\varepsilon > 0$ such that the orbits of $x^{\pm}$ do not enter $(-\varepsilon, \varepsilon)$. As we have already proved, the one-dimensional map f is locally eventually onto and in particular topologically transitive. Hence there exists a point $0 < y < \delta \ll \varepsilon$ with $0 < f(y) - x^{+} < \delta$ whose orbit is dense. Let $n > 0$ be the smallest integer such that $|f^n(y)| < \delta$. Consider $\tilde{f}$ a small perturbation of f such that

- $\tilde{f}$ satisfies all the properties (f2) through (f4);
- $\tilde{f} \mid [-1,1] \setminus (0,\delta) \equiv f$;
- $\tilde{f}(0^+) = y$.

Then $\tilde{f}^k(\tilde{f}(0^+)) = f^k(y)$ for $k = 0, \ldots, n$ and so $\tilde{f}^n(y) \in (-\delta, \delta)$. Now we can perturb $\tilde{f}$ so that $\tilde{f}^n(y)$ changes sign and this would change one of the kneading sequences of $\tilde{f}$. Since δ can be taken arbitrarily small, then we obtain a very small perturbation of f whose kneading sequences are distinct. Since we can build a geometric Lorenz flow from $\tilde{f}$ and from any of its small perturbations, we have shown that we can always find a nearby geometric Lorenz flow Y not topologically conjugated to the given X.

3.3.6 The Geometric Lorenz Attractor Is a Homoclinic Class

Here we prove Proposition 3.17 following Bautista [35].

Observe first that the *geometric Lorenz attractor Λ must contain a hyperbolic periodic orbit*. Indeed since the associated Lorenz transformation f is locally eventually onto, the periodic orbits of f are dense. Let $x_0, \ldots, x_k$ be a periodic orbit of f. Then the leaves $\ell_0, \ldots, \ell_k$ of $\mathscr{F}$ in S which project onto these points form a invariant set under the map P. Since P preserves the leaves of the foliation $\mathscr{F}$ and is a contraction along $\mathscr{F}$, then P^k must send each ℓ_i into itself with a uniform contraction rate. Hence there exists a point p_i which is fixed by P^k on each leaf ℓ_i, i.e., $p_0, \ldots, p_k$ is a periodic orbit of P.

The definition of P shows that the orbit of p_0 by the flow X is periodic and $\mathscr{O}_X(p_0) \cap S = \{p_0, \ldots, p_k\}$.

As already observed every periodic orbit in Λ must be hyperbolic of saddle-type: the expanding and contracting directions can be easily read from the discussion in Sect. 3.3.3. Hence the unstable manifold of p is a disk transverse to S which intersects S in a one-dimensional manifold. The connected component of $W^u(p) \cap S$ which contains p is then a small line transverse to the foliation $\mathscr{F}$.

Now observe that since Λ is an attracting set, that is, $\Lambda = \cap_{t>0} \overline{X^t(U)}$, where U is the trapping ellipsoid, then the unstable manifold $W^u(p)$ of the orbit of $p = p_0$ must be contained in Λ. Indeed if $z \in W^{uu}(p)$ then $\mathrm{dist}(X^{-t}(z), X^{-t}(p)) \xrightarrow[t\to+\infty]{} 0$ and hence $X^{-t}(z) \in U$ for big $t > 0$, and thus $z \in X^t(U)$. This shows that $W^{uu}(p) \subset \Lambda$ and since Λ is X-invariant we also get $W^u(p) \subset \Lambda$.

The definition of homoclinic class and the fact that Λ is closed imply that $H_X(p) \subset \Lambda$. For the converse we need a stronger fact.

Lemma 3.18 *If Λ is the geometric Lorenz attractor and $p \in \Lambda$ is the point of some periodic orbit, then $\Lambda = \overline{W^u(p)}$.*

Proof Let $w \in \Lambda \setminus \{\sigma\}$ be given. Then there exists $t \geq 0$ such that $y = X^t(w) \in S$. Let $\ell = \mathscr{F}(y)$ be the corresponding leaf of $\mathscr{F}$ through y. Then ℓ is not the leaf $S \setminus S^*$. Therefore it projects to a point $x \in (-1/2, 0) \cup (0, 1/2)$. Since the pre-orbit set of every point is dense (because f is locally eventually onto), by definition of f this implies that $\Lambda \cap S = \Lambda \cap \overline{\cup_{n\geq 0} P^{-n}\ell}$.

Hence we have $P^{-n}\ell \cap W^u(p) \neq \emptyset$ for some $n \geq 0$. But this means that $W^s(y) \cap W^u(p) \neq \emptyset$ and so $w, y \in \overline{W^u(p)}$. Thus $\Lambda \setminus \{\sigma\} \subset \overline{W^u(p)}$. □

Finally to prove that $\Lambda \subset H_X(p)$ it is enough to show that $W^u(p) \subset H_X(p)$. Every point $w \in W^u(p)$ admits $t < 0$ such that $q = X^t(w) \in S$. Take a small neighborhood J of q in $W^u(p) \cap S$, which is a small line transverse to $\mathscr{F}$.

Let l be the leaf of $\mathscr{F}$ containing p and let I be the interval inside $(-1/2, 1/2)$ corresponding to J by the projection $S \to S/\mathscr{F} = (-1/2, 1/2)$. Recall that $l \subset W^s(p) \cap S$. Write x for the point corresponding to p under this projection.

Again by Lemma 3.16 there exists $n \geq 0$ such that $f^{-n}\{x\} \cap I \neq \emptyset$. This means that $J \pitchfork P^{-n}(l) \neq \emptyset$, and hence in J there exists a point of the homoclinic class of p. Since J can be taken arbitrarily small near q, we conclude that $q \in H_X(p)$. This concludes the proof that $\Lambda = H_X(p)$.

Chapter 4
Robustness on the Whole Ambient Space

Here we prove that every robustly transitive vector field X in a 3-manifold is an Anosov vector field, i.e., a robustly transitive vector field X is globally hyperbolic. This was first obtained by Doering in [79] and this result is a precursor of the more general results on robustly transitive sets of Chap. 5. In fact we prove an extended version for homogeneous vector fields in compact manifolds of any finite dimension.

The reader should recall Definition 2.32 of homogeneous flow from Sect. 2.6.2. We say that a vector field X is *robustly transitive* if there exists a neighborhood $\mathcal{U}$ of X in $\mathfrak{X}^1(M)$ such that, for all $Y \in \mathcal{U}$, the associated flow Y^t is transitive, i.e., there exits a point $x \in M$ whose past and future orbits are dense: $\overline{\mathcal{O}_Y^-(x)} = M = \overline{\mathcal{O}^+(x)}$.

Theorem 4.1 *Let $X \in \mathfrak{X}^1(M)$ be a robustly transitive homogeneous flow on a n-manifold M. Then X is Anosov.*

We follow Doering from [79] and Vivier from [268]. We use mainly Theorem 2.33 to obtain a dominated splitting for the Linear Poincaré Flow on regular orbits, and a simple growth estimate near linearized saddle equilibria to show that equilibria cannot be in the interior of the non-wandering set of X.

Theorem 4.2 *Let $X \in \mathfrak{X}^1(M)$ be a homogeneous flow on an open subset U of M and assume that*

$$\Lambda_X(U) := \bigcap_{t \in \mathbb{R}} \overline{X^t(U)}$$

is both connected and a subset of $\Omega(X)$. *Then*

1. *there exists a dominated splitting for the Linear Poincaré Flow on $\Lambda_X(U)^* := \Lambda_X(U) \setminus S(X)$;*
2. *there are no hyperbolic linearizable equilibria in the interior of $\Lambda_X(U)$.*

In particular, if $U = M$, then there are no equilibria in M.

V. Araújo, M.J. Pacifico, *Three-Dimensional Flows,* Ergebnisse der Mathematik und ihrer Grenzgebiete. 3. Folge / A Series of Modern Surveys in Mathematics 53,
DOI 10.1007/978-3-642-11414-4_4,

We stress that the connectedness and non-wandering conditions on $\Lambda_X(U)$ are easy consequences if $\Lambda_X(U)$ is transitive.

Here we say that a equilibrium σ of a C^1 vector field X is linearizable if the local dynamics of the flow X^t near σ is conjugated to the dynamics of $DX^t(\sigma)$ on a neighborhood of $\mathbf{0} \in T_\sigma M$ by a C^1 change of coordinates. According to Sternberg [256] this is true depending on certain non-resonance relations between the eigenvalues of $DX(\sigma)$ which amount to conditions satisfied by a dense subset of the family $\mathfrak{X}^1(M)$ of all C^1 vector fields on M.

Having established this, we deduce that, for a robustly transitive vector field in a 3-manifold, every C^1 close vector field Y has no equilibria in M and has a globally defined Linear Poincaré Flow with a dominated splitting. Indeed, we just observe that a robustly transitive vector field in the whole of M is necessarily homogeneous, if M is a 3-manifold.

In dimension 3 this is enough to deduce that X is globally hyperbolic: we apply the Ergodic Closing Lemma (Theorem 2.23) for this.

Theorem 4.3 *Let $X \in \mathfrak{X}^1(M)$ be a robustly transitive flow on a 3-manifold M. Then X is Anosov.*

Vivier in [268] shows that robustly transitive flows on the whole n-dimensional compact boundaryless manifold M must have a dominated splitting for the Linear Poincaré Flow and have no singularities. This is proved showing that, in the absence of dominated splitting for the Linear Poincaré Flow, we can find arbitrarily close vector fields having periodic sinks or sources, contradicting the robust transitivity assumption. However, robust transitivity is not enough to deduce global hyperbolicity in dimensions higher than three, since domination of the Linear Poincaré Flow is compatible with certain bifurcations of periodic orbits and, thus, with non-hyperbolicity. Using the stronger condition of homogeneity together with transitivity enables us to use the same three-dimensional arguments to deduce global hyperbolicity.

The dominated splitting for the Linear Poincaré Flow of a homogeneous vector field in some subset U of M, whose maximal invariant subset is connected and non-wandering, is given by Theorem 2.33, whose proof we present in Sect. 4.2. This proves the first conclusion of Theorem 4.2. Assuming this we deduce the second conclusion in what follows.

4.1 No Equilibria Surrounded by Regular Orbits with Dominated Splitting

Now we explain why the existence of dominated splitting for the Linear Poincaré Flow implies the absence of equilibria in the interior of $\Lambda_X(U)$.

Let us assume then that the Linear Poincaré Flow of X on $\Lambda_X(U)^*$ admits a dominated splitting, that is, there are constants $K, \lambda > 0$ such that the normal bundle N_x decomposes over $x \in \Lambda_X(U)^*$ into $N_x^{cs} \oplus N_x^{cu}$ satisfying

$$\|P^t \mid N_x^{cs}\| \le K e^{-\lambda t} m(P^t \mid N_x^{cu}), \quad \text{for all } t > 0.$$

Arguing by contradiction, we assume also that there exists an equilibrium σ in the *interior* of $\Lambda_X(U)^*$ which is *hyperbolic of saddle-type and the local dynamics of* X *near* σ *is smoothly linearizable of class* C^1.

These conditions are valid on a dense subset of all C^1 vector fields (see e.g. [256]) so that, if the flow is homogeneous on the whole M and we prove that no such equilibria can exist in the interior of M then, for any other type of equilibria σ for a robustly transitive vector field, we obtain some equilibria σ_Y of an arbitrarily C^1 close vector field Y in the previous conditions. Hence it is enough to argue with hyperbolic linearizable equilibria σ in the interior of M to prove Theorem 4.3.

Now we consider the vector subspace E of $T_\sigma M$ spanned by the eigenspace E^{ss} corresponding to the most contracting eigenvalue and by the eigenspace E^{uu} corresponding to the most expanding eigenvalue of $A := DX(\sigma)$, and we show that in this subspace the domination condition on the splitting of the Linear Poincaré Flow leads to a contradiction. We write $A_s := DX(\sigma) \mid E^{ss}$ and $A_u := DX(\sigma) \mid E^{uu}$ in what follows.

The assumption that σ belongs to the interior of the $\Lambda_X(U)$ ensures that we can find regular points of $\Lambda_X(U)$ arbitrarily close to σ and so, by the linearization of the dynamics of X on a neighborhood W of σ and choosing a convenient linear rescaling of coordinates in E and a convenient inner product in E, we can assume that:

- $E = E^{ss} \times E^{uu}$ with $\mathbf{e}_1, \dots, \mathbf{e}_s$ an orthonormal base of E^{ss} and $\mathbf{f}_1, \dots, \mathbf{f}_u$ an orthonormal base of E^{uu} which together form an orthonormal base of E;
- the vector field is given by $q \in W \mapsto A(q) \in E$ and the corresponding flow is $X^t(q) = e^{At}q$ so that $DX^t(q)$ is the linear map $Ae^{At} : E \to E$;
- there are $\lambda, \sigma > 0$ and $K > 1$ so that for $q \in W$ and for all $t \in \mathbb{R}$ such that $e^{At}q \in W$ we have
  - $K^{-1}e^{-\lambda t}\|\mathbf{v}\| \le \|e^{At}\mathbf{v}\| = \|e^{A_s t}\mathbf{v}\| \le Ke^{-\lambda t}\|\mathbf{v}\|$, for $\mathbf{v} \in T_qE$ parallel to E^{ss}; and
  - $K^{-1}e^{\sigma t}\|\mathbf{w}\| \le \|e^{At}\mathbf{w}\| = \|e^{A_u t}\mathbf{w}\| \le Ke^{\sigma t}\|\mathbf{w}\|$, for $\mathbf{w} \in T_qE$ parallel to E^{uu};
- every point in the set $\mathbf{e}_1 \times (B \setminus \{\mathbf{0}\})$, where $B = B(\mathbf{0}, \delta)$ is a small ball around the origin $\mathbf{0}$ in E^{uu}, corresponds to some regular point in $\Lambda_X(U)$.

We note that $\mathbf{e}_1 \times B$ is a disk transversal to the flow and also that the dominated splitting $(N_q^{cs} \oplus N_q^{cu}) \cap E$ of E induced by the splitting of the Linear Poincaré Flow for $q \in E^{ss} \setminus \{\mathbf{0}\}$ is given by $N_q^{ss} = E^{ss} \cap N_q$ and $N_q^{cu} = E^{uu}$. Indeed, this splitting is clearly dominated and by the uniqueness property discussed in Sect. 2.6.2 it must be the P^t invariant and dominated splitting of N_q.

We let $\mathbf{u} \in B \setminus \{\mathbf{0}\}$ be fixed with $\|\mathbf{u}\| = e^{-\sigma T}$ for a big $T > 0$, and we define an initial point $q := (\mathbf{e}_1, \mathbf{u}) \in \mathbf{e}_1 \times (B \setminus \{\mathbf{0}\})$ and consider its forward orbit $q_t := e^{At}q = (e^{A_s t}\mathbf{e}_1, e^{A_u t}\mathbf{u})$ for $t \ge 0$ until the time $t_0 > 0$ such that $\|e^{A_u t_0}\mathbf{u}\| = 1$. We note that by the assumptions above on the linearized flow we have

$$1 \le Ke^{\sigma t_0}\|\mathbf{u}\| = Ke^{\sigma(t_0 - T)}, \quad \text{thus } t_0 \ge T - \frac{\log K}{\sigma} = T\left(1 - \frac{\log K}{\sigma T}\right). \tag{4.1}$$

In this setting, $\tau(t) := \|A_u e^{A_u t}\mathbf{u}\| \, \|A_s e^{A_s t}\mathbf{e}_1\|^{-1}$ is the slope of the vector field at q_t, that is, the tangent of the angle θ_t between $A(q_t)$ and the subspace E^{ss}, see Fig. 4.1.

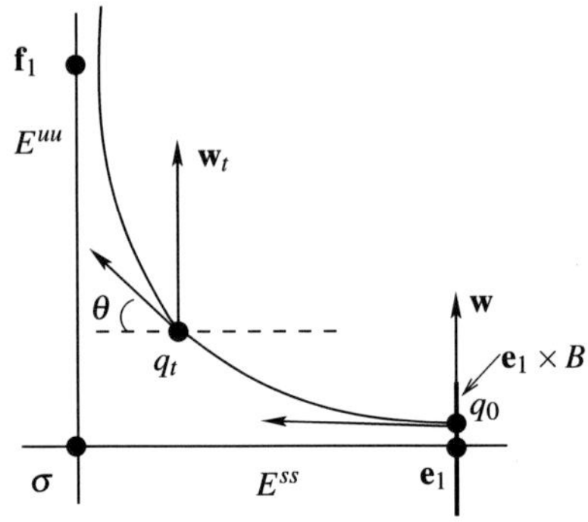

Fig. 4.1 The orbit of q under the linear flow, the vector field at different points of this orbit and the relative position with the image of the vector $\mathbf{w}$ given by the angle θ

We take $\mathbf{w} \in E^{uu}$ with $\|\mathbf{w}\| = 1$. The image $\mathbf{w}$ under the derivative of the flow map is given by $D(e^{At})(q)\mathbf{w} = A_u e^{A_u t}\mathbf{w}$, for $t \in \mathbb{R}$, and so

$$\|P^t\mathbf{w}\| = \|A_u e^{A_u t}\mathbf{w}\| \cos\theta_t = \|A_u e^{A_u t}\mathbf{w}\|(1+\tau(t)^2)^{-1/2}.$$

Now we estimate $\tau(t)$ by

$$K^{-2}\|\mathbf{u}\|e^{(\lambda+\sigma)t} = \frac{K^{-1}e^{\sigma t}\|\mathbf{u}\|}{Ke^{-\lambda t}} \le \frac{\|A_u e^{A_u t}u\|}{\|A_s e^{A_s t}e_1\|} \le \frac{Ke^{\sigma t}\|\mathbf{u}\|}{K^{-1}e^{-\lambda t}} = K^2\|\mathbf{u}\|e^{(\lambda+\sigma)t}$$

and then estimate the growth of $\mathbf{w}$ using these bounds, as follows:

$$\frac{Ke^{\sigma t}}{\sqrt{1+K^4\|\mathbf{u}\|^2e^{2(\lambda+\sigma)t}}} \le \|P^t\mathbf{w}\| \le \frac{Ke^{\sigma t}}{\sqrt{1+K^{-4}\|\mathbf{u}\|^2e^{2(\lambda+\sigma)t}}}. \tag{4.2}$$

On the one hand, we note that for $t \le t_* := \frac{\sigma}{\lambda+\sigma}T$ we have $\|\mathbf{u}\|e^{(\lambda+\sigma)t} \le 1$ and so

$$\|P^t\mathbf{w}\| \ge \frac{Ke^{\sigma t}}{\sqrt{1+K^4}}, \quad 0 \le t \le t_*. \tag{4.3}$$

On the other hand, applying the previous bound (4.2) with $t > t_*$ we get

$$\|P^{t-t_*}(P^{t_*}\mathbf{w})\| = \frac{\|A_u e^{A_u(t-t_*)}(P^{t_*}\mathbf{w})\|}{\sqrt{1+\tau(t)^2}} \le \frac{Ke^{\sigma(t-t_*)}\|P^t\mathbf{w}\|}{\sqrt{1+K^{-4}\|\mathbf{u}\|^2e^{2(\lambda+\sigma)t_*}e^{2(\lambda+\sigma)(t-t_*)}}}$$

and since $\|\mathbf{u}\|^2e^{2(\lambda+\sigma)t_*} = 1$ we obtain

$$\|P^{t-t_*}(P^{t_*}\mathbf{w})\| \le \frac{Ke^{-\lambda(t-t_*)}\|P^t\mathbf{w}\|}{\sqrt{K^{-4}+e^{-2(\lambda+\sigma)(t-t_*)}}} \le \frac{e^{-\lambda(t-t_*)}}{K^3}\|P^t\mathbf{w}\|, \quad t > t_*. \tag{4.4}$$

Hence from (4.3) $P^t\mathbf{w}$ grows at a rate σ for $0 < t < t_*$, and then from (4.4) we see that the Linear Poincaré Flow shrinks $P^{t_*}\mathbf{w}$ at a rate λ for $t_* < t < t_0$. Moreover (4.1) shows that $t_0 \gg t_*$ if T is big enough, that is, we can make t_* and $t_0 - t_*$ arbitrarily big by letting T grow.

This contradicts the domination of the splitting $N^{cs} \oplus N^{cu}$ of the normal bundle on $M \setminus S(X)$ because, as explained in Sect. 2.6.2, vectors in the normal bundle in this setting will converge to the direction of the dominating bundle, N^{cu}. Hence their growth rate will tend to be at least the minimal growth rate along N^{cu}. We have shown that, on the contrary, the growth rate of a certain vector changes from the maximal growth rate of vectors along an orbit of the flow to the minimum growth rate along the same orbit, and the pieces of orbit with this behavior can be made arbitrarily long. This concludes the proof of item 2 and the last statement of Theorem 4.2.

4.2 Homogeneous Flows and Dominated Splitting

Here we prove the following central result in this book.

Theorem 4.4 *Let $X \in \mathfrak{X}^1(M)$ be a homogeneous vector field in an open subset U of M and let us assume that $\Lambda_X(U)^* \subset \Omega(X)$ and that $\Lambda_X(U)^*$ is connected. Then there exists an invariant, continuous and dominated splitting $N_{\Lambda_X^*(U)} = N^{cs} \oplus N^{cu}$ for the Linear Poincaré Flow P^t on $\Lambda_X^*(U)$.*

In the particular case of a homogeneous vector field in the whole M we get

Corollary 4.5 *Let a homogeneous vector field $X \in \mathfrak{X}^1(M)$ on M be given such that $M \setminus S(X) \subset \Omega(X)$ and $M \setminus S(X)$ is connected. Then there exists an invariant, continuous and dominated splitting $N_{\Lambda_X^*(U)} - N^{cs} \oplus N^{cu}$ for the Linear Poincaré Flow P^t on $M \setminus S(X)$.*

These results follow from Pugh's Closing Lemma together with two estimates on the eigenvalues and splittings of periodic orbits with high period for all C^1 flows close to X and from Theorems 4.7 and 4.8, presented in Sect. 4.2.1. Some of the ideas can be traced back to several works of Pliss, for example [203, 204], and also of Liao [133, 134].

We explain how these theorems imply Theorem 4.4 in Sect. 4.2.2 and prove Theorems 4.7 and 4.8 in Sects. 4.2.3 and 4.2.4.

4.2.1 Dominated Splitting over the Periodic Orbits

Let $\Lambda_Y(U)$ be an isolated set of a homogeneous C^1 vector field in U. Since every $p \in \mathrm{Per}(Y) \cap \Lambda_Y(U)$ is hyperbolic of saddle type, we have that the tangent bundle of M over p can be written as

$$T_pM = E_p^s \oplus E_p^Y \oplus E_p^u,$$

where E^s_p is the eigenspace associated to the contracting eigenvalues of $DY^{t_p}(p)$, E^u_p is the eigenspace associated to the expanding eigenvalues of $DY^{t_p}(p)$, and we write t_p for the (minimal) period of p. We recall from Sect. 2.6.1 that $E^s_p \subset N^s_p \oplus E^Y_p$ and $E^u_p \subset N^u_p \oplus E^Y_p$, where $N^s \oplus N^u$ is any dominated splitting for the linear Poincaré Flow over regular orbits.

To obtain a splitting for the Linear Poincaré Flow we start with a splitting for the tangent bundle over periodic orbits of every nearby flow to X inside U. Let $\mathscr{U}$ be a C^1 neighborhood of X where all $Y \in \mathscr{U}$ are homogeneous vector fields.

Definition 4.6 Given $Y \in \mathscr{U}$ define for any $p \in \mathrm{Per}(Y) \cap \Lambda_Y(U)$ the subspaces

$$E^{cs,Y}_p := E^{s,Y}_p \oplus E^Y_p \quad \text{and} \quad E^{cu,Y}_p := E^Y_p \oplus E^u_p.$$

which give a pair of subbundles over $\mathrm{Per}(Y) \cap \Lambda_Y(U)$.

From these definitions we will show that $N^{cs,Y}_p := \mathscr{O}_Y(E^{cs,Y}_p)$ and $N^{cu,Y}_p := \mathscr{O}_Y(E^{cu,Y}_p)$ is a splitting of N_p at every $p \in \mathrm{Per}(Y) \cap \Lambda_Y(U)$ and that the splitting is dominated.

The following result establishes, first, that the periodic points are uniformly hyperbolic, i.e., the periodic points are of saddle-type and the Lyapunov exponents are uniformly bounded away from zero. Secondly, the angle between the stable and the unstable eigenspaces at periodic points are uniformly bounded away from zero.

Theorem 4.7 *There are a neighborhood $\mathscr{V} \subset \mathscr{U}$ of X and constants $0 < \lambda < 1$ and $c > 0$, such that, for every $Y \in \mathscr{V}$, if $p \in \mathrm{Per}(Y) \cap \Lambda_Y(U)$ and t_p is the period of p then*

1. (a) $\|DY^{t_p} \mid E^s_p\| < \lambda^{t_p}$ *(uniform contraction on the period)*
   (b) $\|DY^{-t_p} \mid E^u_p\| < \lambda^{t_p}$ *(uniform expansion on the period).*
2. $\angle(N^{cs,Y}_p, N^{cu,Y}_p) > c$ *(angle uniformly bounded away from zero).*

The proof of this result, presented in Sect. 4.2.3, is an argument of perturbation: if the conclusions fail, then we can, by an arbitrarily small C^1 perturbation of the flow, obtain a hyperbolic periodic orbit for the nearby flow which is either a sink or a source, in the three-dimensional case, or with a different index (dimension of the unstable subspace) from the original one if the dimension of M is higher than three. In any case, we contradict the homogeneity assumption.

Before stating the next theorem, we observe that from Theorem 4.7 we have naturally defined over the set of periodic orbits of $Y \in \mathscr{V}$, whose period is big enough, an invariant splitting $N_p = N^{cs}_p \oplus N^{cu}_p$ for the Linear Poincaré Flow.

We also note that Theorem 4.7 already ensures that the angle between N^{cs} and N^{cu} is uniformly bounded away from zero. Moreover, we also obtain (see Sect. 2.6)

$$\|P^{t_p} \mid N^{cs}_p\| \le \lambda^{t_p} \quad \text{and} \quad \|P^{t_p} \mid N^{cu}_p\| \ge \lambda^{-t_p}$$

so that we have *dominated splitting at multiples of the period of the periodic orbits* for the Linear Poincaré Flow $(P^t)_{t\in\mathbb{R}}$ associated to Y.

This result is to be combined with the following main theorem, first obtained by Mañé for diffeomorphisms [145] and by Liao [134] for flows, which provides dominated splitting of the Linear Poincaré Flow over the periodic orbits of flows whose critical elements are robustly hyperbolic.

Theorem 4.8 *Let $X \in \mathfrak{X}^1(M)$ be such that on a neighborhood $\mathcal{U}$ of X in the C^1 topology every critical element of $Y \in \mathcal{U}$ inside U is hyperbolic* (*this is known as a* star flow). *Then there is a C^1 neighborhood $\mathcal{V}$ of X and numbers $\lambda, T > 0$ such that, for all $Y \in \mathcal{V}$, each $p \in \mathrm{Per}(Y) \cap \Lambda_Y(U)$ with $t_p > T$ and every $t \geq T$, we have*

$$\|P^t \mid N_p^{cs}\| \leq e^{-\lambda t} m(P^t \mid E_p^{cu}).$$

The proof of this result, presented in Sect. 4.2.4, is again by contradiction via perturbation. If the conclusion is not true, then after a small C^1 perturbation we can find a hyperbolic periodic orbit whose Linear Poincaré Flow has a splitting into stable and unstable subspaces with an arbitrary small angle, contradicting the previous Theorem 4.7.

4.2.2 Dominated Splitting over Regular Orbits from the Periodic Ones

Here we explain how Theorem 4.4 follows from Theorems 4.7 and 4.8.

We induce a dominated splitting over the normal bundle of $\Lambda_X(U)^*$ using the dominated splitting over the normal bundle to the orbits of $\{p \in \mathrm{Per}(Y) \cap \Lambda_Y(U) : t_p \geq T_0\}$ for flows $Y \in \mathcal{V}$ near X, provided by Theorem 4.8.

On the one hand, since $\Lambda_Y(U)$ is an attracting set for every Y close to X in $\mathfrak{X}^1$, we can assume without loss of generality that, for all $Y \in \mathcal{V}$ and $x \in \mathrm{Per}(Y)$ with $\mathcal{O}_Y(x) \cap U \neq \emptyset$, we have

$$\mathcal{O}_Y(x) \subset \Lambda_Y(U). \tag{4.5}$$

On the other hand, since $\Lambda = \Lambda_X(U)$ is assumed to be connected, we get that

$$\Lambda(T_0) := \Lambda \setminus \{p \in \mathrm{Per}(X) \cap U : t_p < T_0\} \text{ is dense in } \Lambda. \tag{4.6}$$

Indeed, the family of hyperbolic periodic orbits with period bounded from above in U forms a set of isolated simple curves in U. Arguing by contradiction, if $\Lambda(T_0)$ is not dense in Λ, then $A := \Lambda \setminus \overline{\Lambda(T_0)} \neq \emptyset$ is a nonempty open invariant subset of Λ in the relative topology and $A \subset \Lambda(T_0)$. Thus A contains some periodic orbit, which is then a connected simple curve γ in the interior of Λ, that is, γ is a closed and open subset in Λ, contradicting the assumption that Λ is connected. This proves (4.6).

From property (4.6), to induce an invariant splitting over $\Lambda_X(U)$ it is enough to do so over $\Lambda(T_0)$ (the reader can consult [144] and references therein for another instance of this argument). For this we proceed as follows.

Given $X \in \mathscr{V}$, let $K(X) \subset \Lambda(T_0)$ be such that $X^t(x) \notin K(X)$ for all $x \in K(X)$ if $t \neq 0$. In other words, $K(X)$ is a set of representatives of the quotient $\Lambda(T_0)/\sim$ by the equivalence relation $x \sim y \iff x \in \mathscr{O}_X(y)$. Since we are also assuming that $\Lambda_X(U) \subset \Omega(X)$, then we can use the Closing Lemma (Theorem 2.19): for any $x \in K(X)$ and $x \notin S(X)$ there exist $Y_n \to X$ in $\mathfrak{X}^1$ and $y_n \to x$ such that $y_n \in \mathrm{Per}(Y_n)$. We can assume that $Y_n \in \mathscr{V}$ for all n. In particular, inclusion (4.5) holds for all $Y = Y_n$, that is, $\mathscr{O}_{Y_n}(y_n) \subset \Lambda_{Y_n}(U)$.

Moreover, since the periods of the periodic points in $K(X)$ are larger than T_0, we can also assume that the periods of y_n are $t_{y_n} > T_0$ for all n. Thus the λ-dominated splitting $N^{cs,Y_n} \oplus N^{cu,Y_n}$ over $\{p \in \mathrm{Per}(Y_n) \cap \Lambda_{Y_n}(U)^* : t_p > T_0\}$, provided by Theorem 4.8, is well defined.

Let us take a converging subsequence $N^{cs,Y^{n_k}}_{y_{n_k}} \oplus N^{cu,Y^{n_k}}_{y_{n_k}} = N_{y_{n_k}}$ and define

$$N^{cs,X}_x = \lim_{k\to\infty} N^{cs,Y^{n_k}}_{y_{n_k}}, \qquad N^{cu,X}_x = \lim_{k\to\infty} N^{cu,Y^{n_k}}_{y_{n_k}}.$$

Since $N^{cs,Y_n} \oplus N^{cu,Y_n} = N$ is a splitting at y_n with angle bounded away from zero uniformly for all n, then this property is also true for the limit $N^{cs,X}_x \oplus N^{cu,X}_x$. Moreover $\dim(N^{cs,X}_x) = s$ and $\dim(N^{cu,X}_x) = u$ for all $x \in K(X)$ with $s+u+1 = n$ by the homogeneity assumption on the neighborhood $\mathscr{V}$.

Define the following eigenspaces along $X^t(x)$ for $t \in \mathbb{R}$:

$$N^{cs,X}_{X^t(x)} := P^t_X(N^{cs,X}_x) \quad \text{and} \quad N^{cu,X}_{X^t(x)} := P^t_X(N^{cu,X}_x).$$

Since for every n the splitting over $\{p \in \mathrm{Per}(Y_n) \cap \Lambda_{Y_n}(U)^* : t_p > T_0\}$, is λ-dominated, it follows that the splitting defined above along X-orbits of points in $K(X)$ is also λ-dominated. Moreover we also have that $N^{cs,X}_{X^t(x)}$ is s-dimensional and $N^{cu,X}_{X^t(x)}$ is u-dimensional, for all $t \in \mathbb{R}$. This provides the desired extension of a dominated splitting to $\Lambda_X(U)^*$.

This concludes the proof of Theorem 4.4, assuming Theorems 4.7 and 4.8.

4.2.3 Bounded Angles on the Splitting over Hyperbolic Periodic Orbits

Here we prove Theorem 4.7. We divide the proof into two parts.

Item 1 of Theorem 4.7 We suppose, by contradiction, that given $\delta > 0$ small, there is $Y \in \mathfrak{X}^\infty(M)$ arbitrarily C^1 close to X, and a periodic orbit y of Y with period t_y, such that $\|DY^{t_y} \mid E^s_y\| \geq (1-\delta)^{t_y}$.

Let A_t be the one-parameter family of linear maps

$$A_t = (1-2\delta)^{-t} \cdot DY^t(y), \quad 0 \leq t \leq t_y.$$

By construction A_t preserves the flow direction and the eigenspaces of DY^{t_y}. Moreover

$$\|\partial_h A_{t+h} A_t^{-1}|_{h=0} - DY(Y^t(y))\| < -\log(1-\delta).$$

Since we can take δ as close to 0 as needed, the inequality above together with $Y \in C^\infty$ imply that A_t satisfies Frank's Lemma 2.24. Hence there exists $Z \in \mathfrak{X}^1$, which is C^1 near Y, such that y is a periodic point of Z with period t_y, and $DZ^t(Z^t(y)) = A_t$ for $0 \le t \le t_y$. By definition of A_t we get $\|DZ^{t_y} \mid E_y^s\| > 1$, implying that y is a periodic orbit for Z with a different index (dimension of the stable manifold) from the index for Y, which contradicts the homogeneous assumption on X. This proves subitem (1a).
A similar argument proves subitem (1b) are we are done with the first item of Theorem 4.7.

Item 2 of Theorem 4.7 Again, by contradiction, we assume that for every $\theta > 0$ small there exist $Y \in \mathfrak{X}^\infty$, which is C^1 close to X, and $p \in \mathrm{Per}(Y) \cap \Lambda_Y(U)$ such that $\angle(N_p^{cs}, N_p^{cu}) < \theta$ and $t_p > 1/\theta$. The proof follows Mañé [145, Lemma II.9]. Consider the splitting $(N_p^{cs})^\perp \oplus N_p^{cs}$ of N_p and the (orthogonal) splitting $E_p^X \oplus (N_p^{cs})^\perp \oplus N_p^{cs}$ of T_pM. The transformation $DY^{t_p}(p)$ on T_pM has a matrix which can be written with respect to this splitting as

$$\begin{bmatrix} 1 & \star & \star \\ 0 & A & 0 \\ 0 & P & B \end{bmatrix}$$

where $A : (N_p^{cs})^\perp \to (N_p^{cs})^\perp$, $P : N_p^{cs} \to (N_p^{cs})^\perp$ and $B : N_p^{cs} \to N_p^{cs}$ are linear maps with

$$\|A^{-1}\| \le \lambda^{t_p} \quad \text{and} \quad \|B\| \le \lambda^{t_p}$$

according to the first item of the theorem, already proved.
Let N_p^{cs} be given by $\{v + Tv : v \in (N_p^{cs})^\perp\}$ for a linear map $T : (N_p^{cs})^\perp \to N_p^{cs}$ in the coordinates provided by the chosen splitting. The invariance ensures that there exists $u \in (N_p^{cs})^\perp$ such that

$$\begin{bmatrix} A & 0 \\ P & B \end{bmatrix} \cdot \begin{bmatrix} v \\ Tv \end{bmatrix} = \underbrace{Av}_{(N_p^{cs})^\perp} + \underbrace{Pv + BTv}_{N_p^{cs}} = u + Tu$$

and so we get $TA = P + BT$ or $T = PA^{-1} + BTA^{-1}$. Hence by the first item

$$\|T\| \le \|PA^{-1}\| + \lambda^{2t_p}\|T\| \quad \text{and} \quad \|T\| \le 2\|PA^{-1}\|$$

by letting t_p be so big that $\lambda^{2t_p} \le 1/2$. Then the following can be made arbitrarily small:

$$\frac{1}{2}\|PA^{-1}\| \le \|T\|^{-1} = o(\theta) \xrightarrow[\theta \to 0]{} 0.$$

Let us assume that there exist a map $C : (N_p^{cs})^\perp \to N_p^{cs}$ with small norm so that the product

$$\begin{bmatrix} A & 0 \\ P & B \end{bmatrix} \cdot \begin{bmatrix} I & C \\ 0 & I \end{bmatrix} = \begin{bmatrix} A & AC \\ P & PC + B \end{bmatrix} =: M$$

has some eigenvalue with norm 1, where I is the identity map of the corresponding subspace of the decomposition. We consider a non-negative C^2 real function $\delta : [0, t_p] \to [0, 1]$ such that $\delta(0) = 0$, $\delta(t_p) = 1$ and $|\delta'(t)| < \theta$ for $0 \le t \le t_p$ (we can take $\delta(t) = t/t_p$ so that $\delta' = 1/t_p < \theta$). We now define the one-parameter family of linear maps

$$A_t := \begin{bmatrix} 1 & 0 & 0 \\ 0 & I & \delta(t)C \\ 0 & 0 & I \end{bmatrix}, \quad 0 \le t \le t_p,$$

and set $C_t := A_t \cdot DY^t(p)$ for $0 \le t \le t_p$. By construction, the transformation C_t preserves the flow direction along the Y-orbit of p. The choice of $\delta(t)$ implies that A_t is a small perturbation of the identity map $I_t : T_{Y^t(p)}M \to T_{Y^t(p)}M$ for $0 \le t \le t_p$ and so C_t is in the setting of Frank's Lemma 2.24.

Hence we can find a vector field Z which is C^1 near Y, and a periodic point $p \in \mathrm{Per}(Z) \cap \Lambda_Z(U)$ such that $DZ^t(p) = C_t = A_t \cdot DY^t(p)$, for $0 \le t \le t_p$. Moreover $DZ^{t_p} = A_{t_p} \cdot DY^{t_p}(p) = M$. Thus, taking θ small enough, we get a C^1 vector field Z near to Y exhibiting a periodic orbit through p which is not hyperbolic.

This contradicts the homogeneity condition on X and concludes the proof, under the assumption that we can find C with very small norm such that M has some eigenvalue with norm 1.

Now we explain how to find C. We need a vector $(x, y) \in N_p$ so that

$$M \begin{bmatrix} x \\ y \end{bmatrix} = \begin{bmatrix} x \\ y \end{bmatrix} \quad \text{or} \quad \begin{cases} Ax + ACy = x \\ Px + (PC + B)y = y \end{cases}.$$

But this is the same as

$$\begin{cases} x = (I - A)^{-1}ACy = (A^{-1} - I)^{-1}Cy = -(I - A^{-1})^{-1}Cy \\ -P(I - A^{-1})^{-1}Cy + PCy = (I - B)y \end{cases}$$

and so $y = (I - B)^{-1}P(I - (I - A^{-1})^{-1})Cy$. Since $I - (I - A^{-1})^{-1} = -A^{-1}(I - A^{-1})^{-1}$ we deduce

$$-(I - B)^{-1}PA^{-1}(I - A^{-1})^{-1}Cy = y.$$

We can take v such that[1] $\|PA^{-1}v\| = 1$ and $\|v\| = \|PA^{-1}\|^{-1}$. We can now define $y = -(I - B)^{-1}PA^{-1}v$.

The fact that $\|B\|$ can be made arbitrarily small ensures that we can assume without loss of generality that $\|I - B\| \leq 2$ and so

$$\|y\| = \|(I - B)^{-1}PA^{-1}v\| \geq \|(I - B)^{-1}\|^{-1}\|PA^{-1}v\| = \|(I - B)^{-1}\|^{-1}.$$

Thus $\|y\|^{-1} \leq \|I - B\| \leq 2$.

We can now choose w such that $(I - A^{-1})^{-1}w = v$. Again, since $\|A^{-1}\|$ is very small, we can assume that $\|(I - A^{-1})^{-1}\| \geq 1/2$ and so $\|w\| \leq 2\|v\|$.

Now we choose C such that $Cy = w$ and $\|C\| = \|w\|/\|y\|$.

This implies that $\|C\| \leq 2\|v\|/\|y\| \leq 4\|v\| = 4\|PA^{-1}\| = o(\theta)$ which can be made arbitrarily small by taking θ small and t_p big. For this transformation C the vector y provides a fixed vector for the transformation M, that is, M has an eigenvalue 1, as we wanted to prove.

This concludes the proof of Theorem 4.7.

4.2.4 Dominated Splitting for the Linear Poincaré Flow Along Regular Orbits

Here we prove Theorem 4.8 following Mañé [145, Lemma II.3]. For this it is enough to show that there exist a C^1 neighborhood $\mathcal{V}$ of X and $T_0 > 0$ such that, for every vector field $Y \in \mathcal{V}$, if $p \in \mathrm{Per}(Y) \cap \Lambda_Y(U)$ and $t_p > T_0$ then

$$\|P^{T_0} \mid N_p^{cs}\| \cdot \|P^{-T_0} \mid N^{cu}_{Y^{T_0}(p)}\| \leq \frac{1}{2}. \tag{4.7}$$

We prove (4.7) arguing by contradiction. If (4.7) is not true then, for any given $T_0 > 0$, we can find $Y \in \mathfrak{X}^1$ arbitrarily C^1 close to X and a point $y \in \mathrm{Per}(Y) \cap \Lambda_Y(U)$ with $t_y > T_0$ satisfying

$$\|P^{T_0} \mid N_y^{cs}\| \cdot \|P^{-T_0} \mid N^{cu}_{Y^{T_0}(y)}\| > \frac{1}{2}. \tag{4.8}$$

Under this assumption, we have

$$\|P^{T_0} \mid N_y^{cs}\| > \frac{1}{2}\|P^{-T_0} \mid N^{cu}_{Y^{T_0}(y)}\|^{-1} = \frac{1}{2}\|(P^{T_0} \mid N_y^{cu})^{-1}\|^{-1} = \frac{1}{2}m(P^{T_0} \mid N_y^{cu})$$

[1]For an invertible linear map L and each $\mathbf{w}$ with $\|\mathbf{w}\| = 1$ we have $\|LL^{-1}\mathbf{w}\| = 1$; and if $\|L^{-1}w\| > \|L\|^{-1}$ for all $\|\mathbf{w}\| = 1$, then $1 = \|L^{-1}L\mathbf{w}\| > \|L\|^{-1}\|L\mathbf{w}\| = 1$ for some $\mathbf{w}$ with $\|\mathbf{w}\| = 1$. Since $\|L^{-1}\| \geq \|L\|^{-1}$ we see that there exists $\mathbf{w}$ with $\|\mathbf{w}\| = 1$ and $\|L^{-1}\mathbf{w}\| = \|L\|^{-1}$ and we just have to take $v = L^{-1}\mathbf{w}$ and $L = PA^{-1}$.

and the compactness of the unit ball in finite dimensional vector spaces ensures the existence of $w_0 \in N_y^{cs}$, $v_0 \in N_y^{cu}$ such that

$$\|P^{T_0}(y)\cdot w_0\| > \frac{1}{2}\|P^{T_0}(y)\cdot v_0\|. \tag{4.9}$$

We define now a linear map $L : N_y^{cu} \to N_y^{cs}$ such that $Lv_0 = \varepsilon_1 w_0$ and $\|L\| = \varepsilon_1$ for a small $\varepsilon_1 > 0$ to be determined in what follows. Next we set

$$\hat{L} := (1+\varepsilon_1)^{t_y} P^{t_y}(y)\cdot L\cdot \left(P^{t_y} \mid N_y^{cu}\right)^{-1},$$

so that $\hat{L}$ is the transformation whose graph is the image of the graph of L by P^{t_y} after stretching, that is, $\hat{L}$ represents the "graph transform" of L:

$$\mathrm{Graph}(\hat{L}) = (1+\varepsilon_1)^{t_y} P^{t_y}(y)\cdot \mathrm{Graph}(L).$$

Starting with $\|L\|$ small just means that the graph of L is a small perturbation of N_y^{cu}. Then $\hat{L}$ is the result of flowing along the periodic orbit through y by the action of the Linear Poincaré Flow.

Now we choose ε_0 small and a big enough T_0 so that, for $\alpha = \alpha(N_y^{cs}, N_y^{cu}) > c$ (from Theorem 4.7),

$$\varepsilon_1 := \frac{\alpha}{1+\alpha}\varepsilon_0, \quad \varepsilon_1(1+\varepsilon_1)^{T_0} \geq 4 + 2/c \quad \text{and} \quad [(1+\varepsilon_1)\lambda]^{T_0} \leq \varepsilon_1.$$

From Theorem 4.7 we deduce

$$\|\hat{L}\| \leq (1+\varepsilon_1)^{t_y}\lambda^{2t_y}\|L\| = [(1+\varepsilon_1)^2\lambda^2]^{t_y}\|L\| \leq \varepsilon_1\|L\|$$

where we have used that $t_y \gg T_0$ and $[(1+\varepsilon_1)^2\lambda^2]^{t_y} \leq [(1+\varepsilon_1)\lambda]^{T_0} \leq \varepsilon_1$. Since λ is uniform for all periodic orbits in a C^1 neighborhood of X, we can take T_0 big enough and ε_1 close enough to zero to achieve this. This shows that the angle between the graph of L and N_y^{cu} is contracted after a full turn around the orbit, as it should be since N_y^{cu} dominates N_y^{cs}, see Sect. 2.6.2.

Now we build up a perturbation of the flow along the orbit through y so that the perturbed flow is arbitrarily C^1 close to Y, with a periodic orbit through y having a hyperbolic splitting for its Linear Poincaré Flow with an arbitrarily small angle between the contracting and expanding subspace. This provides the contradiction we need to conclude the argument.

To build the perturbation we need the following linear algebra result.

Lemma 4.9 *Let us assume that* $\mathbb{R}^n = E \oplus F$ *with* E, F *non-trivial subspaces and let us write* $\alpha(E,F) := \|H\|^{-1}$ *for the linear operator* $H : E^{\perp} \to E$ *such that* $F = \mathrm{Graph}(H) = \{u + Hu : u \in E^{\perp}\}$. *Then*

1. $\|v - u\| \geq \frac{\alpha(E,F)}{1+\alpha(E,F)}\|v\|$, $u \in E$, $v \in F$;
2. $\|T\| \leq \frac{1+\alpha(E,F)}{\alpha(E,F)}(\|T \mid E\| + \|T \mid F\|)$, *for any linear transformation* T *of* $\mathbb{R}^n$;

3. *Given a linear transformation* $A : F \to E$ *and* $G = \text{Graph}(A)$, *there are linear maps* $T_i : \mathbb{R}^n \to \mathbb{R}^n$ *satisfying*
   a. $\|T_i\| \le \frac{1+\alpha(E,F)}{\alpha(E,F)}\|A\|$ *and* $T_i \mid E \equiv 0$ *for* $i = 1, 2$;
   b. $(T_1 + I) \cdot G = F$ *and* $(T_2 + I) \cdot F = G$.

Proof

1. We can write $v \in F$ as $w + Hw$ for some $w \in E^{\perp}$, and so for $u \in E$ we get

$$\|v-u\| = \|w+(Hw-u)\| \ge \|w\| \quad \text{and} \quad \|v\| \le \|w\| + \|Hw\| \le (1+\|H\|)\|w\|.$$

   Hence $\|v - u\| \ge \|w\| \ge \frac{\|v\|}{1+\|H\|} \ge \frac{\alpha(E,F)}{1+\alpha(E,F)}\|v\|$.
2. For each $u \in \mathbb{R}^n$ we can write $u = v + w$ with $v \in E$ and $w \in F^{\perp}$, and

$$\begin{aligned}\|Tu\| &= \|T(v+w)\| = \|T(\underbrace{v+Hv}_{F}) + T(\underbrace{w-Hv}_{E})\| \\ &\le \|T \mid F\|(1+\|H\|)\|v\| + \|T \mid E\|(\|w\| + \|H\|\|v\|) \\ &\le \|T \mid F\|(1+\|H\|)\|u\| + \|T \mid E\|(1+\|H\|)\|u\|\end{aligned}$$

   since $\|v\|, \|w\| \le \|u\|$. This proves this item because $\|H\| = \alpha(E, F)^{-1}$.
3. Let $\pi : \mathbb{R}^n \to F$ be the projection parallel to E, and $T_1 := -A \cdot \pi$, $T_2 = A \cdot \pi$. Clearly $T_i \mid E \equiv 0$ and $T_i \mid F = \pm A$. Then item (a) follows from item 2. For $v \in F$ we have

$$\begin{aligned}(I + T_1)(v + Av) &= v + Av - A \cdot \pi v - A \cdot \pi \cdot Av = v \quad \text{and} \\ (I + T_2)v &= v + A \cdot \pi v = v + Av,\end{aligned}$$

   which completes the proof of (b). □

Now we define for each $0 \le \xi \le 1$ the subspaces

$$\begin{aligned}G_\xi &= \text{Graph}(\xi L) = \{u + \xi Lu : u \in N_y^{cu}\} \\ \hat{G}_\xi &= \text{Graph}(\xi \hat{L}) = \{u + \xi \hat{L}u : u \in N_y^{cu}\},\end{aligned}$$

and use Lemma 4.9 to find the maps

$$\begin{cases} T_\xi \mid N_y^{cs} \equiv 0 \\ (I + T_\xi) \cdot N_y^{cu} = G_\xi \end{cases} \quad \text{and} \quad \begin{cases} S_\xi \mid N_y^{cs} \equiv 0 \\ (I + S_\xi) \cdot \hat{G}_\xi = N_y^{cu} \end{cases},$$

which satisfy (with $\alpha = \alpha(N_y^{cs}, N_y^{cu})$)

$$\begin{aligned}\|T_\xi\| &\le \frac{1+\alpha}{\alpha}\|\xi L\| \le \frac{1+\alpha}{\alpha}\varepsilon_1 = \varepsilon_0 \quad \text{and} \\ \|S_\xi\| &\le \frac{1+\alpha}{\alpha}\|\xi \hat{L}\| \le \frac{1+\alpha}{\alpha}\varepsilon_1 = \varepsilon_0.\end{aligned}$$

For $0 \le t \le t_y$ and t_y big enough we consider also the projection $\pi_t^{cs} : N_{Y^t(y)} \to N_{Y^t(y)}^{cs}$ parallel to $N_{Y^t(y)}^{cu}$ and define the function

$$\varepsilon_1(t) := [(1+\varepsilon_1)^t - 1](1-\delta(t)),$$

where $\delta : [0, t_y] \to [0, 1]$ is a smooth function so that

$$\begin{cases} \delta \equiv 0 & \text{for } 0 \le t \le [t_y] - 2 \\ \delta \equiv 1 & \text{for } [t_y] - 1/2 \le t \le t_y \end{cases} \quad \text{and} \quad |\delta'(t)| \le 1 \quad \text{for all } 0 \le t \le t_y.$$

Now we are ready to define the perturbation of the flow: for each fixed $\xi \in [0, 1]$ consider the family of maps $A_t : N_y \to N_{Y^t(y)}$ (which can be extended to maps $A_t : T_y M \to T_{Y^t(y)} M$ preserving the flow direction) as follows:

$$A_t := (I + \delta(t) S_\xi) \cdot (I + \varepsilon_1(t)\pi_t^{cs}) \cdot P^t(y) \cdot (I + T_\xi).$$

It is easy to see that for $0 < t < t_y$ and small s the map $A_{t+s} A_t^{-1}$ equals

$$(I+\delta(t+s)S_\xi)\cdot(I+\varepsilon_1(t+s)\pi_t^{cs})\cdot P^s(Y^t(y))\cdot(I+\varepsilon_1(t)\pi_t^{cs}))^{-1}\cdot(I+\delta(t)S_\xi)^{-1}.$$

Hence, by the previous choices of the maps T_ξ, S_ξ and the functions $\delta(t), \varepsilon_1(t)$, taking ε_0 (and so also ε_1) small enough, a straightforward calculation shows that $\partial_s(A_{t+s}A_t^{-1}) \mid_{s=0}$ equals $P_Y^t(Y^t(y)) + L_0$ for some linear map L_0 whose norm $\|L_0\| = o(\varepsilon_0)$ tends to zero as $\varepsilon_0 \to 0$.

Applying Frank's Lemma for vector fields we obtain a vector field Z arbitrarily C^1 close to Y with a periodic orbit through y with the same period and such that $DZ^t(y) = (I_t, A_t) : E_y^X \oplus N_y \to E_{Z^t(y)}^X \oplus N_{Z^t(y)}$ for $0 \le t \le t_y$, where I_t is the derivative of Y^t along the flow direction at $Y^t(p) = Z^t(p)$.

Now we check the resulting change in the angles between the stable and unstable direction for the Linear Poincaré Flow. First we show that the stable manifold is unchanged: by the choice of A_t we obtain for each $0 \le \xi \le 1$

$$\begin{aligned} \|P_Z^{t_y} \mid N_y^{cs,Y}\| &\le \|(I+S_\xi)(I+\varepsilon_1(t_y)\pi_0^{cs})P_Y^{t_y}(I+T_\xi)\| \\ &\le (1+\varepsilon_1)^{t_y}\|P_Y^{t_y} \mid N_y^{cs,Y}\| \le [(1+\varepsilon_1)\lambda]^{t_y} < 1. \end{aligned}$$

This shows that $N_y^{cs,Y} \subseteq N_y^{cs,Z}$ and it is enough to conclude that $N_y^{cs,Y} = N_y^{cs,Z}$ since, by assumption, we have $Y, Z \in \mathscr{V}$ so that the orbit of y with respect to Z is hyperbolic with the same index with respect to Y.

Now we show that the unstable manifold for the transformation $P_Z^{t_y}$ is also unchanged. Indeed, by definition of the perturbation we have $\varepsilon_1(t_y) = 0$ and

$$v \in N_y^{cu,Y} \xrightarrow[I+T_\xi]{} v_1 \in G_\xi \xrightarrow[P_Y^{T_0}]{} v_2 \in \hat{G}_\xi \xrightarrow[I+S_\xi]{} P_Z^{t_y}(y)\cdot v \in N_y^{cu,Y}$$

so that $P_Z^{t_y} \cdot N_y^{cu,Y} \subset N_y^{cu,Y}$. The fact that $N_y^{cu,Y}$ and $N_y^{cs,Y}$ form a splitting of N_y ensures that we have in fact equality. Moreover the norm of $P_Z^{-t_y} \mid N_y^{cu,Y}$ depends continuously on ξ and is strictly smaller than 1 for $\xi = 0$ (since in this case $P_Z^{t_y} = P_Y^{t_y}$). Also the homogeneous assumption on $\mathcal{V}$ ensures that this norm is never close to 1 for all $\xi \in (0,1]$. Therefore $P_Z^{-t_y} \mid N_y^{cu,Y}$ is a contraction for each $0 \le \xi \le 1$ and so $N_y^{cu,Z} = N_y^{cu,Y}$.

Finally we estimate $\angle(N^{cs,Z}_{Z^{T_0}(y)}, N^{cu,Z}_{Z^{T_0}(y)})$ in terms of $\beta := \alpha(N^{cs,Z}_{Z^{T_0}(y)}, N^{cu,Z}_{Z^{T_0}(y)})$. For this we take $\xi = 1$ and the vectors $w_0 \in N_y^{cs}$ and $v_0 \in N_y^{cu}$ fixed earlier in (4.9), and note that

$$w_1 := P_Z^{T_0}(y) \cdot w_0 \in N^{cs,Z}_{Z^{T_0}(y)} \quad \text{and} \quad v_1 := P_Z^{T_0}(y) \cdot v_0 \in N^{cu,Z}_{Z^{T_0}(y)}.$$

Since $\delta(T_0) = 0$ (we just take $t_y \gg T_0$) we have

$$w_1 = (1+\varepsilon_1)^{T_0} P^{T_0}(y) \cdot w_0 \quad \text{and}$$
$$v_1 = P^{T_0}(y) \cdot v_0 + \varepsilon_1 (1+\varepsilon_1)^{T_0} P^{T_0}(y) \cdot w_0.$$

We use Lemma 4.9 and the choice of v_0, w_0 in (4.9) to bound

$$\begin{aligned} \|P^{T_0}(y) \cdot v_0\| = \|v_1 - \varepsilon_1 w_1\| &\ge \frac{\beta}{1+\beta} \|v_1\| \\ &\ge \frac{\beta}{1+\beta} \Big| \|\varepsilon_1 (1+\varepsilon_1)^{T_0} P^{T_0}(y) \cdot w_0\| - \|P^{T_0}(y) \cdot v_0\| \Big| \\ &\ge \frac{\beta}{1+\beta} \left| \frac{\varepsilon_1}{2} (1+\varepsilon_1)^{T_0} - 1 \right| \|P^{T_0}(y) \cdot v_0\| \end{aligned}$$

which means that

$$\frac{1+\beta}{\beta} \ge \left| \frac{\varepsilon_1}{2} (1+\varepsilon_1)^{T_0} - 1 \right| \iff \beta \le \frac{2}{\varepsilon_1 (1+\varepsilon_1)^{T_0} - 4} < c$$

by the choice of T_0 and ε_1 made earlier.

This contradicts Theorem 4.7 and this contradiction proves Theorem 4.8.

4.3 Uniform Hyperbolicity for the Linear Poincaré Flow

Here we complete the proof of Theorem 4.1. We state the assumptions and the conclusion in the following theorem.

Theorem 4.10 *Let a vector field $X \in \mathfrak{X}^1(M)$ be given with an attracting set $\Lambda_X(U)$ on some isolating neighborhood $U \subset M$ satisfying*

- *$\Lambda_X(U) \subset \Omega(X)$ and*
- *for every vector field Y in a C^1 neighborhood $\mathcal{V}$ of X the critical elements of Y within U do not change index, i.e., X is homogeneous in U.*

Then there exists a hyperbolic splitting for the Linear Poincaré Flow of X over every compact invariant subset Γ *of* $\Lambda_X(U)^* = \Lambda_X(U) \setminus S(X)$.

From Theorem 2.27 the existence of a hyperbolic splitting for the Linear Poincaré Flow over a compact invariant set without singularities is equivalent to hyperbolicity of this invariant subset.

In the above setting, we note in particular that, *if there are no singularities in* U, *then* $\Lambda_X(U)$ *has a globally defined Linear Poincaré Flow with a hyperbolic splitting*, that is, $\Lambda_X(U)$ is hyperbolic.

The proof is adapted from the work [145] of Mañé adapted to the flow setting. The main idea is to show that, if the splitting of the Linear Poincaré Flow provided by Theorem 4.4 is not hyperbolic, then we can see the lack of hyperbolicity on some periodic orbit for a C^1 nearby flow, which contradicts Theorems 4.7 and 4.8. This "transfer of non-hyperbolicity" is provided by the Ergodic Closing Lemma for flows (Theorem 2.23), which enables us to approximate a full measure subset of points by periodic orbits for nearby flows.

In Sect. 4.3.1 we first state an abstract result that captures the conditions we use in our setting. Then in Sect. 4.3.2 we then apply this abstract result twice, for each subbundle of the dominated decomposition of the Linear Poincaré Flow, to complete the proof.

4.3.1 Subadditive Functions of the Orbits of a Flow and Exponential Growth

Let U be an open subset of M such that $\overline{X^t(U)} \subset U$ for all $t > 0$ and $\Lambda_Y(U)$ the corresponding attracting set for $Y \in \mathscr{V}$, where $\mathscr{V}$ is a C^1 neighborhood of $X \in \mathfrak{X}^1(M)$. We say that a family of functions $\phi_Y : \mathbb{R} \times \Lambda_Y(U) \to \mathbb{R}$ is *subadditive* if

$$\phi_Y(t+s,x) \le \phi_Y(s, X^t(x)) + \phi_Y(t,x), \quad \text{for all } t,s \in \mathbb{R}, x \in \Lambda \quad \text{and} \quad Y \in \mathscr{V}.$$

Theorem 4.11 *Let* ϕ_Y *be a family of continuous subadditive functions on the subsets* Λ_Y *for* $Y \in \mathscr{V}$ *such that for all* $x \in \Lambda_Y$ *and every* $Y \in \mathscr{V}$

- $\phi_Y(0,x) = 0$;
- $D_Y(x) = \limsup_{h\to 0}(\phi_Y(h,x)/h) < \infty$ *depends continuously on* x;
- $\Lambda_X(U) \cap S(X)$ *is discrete*;
- $D(\sigma) < 0$ *for each* $\sigma \in \Lambda_X(U) \cap S(X)$.

Moreover $\phi_Y(s,y)$ *depends continuously on* (s,y,Y) *as follows: if* $Y_n \xrightarrow[n\to\infty]{C^1} X$, $y_n \in \Lambda_{Y_n}(U)$, $s_n \in \mathbb{R}$ *are such that* $y_n \xrightarrow[n\to\infty]{} x \in \Lambda_X(U)$ *and* $s_n \xrightarrow[n\to\infty]{} t \in \mathbb{R}$, *then*

$$\phi_{Y_n}(s_n, y_n) \xrightarrow[n\to\infty]{} \phi_X(t,x).$$

Let us assume that

1. *there exists $T_0 > 0$ and $a > 0$ such that, for each $Y \in \mathscr{V}$ and for every periodic point for Y whose minimal period is bigger than T_0, that is, $p \in \{y \in \mathrm{Per}(Y) \cap \Lambda_Y(U) : t_y \geq T_0\}$, we have*

$$\phi_Y(t_p, p) \leq -a t_p,$$

where t_p is the period of p; and

2. *$\phi_Y(t_p, p) < 0$ for all $p \in \mathrm{Per}(Y) \cap \Lambda_X(U)$.*

Then for every compact invariant subset Γ of $\Lambda_X(U)$

$$\phi_X(t, x) \leq -at \quad \textit{for all } t > 0, \ \textit{every } x \in \Gamma \textit{ and all } X \in \mathscr{V}.$$

We need some technical observations about subadditive functions before we prove this result.

4.3.1.1 Differentiability of Subadditive Functions

Lemma 4.12 *Let $\phi_X : \mathbb{R} \times M \to \mathbb{R}$ be a subadditive function for the flow $X \in \mathfrak{X}^1(M)$ satisfying*

- *$\phi(0, x) = 0$;*
- *$D(x) := \limsup_{h \to 0}(\phi(h, x)/h) < \infty$.*

Then $\partial_h \phi(h, x) \mid_{h=0} = D(x) = \lim_{h \to 0} \phi(h, x)/h$.

Proof Let us define $\omega_\delta(x) := \sup_{0 \leq \eta \leq \delta}(\phi(\eta, x)/\eta)$ for all $x \in M$. Then we see that $D(x) = \lim_{\delta \to 0} \omega_\delta(x)$. We fix $x \in M$ and $\varepsilon > 0$ and consider $0 < h < \varepsilon$ and $\delta_n = 2^{-n}\varepsilon$, for some big $n \in \mathbb{N}$, and obtain

$$D(x) - \varepsilon \leq \frac{\phi(h, x)}{h} = \frac{\phi(2^n \delta_n, x)}{h} \leq \frac{1}{h} \sum_{j=0}^{2^n - 1} \phi(\delta_n, X^{j\delta_n}(x))$$

$$\leq \frac{1}{h} \sum_{j=0}^{2^n - 1} \delta_n \omega_{\delta_n}(X^{j\delta_n}(x)).$$

In this way we are dividing the interval $[0, h]$ into dyadic subintervals with length δ_n, evaluating the function ω_{δ_n} at these points and making a Riemann sum. We note that, fixing a point τ of the dyadic subdivision of level n, we have $\omega_{\delta_n}(X^\tau(x)) \xrightarrow[n \to \infty]{} D(X^\tau(x))$. Therefore

$$\frac{1}{h} \sum_{j=0}^{2^n - 1} \delta_n \omega_{\delta_n}(X^{j\delta_n}(x)) \xrightarrow[n \to \infty]{} \frac{1}{h} \int_0^h D(X^t(x))\, dt.$$

Hence we get

$$D(x) - \varepsilon \le \liminf_{h\to 0+} \frac{\phi(h,x)}{h} \le \limsup_{h\to 0+} \frac{\phi(h,x)}{h}$$
$$\le \limsup_{h\to 0+} \frac{1}{h}\int_0^h D(X^t(x))\,dt = D(x).$$

Since $\varepsilon > 0$ was arbitrarily chosen and $\phi(0,x) = 0$, we have

$$D(x) = \lim_{h\to 0+} \frac{\phi(h,x)}{h}.$$

For $-\varepsilon < h < 0$ we can write

$$D(x) + \varepsilon \ge \frac{\phi(h,x)}{h} = \frac{\phi(2^n\delta_n, x)}{h} \ge \frac{1}{h}\sum_{j=0}^{2^n-1} \phi(\delta_n, X^{j\delta_n}(x))$$
$$\ge \frac{1}{h}\sum_{j=0}^{2^n-1} \delta_n \omega_{\delta_n}(X^{j\delta_n}(x)),$$

and the same argument with the reverse inequalities shows that

$$D(x) = \lim_{h\to 0-} \frac{\phi(h,x)}{h} = \partial_h \phi(h,x)\,|_{h=0}\,.$$

The proof is complete. □

We remark that for all $x \in M$ and $T > 0$ we have on the one hand

$$\left.\frac{\partial}{\partial t}\phi(t,x)\right|_{t=T} = \lim_{h\to 0+} \frac{\phi(T+h,x) - \phi(T,x)}{h}$$
$$\le \lim_{h\to 0+} \frac{\phi(T,x) + \phi(h, X^T(x)) - \phi(T,x)}{h} = \left.\frac{\partial}{\partial h}\phi(h, X^T(x))\right|_{h=0}.$$

On the other hand

$$\left.\frac{\partial}{\partial t}\phi(t,x)\right|_{t=T} = \lim_{h\to 0-} \frac{\phi(T+h,x) - \phi(T,x)}{h}$$
$$\ge \lim_{h\to 0-} \frac{\phi(T,x) + \phi(h, X^T(x)) - \phi(T,x)}{h} = \left.\frac{\partial}{\partial h}\phi(h, X^T(x))\right|_{h=0}.$$

Therefore we deduce the following useful bound

$$\phi(T,x) = \int_0^T \left.\frac{\partial}{\partial t}\phi(t,x)\right|_{t=s} ds = \int_0^T \left.\frac{\partial}{\partial h}\phi(h, X^s(x))\right|_{h=0} ds. \tag{4.10}$$

4.3.1.2 Subadditivity and Exponential Bounds

Here we obtain sufficient conditions for linear bounds on a subadditive function. Then we present a proof of Theorem 4.11.

Lemma 4.13 *Let $\phi = \phi_X : \mathbb{R} \times \Gamma \to \mathbb{R}$ be a continuous subadditive function, where $\Gamma \subset \Lambda_X(U)$ is a compact invariant set.*

If $\liminf_{t\to\infty} \phi(t,x) = -\infty$ for all $x \in \Gamma$, then there exists $T_0 > 0$ such that $\phi(T_0, x) < -\log 2$ for every $x \in \Gamma$.

Proof For each $x \in \Gamma$ there exists t_x such that $\phi(t_x, x) < -\log 3$. Hence for each $x \in \Gamma$ there is a neighborhood $B(x)$ such that for all $y \in B(x)$ we have $\phi(t_x, y) < -\log 2$. Since Γ is compact, there are $B(x_i)$, $i = 1, \dots, n$, such that $\Gamma \subset B(x_1) \cup \dots \cup B(x_n)$.

Let $K_0 = \sup\{\exp(\phi(t, y)), y \in B(x_i), 0 \le t \le t_{x_i}, i = 1, \dots, n\}$, let j_0 be such that $2^{-j_0} \cdot K_0 < 1/2$ and fix $T_0 > j_0 \cdot \sup\{t_{x_i}, i = 1, \dots, n\}$. We claim that T_0 satisfies the statement of the lemma.

Indeed, given $y \in \Gamma$ we have $y \in B(x_{i_1})$ for some $1 \le i_1 \le n$. Let $t_{i_1}, \dots, t_{i_k}, t_{i_{k+1}}$ satisfy

- $X^{t_{i_1}+\dots+t_{i_j}}(y) \in B(x_{i_{j+1}})$, $1 \le j \le k$, and
- $t_{i_1} + \dots + t_{i_k} \le T_0 \le t_{i_1} + \dots + t_{i_{k+1}}$.

Observe that $k \ge j_0$. Then for $\ell_j = t_{i_1} + \dots + t_{i_j}$, $j = 1, \dots, k+1$, we have

$$\phi(T_0, y) \le \phi(T_0 - \ell_k, X^{\ell_k}(y)) + \sum_{j=1}^{k} \phi(t_{i_j}, X^{\ell_{j-1}}(y))$$

$$< \log K_0 - j_0 \log 2 = \log(2^{-j_0} K_0) < -\log 2,$$

and the proof is complete. □

Lemma 4.14 *Let $\phi = \phi_X : \mathbb{R} \times \Gamma \to \mathbb{R}$ be a continuous subadditive function, where $\Gamma \subset \Lambda_X(U)$ is a compact invariant set.*

If there exists $T_0 > 0$ such that $\phi(T_0, x) < -\log 2$ for all $x \in \Gamma$, then there are $c \in \mathbb{R}$ and $0 < \lambda < 1$ such that $\phi(T, x) < c + T \log \lambda$ for all $x \in \Gamma$ and $T > 0$.

Proof Let $K_1 = \sup\{\exp(\phi(t, x)), 0 \le t \le T_0, x \in \Gamma\}$. Choose $0 < \lambda < 1$ such that $1/2 < \lambda^{T_0}$ and $c \in \mathbb{R}$ such that $\log K_1 < c + r \log \lambda$ for all $0 \le r \le T_0$. Then for any $x \in \Gamma$ and all $T > 0$ we have $T = nT_0 + r$ with $n = [T/T_0] = \max\{k \in \mathbb{Z} : k \le T/T_0\}$ and $0 \le r = T - nT_0 < T_0$. Consequently

$$\phi(T, x) \le \phi(r, X^{nT_0}(x)) + \sum_{j=0}^{n-1} \phi(T_0, X^{jT_0}(x))$$

$$< \log K_1 - n \log 2 < c + r \log \lambda + nT_0 \log \lambda < c + T \log \lambda,$$

concluding the proof. □

4.3.1.3 Uniform Linear Bound on All Orbits

Now we are ready to prove Theorem 4.11. Let $\Gamma \subset \Lambda_X(U)$ be a compact invariant set. From Lemmas 4.13 and 4.14 it is enough to show that $\liminf_{t\to+\infty} \phi(t,x) = -\infty$ for each $x \in \Gamma$.

We argue by contradiction and assume that there exists $x \in \Gamma$ such that

$$\liminf_{t\to+\infty} \phi(t,x) = L \in \mathbb{R}.$$

Then there exists $s_n \xrightarrow[n\to\infty]{} \infty$ such that

$$\lim_{n\to\infty} \frac{1}{s_n}\phi(s_n,x) \geq 0. \tag{4.11}$$

Let $C^0(\Gamma)$ be the set of real continuous functions defined on Γ with the topology of uniform convergence, and define the sequence of continuous operators

$$\Psi_n : C^0(\Gamma) \to \mathbb{R}, \qquad \varphi \in C^0(\Gamma) \mapsto \frac{1}{s_n}\int_0^{s_n} \varphi(X^s(x))\,ds.$$

Since in the C^0 norm this sequence is bounded, $\|\Psi_n\| \leq 1$, and the unit ball of the dual $C^0(\Gamma)^*$ is weak*-compact (see any standard reference on Functional Analysis e.g. [234]), there exists a subsequence of Ψ_n, which we still denote by Ψ_n, converging to a continuous map $\Psi \in C^0(\Gamma)^*$ in the weak* topology. Let $\mathcal{M} = \mathcal{M}(\Gamma)$ be the space of measures with support on Γ. By the Riesz Representation Theorem (see e.g. [233]) there exists a probability measure $\mu \in \mathcal{M}$ such that

$$\int_\Gamma \varphi\,d\mu = \lim_{n\to+\infty} \frac{1}{s_n}\int_0^{s_n} \varphi(X^s(x))\,ds = \Psi(\varphi), \tag{4.12}$$

for every continuous function $\varphi : \Gamma \to \mathbb{R}$. Note that such μ is invariant under the flow since for all $t \in \mathbb{R}$

$$\begin{aligned}\Psi(\varphi\circ X^t) &= \lim_{n\to+\infty} \frac{1}{s_n}\int_0^{s_n} \varphi(X^{s+t}(x))\,ds \\ &= \lim_{n\to+\infty} \frac{s_n+t}{s_n}\cdot\frac{1}{s_n+t}\Big(\int_0^{s_n+t}\varphi(X^s(x))\,ds - \int_0^t \varphi(X^s(x))\,ds\Big) = \Psi(\varphi).\end{aligned}$$

Now we use the fact that $D(x) = \partial_h\phi(h,x)\mid_{h=0}$ is continuous by assumption. This together with (4.10) ensures that

$$\int_0^T D(X^s(x))\,ds = \phi(T,x), \quad \text{for all } T > 0, \tag{4.13}$$

and so, by (4.11) and (4.12),

$$\int_\Gamma D\,d\mu \geq 0. \tag{4.14}$$

The Ergodic Theorem now implies that

$$\int_{\Gamma} D\,d\mu = \int_{\Gamma} \lim_{T\to\infty} \frac{1}{T}\int_0^T D(X^s(x))\,ds\,d\mu(x). \tag{4.15}$$

Let $\Sigma(X)$ be the set of strongly closed points see Sect. 2.5.8. Since μ is X-invariant and $\operatorname{supp}(\mu) \subset \Gamma$, the Ergodic Closing Lemma (Theorem 2.23) ensures that $\mu(\Gamma \cap (S(X) \cup \Sigma(X))) = 1$.

We claim that $\mu(\Gamma \cap \Sigma(X)) > 0$. For otherwise $\mu(\Gamma \cap S(X)) = 1$ and, since $S(X)$ is X-invariant and discrete, we would get that μ is a finite convex linear combination of point masses in $S(X)$: $\mu = \sum_{\sigma \in S(X)} a_i \delta_\sigma$. But by assumption

$$\int_{\Gamma} D\,d\mu = \int_{S(X)} D\,d\mu = \sum_{\sigma\in S(X)} a_i D(\sigma) < 0$$

contradicting (4.14). This contradiction proves the claim.

The Ergodic Decomposition Theorem (see Sect. 2.7) enables us to assume without loss of generality that μ is ergodic. Hence $\mu(\Gamma \cap \Sigma(X)) = 1$. Therefore by (4.14) and (4.15) there exists $y \in \Gamma \cap \Sigma(X)$ such that

$$\lim_{T\to\infty} \frac{1}{T}\int_0^T D(X^s(y))\,ds \geq 0. \tag{4.16}$$

Since $y \in \Sigma(X)$, there are $\delta_n \xrightarrow[n\to+\infty]{} 0$, $Y_n \in \mathcal{V}$, and $p_n \in \operatorname{Per}(Y_n) \cap \Lambda_{Y_n}(U)$ with period t_n such that

$$\|Y_n - X\| < \delta_n \quad \text{and} \quad \sup_{0\leq s\leq t_n} \operatorname{dist}(Y_n^s(p_n), X^s(y)) < \delta_n.$$

We must have $t_n \xrightarrow[n\to\infty]{} \infty$. For otherwise y would be periodic for X and, if t_y is its period, then (4.13) together with (4.16) imply that

$$\frac{1}{t_y}\phi_X(t_y, y) \geq 0$$

contradicting the negative bound assumption on periodic orbits of item 2 of Theorem 4.11. We have arrived at a contradiction and so t_n grows without bound.

Let $\gamma < 0$ be arbitrarily small. By (4.16) again, there exists T_γ such that for $t \geq T_\gamma$ we have

$$\frac{1}{t}\int_0^t D(X^s(y))ds = \frac{1}{t}\phi(t, y) \geq \gamma. \tag{4.17}$$

Since $t_n \xrightarrow[n\to\infty]{} \infty$, we can assume that $t_n > T_\gamma$ for every n. The continuity of the family ϕ_Y together with (4.17) implies that for n big enough

$$\frac{1}{t_n}\phi(t_n, p_n) \geq \gamma \quad \text{or} \quad \phi(t_n, p_n) \geq \gamma t_n.$$

Taking n sufficiently large and $\gamma < 0$ sufficiently small, this last inequality contradicts the assumption stated as item 1 in Theorem 4.11.

This contradiction shows that our assumption (4.11) cannot be true and this completes the proof of Theorem 4.11.

4.3.1.4 Uniform Hyperbolic Splitting for the Linear Poincaré Flow

Here we present the proof of Theorem 4.10. The assumptions imply, by Theorem 4.4, that there exists a dominated splitting for the Linear Poincaré Flow over $\Lambda_Y(U)^*$, for every vector field Y in some C^1 neighborhood $\mathscr{V}$ of X. Here U is a trapping region for $\Lambda_X(U)$. We can define the family of subadditive functions

$$\phi_Y(t,y) := \log \| P_Y^t \mid N_y^{cs,Y} \|, \quad t \in \mathbb{R},\ y \in \Lambda_Y(U)^*, Y \in \mathscr{V}.$$

We note that this family of functions satisfies all the conditions in the statement of Theorem 4.11 restricted to non-equilibrium points. Indeed, the Chain Rule together with the elementary properties of the norm shows that ϕ is subadditive; it is obvious that $\phi_Y(0,y) = 0$ for all $y \in \Lambda_Y(U)^*$; the boundedness of the derivative follows from the boundedness of Y on a bounded subset of a manifold; there are no equilibria to consider in this setting and the dominated splitting implies the continuity of the family $\{\phi_Y\}_{Y\in\mathscr{V}}$ see Theorem 2.33. Theorem 4.7 provides the main assumption on the behavior of ϕ_Y on periodic orbits in $\Lambda_Y(U)^*$ for $Y \in \mathscr{V}$.

Hence, given a compact invariant subset Γ of $\Lambda_X(U)^*$, from Theorem 4.11 we have that there exists $c \in \mathbb{R}$ and $\lambda_0 \in (0,1)$ such that

$$\phi_Y(t,x) \le c + t\log\lambda \quad \text{for all } x \in \Gamma, t \in \mathbb{R}, Y \in \mathscr{V}$$

which implies that $\| P_X^t \mid N_x^{cs,X} \| \le K\lambda_0^t = Ke^{-\lambda t}$ for every $x \in \Gamma$ and $t \in \mathbb{R}$, where $K = e^c > 0$ and $\lambda < 0$ is given by $e^{-\lambda} = \lambda_0$. Thus $N^{cs,X}$ is a uniformly contracted subbundle of the normal bundle over Γ.

Analogously, we show that $N^{cu,X}$ is a uniformly expanded subbundle of the normal bundle for X, by reasoning with the family of subadditive functions

$$\psi_Y(t,y) = \log \| (P_Y^t \mid N_y^{cu,Y})^{-1} \|, \quad Y \in \mathscr{V}, y \in \Gamma, t \in \mathbb{R}.$$

Altogether this shows that $X \in \mathscr{V}$ has a hyperbolic splitting for the Linear Poincaré Flow over Γ. The proof of Theorem 4.10 is complete.

4.3.2 Uniform Hyperbolicity for the Linear Poincaré Flow on the Whole Manifold

Here we prove Theorem 4.1 using Theorem 4.10.

Let $\mathscr{V}$ be a C^1 neighborhood of $X \in \mathfrak{X}^1(M)$ such that every critical element of $Y \in \mathscr{V}$ is hyperbolic with the same index (dimension of the stable manifold) and there exists w such that the closure of $\{Y^t(w) : t \ge 0\}$ is dense.

From Theorem 4.2, already proved, with $U = M$ we see that there are no equilibria for $Y \in \mathscr{V}$, and there exists a dominated splitting $N^{cs,Y} \oplus N^{cu,Y}$ for the Linear Poincaré Flow P^t_Y of Y defined on the whole M. From Theorem 4.10 applied with $U = M$, we deduce that M admits a hyperbolic splitting for the Linear Poincaré Flow with respect to the vector field X. Thus, by Theorem 2.27, M is a hyperbolic set for X, i.e., X is an Anosov flow.

This completes the proof of Theorem 4.1.

Chapter 5
Robust Transitivity and Singular-Hyperbolicity

In the theory of differentiable dynamics for flows, i.e., in the study of the asymptotic behavior of orbits $\{X^t(x)\}_{t\in\mathbb{R}}$ for $X \in \mathfrak{X}^r(M)$, $r \geq 1$, a fundamental problem is to understand how the behavior of the tangent map DX controls or determines the dynamics of the flow X^t. Since the 1970s there is a complete description of the dynamics of a system under the assumption that the tangent map has a hyperbolic structure.

The spectral decomposition theorem, presented in Sect. 2.3 and first proved in [252], provides a description of the non-wandering set of a structural stable system as a finite number of disjoint *compact maximal invariant and transitive sets*, each of these pieces being well understood from both the deterministic and from statistical viewpoints. Moreover such a decomposition persists under small C^1 perturbations. This naturally leads to the study of isolated transitive sets that remain transitive for all nearby systems (robustness).

The Lorenz equations (2.2) provide an example of a robust attractor containing an equilibrium point at the origin and periodic points accumulating on it. This is a non-uniformly hyperbolic attractor which cannot be destroyed by any small perturbation of the parameters. See Sect. 2.2.3 for more on this.

The existence of robust non-hyperbolic attractors for flows was first proved rigorously through the study of *geometric models for Lorenz attractors*, see Sect. 3.3. In particular, they exhibit in a robust way an attracting transitive set with an equilibrium (singularity) whose eigenvalues λ_i, $1 \leq i \leq 3$, are real and satisfy $\lambda_2 < \lambda_3 < 0 < -\lambda_3 < \lambda_1$. In the definition of geometrical models, another key requirement was the existence of an invariant foliation whose leaves are forward contracted by the flow. These features enable us to extract very complete topological, dynamical and ergodic information about these geometrical Lorenz models, as explained in Sect. 3.3. We prove now that these features are present for any robustly transitive set.

Hence the main properties of the Lorenz attractor and geometric Lorenz models are consequences of their robust transitivity. Building on this characterization, in Chap. 7 we elaborate on the ergodic properties of singular-hyperbolic attractors.

V. Araújo, M.J. Pacifico, *Three-Dimensional Flows*, Ergebnisse der Mathematik und ihrer Grenzgebiete. 3. Folge / A Series of Modern Surveys in Mathematics 53,
DOI 10.1007/978-3-642-11414-4_5,

5.1 Definitions and Statement of Results

Definition 5.1 An isolated set Λ of a C^1 vector field X is robustly transitive if it has an open neighborhood U such that

$$\Lambda_Y(U) = \bigcap_{t \in \mathbb{R}} Y^t(U)$$

is both transitive and non-trivial (i.e., *it is neither a singularity nor a periodic orbit*) for any vector field Y C^1-close to X.

First we recall the following simpler result for global transitive flows on 3-manifolds which was first proved by Doering in [79] and whose proof is presented in Chap. 4.

Theorem 5.2 *Assume that $\Lambda = M$ is a robustly transitive set (on a three dimensional manifold). Then the flow is Anosov. In particular the flow has no singularities.*

In general, when Λ is a proper subset of M and contains singularities, we have the following characterization.

Theorem 5.3 *A robustly transitive set* containing singularities *of a flow on a closed 3-manifold is either a proper attractor or a proper repeller.*

We present now some examples which help to make the meaning of Theorem 5.3 more precise.

Example 5.4 We recall that an example of an isolated and invariant compact subset of a three-dimensional vector field, robustly transitive but not an attractor nor a repeller, is the suspension of a horseshoe for a surface diffeomorphism. *In this case the set has no singularities.* This is the case of all transitive hyperbolic isolated subsets for three-dimensional closed manifolds.

Example 5.5 Theorem 5.3 is false for dimensions bigger than three. Indeed consider the vector field $Y : (z, w) \in \mathbb{S}^3 \times \mathbb{S}^1 \mapsto (X(z), N(w))$ in $\mathbb{S}^3 \times \mathbb{S}^1$, where

- X is the vector field given by the Lorenz equations (2.2) or the vector field obtained after the construction of any geometric Lorenz attractor (see Sect. 3.3.2) suitably embedded in $\mathbb{S}^3$, for example with a hyperbolic unstable equilibrium at infinity;
- N is the "North-South" vector field on the circle $\mathbb{S}^1 = \{(x, y) : x^2 + y^2 = 1\}$ given by $-k \cdot \nabla(\pi \mid \mathbb{S}^1)$ where π is the projection on the second coordinate and $k > 0$ is big enough so that the expansion rate e^k at the "North" $(0, 1)$ and the contraction rate e^{-k} at the "South" $(0, -1)$ dominate every eventual expansion or contraction along the directions of $T\mathbb{S}^3 \times \{0\}$.

Then $\Lambda_Y = \cap_{t>0} Y^t(U \times V)$, which is the maximal invariant subset of $U \times V$ with respect to Y, equals $\Lambda_X \times \{(0,1)\}$ and is transitive, where

- U is an isolating neighborhood of the (geometric) Lorenz attractor for the X-flow;
- V is a small neighborhood of $(0,1)$ in $\mathbb{S}^1$; and
- Λ_X is the (geometric) Lorenz attractor.

Notice that $\mathbb{S}^3 \times \{(0,1)\}$ is an invariant and normally hyperbolic compact submanifold of $\mathbb{S}^3 \times \mathbb{S}^1$, see [110]. It follows that, for all vector fields Z C^1-close to Y, there exists an "analytic continuation" $\tilde{M}$ of the submanifold $\mathbb{S}^3 \times \{(0,1)\}$ such that

- $\tilde{M}$ is a Z^t-invariant, compact and normally hyperbolic submanifold of $\mathbb{S}^3 \times \mathbb{S}^1$; in particular the length of any smooth curve transverse to $\tilde{M}$ inside $U \times V$ is exponentially expanded by Z^t, $t > 0$.
- $\tilde{M}$ is C^1-close to $\mathbb{S}^3 \times \{(0,1)\}$ as embeddings in $\mathbb{S}^3 \times \mathbb{S}^1$.

Hence there exists a diffeomorphism $\phi : \tilde{M} \to \mathbb{S}^3$, close to the identity, and the restriction of the vector field Z to $\tilde{M}$ can be seen as a vector field C^1-close to X under a global change of coordinates extending ϕ. Therefore the maximal invariant subset of $U \times V$ for Z is $\cap_{t>0} Z^t(\phi^{-1}(U \times V)) \subset \tilde{M}$, which is transitive by the robustness of the (geometric) Lorenz attractor. In this way we get a robustly transitive set Λ_Y which is neither an attractor nor a repeller.

Example 5.6 In the setting of boundary-preserving vector fields, on 3-manifolds with boundary, the singular-horseshoe provides another counter-example (see Remark 3.3) since it is robustly transitive in the space of vector fields preserving the boundary, but it is not an attractor nor a repeller.

Example 5.7 The converse to Theorem 5.3 is also not true: proper attractors (or repellers) with singularities are not necessarily robustly transitive, even if their periodic points and singularities are hyperbolic in a robust way. For examples, see e.g. Morales and Pujals [171], where we can find the following construction which we detail here.

In the construction of the geometric Lorenz attractor presented in Sect. 3.3, replace the one-dimensional map of Fig. 3.24 with the map represented on the left hand side of Fig. 5.1. The fixed point p of this map represents a saddle-type hyperbolic periodic orbit in the maximal invariant subset of a neighborhood of the attractor.

However, by an arbitrarily small perturbation of the one-dimensional map, we can construct a C^1-close vector field with a geometric Lorenz attractor whose one-dimensional quotient map is represented on the right hand side of Fig. 5.1. The hyperbolic periodic orbit persists, and the new orbit corresponding to p' still belongs to the maximal invariant set of the same neighborhood, but does not belong to the new attractor corresponding to the maximal invariant subset inside the interval $[-1/2, b]$ in Fig. 5.1. Hence the original attractor is transitive but not robustly transitive.

This last example prompts for a search for sufficient conditions ensuring robustness of a singular-hyperbolic attractor. We shall say more about this in Chap. 6.

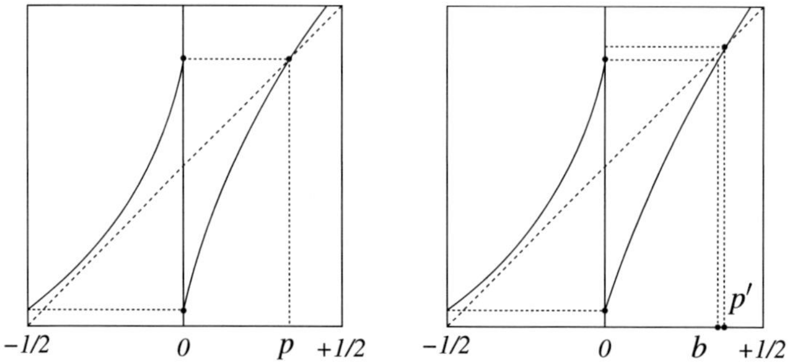

Fig. 5.1 A non-robustly transitive Lorenz transformation

Theorem 5.3 is obtained from a general result on n-manifolds, $n \geq 3$, which shows that the next conditions are sufficient for an isolated set to be an attracting set:

1. all its periodic points and singularities are hyperbolic, and
2. it robustly contains the unstable manifold either of a periodic point or of an equilibrium point.

Before giving the proofs let us describe a global consequence of Theorem 5.3 which improves Theorem 5.2 (or Theorem 4.3).

Theorem 5.8 *A C^1 vector field on a 3-manifold having a robustly transitive non-wandering set is Anosov.*

Proof Let X be a C^1 vector field satisfying the conditions of the statement above, that is: $\Omega(X)$ is (an isolated set and) robustly transitive.

If $\Omega(X)$ has singularities, then $\Omega(X)$ is either a proper attractor or a proper repeller of X by Theorem 5.3, which is impossible by Lemma 2.1 from Sect. 2. Then $\Omega(X)$ is a robustly transitive set without singularities. By [79, 270] we conclude that $\Omega(X)$ is hyperbolic and so X is Axiom A with a unique basic set in its spectral decomposition. Since Axiom A vector fields always exhibit at least one attractor and $\Omega(X)$ is the unique basic set of X, it follows that $\Omega(X)$ is an attractor. By Lemma 2.1 again this implies that $\Omega(X)$ is the whole manifold.

Hence we are in the setting of Theorem 5.2 and we conclude that X is Anosov as desired. □

Remark 5.9 As observed after the proof of Lemma 2.1 in Sect. 2, the same argument shows that Theorem 5.8 remains true if one exchanges non-wandering set by limit set in its statement.

5.1.1 Equilibria of Robust Attractors Are Lorenz-Like

We say that an isolated set $\Lambda \subset M$ is *robustly singular* for $X \in \mathfrak{X}^1(M)$ if there is a neighborhood U of Λ in M and a C^1-neighborhood $\mathcal{U}$ of X in $\mathfrak{X}^1(M)$ such that $\Lambda_Y(U)$ contains a singularity for all $Y \in \mathcal{U}$.

Theorem 5.10 *Let Λ be a robustly singular transitive set of $X \in \mathfrak{X}^1(M)$. Then, for either $Y = X$ or $Y = -X$, every $\sigma \in S(Y) \cap \Lambda$ is Lorenz-like and satisfies $W_Y^{ss}(\sigma) \cap \Lambda = \{\sigma\}$.*

As a consequence, considering *robust attractors*, that is, attractors which persist for all C^1-nearby vector fields and remain transitive, we get

Theorem 5.11 *Every singularity of a robust attractor on a closed 3-manifold is Lorenz-like.*

5.1.2 Robust Attractors Are Singular-Hyperbolic

A compact invariant set Λ of X is *partially hyperbolic* if there are a continuous invariant tangent bundle decomposition $T_\Lambda M = E^s_\Lambda \oplus E^c_\Lambda$ and constants $\lambda, K > 0$ such that

- E^c_Λ (K,λ)*-dominates* E^s_Λ, i.e., for all $x \in \Lambda$ and for all $t \geq 0$

$$\|DX^t(x) \mid E^s_x\| \leq \frac{e^{-\lambda t}}{K} \cdot m(DX^t(x) \mid E^c_x); \tag{5.1}$$

- E^s_Λ is (K,λ)-contracting (see Sect. 2.3).

We shall say that $T_\Lambda M = E^s_\Lambda \oplus E^c_\Lambda$ is a (K,λ)*-splitting* for short. For $x \in \Lambda$ and $t \in \mathbb{R}$ we let $J^c_t(x)$ be the absolute value of the determinant of the linear map $DX^t(x) \mid E^c_x : E^c_x \to E^c_{X^t(x)}$. We say that the sub-bundle E^c_Λ of the partial hyperbolic set Λ is *volume expanding* if

$$J^c_t(x) = \left|\det(DX^t \mid E^c_x)\right| \geq K e^{\lambda t},$$

for every $x \in \Lambda$ and $t \geq 0$ (in this case we say that E^c_Λ is (K,λ)*-volume expanding* to indicate the dependence on (K,λ)).

It is known, see Theorem 2.27, that a non-singular partially hyperbolic set for a three-dimensional flow, with volume expanding central direction, is uniformly hyperbolic.

Definition 5.12 A partially hyperbolic set is *singular-hyperbolic* if its singularities are hyperbolic and it has volume expanding central direction.

A *singular-hyperbolic attractor* is a singular-hyperbolic set which is an attractor as well: an example is the (geometric) Lorenz attractor presented in Sect. 3.3. A *singular-hyperbolic repeller* of X is a singular-hyperbolic attractor of $-X$. An example of a singular-hyperbolic set which is neither an attractor nor a repeller is the singular horseshoe presented in Sect. 3.1.

We note that we can also say, in a 3-manifold, that an invariant subset is singular-hyperbolic if it is *volume hyperbolic*: that is, it admits a DX^t-invariant dominated splitting $E^{cs}_{\Lambda} \oplus E^{cu}_{\Lambda}$ such that the volume element in E^{cs}_{Λ} is uniformly contracted and the volume element in E^{cu}_{Λ} is uniformly expanded. Since one of these subbundles is one-dimensional, we recover the above notion of singular-hyperbolicity, either for X or for $-X$.

The following result characterizes robust attractors for three-dimensional flows.

Theorem 5.13 *Robust attractors of C^1 flows containing singularities are singular-hyperbolic sets for X.*

We remark that robust attractors cannot be C^1 approximated by vector fields presenting either attracting or repelling periodic points. This implies that, on 3-manifolds, any periodic point lying in a robust attractor is hyperbolic of saddle-type. Thus, as in Theorem 4.10 (see also Liao [135, Theorem A]), we conclude that robust attractors *without singularities* on closed 3-manifolds are hyperbolic. We can also use Theorem 2.27 and show that a singular-hyperbolic set without singularities admits a hyperbolic splitting for the Linear Poincaré Flow: this is done in Proposition 6.2 in Chap. 6. Therefore we obtain a dichotomy as follows.

Theorem 5.14 *Let Λ be a robust attractor of $X \in \mathfrak{X}^1(M)$. Then Λ is either hyperbolic or singular-hyperbolic.*

5.1.3 Brief Sketch of the Proofs

To prove Theorem 5.3 we first obtain a sufficient condition for a transitive isolated set with hyperbolic critical elements of a C^1 vector field on an n-manifold, $n \geq 3$, to be an attractor (Theorem 5.17). We use this to prove that a robustly transitive set whose critical elements are hyperbolic is an attractor if it contains a singularity whose unstable manifold has dimension one (Theorem 5.18). This implies that C^1 robustly transitive sets with singularities on closed 3-manifolds are either proper attractors or proper repellers (Theorem 5.3).

The characterization of singularities in a robust transitive set (Theorem 5.10) is obtained by contradiction. Using the Connecting Lemma (see Sect. 2.5.7), we can produce special types of cycles (inclination-flip or Shil'nikov, see Chap. 3) associated to a singularity leading to nearby vector fields which exhibit attracting or repelling periodic points. This contradicts the robustness of the transitivity condition.

Theorem 5.13 is proved in Sect. 5.4. We start by proposing an invariant splitting over the periodic points lying in Λ and prove uniform estimates on angles between stable, unstable, and central unstable bundles for periodic points. Roughly speaking, if such angles are not uniformly bounded away from zero, we construct a new vector field near the original one exhibiting either a sink or a repeller, yielding a contradiction. Such a perturbation is obtained using the extension for flows of a perturbation lemma of Franks, given by Theorem 2.24. This allows us to prove that the splitting proposed for the periodic points is partially hyperbolic with volume expanding central direction. We then extend this splitting to the closure of the periodic points. We show that the splitting proposed for the periodic points is compatible with the local partial hyperbolic splitting at the singularities (Proposition 5.41) using the fact that the Linear Poincaré Flow has a dominated splitting outside the singularities ([270, Theorem 3.8] stated as Theorem 2.33 in Sect. 2.6); and that the non-wandering set outside a neighborhood of the singularities is hyperbolic (Lemma 5.44). We next extend this splitting to all of Λ, obtaining Theorem 5.13.

5.2 Higher Dimensional Analogues

An example of a higher-dimensional invariant robust attractor with multidimensional expanding directions was given by Bonatti, Pumariño and Viana in [58], which we present below.

5.2.1 *Singular-Attractor with Arbitrary Number of Expanding Directions*

Consider a "solenoid" constructed over a uniformly expanding map $f : \mathbb{T}^k \to \mathbb{T}^k$ of the k-dimensional torus, for some $k \geq 2$. That is, let $\mathbb{D}$ be the unit disk on $\mathbb{R}^2$ and consider a smooth embedding $F : \mathbb{T}^k \times \mathbb{D} \to \mathbb{T}^k \times \mathbb{D}$ of $N = \mathbb{T}^k \times \mathbb{D}$ into itself, which preserves and contracts the foliation

$$\mathcal{F}^s = \left\{ \{z\} \times \mathbb{D} : z \in \mathbb{T}^k \right\},$$

and moreover the natural projection $\pi : N \to \mathbb{T}^k$ on the first factor conjugates F to f: $\pi \circ F = f \circ \pi$.

Now consider the linear flow over $M = N \times [0,1]/\sim$ given by the vector field $X = (0,1)$ on $TN \times \mathbb{R}$ where we make the identification $(x,0) \sim (x,1)$ for all $x \in N$. Modify the flow on a cylinder $U \times \mathbb{D} \times [0,1]$ around the orbit of a point $p = (z,0) \in N$, where U is a neighborhood of z in $\mathbb{T}^k$, in such a way as to create a hyperbolic singularity σ of saddle-type with k-expanding and 3 contracting eigenvalues, as depicted in Fig. 5.2.

This modified flow defines a transition map L from $\Sigma_0 = \mathbb{T}^k \times \{0\}$ to $\Sigma_1 = \mathbb{T}^k \times \{1\}$ which through the identification given by $(w,1) \sim_F (F(w),0)$ defines the return map to the global cross-section Σ_0 of a flow Y on the space $M^F = M/\sim_F$.

In [58] it is shown that, if the expanding rate of f is sufficiently big, then the set

$$\Lambda = \bigcup_{T>0} \overline{\bigcap_{t>T} Y_t(\Sigma_0)}$$

is a robust partially hyperbolic attractor with singularities.

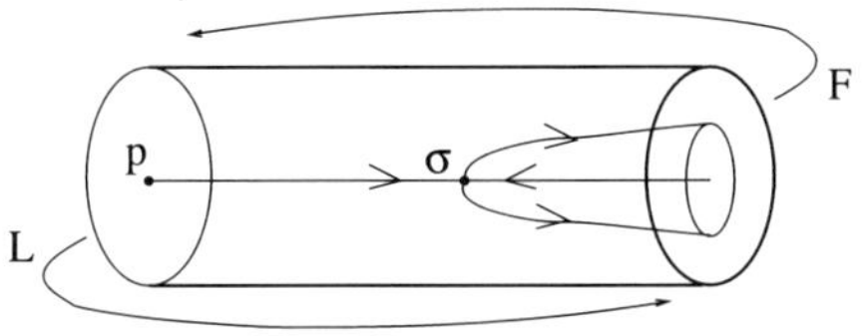

Fig. 5.2 A sketch of the construction of a robust singular-attractor in higher dimensions

5.2.2 The Notion of Sectionally Expanding Sets

Metzger and Morales in [156] introduced the notion of *sectionally expanding set* in a manifold of arbitrary finite dimension. This notion encompasses that of singular-hyperbolic sets in 3-manifolds as a particular case.

We say that a compact invariant set Λ for a flow, generated by a vector field $X \in \mathfrak{X}^1(M)$ on a compact finite dimensional manifold M, is *sectionally expanding* if it is partially hyperbolic and the central direction expands uniformly the area along any two-dimensional subspace. More precisely, the tangent bundle over Λ admits a DX^t-invariant and dominated splitting $T_\Lambda M = E^s \oplus E^c$, such that there are $C, \lambda > 0$ satisfying for every $x \in \Lambda$ and $t > 0$

- E^s is uniformly contracted: $\|DX^t \mid E^s_x\| \le Ce^{-\lambda t}$;
- E^c is sectionally expanded: for every bidimensional subspace F_x contained in E^c_x we have $|\det(DX^t \mid F_x)| \ge Ce^{\lambda t}$.

Similarly to the notion of singular-hyperbolicity, robust attractors in higher dimensional manifolds need not be sectionally expanding, as the example of Turaev and Shil'nikov is [263] shows.

5.2.3 Homogeneous Flows and Sectionally Expanding Attractors

However, strengthening the robust transitivity assumption with homogeneity (the definition of homogeneous flow on a neighborhood is Definition 2.32 in Sect. 2.6). enables us to essentially apply the same arguments of the proof of Theorem 5.13.

Theorem 5.15 *Let $\Lambda_X(U)$ be a homogeneous robust attractor for a vector field $X \in \mathfrak{X}^1(M^n)$ with $n \ge 3$ and index $i < n-1$ (the dimension of the stable manifold of periodic orbits in U for all nearby vector fields). Then any equilibrium in Λ is generalized Lorenz-like, with index $i+1$, and Λ is a sectionally expanding set.*

A proof of this result can be seen as an adaptation of the proof of Theorem 5.13, explained in several remarks during the rest of this chapter, following the proof of the three-dimensional case. For a partial result in this direction we mention Li, Gan and Wen, who have shown in [132] that *robustly transitive singular sets for homogeneous vector fields, with index i, having all hyperbolic singularities with index i, are partially hyperbolic sets with contracting bundle with dimension i.*

For a proof of Theorem 5.15 the reader can consult Metzger and Morales in [156], which refers to [132] for some technical higher-dimensional arguments.

5.3 Attractors and Isolated Sets for C^1 Flows

Here we prove Theorems 5.3 and 5.10. We start by focusing on isolated sets, obtaining the following sufficient conditions for an isolated set of a C^1 flow on an n-manifold, $n \ge 3$, to be an attractor:

- all its periodic points and singularities are hyperbolic, and
- it contains, in a robust way, the unstable manifold of some critical element.

Using this we prove that isolated sets Λ satisfying the following conditions are attractors:

- the critical elements $C(X) \cap \Lambda$ are hyperbolic;
- Λ contains a singularity with one-dimensional unstable manifold, and
- Λ is
  - either robustly non-trivial and transitive (robustly transitive),
  - or $\Lambda = \overline{C(X) \cap \Lambda}$ is robustly the closure of its periodic points (C^1 robustly periodic).

In particular robustly transitive sets with singularities on closed 3-manifolds are either proper attractors or proper repellers, proving Theorem 5.3. Then we characterize the singularities on robustly transitive sets on 3-manifolds, obtaining Theorem 5.10.

Elementary topological dynamics ensures that an attractor containing a hyperbolic critical element contains the unstable manifold of this critical element. The converse, although false in general, is true for a residual subset of C^1 vector fields, as shown in [67]. We derive a sufficient condition for the converse to hold inspired by the following property of uniformly hyperbolic attractors (see e.g. [193]): if Λ is a uniformly hyperbolic attractor of a vector field X, then there is an isolating block U of Λ and $x_0 \in C(X) \cap \Lambda$ such that $W^u_Y(x_0(Y)) \subset U$ for every Y close to X, where $x_0(Y)$ is the hyperbolic continuation of x_0 for Y. This property motivates the following definition.

Definition 5.16 Let Λ be an isolated set of $X \in \mathfrak{X}^r(M)$, $r \geq 1$. We say that Λ *robustly contains the unstable manifold of a critical element* if there are $x_0 \in C(X) \cap \Lambda$ hyperbolic, an isolating block U of Λ and a neighborhood $\mathcal{U}$ of X in $\mathfrak{X}^r(M)$ such that $W^u_Y(x_0(Y)) \subset U$, for all $Y \in \mathcal{U}$.

With this definition in mind we are able to prove

Theorem 5.17 *Let Λ be a transitive isolated set of $X \in \mathfrak{X}^1(M)$ where M is a compact n-manifold, $n \geq 3$, and suppose that every $x \in C(X) \cap \Lambda$ is hyperbolic. If Λ robustly contains the unstable manifold of a critical element, then Λ is an attractor.*

Now we derive an application of Theorem 5.17.

The geometric Lorenz attractor is a robustly transitive (periodic) set, and it is an attractor satisfying (see Sect. 3.3)

- all its periodic points are hyperbolic, and
- it contains a singularity whose unstable manifold has dimension one.

We say that an isolated set $\Lambda_X(U)$ is C^1 *robustly periodic* if for all vector fields Y in a C^1-neighborhood of X the periodic orbits of $\Lambda_Y(U)$ are dense; see Definition 6.4.

The result below shows that such conditions are enough for a robustly transitive (periodic) set to be an attractor.

Theorem 5.18 *Let Λ be either a robustly transitive or a transitive C^1 robustly periodic set of $X \in \mathfrak{X}^1(M)$, where M is an n-dimensional compact manifold, $n \geq 3$. If*

1. *every $x \in C(X) \cap \Lambda$ is hyperbolic and*
2. *Λ has a singularity whose unstable manifold is one-dimensional,*

then Λ is an attractor of X.

Theorem 5.18 follows from Theorem 5.17 showing that Λ robustly contains the unstable manifold of the singularity provided by condition 2 above.

5.3.1 Proof of Sufficient Conditions to Obtain Attractors

The proof of Theorem 5.17 is based on the following lemma.

Lemma 5.19 *Let Λ be a transitive isolated set of $X \in \mathfrak{X}^1(M)$ such that every $x \in C(X) \cap \Lambda$ is hyperbolic. Suppose that the following condition holds:*

(H3) *There are $x_0 \in C(X) \cap \Lambda$, an isolating block U of Λ and a neighborhood $\mathscr{U}$ of X in $\mathfrak{X}^1(M)$ such that*

$$W^u_Y(x_0(Y)) \subset U, \quad \forall Y \in \mathscr{U}.$$

Then $W^u_X(x) \subset \Lambda$ for every $x \in C(X) \cap \Lambda$.

Proof Let x_0, U and $\mathscr{U}$ as in (H3). By assumption $\mathscr{O}_X(x_0)$ is hyperbolic. If $\mathscr{O}_X(x_0)$ is attracting then $\Lambda = \mathscr{O}_X(x_0)$ since Λ is transitive and we are done. We can then assume that $\mathscr{O}_X(x_0)$ is not attracting. Thus, $W^u_X(x_0) \setminus \mathscr{O}_X(x_0) \neq \emptyset$.

By contradiction, suppose that there is $x \in C(X) \cap \Lambda$ such that $W^u_X(x)$ is not contained in Λ. Then $W^u_X(x)$ is not contained in $\overline{U}$. As $M \setminus \overline{U}$ is open there is a cross-section $\Sigma \subset M \setminus \overline{U}$ of X such that $W^u_X(x) \cap \Sigma \neq \emptyset$ is transverse. Shrinking $\mathscr{U}$ if necessary we can assume that $W^u_Z(x(Z)) \cap \Sigma \neq \emptyset$ is transverse for every $Z \in \mathscr{U}$.

Now $W^u_X(x_0) \subset \Lambda$ by (H3) applied to $Y = X$. Choose $p \in W^u_X(x_0) \setminus \mathscr{O}_X(x_0)$. As Λ is transitive and $p, x \in \Lambda$, there is $q \in W^s_X(x) \setminus \mathscr{O}_X(x)$ such that p, q satisfy (H1) in Theorem 2.21 (the Connecting Lemma). Indeed, the dense orbit of Λ accumulates both p and x. Then, by Theorem 2.21, there are $Z \in \mathscr{U}$ and $T > 0$ such that $p \in W^u_Z(x(Z))$, $q \in W^s_Z(x(Z))$ and $Z_T(p) = q$. In other words, $\mathscr{O}_Z(q)$ is a saddle connection between $x_0(Z)$ and $x(Z)$. On the other hand, as $Z \in \mathscr{U}$, we have that $W^u_Z(x(Z)) \cap \Sigma \neq \emptyset$ is transverse. It follows from the λ-Lemma (see Sect. 2.5.4 of Chap. 1) that $Z^t(\Sigma)$ accumulates on q as $t \to \infty$. This allows us to break the saddle-connection $\mathscr{O}_Z(q)$ in the standard way in order to find $Z' \in \mathscr{U}$ such that $W^u_{Z'}(x_0(Z')) \cap \Sigma \neq \emptyset$ (see e.g. [190] or the proof of Theorem 3.6 in [167]). In particular, $W^u_{Z'}(x_0(Z'))$ is not contained in U. This contradicts (H3) and the lemma follows. □

Proof of Theorem 5.17 Let Λ and X be as in the statement of Theorem 5.17. It follows that there are $x_0 \in C(X) \cap \Lambda$, U and $\mathcal{U}$ such that (H3) holds.

Next we prove that Λ satisfies the hypothesis of Lemma 2.5, that is, there is an open set W containing Λ such that $X^t(W) \subset U$ for every $t \geq 0$.

Indeed, suppose that such a W does not exist. Then there are sequences $x_n \to x \in \Lambda$ and $t_n > 0$ such that $X^{t_n}(x_n) \in M \setminus U$. By compactness we can assume that $X^{t_n}(x_n) \to q$ for some $q \in \overline{M \setminus U}$.

Fix an open set $V \subset \overline{V} \subset U$ containing Λ. As $q \in \overline{M \setminus U}$,

$$\overline{M \setminus U} \subset M \setminus \text{int}(U), \quad \text{and} \quad M \setminus \text{int}(U) \subset M \setminus \overline{V}$$

we have $q \notin \overline{V}$. By Lemma 2.3 there is a neighborhood $\mathcal{U}_0 \subset \mathcal{U}$ of X such that

$$\Lambda_Y(U) \subset V, \quad \text{for all } Y \in \mathcal{U}_0. \tag{5.2}$$

Then condition (H3), the invariance of $W^u_Y(x_0(Y))$ and the relation (5.2) imply that

$$W^u_Y(x_0(Y)) \subset V \subset \overline{V}, \quad \text{for every } Y \in \mathcal{U}_0. \tag{5.3}$$

Now we have two cases:

1. either $x \notin C(X)$;
2. or $x \in C(X)$.

In Case 1 we obtain a contradiction as follows. Let $\mathcal{O}_X(z)$ be the dense orbit of Λ, i.e., $\Lambda = \omega_X(z)$. Fix $p \in W^u_X(x_0) \setminus \mathcal{O}_X(x_0)$. Then $p \in \Lambda$ by (H3) applied to $Y = X$. As $x \in \Lambda$ we can choose sequences $z_n \in \mathcal{O}_X(z)$ and $t'_n > 0$ such that

$$z_n \to p \quad \text{and} \quad X^{t'_n}(z_n) \to x.$$

It follows that p, q, x satisfy (H2) of Theorem 2.22 for $Y = X$. Then from Theorem 2.22 there is $Z \in \mathcal{U}_0$ such that $q \in W^u_Z(x_0(Z))$. As $q \notin \overline{V}$ we have that $W^u_Z(x_0(Z))$ is not contained in U. And this is a contradiction by (5.3) since $Z \in \mathcal{U}_0$.

In Case 2 we use (H3) to obtain a contradiction as follows. By assumption $\mathcal{O}_X(x)$ is a hyperbolic closed orbit. Clearly $\mathcal{O}_X(x)$ is neither attracting nor repelling. In particular $W^u_X(x) \setminus \mathcal{O}_X(x) \neq \emptyset$. But $x_n \notin W^s_X(x)$ since $x_n \to x$ and $X^{t_n}(x_n) \notin U$. Then, using linearizing coordinates given by the Grobman-Hartman Theorem around $\mathcal{O}_X(x)$ (see Sect. 2.5.3), we can find x'_n in the positive orbit of x_n such that $x'_n \to r \in W^u_X(x) \setminus \mathcal{O}_X(x)$. Note that $r \notin C(X)$ and that there are $t'_n > 0$ such that $X^{t'_n}(x'_n) \to q$.

Since (H3) holds, by Lemma 5.19 we have $W^u_X(x) \subset \Lambda$. This implies that $r \in \Lambda$. Then we have Case 1 replacing x by r, t_n by t'_n and x_n by x'_n. As Case 1 results in a contradiction, we conclude that Case 2 also results in a contradiction.

Hence Λ satisfies the hypothesis of Lemma 2.5, and Theorem 5.17 follows. □

Proof of Theorem 5.18 Let Λ be either a robust transitive set or a transitive C^1 robust periodic set of $X \in \mathfrak{X}^1(M)$ satisfying the following conditions:

1. Every critical element of X in Λ is hyperbolic.
2. Λ contains a singularity σ with $\dim(W^u_X(\sigma)) = 1$.

On the one hand, if Λ is robustly transitive, we can by Definition 5.1 fix a neighborhood $\mathcal{U}$ of X and an isolating block U of Λ such that $\Lambda_Y(U)$ is a non-trivial transitive set of Y, for every $Y \in \mathcal{U}$. Clearly we can assume that the continuation $\sigma(Y)$ is well defined for all $Y \in \mathcal{U}$. Since transitive sets are connected sets, we have:

(C) $\Lambda_Y(U)$ is connected for each $Y \in \mathcal{U}$.

On the other hand, if Λ is C^1 robustly periodic, we can fix by Definition 6.4 a neighborhood $\mathcal{U}$ of X and an isolating block U of Λ such that for each $Y \in \mathcal{U}$ we have $\Lambda_Y(U) = \overline{\mathrm{Per}(Y) \cap \Lambda_Y(U)}$. Assuming that $\sigma(Y)$ is well defined for $Y \in \mathcal{U}$ we have

(C') $\sigma(Y) \in \overline{\mathrm{Per}(Y) \cap \Lambda_Y(U)}$, for every $Y \in \mathcal{U}$.

Claim The set Λ robustly contains the unstable manifold of a critical element.

By Definition 5.16, if $\mathcal{U}$ is the neighborhood of X described in either Property (C) or (C'), then it suffices to prove $W^u_Y(\sigma(Y)) \subset \overline{U}$ for all $Y \in \mathcal{U}$.

Arguing by contradiction, suppose that there exists $Y \in \mathcal{U}$ such that $W^u_Y(\sigma(Y))$ *is not contained in* U.

From Condition 2 above it follows that $W^u_X(\sigma) \setminus \{\sigma\}$ has two branches which we denote by w^+ and w^- respectively. Fix $q^+ \in w^+$ and $q^- \in w^-$. Denote by $q^\pm(Y)$ the continuation of $q^\pm$ for Y close to X. We can assume that the $q^\pm(Y)$ are well defined for all $Y \in \mathcal{U}$.

As $q^\pm(Y) \in W^u_Y(\sigma(Y))$, the negative orbit of $q^\pm(Y)$ converges to $\sigma(Y) \in \mathrm{int}(U) \subset U$. If the positive orbit of $q^\pm(Y)$ is in U, then $W^u_Y(\sigma(Y)) \subset U$, which is a contradiction. Consequently the positive orbit of either $q^+(Y)$ or $q^-(Y)$ leaves U. It follows that there is $t > 0$ such that either $Y^t(q^+(Y))$ or $Y^t(q^-(Y)) \notin U$. Assume the first case. The other case is analogous. As $M \setminus U$ is open, the continuous dependence of the unstable manifolds implies that there is a neighborhood $\mathcal{U}' \subset \mathcal{U}$ of Y such that

$$Z^t(q^+(Z)) \notin U, \quad \text{for every } Z \in \mathcal{U}'. \tag{5.4}$$

Now we split the proof of the claim into two cases.

Case I: Λ is robustly transitive.

In this case $\Lambda_Y(U)$ is a non-trivial transitive set of Y. Fix $z \in \Lambda_Y(U)$ such that $\omega_Y(z) = \Lambda_Y(U)$. As $\sigma(Y) \in \Lambda_Y(U)$ it follows that either $q^+(Y)$ or $q^-(Y) \in \omega_Y(z)$. As $Y \in \mathcal{U}'$, the relation (5.4) implies that $q^-(Y) \in \omega_Y(z)$. Thus, there is a sequence $z_n \in \mathcal{O}_Y(z)$ converging to $q^-(Y)$. Similarly there is a sequence $t_n > 0$ such that $Y^{t_n}(z_n) \to q$ for some $q \in W^s_Y(\sigma(Y) \setminus \{\sigma(Y)\}$. Define $p = q^-(Y)$.

It follows that p, q, Y satisfy (H1) in Theorem 2.21, and so there is $Z \in \mathscr{U}'$ such that $q^-(Z) \in W^s_Z(\sigma(Z))$. This gives a homoclinic connection associated to $\sigma(Z)$. Breaking this connection as in the proof of Lemma 5.19, we can find $Z' \in \mathscr{U}'$ close to Z and $t' > 0$ such that

$$Z'^{t'}(q^-(Z')) \notin U. \tag{5.5}$$

Now, (5.4) and (5.5) together with the Grobman-Harman Theorem 2.14 imply that the set $\{\sigma(Z')\}$ is isolated in $\Lambda_Z(U)$. But $\Lambda_{Z'}(U)$ is connected by Property (C) since $Z' \in \mathscr{U}' \subset \mathscr{U}$. Then $\Lambda_{Z'}(U) = \{\sigma(Z')\}$, a contradiction since $\Lambda_{Z'}(U)$ is non-trivial. This proves the claim in this case.

Case II: Λ is C^1 robustly periodic.

The proof is similar to the previous one. In this case $\Lambda_Y(U)$ is the closure of its periodic orbits and $\dim(W^u_Y(\sigma(Y)) = 1$. As the periodic points of $\Lambda_Y(U)$ do accumulate either $q^+(Y)$ or $q^-(Y)$, relation (5.4) implies that there is a sequence $p_n \in \mathrm{Per}(Y) \cap \Lambda_Y(U)$ such that $p_n \to q^-(Y)$. Clearly there is another sequence $p'_n \in \mathscr{O}_Y(p_n)$ now converging to some $q \in W^s_Y(\sigma(Y) \setminus \{\sigma(Y)\}$. Set $p = q^-(Y)$. Again p, q, Y satisfy (H1) in Theorem 2.21, and so there is $Z \in \mathscr{U}'$ such that $q^-(Z) \in W^s_Z(\sigma(Z))$. As before we have a homoclinic connection associated to $\sigma(Z)$. Breaking this connection we can find $Z' \in \mathscr{U}'$ close to Z and $t' > 0$ such that

$$Z'^{t'}(q^-(Z')) \notin U.$$

Again this relation together with the Grobman-Harman Theorem 2.14 and the relation (5.4) would imply that every periodic point of Z' passing close to $\sigma(Z')$ is not contained in $\Lambda_{Z'}(U)$. But this contradicts Property (C') since $Z' \in \mathscr{U}' \subset \mathscr{U}$. This completes the proof of the claim in this case.

It follows that Λ is an attractor by condition (1) above, Theorem 5.17 and by the claim. This completes the proof of Theorem 5.18. □

5.3.2 *Robust Singular Transitivity Implies Attractors or Repellers*

In this section M is a closed 3-manifold and Λ is a robustly transitive set of $X \in \mathfrak{X}^1(M)$.

According to Definition 5.1 we can fix an isolating block U of Λ and a neighborhood $\mathscr{U}_U$ of X such that $\Lambda_Y(U) = \cap_{t\in\mathbb{R}} Y^t(U)$ is a non-trivial transitive set of Y, for every $Y \in \mathscr{U}_U$. Robustness of transitivity implies that $X \in \mathscr{U}_U$ cannot be C^1-approximated by vector fields exhibiting either sinks or sources in U. And since $\dim(M) = 3$ this easily implies the following.

Lemma 5.20 *Let $X \in \mathscr{U}_U$. Then X has neither sinks nor sources in U, and any $p \in \mathrm{Per}(X) \cap \Lambda_X(U)$ is hyperbolic.*

The following result shows that singularities in this setting are Lorenz-like, either for the given flow X or for the reversed flow $-X$. We need the following definition.

Definition 5.21 A hyperbolic singularity σ belonging to a non-trivial compact invariant set Λ for a C^1 flow X^t is *properly accumulated by regular orbits* in Λ if there exists a sequence $(\gamma_n)_{n\in\mathbb{N}}$ of distinct regular orbits contained in Λ such that

- $\alpha(\gamma_n) \neq \{\sigma\}$ and $\omega(\gamma_n) \neq \{\sigma\}$ for all $n \in \mathbb{N}$;
- for every neighborhood V of σ there are points $p_n, q_n \in V \cap \gamma_n$ such that $q_n = X^{t_n}$ for some $t_n > 0$ and $n \in \mathbb{N}$, and $X^{[0,t_n]}(p_n)$ is not contained in V.

Observe that a hyperbolic singularity σ properly accumulated by regular orbits will have points in $\Lambda \cap W^s(\sigma) \setminus \{\sigma\}$ and in $\Lambda \cap W^u(\sigma) \setminus \{\sigma\}$ accumulated by distinct regular orbits of Λ.

For the next result the reader should recall Definition 2.32 of a homogeneous vector field from Chap. 2.

Lemma 5.22 *Let $Y \in \mathfrak{X}^1(M)$ be a homogeneous vector field in U on a 3-manifold and $\sigma \in S(Y) \cap \Lambda_Y(U)$ be properly accumulated by regular orbits in $\Lambda = \Lambda_Y(U)$. Then*

1. *the eigenvalues of σ are real;*
2. *if $\lambda_2 \le \lambda_3 \le \lambda_1$ are the eigenvalues of σ, then $\lambda_2 < 0 < \lambda_1$;*
3. *for λ_i as above we have*
   (a) $\lambda_3(\sigma) < 0 \Longrightarrow -\lambda_3(\sigma) < \lambda_1(\sigma)$;
   (b) $\lambda_3(\sigma) > 0 \Longrightarrow -\lambda_3(\sigma) > \lambda_2(\sigma)$.

Note that if $\Lambda(U) = \Lambda_X(U)$ is robustly transitive, then the homogeneity condition is satisfied for all Y C^1-close to X; this is just Lemma 5.20. In addition if $\Lambda_Y(U)$ is transitive, then every equilibrium of $\Lambda_Y(U)$ is properly accumulated by regular orbits. Thus $Y \in \mathcal{U}_U$ satisfies the condition of the lemma.

Remark 5.23 For an example of a singular-hyperbolic isolated set of a flow X with non-Lorenz-like singularities, consider the maximal invariant set L inside the ellipsoid E of the flow described in Fig. 3.16. Note that the singularities σ_1, σ_2 with complex eigenvalues inside the ellipsoid belong to L together with orbits inside their unstable manifolds connecting them with the Lorenz attractor. Hence σ_1, σ_2 are accumulated by regular orbits of L but are *not properly accumulated* since there are no orbits γ_n in L approaching σ_i with $\alpha(\gamma_n) \neq \{\sigma_i\}$ for $i = 1, 2$.

Remark 5.24 The example give in Fig. 9.1, in Chap. 9, provides an example of a singular-hyperbolic attractor Λ, obtained through a modification of the construction of the geometric Lorenz attractor, *with a singularity properly accumulated by regular orbits which is not hyperbolic*. This is explained in Remark 9.10. The vector field in this example is non-homogeneous and is not Kupka-Smale. We note that this does not contradict item 3 of Lemma 5.22, since this item does not specify what happens when $\lambda_3(\sigma) = 0$.

Remark 5.25 Assume that we are given a nonempty compact invariant isolated set $\Lambda = \Lambda_X(U)$ under a flow on a 3-manifold, which is partially hyperbolic with volume expanding central direction. Since partial hyperbolicity is a robust property, then for every close flow Y we have that $\Lambda_Y(U)$ is also partially hyperbolic. This implies that there are no sources in $\Lambda_Y(U)$. The uniform volume expansion along the central direction of $T_\Lambda M$ for X implies that there are no sinks in $\Lambda_Y(U)$, for otherwise we would get volume contraction along the central direction for points and flows arbitrary close to Λ and X. This is a contradiction since dominated splittings depend continuously on the base point and on the dynamics, thus taking limits we obtain a point in Λ with central direction whose volume is contracted by the X flow.

Hence in this setting the flow is homogeneous in U and we conclude that *every hyperbolic singularity properly accumulated by regular orbits of a singular-hyperbolic isolated set of a flow X is either Lorenz-like for X, or Lorenz-like for $-X$*. Note that we need to assume hyperbolicity of the singularity by Remark 5.24.

Proof of Lemma 5.22 Let us prove the first item by contradiction. Suppose that there are $Y \in \mathscr{U}_U$ and $\sigma \in S(Y) \cap \Lambda_Y(U)$ with a complex eigenvalue ω. We can assume that σ is hyperbolic by Lemma 5.20 (or, in a higher dimensional setting, by the homogeneity assumption on U). As $\dim(M) = 3$ the remaining eigenvalue λ of σ is real. We have either $Re(\omega) < 0 < \lambda$ or $\lambda < 0 < Re(\omega)$. Reversing the flow direction if necessary we can assume that we are in the first case. We can further assume, by a small perturbation keeping the vector field inside $\mathscr{U}_U$, that Y is C^∞ and

$$\lambda \neq -Re(\omega). \tag{5.6}$$

According to a form of the Connecting Lemma stated in Theorem 2.21, we can assume that there is a homoclinic loop $\Gamma \subset \Lambda_Y(U)$ associated to σ. Then Γ is a *Shil'nikov bifurcation*, see Sect. 3.2, and thus there is a vector field Z arbitrarily C^1 close to Y exhibiting a sink or a source in $\Lambda_Z(U)$. This contradicts Lemma 5.20 and concludes the proof of the first item.

Thus we can arrange the eigenvalues λ_1, λ_2, λ_3 of σ in such a way that $\lambda_2 \le \lambda_3 \le \lambda_1$. By Lemma 5.20 we have that $\lambda_2 < 0$ and $\lambda_1 > 0$. This proves the second item in the statement.

To prove the third item we can apply Theorem 3.4 from Sect. 3.2. This shows that if item (a) fails, there is Z arbitrarily C^1 close to Y exhibiting a sink in $\Lambda_Z(U)$; or if item (b) fails, we can find a source in $\Lambda_Z(U)$. Either case is a contradiction as before, concluding the proof of the lemma. □

Remark 5.26 The same proof above extends to homogeneous attracting sets in higher dimensions, since we can always argue with the least contracting/expanding eigenvalues of σ through a reduction to the corresponding central manifold, see the discussion in Sect. 3.2.3. We argue in a slightly different way from Metzger and Morales in [156]. We first recall that in a homogeneous vector field equilibria must be hyperbolic by definition. Keeping the notations introduced in Definition 2.30 of generalized Lorenz-like singularity properly accumulated by regular orbits in Λ, we can show that:

1. the contracting and expanding eigenvalues with smallest real part are real and $\lambda < 0 < \mu$.

   Indeed, on the one hand, if both the contracting eigenvalue with smallest real part and the expanding eigenvalue with smallest real part are complex, then we can, after a small perturbation of X using that σ is properly accumulated by regular orbits in Λ, obtain a double-focus connection for σ such that the eigenvalues with smallest real part are simple (they are not multiple zeroes of the characteristic polynomial of $DX(\sigma)$). Moreover, since regular orbits accumulating on σ must do so along the central direction given by the sum of the eigenspaces of these weak eigenvalues, we can obtain such a double-focus saddle-connection along a central manifold corresponding to the weakest eigenvalues. This type of connection bifurcates in a way that new periodic orbits appear with different indexes, which contradicts the homogeneous assumption on X, see Sect. 3.2.3.

   On the other hand, if one of these eigenvalues ω is complex and the other λ is real, we can argue as in the three-dimensional proof of Lemma 5.22 since their real parts have opposite signs, restricting the dynamics to the central manifold corresponding to λ, ω.

   Hence both such eigenvalues are real, as claimed.
2. a. either $\lambda^-(\sigma) \neq \emptyset \implies -\lambda < \mu$;
   b. or $\lambda^-(\sigma) = \emptyset \implies \lambda < -\mu$.

   Indeed, if $\lambda^-(\sigma) \neq \emptyset$, then we can obtain a saddle-connection associated to σ by a perturbation through the Connecting Lemma, using the fact that σ is properly accumulated by regular orbits, connecting the one-dimensional unstable manifold with the one-dimensional central-stable (weak-stable) manifold. The bifurcation of this saddle-connection provides periodic orbits with different indices for nearby flows if $\lambda + \mu \leq 0$, by the same arguments proving Theorem 3.4.

   For the other case, just note that if $\lambda^-(\sigma) = \emptyset$, then $\mu^+(\sigma) \neq \emptyset$ and we find ourselves in the same setting as above after reversing time.

Summing up: *given an attracting set Λ of a homogeneous vector field in a finite dimensional manifold, then every equilibrium σ properly accumulated by regular orbits within Λ is generalized Lorenz-like either for X or for $-X$.* See also Lemma 5.32 in a 3-dimensional setting.

The following fact is extremely useful.

Lemma 5.27 *There is no $Y \in \mathscr{U}_U$ exhibiting two hyperbolic singularities in $\Lambda_Y(U)$ with different unstable manifold dimensions.*

Proof Suppose by contradiction that there is $Y \in \mathscr{U}_U$ exhibiting two hyperbolic singularities with different unstable manifold dimensions in $\Lambda_Y(U)$. Note that $\Lambda' = \Lambda_Y(U)$ is a robust transitive set of Y and $-Y$ respectively. Since Kupka-Smale vector fields are generic (by the results in Sect. 2.5.10) we can assume that all the critical elements of Y in Λ' are hyperbolic.

As $\dim(M) = 3$ and Y has two hyperbolic singularities with different unstable manifold dimensions, it follows that both Y and $-Y$ have a singularity in Λ' whose unstable manifold has dimension one. Then, by Theorem 5.18 applied to Y and $-Y$ respectively, Λ' is a proper attractor and a proper repeller of Y. In particular, Λ' is an attracting set and a repelling set of Y. It would follow from Lemma 2.4 that $\Lambda' = M$. But this is a contradiction since Λ' is proper. □

From this we can derive the following.

Corollary 5.28 *If $Y \in \mathscr{U}_U$, then every critical element of Y in $\Lambda_Y(U)$ is hyperbolic.*

Proof By Lemma 5.20 every periodic point of Y in $\Lambda_Y(U)$ is hyperbolic, for all $Y \in \mathscr{U}$. It remains to prove that every $\sigma \in S(Y) \cap \Lambda_Y(U)$ is hyperbolic, for all $Y \in \mathscr{U}_U$. By Lemma 5.22 the eigenvalues $\lambda_1, \lambda_2, \lambda_3$ of σ are real and satisfy $\lambda_2 < 0 < \lambda_1$. Then, to prove that σ is hyperbolic, we only have to prove that $\lambda_3 \neq 0$. If $\lambda_3 = 0$, then σ is a generic saddle-node singularity (after a small perturbation if necessary). Unfolding this saddle-node we obtain $Y' \in \mathscr{U}_U$ close to Y having two hyperbolic singularities with different unstable manifold dimensions in $\Lambda_{Y'}(U)$. This contradicts Lemma 5.27 and the proof follows. □

Finally we can start proving the main theorems.

Proof of Theorem 5.3 Let Λ be a robustly transitive set with singularities of $X \in \mathscr{X}^1(M)$ with $\dim(M) = 3$. By Corollary 5.28 applied to $Y = X$ we have that every critical element of X in Λ is hyperbolic. So Λ satisfies condition (1) of Theorem 5.18. As $\dim(M) = 3$ and Λ is non-trivial, if Λ has a singularity, then this singularity has unstable manifold dimension equal to one, for either X or $-X$. So Λ also satisfies condition (2) of Theorem 5.18, for either X or $-X$. Applying Theorem 5.18 we have that Λ is an attractor (in the first case) or a repeller (in the second case).

We shall prove that Λ is proper in the first case. The proof is similar in the second case. If $\Lambda = M$ then we would have $U = M$. From this it would follow that $\Omega(X) = M$ and, moreover, that X cannot be C^1 approximated by vector fields exhibiting attracting or repelling critical elements. It would follow from Theorem 4.3 that X is Anosov. But this is a contradiction since Λ (and so X) has a singularity and Anosov vector fields do not. This finishes the proof of Theorem 5.3. □

Now we prove Theorem 5.10. We start with the following corollary.

Corollary 5.29 *If $Y \in \mathscr{U}_U$ then, for either $Z = Y$ or $Z = -Y$, every singularity of Z in $\Lambda_Z(U)$ is Lorenz-like.*

Proof Apply Lemmas 5.22, 5.27 and Corollary 5.28. We deduce that any singularity, for $Y = Z$ or $Y = -Z$, has real eigenvalues satisfying $\lambda_2 \leq \lambda_3 < 0 < -\lambda_3 < \lambda_1$. To conclude that $\lambda_2 < \lambda_3$ for such σ we argue by contradiction. If $\lambda_2 = \lambda_3$, then by an arbitrarily small perturbation of Y, we obtain a vector field Y_1 for which the continuation σ_1 of σ has eigenvalues λ and $-\rho \pm i\omega$ with $\lambda > 0$, $\rho > 0$ and $|\omega|$ arbitrarily small. But we can ensure that $Y_1 \in \mathscr{U}_U$, and thus the transitivity of $\Lambda_{Y_1}(U)$

enables us to use the Connecting Lemma to get another C^1-close vector field Y_2 having a saddle-connection for the continuation σ_2 of σ_1 in the setting of the Shilnikov bifurcation (see Sect. 3.2). In this way we obtain either a sink or a source for a vector field $X \in \mathcal{U}_U$, a contradiction. □

Now we use the existence of dominated splitting for the Linear Poincaré Flow with respect to $X \in \mathcal{U}_U$; see Sect. 2.6 for the relevant results and definitions.

Given $X \in \mathcal{U}_U$ define $\Lambda_X^*(U) = \Lambda_X(U) \setminus S(X)$. According to Theorem 5.3 we can assume that $\Lambda_X(U)$ is a proper and isolated attractor of X. Using Lemma 5.20 and the fact that $\Lambda_X^*(U) \subset \Omega(X)$, we see that we are in the setting of Theorem 2.33. Then we conclude that the Linear Poincaré Flow P^t on $\Lambda_X^*(U)$ admits a partially hyperbolic splitting: $N_{\Lambda_X^*(U)} = N^{s,X} \oplus N^{u,X}$.

The following consequence of this is used in a crucial way for the proof of expansiveness in Chap. 7.

Lemma 5.30 *Let Λ be a compact isolated and transitive invariant set for X, with isolating neighborhood U such that every C^1-close vector field admits a dominated splitting for the corresponding Linear Poincaré Flow on U away from singularities. Fix $\sigma \in S(X) \cap \Lambda$ and write $\lambda_2 < \lambda_3 < \lambda_1$ for its eigenvalues.*

1. *If $\lambda_2 < \lambda_3 < 0$, then σ is Lorenz-like for X and $W_X^{ss}(\sigma) \cap \Lambda = \{\sigma\}$.*
2. *If $0 < \lambda_3 < \lambda_1$, then σ is Lorenz-like for $-X$ and $W_X^{uu}(\sigma) \cap \Lambda = \{\sigma\}$.*

Remark 5.31 If we are given a singular-hyperbolic isolated set Λ for a flow X with isolating neighborhood U then, by Remark 5.25, the singularities of Λ, which are *properly accumulated by regular orbits* in Λ, are Lorenz-like for either X or $-X$. Moreover the Linear Poincaré Flow on $\Lambda^* = \Lambda \setminus S(X)$ admits a partially hyperbolic splitting naturally. Indeed the Linear Poincaré Flow is dominated by Theorem 2.33 (since singular-hyperbolicity prevents sinks and sources for nearby flows on U and guarantees hyperbolicity of all critical elements in U) and its central-stable bundle is uniformly contracted by the same argument as in the proof of Proposition 6.2.

In addition, for all close enough vector fields Y the corresponding locally maximal invariant subset $\Lambda_Y(U)$ is also partially hyperbolic with volume expanding central direction, and so the domination property of the splitting for the Linear Poincaré Flow of X on Λ is robust.

Hence we have the same properties used in the proof of Lemma 5.30. We conclude that *every singularity properly accumulated by regular orbits in a singular-hyperbolic isolated set satisfies either item* 1 *or item* 2 *of Lemma* 5.30 *above.*

We can improve and eliminate the possible Lorenz-like singularities for $-X$ if we assume from the beginning that Λ is an attractor. This lemma will be very useful in the next chapters.

Lemma 5.32 *Let $X \in \mathfrak{X}^1(M)$, with M a 3-manifold, and let Λ be an attractor for X such that the Linear Poincaré Flow over Λ^* admits a dominated splitting. Then the only possible singularities of X in Λ are Lorenz-like for X.*

Proof We recall that the dominated splitting for the Linear Poincaré Flow extends to periodic orbits on a small neighborhood U of Λ and for every vector field Y sufficiently C^1-close to X. Now we argue by contradiction: we know from Lemma 5.30 that if $\sigma \in \Lambda \cap S(X)$ then σ is a Lorenz-like singularity, either for X or for $-X$, and we assume it is Lorenz-like for $-X$ and derive a contradiction.

The bidimensional unstable manifold $W^u_X(\sigma)$ is inside Λ because Λ is an attracting set. The transitivity of Λ ensures that there exists some point p of $\Lambda \setminus \{\sigma\}$ in $W^s_X(\sigma)$, because regular orbits accumulating on σ must accumulate also arbitrarily near $W^s_X(\sigma)$ by the linearization of the flow near σ. Considering a point q of $\Lambda \setminus \{\sigma\}$ in $W^{uu}_X(\sigma)$ (the strong-unstable manifold of σ), then there are regular orbits in Λ passing arbitrarily close to p and then to q, since $W^{uu}(\sigma) \subset \Lambda$ and Λ is transitive. The Connecting Lemma ensures that we can find an arbitrarily close C^1 vector field Y admitting a regular orbit in a neighborhood U of Λ connecting $W^s_Y(\sigma_Y)$ and $W^{uu}_Y(\sigma_Y)$. This saddle-connection can be transformed into an inclination-flip type saddle-connection as in the proof of Lemma 5.30, by an arbitrarily small C^1 perturbation of Y. This would contradict the robustness of dominated decomposition of the Linear Poincaré Flow. □

Proof of Lemma 5.30 To prove the first item we assume that $\lambda_2 < \lambda_3 < 0$. Then σ is Lorenz-like for X by Corollary 5.29. Assume by contradiction that $W^{ss}_X(\sigma) \cap \Lambda \neq \{\sigma\}$.

Since Λ is transitive, by Theorem 2.21 there is $Z \in \mathcal{U}_U$ exhibiting a homoclinic connection $\Gamma \subset W^u_Z(\sigma(Z)) \cap W^{ss}_Z(\sigma(Z))$. This saddle-connection is an *orbit-flip* type connection, see Sect. 3.2.2. Using Theorem 3.6 we can approximate Z by $Y \in \mathcal{U}_U$ with a homoclinic connection

$$\Gamma' \subset W^u_Y(\sigma(Y)) \cap (W^s_Y(\sigma(Y)) \setminus W^{ss}_Y(\sigma(Y))).$$

Hence there exists a center-unstable manifold $W^{cu}_Y(\sigma(Y))$ containing Γ' and tangent to $W^s_Y(\sigma(Y))$ along Γ'. This saddle-connection is an *inclination-flip* type connection.

The existence of inclination-flip connections contradicts the existence of the dominated splitting for the Linear Poincaré Flow from Theorem 2.33, as a direct consequence of Theorem 3.6 in Sect. 3.2.2. This contradiction proves the first item.

The proof of the second item follows from the above argument applied to $-X$. □

Proof of Theorem 5.10 Let Λ be a robust transitive set of $X \in \mathfrak{X}^1(M)$ with $\dim(M) = 3$. By Corollary 5.29, if $\sigma \in \sigma_X(\Lambda)$, then σ is Lorenz-like for either X or $-X$. If σ is Lorenz-like for X we have $W^{ss}_X(\sigma) \cap \Lambda = \{\sigma\}$ by Lemma 5.30-(1) applied to $Y = X$. If σ is Lorenz-like for $-X$ we have that $W^{uu}_X(\sigma) \cap \Lambda = \{\sigma\}$ by Lemma 5.30-(2) again applied to $Y = X$. As $W^{ss}_{-X}(\sigma) = W^{uu}_X(\sigma)$ the proof is complete. □

Remark 5.33 In higher dimensions, given a homogeneous attracting set $\Lambda = \Lambda_X(U)$, the possible equilibria belonging to Λ are generalized Lorenz-like (recall Definition 2.30) either for X or for $-X$, by Remark 5.26.

In this setting, a periodic orbit contained in U has index (dimension of the stable manifold) equal to a fixed integer $s(X)$, the index of the vector field in a neighborhood of Λ for all C^1 close vector fields.

The index $s(\sigma)$ of a generalized Lorenz-like equilibrium, i.e., the dimension of the stable eigenspace of σ, properly accumulated by regular orbits in an attracting set Λ with respect to a homogeneous vector field, is one higher than the index $s(X)$ of the homogeneous vector field. Indeed, a generalized Lorenz-like equilibrium properly accumulated by regular orbits in Λ can be turned into a saddle-connection in the setting of Theorem 3.4 along a central manifold, by a small perturbation using the Connecting Lemma. The unfolding of this connection induces the creation of periodic orbits with index $s(\sigma)-1$, so that, to respect the homogeneous assumption on X inside U, we must have $s(\sigma)-1=s(X)$.

A similar result for robustly transitive attractors in homogeneous vector fields is given by [132, Corollary B].

If Λ also admits a dominated splitting with the dimension of the dominated bundle equal to the index of the vector field (which is a C^1 robust property), then we can argue as in the proof of Lemma 5.30. Here the strong-stable manifold of a generalized Lorenz-like singularity σ is the invariant manifold tangent at σ to the directions corresponding to the eigenvalues in $\lambda^-(\sigma)$.

Indeed, if σ is generalized Lorenz-like, properly accumulated by regular orbits in Λ and $W^{ss}(\sigma)\cap\Lambda\neq\{\sigma\}$, then we can use the Connecting Lemma to obtain an orbit-flip type saddle-connection along the central manifold corresponding to the weakest contracting and expanding eigenvalues. A small perturbation provides an inclination-flip type saddle-connection, contradicting the robust domination assumption. The same argument holds for a generalized Lorenz-like singularity after time reversal. Therefore Lemma 5.30 extends to the setting of homogeneous attracting sets.

Again, Li, Gan and Wen gave a similar result for robustly transitive singular sets of a homogeneous vector field in [132, Lemma 4.3].

In addition, Lemma 5.32 also extends to this setting with the same argument: *given a homogeneous attractor Λ with dominated splitting, then all its equilibria must be generalized Lorenz-like.* Otherwise the same type of perturbation arguments builds an inclination-flip type saddle-connection contradicting the domination.

In particular, *every equilibria in a sectionally expanding attractor is generalized Lorenz-like.* All these results were obtained by Metzger and Morales in [156].

5.4 Attractors and Singular-Hyperbolicity

Here we present a proof of Theorem 5.13.

Let Λ be a robust attractor of $X\in\mathfrak{X}^1(M)$ with $\dim(M)=3$, U an isolating block of Λ, and $\mathcal{U}_U$ a neighborhood of X such that $\Lambda_Y(U)=\cap_{t\in\mathbb{R}}Y^t(U)$ is transitive for all $Y\in\mathcal{U}_U$. By definition $\Lambda=\Lambda_X(U)$.

As we have already proved (in Lemma 5.20 and Corollary 5.28), for all $Y\in\mathcal{U}_U$, all the singularities of $\Lambda_Y(U)$ are Lorenz-like and all the critical elements in $\Lambda_Y(U)$ are hyperbolic of saddle type.

For future reference we state precisely the technical conditions for the arguments that follow.

Theorem 5.34 *Let $\Lambda = \Lambda_X(U)$ be a compact proper isolated invariant subset of $X \in \mathfrak{X}^1(M)$ such that*

1. *one of the following conditions holds*
   - *either Λ contains a dense regular orbit;*
   - *or Λ is connected and contains a dense subset of periodic orbits;*
2. *for every vector field C^1-close to X, all critical elements in U are hyperbolic of saddle-type (i.e., there are no sinks or sources nearby);*
3. *for all C^1-close vector fields every equilibrium in U is Lorenz-like.*

Then Λ is singular-hyperbolic.

The strategy to prove Theorem 5.13 is the following: given $X \in \mathscr{U}_U$ we show that there exists a neighborhood $\mathscr{V}$ of X, $c > 0$, $0 < \lambda < 1$ and $T_0 > 0$ such that, for all $Y \in \mathscr{V}$, the set

$$\{y \in \mathrm{Per}(Y) \cap U : (\text{minimal period of } y) \geq T_0\}$$

has a continuous invariant (c, λ)-dominated splitting $E^s \oplus E^{cu}$, with the dimension of E^s equal to 1.

Using the Closing Lemma of Pugh (Theorem 2.19) and the absence of sinks and sources, we obtain a dominated splitting for the Linear Poincaré Flow with uniform bounds for all C^1 close maximal invariant subsets, and this induces a dominated splitting over $\Lambda_X(U)$. The natural difficulty is to obtain the splitting around the singularities. The singularities are Lorenz-like and, consequently, they carry the local hyperbolic bundle $\hat{E}^{ss}$ associated to the strongest contracting eigenvalue of $DX(\sigma)$, and the central bundle $\hat{E}^{cu}$ associated to the remaining eigenvalues of $DX(\sigma)$. These bundles induce a local partial hyperbolic splitting $\hat{E}^{ss} \oplus \hat{E}^{cu}$ around the singularities.

The main step now is to prove that the splitting proposed for the periodic points is compatible with the local partial hyperbolic splitting at the singularities. Proposition 5.41 expresses this fact. Finally we prove that E^s is contracting and that the central direction E^{cu} is volume expanding, using the Ergodic Closing Lemma through Theorem 4.11 (on subadditive functions along orbits of a flow), concluding the proof of Theorem 5.34 and of Theorem 5.13.

We point out that the splitting for the Linear Poincaré Flow P^t obtained in Theorem 2.33 is not necessarily invariant under DX^t. When $\Lambda_X^*(U) = \Lambda_X(U) \setminus S(X)$ is closed, this splitting induces a hyperbolic splitting for X; see Theorem 2.27. However; the arguments used there do not apply here since $\Lambda_X^*(U)$ is not closed. We also note that a hyperbolic splitting for X *over* $\Lambda_X^*(U)$ cannot be automatically extended to a hyperbolic one over $\overline{\Lambda_X^*(U)}$: the presence of a singularity is a natural obstruction for this. On the other hand, Theorem 5.34 shows that this can be circumvented to get a partially hyperbolic structure for X over the whole of $\Lambda_X(U)$.

Let us present now each step of the proof. We keep in mind the original setting of a robustly transitive attractor for clarity, but the only properties used in the arguments are those stated in Theorem 5.34.

Remark 5.35 It is not difficult to see that these steps can be performed for vector fields X in compact manifolds of any dimension, since a homogeneous attractor $\Lambda_X(U)$ such that every equilibrium close to Λ for all C^1 nearby vector fields is generalized Lorenz-like. We comment on some of the adapted arguments along the several steps of the proof in a number of remarks in what follows.

5.4.1 Uniformly Dominated Splitting over the Periodic Orbits

Let $\Lambda_Y(U)$ be a robust attractor of $Y \in \mathscr{U}_U$, where U and $\mathscr{U}_U$ are as in the previous section, that is: for every $Y \in \mathscr{U}_U$ every critical element inside U is hyperbolic of saddle-type. Moreover the equilibria are all Lorenz-like.

Since every $p \in \mathrm{Per}_Y(\Lambda_Y(U))$ is hyperbolic of saddle type, we have that the tangent bundle of M over p can be written as

$$T_pM = E^s_p \oplus E^Y_p \oplus E^u_p,$$

where E^s_p is the eigenspace associated to the contracting eigenvalue of $DY^{t_p}(p)$, E^u_p is the eigenspace associated to the expanding eigenvalue of $DY^{t_p}(p)$, and we write t_p for the (minimal) period of p.

Note that $E^s_p \subset N^s_p \oplus E^Y_p$ and $E^u_p \subset N^u_p \oplus E^Y_p$, where $N^s \oplus N^u$ is the splitting for the Linear Poincaré Flow over regular orbits.

Observe that, if we consider the previous splitting over all $\mathrm{Per}(Y) \cap \Lambda_Y(U)$, the presence of a singularity in $\overline{\mathrm{Per}(Y) \cap \Lambda_Y(U)}$ is an obstruction for the extension of the stable and unstable bundles E^s and E^u to $\overline{\mathrm{Per}(Y \cap \Lambda_Y(U)}$. Indeed, near a singularity, the angle between either E^u and E^X, or E^s and E^X, goes to zero. To bypass this difficulty, we introduce the following notion.

Definition 5.36 Given $Y \in \mathscr{U}_U$ define for any $p \in \mathrm{Per}(Y) \cap \Lambda_Y(U)$ the splitting

$$T_pM = E^{s,Y}_p \oplus E^{cu,Y}_p, \quad \text{where } E^{cu,Y}_p := E^Y_p \oplus E^u_p.$$

Moreover we define a splitting over $\mathrm{Per}(Y) \cap \Lambda_Y(U)$ by

$$T_{\mathrm{Per}(Y)\cap\Lambda_Y(U)}M = \bigcup_{p\in\mathrm{Per}(Y)\cap\Lambda_Y(U)} (E^{s,Y}_p \oplus E^{cu,Y}_p).$$

In addition, we define the subspace $E^{cs,Y}_p := E^{s,Y}_p \oplus E^Y_p$ of the tangent space at $p \in \mathrm{Per}(Y) \cap \Lambda_Y(U)$ which gives another bundle over $\mathrm{Per}(Y) \cap \Lambda_Y(U)$.

When no confusion arises we drop the Y-dependence on the notation just defined. To simplify notation we denote the restriction of $DY^t(p)$ to $E_p^{s,Y}$ (respectively $E_p^{cu,Y}$) simply by $DY^t \mid E_p^s$ (respectively $DY^t \mid E_p^{cu}$) for $t \in \mathbb{R}$ and $p \in \mathrm{Per}(Y) \cap \Lambda_Y(U)$.

We now prove that the splitting over $\mathrm{Per}(Y) \cap \Lambda_Y(U)$ given by Definition 5.36 is a DY^t-invariant and uniformly dominated splitting along periodic points with large period.

Theorem 5.37 *Given $X \in \mathscr{U}_U$, there are a neighborhood $\mathscr{V} \subset \mathscr{U}_U$ and constants $0 < \lambda < 1$, $c > 0$, and $T_0 > 0$ such that, for every $Y \in \mathscr{V}$, if $p \in \mathrm{Per}(Y) \cap \Lambda_Y(U)$, $t_p > T_0$ and $T > 0$, then*

$$\|DY^T \mid E_p^s\| \cdot \|DY^{-T} \mid E_{Y^T(p)}^{cu}\| < c \cdot \lambda^T.$$

This result is similar to Theorem 4.8 but, since the angle between E_p^s and N_p can be very close to a right angle, we cannot use estimates for the Linear Poincaré Flow directly to obtain this result.

Theorem 5.37 will be proved in Sect. 5.4.3.3, with the help of Theorems 5.38 and 5.39 below, corresponding to Theorem 4.7, proved in Chap. 4.

Theorem 5.38 establishes, first, that the periodic points are uniformly hyperbolic, i.e., the periodic points are of saddle-type and the Lyapunov exponents are uniformly bounded away from zero. Secondly, the angle between the stable and the unstable eigenspaces at periodic points are uniformly bounded away from zero.

Theorem 5.38 *Given $X \in \mathscr{U}_U$, there are a neighborhood $\mathscr{V} \subset \mathscr{U}_U$ of X and constants $0 < \lambda < 1$ and $c > 0$, such that for every $Y \in \mathscr{V}$, if $p \in \mathrm{Per}(Y) \cap \Lambda_Y(U)$ and t_p is the period of p then*

a) (a1) $\|DY^{t_p} \mid E_p^s\| < \lambda^{t_p}$ *(uniform contraction on the period)*
   (a2) $\|DY^{-t_p} \mid E_p^u\| < \lambda^{t_p}$ *(uniform expansion on the period).*
b) $\angle(\mathscr{O}_p(E_p^s), \mathscr{O}_p(E_p^u)) > c$ *(angle uniformly bounded away from zero between center-stable and center-unstable directions).*

This result is exactly the same as Theorem 4.7, proved in Chap. 4.

Theorem 5.39 is a strong version of Theorem 5.38-b). It establishes that, at periodic points, the angle between the stable and the central unstable bundles is uniformly bounded away from zero.

Theorem 5.39 *Given $X \in \mathscr{U}_U$ there are a neighborhood $\mathscr{V} \subset \mathscr{U}_U$ of X and a positive constant C such that for every $Y \in \mathscr{V}$ and $p \in \mathrm{Per}(Y) \cap \Lambda_Y(U)$ we have angles uniformly bounded away from zero: $\angle(E_p^s, E_p^{cu}) > C$.*

This result is proved in Sect. 5.4.3.4.

We shall prove that, if Theorem 5.37 fails, then we can create a periodic point for a nearby flow with the angle between the stable and the central unstable bundles arbitrarily small. This yields a contradiction to Theorem 5.39. In proving the existence of such a periodic point for a nearby flow we use Theorem 5.38.

Assuming Theorem 5.37, we establish in the following section the extension of the splitting given in Definition 5.36 to all of $\Lambda_X(U)$. Afterward, with the help of Theorem 5.38, we show that E^s is uniformly contracting and that E^{cu} is volume expanding.

In the proof that E^s is uniformly contracted (respectively E^{cu} is volume expanding) we show that the opposite assumption leads to the creation of periodic points for flows near to the original one with contraction (respectively expansion) along the stable (respectively unstable) bundle arbitrarily small, contradicting the first part of Theorem 5.38.

All of these facts together imply Theorem 5.13.

5.4.2 Dominated Splitting over a Robust Attractor

Here we induce a dominated splitting over $\Lambda_X(U)$ using the dominated splitting over $\{p \in \mathrm{Per}(Y) \cap \Lambda_Y(U) : t_p \geq T_0\}$ for flows near X, given by Definition 5.36. The method is the same as in the proof of Theorem 4.4.

On the one hand, since $\Lambda_Y(U)$ is an attracting set for every Y close to X in $\mathfrak{X}^1$, we can assume without loss of generality, that for all $Y \in \mathscr{V}$ and $x \in \mathrm{Per}(Y)$ with $\mathscr{O}_Y(x) \cap U \neq \emptyset$, we have $\mathscr{O}_Y(x) \subset \Lambda_Y(U)$.

On the other hand, since $\Lambda = \Lambda_X(U)$ is assumed to be connected (this property is also a consequence of the existence of a dense forward orbit in Λ), we get

$$\Lambda(T_0) := \Lambda \setminus \{p \in \mathrm{Per}(X) \cap U : t_p < T_0\} \text{ is dense in } \Lambda.$$

From this, to induce an invariant splitting over $\Lambda_X(U)$ it is enough to do it over $\Lambda(T_0)$, as in Sect. 4.2.2.

Let us take a converging subsequence $E^{s,Y^{n_k}}_{y_{n_k}} \oplus E^{cu,Y^{n_k}}_{y_{n_k}}$ and define

$$E^{s,X}_x = \lim_{k\to\infty} E^{s,Y^{n_k}}_{y_{n_k}}, \qquad E^{cu,X}_x = \lim_{k\to\infty} E^{cu,Y^{n_k}}_{y_{n_k}}.$$

Since $E^{s,Y_n} \oplus E^{cu,Y_n}$ is a (c,λ)-dominated splitting for all n, then this property is also true for the limit $E^{s,X}_x \oplus E^{cu,X}_x$. Moreover $\dim(E^{s,X}_x) = 1$ and $\dim(E^{cu,X}_x) = 2$ for all $x \in \Lambda_X(U)$.

Define the following eigenspaces along $X^t(x)$ for $t \in \mathbb{R}$

$$E^{s,X}_{X^t(x)} := DX^t(E^{s,X}_x) \quad \text{and} \quad E^{cu,X}_{X^t(x)} := DX^t(E^{cu,X}_x).$$

Since for every n the splitting over $\{p \in \mathrm{Per}(Y_n) \cap \Lambda_{Y_n}(U) : t_p \geq T_0\}$ is (c,λ)-dominated, it follows that the splitting defined above along X-orbits of points in $K(X)$ is also (c,λ)-dominated. Moreover we also have that $E^{s,X}_{X^t(x)}$ is unidimensional and $E^{cu,X}_{X^t(x)}$ is bidimensional, for all $t \in \mathbb{R}$. This provides the desired extension of a dominated splitting to $\Lambda_X(U)$.

We denote by $E^s \oplus E^{cu}$ the splitting over $\Lambda_X(U)$ obtained in this way. Since this splitting is uniformly dominated we deduce that $E^s \oplus E^{cu}$ depends continuously on the points of $\Lambda_X(U)$ and also on the vector field X; see Sect. 2.6 and also [110].

When necessary we denote by $E^{s,Y} \oplus E^{cu,Y}$ the above splitting for Y near X.

Remark 5.40 If $\sigma \in S(X) \cap \Lambda_X(U)$ then E^s_σ is the eigenspace E^{ss}_σ associated to the strongest contracting eigenvalue of $DX(\sigma)$, and E^{cu}_σ is the bidimensional eigenspace associated to the remaining eigenvalues of $DX(\sigma)$. This follows from the uniqueness of dominated splittings; see Sect. 2.6 and also [79, 148].

5.4.3 Robust Attractors Are Singular-Hyperbolic

Next we prove that the splitting $E^s \oplus E^{cu}$ over $\Lambda_X(U)$ is partially hyperbolic with volume expanding central-unstable direction. For this we use Theorem 4.11.

5.4.3.1 E^s Is Uniformly Contracting

Consider the family of subadditive functions

$$\phi_Y(t,x) := \log \| DY^t \mid E^{s,Y}_x \|, \quad Y \in \mathcal{V},\ x \in \Lambda_Y(U),\ t \in \mathbb{R}.$$

This family of functions satisfies the conditions of Theorem 4.11, assuming Theorems 5.37, 5.37 and 5.39, and using the fact that the equilibria are Lorenz-like, and thus isolated.

Therefore there exist $c \in \mathbb{R}$ and $\lambda_0 \in (0,1)$ such that $\phi_X(t,x) \le c + t\log\lambda_0$ for all $X \in \mathcal{V}$, $x \in \Lambda_X(U)$, $t \ge 0$. This means that $\|DX^t \mid E^s_x\| \le Ke^{-\lambda t}$ with $K = e^c$ and $e^{-\lambda} = \lambda_0$.

5.4.3.2 E^{cu} Is Uniformly Volume Expanding

Analogously, we consider the family of subadditive functions

$$\psi_Y(t,x) := \log \| \det(DY^{-t} \mid E^{cu,Y}_x) \|, \quad Y \in \mathcal{V},\ x \in \Lambda_Y(U),\ t \in \mathbb{R}.$$

This family of functions satisfies the conditions of Theorem 4.11, assuming Theorems 5.37, 5.37 and 5.39, and using the fact that the equilibria are Lorenz-like, and thus isolated and their central-unstable subbundle expands area.

Again there exist $c \in \mathbb{R}$ and $\lambda_0 \in (0,1)$ such that $\psi_X(t,x) \le c + t\log\lambda_0$ for all $X \in \mathcal{V}, x \in \Lambda_X(U)$, $t \ge 0$. This means that $\| \det(DX^{-t} \mid E^{cu}_x \| \le Ke^{-\lambda t}$ with $K = e^c$ and $e^{-\lambda} = \lambda_0$, as needed.

5.4.3.3 Uniform Dominated Splitting on Periodic Orbits

Let us assume Theorems 5.38 and 5.39 and show how we obtain Theorem 5.37. The central idea of the proof is to show that, if Theorem 5.37 fails, then we can obtain a flow near X exhibiting a periodic point with an arbitrarily small angle between the central stable and the central-unstable bundles of the Linear Poincaré Flow, leading to a contradiction to Theorem 5.39.

For this we reduce the problem to the proof of Theorem 4.8 in Sect. 4.2.4, as follows.

As in the proof of Lemma 4.14, to obtain Theorem 5.37 it is enough to show that there exist a neighborhood $\mathcal{V} \subset \mathcal{U}_U$ of X and $T_0 > 0$ such that, for every vector field $Y \in \mathcal{V}$, if $p \in \mathrm{Per}_Y^{T_0}(\Lambda_Y(U))$ then

$$\|DY^{T_0} \mid E_p^s\| \cdot \|DY^{-T_0} \mid E_{Y^{T_0}(p)}^{cu}\| \le \frac{1}{2}. \tag{5.7}$$

We prove (5.7) arguing by contradiction. If (5.7) fails then given $X \in \mathcal{U}_U$ and $T_0 > 0$, we can find $Y \in \mathcal{U}_U$ arbitrarily close to X, $y \in \mathrm{Per}(Y) \cap \Lambda_Y(U)$ with $t_y \ge T_0$ and $w_0 \in E_y^s$ and $v_0 \in E_y^{cu}$ satisfying (see Sect. 4.2.4 for a similar argument)

$$\|DY^{T_0}(y) \cdot w_0\| > \frac{1}{2}\|DY^{T_0}(y) \cdot v_0\| \ge \frac{1}{2}\|P_Y^{T_0} \cdot v_0\|. \tag{5.8}$$

If we show that the angle between $E^s_{X^{T_0}(y)}$ and $N_{X^{T_0}(y)}$ is bounded above by a function of $\angle(E^s, E^{cu})$, then we find $\eta > 0$ so that

$$\|P_Y^{T_0} \mid E_y^s\| \ge \eta\|DY^{T_0} \mid E_y^s\|$$

since the orthogonal projection on N restricted to E^s is close to the identity, and thus

$$\|P_Y^{T_0}(y) \cdot w_0\| > \eta\frac{1}{2}\|P_Y^{T_0} \cdot v_0\|. \tag{5.9}$$

From (5.9) we can argue just as in Sect. 4.2.4 to obtain a perturbation of Y, as close as we want to Y, with arbitrarily small angles between $N^{cs}_{X^{T_0}(y)}$ and $N^{cu}_{X^{T_0}(y)}$. This contradicts Theorem 5.39 and proves Theorem 5.37.

Now we find the upper bound for the angle between the stable and the normal direction to the flow. Let $L : (E^{cu}_{X^{T_0}(y)})^{\perp} \to E^{cu}_{X^{T_0}(y)}$ be a linear map such that $E^s_{X^{T_0}(y)} = \mathrm{Graph}(L)$. Then $\alpha(E^s, E^{cu}) = \|L\|^{-1} \ge c$. Moreover, if $\theta = \angle(E^s, N)$, then

$$\cos\theta = \frac{\|O_{X^{T_0}(y)}(w + Lw)\|}{\|w + Lw\|} = \frac{\|w + O_{X^{T_0}(y)}Lw\|}{\|w + O_{X^{T_0}(y)}Lw + (Lw - O_{X^{T_0}(y)}Lw)\|}.$$

We note that $Lw - O_{X^{T_0}(y)}Lw = \pi(Lw)$, where $\pi : T_{X^{T_0}(y)} \to E^X_{X^{T_0}(y)}$ is the orthogonal projection onto the flow direction, and that the vectors w, $O_{X^{T_0}(y)}Lw$ and $\pi(Lw)$ are mutually orthogonal. Thus

$$\cos\theta = \left(1 + \frac{\|\pi(Lw)\|^2}{\|w\|^2 + \|O_{X^{T_0}(y)}Lw\|^2}\right)^{-1/2} \geq \frac{1}{\sqrt{1+\|L\|^2}} \geq \frac{1}{\sqrt{1+c^{-2}}}.$$

Hence the angle θ is uniformly bounded from above, which is what we wanted to prove.

This completes the proof of Theorem 5.37 using Theorems 5.38 and 5.39.

5.4.3.4 Uniformly Bounded Angles Between Stable and Unstable Directions at Periodic Orbits

Here we prove Theorem 5.39, used in the proofs of the results in the previous section.

Proof of Theorem 5.39 Arguing by contradiction, we show that, if Theorem 5.39 fails, then we can create periodic points with arbitrarily small angle between the stable and unstable direction, leading to a contradiction to the second part of Theorem 5.38, already proved.

Theorem 5.39 is a consequence of Propositions 5.41 and 5.42 below. The first one establishes that, for periodic points close to a singularity, the stable direction remains close to the strong-stable direction of the singularity, and the central-unstable direction is close to the central-unstable direction of the singularity. This result gives the compatibility between the splitting proposed for the periodic points in Definition 5.36 and the local partially hyperbolic splitting at the singularities.

Proposition 5.41 *Given* $X \in \mathcal{U}_U$, $\varepsilon > 0$ *and* $\sigma \in S(X) \cap \Lambda_X(U)$, *there exist a neighborhood* $\mathcal{V} \subset \mathcal{U}_U$ *of* X *and* $\delta > 0$ *such that for all* $Y \in \mathcal{V}$ *and* $p \in \mathrm{Per}_Y(\Lambda_Y(U))$ *satisfying* $\mathrm{dist}(p, \sigma_Y) < \delta$ *we have*

(a) $\angle(E_p^{s,Y}, \hat{E}_{\sigma_Y}^{ss,Y}) < \varepsilon$, *and*
(b) $\angle(E_p^{cu,Y}, \hat{E}_{\sigma_Y}^{cu,Y}) < \varepsilon$.

The second statement says that, *far from singularities*, the angle between the stable direction and the central unstable direction of any periodic point, inside the maximal invariant set, is uniformly bounded away from zero.

Proposition 5.42 *Given* $X \in \mathcal{U}_U$ *and* $\delta > 0$, *there are a neighborhood* $\mathcal{V} \subset \mathcal{U}$ *of* X *and a positive constant* $C = C(\delta)$ *such that if* $Y \in \mathcal{V}$ *and* $p \in \mathrm{Per}_Y(\Lambda_Y(U))$ *satisfies* $\mathrm{dist}(p, S(Y) \cap \Lambda_Y(U)) > \delta$ *then*

$$\angle\big(E_p^{s,Y}, E_p^{cu,Y}\big) > C.$$

Theorem 5.39 follows from these propositions since

- away from equilibria, the uniform domination between the stable and center-unstable directions at periodic orbits is a consequence of the uniform growth rates provided by Theorem 5.38 together with the angle estimate of Proposition 5.42;

- for orbits passing close to equilibria, Proposition 5.41 ensures that the stable and center-unstable directions are essentially the same as the strong-stable and center-unstable direction at the singularity. The angle between these is bounded away from zero since each equilibrium is Lorenz-like, by Theorem 5.10, and the set $S(X) \cap \Lambda_Y(U)$ is finite because each singularity is hyperbolic. This, together with the uniform growth rates provided by Theorem 5.38, ensures the uniform domination between the stable and center-unstable directions.

The proof of Theorem 5.39 is complete depending only on Propositions 5.41 and 5.42. □

Remark 5.43 This same reasoning above is also valid to prove that a homogeneous attractor with generalized Lorenz-like singularities admits a dominated splitting.

We present the proofs in the three-dimensional case but for higher dimensions case the adaptations are straightforward.

5.4.4 Flow-Boxes Near Equilibria

Since the equilibria σ in our setting are all Lorenz-like, the unstable manifold $W^u(\sigma)$ is one-dimensional, and there is a one-dimensional strong-stable manifold $W^{ss}(\sigma)$ contained in the two-dimensional stable manifold $W^s(\sigma)$. Using the linearization given by the Hartman-Grobman Theorem 2.14 or, in the absence of resonances, the smooth linearization results provided by Sternberg [256], orbits of the flow in a small neighborhood U_0 of the equilibrium are solutions of the linear system (3.3), modulo a continuous change of coordinates.

Then for some $\delta > 0$ we may choose cross-sections contained in U_0

- $\Sigma^{o,\pm}$ at points $y^\pm$ in different components of $W^u_{loc}(\sigma) \setminus \{\sigma\}$;
- $\Sigma^{i,\pm}$ at points $x^\pm$ in different components of $W^s_{loc}(\sigma) \setminus W^{ss}_{loc}(\sigma)$

and Poincaré first hitting time maps $R^\pm : \Sigma^{i,\pm} \setminus \ell^\pm \to \Sigma^{o,-} \cup \Sigma^{o,+}$, where $\ell^\pm = \Sigma^{i,\pm} \cap W^s_{loc}(\sigma)$, satisfying (see Fig. 5.3)

1. every orbit in the attractor passing through a small neighborhood of the equilibrium σ intersects some of the incoming cross-sections $\Sigma^{i,\pm}$;
2. $R^\pm$ maps each connected component of $\Sigma^{i,\pm} \setminus \ell^\pm$ diffeomorphically inside a different outgoing cross-section $\Sigma^{o,\pm}$, preserving the corresponding stable foliations.

These cross-sections may be chosen to be planar relative to some linearizing system of coordinates near σ, e.g., for a small $\delta > 0$

$$\Sigma^{i,\pm} = \{(x_1, x_2, \pm 1) : |x_1| \le \delta, |x_2| \le \delta\} \quad \text{and}$$

$$\Sigma^{o,\pm} = \{(\pm 1, x_2, x_3) : |x_2| \le \delta, |x_3| \le \delta\},$$

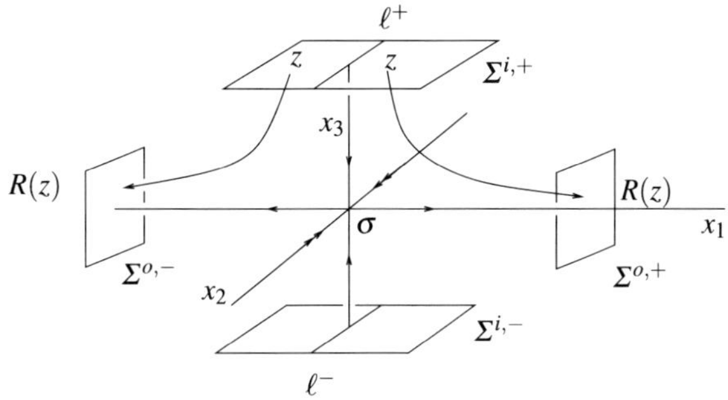

Fig. 5.3 Cross-sections near a Lorenz-like equilibrium

where the x_1-axis corresponds to the unstable manifold near σ, the x_2-axis to the strong-stable manifold and the x_3-axis to the weak-stable manifold of the equilibrium which, in turn; is at the origin, see Fig. 5.3.

The equilibrium is hyperbolic for the vector field X. Hence for every C^1 nearby vector field Y there exists a unique Lorenz-like equilibrium σ_Y in U_0. Moreover the submanifolds $\Sigma^{i,\pm}$ and $\Sigma^{o,\pm}$ remain transverse to Y. So all local properties of these cross-sections are robust under small C^1 perturbations of the flow.

5.4.5 Uniformly Bounded Angle Between Stable and Center-Unstable Directions on Periodic Orbits

Let us recall some facts and notation before starting the proof of Propositions 5.41 and 5.42.

Given a singularity σ of $X \in \mathcal{U}_U$, we know that σ is hyperbolic. So for Y close to X there exists a unique continuation of σ, which we write σ_Y. Since every singularity of X in $\Lambda_X(U)$ is hyperbolic, we conclude that the singularities of Y near to X are the continuations of the singularities of X. Hence we can assume that, for any Y close to X, the singularities of Y in $\Lambda_Y(U)$ coincide with those of X in $\Lambda_X(U)$.

According to Theorem 5.10, for all $Y \in \mathcal{U}_U$ the eigenvalues $\lambda_i = \lambda_i(Y)$, $i = 1, 2, 3$ of $DY(\sigma_Y)$ are real and satisfy $\lambda_2 < \lambda_3 < 0 < -\lambda_3 < \lambda_1$. We write $\hat{E}^{ss,Y}_{\sigma_Y}$ for the eigenspace associated to the strongest contracting eigenvalue λ_2 and $\hat{E}^{cu,Y}_{\sigma_Y}$ for the bidimensional eigenspace associated to $\{\lambda_3, \lambda_1\}$. Without loss of generality we can assume that, for Y close to X, the eigenvalues of $DY(\sigma_Y)$ are the same as the ones of $DX(\sigma)$.

Since M is a Riemannian manifold, for any $x \in M$ and every neighborhood U of x there exists a *normal neighborhood* $V \subset U$ of x, i.e., for any pair of points in V there is a unique geodesic contained in V connecting them. Thus using parallel transport in V we can define angles between any pair of tangent vectors at points of V. We will use this in what follows to compare angles of tangent vectors at nearby points. Alternatively, we can simply take a local coordinate system at x and compare angles in those coordinates: the distortion will be bounded near x.

We reduce the proof of Propositions 5.41 and 5.42 to the following results.

The first one is the next lemma (which is Theorem 4.10, already proved) establishing that any compact invariant set $\Gamma \subset \Lambda_X(U)$ *containing no singularities* is uniformly hyperbolic.

Lemma 5.44 *Let $X \in \mathscr{U}_U$ and $\Gamma \subset \Lambda_X(U)$ be a compact invariant set without singularities. Then Γ is uniformly hyperbolic.*

Given $X \in \mathscr{U}_U$ and $\delta > 0$ we define

$$C_\delta = \bigcup_{\sigma \in S(X) \cap \Lambda_X(U)} B_\delta(\sigma),$$

the δ-neighborhood around the singularities of X in Λ. Write $U_\delta = \overline{U \setminus C_\delta}$ for the closure of the complement of C_δ in U and define

$$\Omega_X(U_\delta) = \{x \in \Omega(X) : \mathscr{O}_X(x) \subset U_\delta\}.$$

We use the following application of Lemma 5.44.

Corollary 5.45 *For any $\delta > 0$, $\Omega_X(U_\delta)$ is hyperbolic.*

Recall that, given a regular point $x \in M$, we define N_x^Y as the orthogonal complement of E_x^Y in T_xM, $\Lambda_Y(U)^* = \Lambda_Y(U) \setminus S(Y)$, and

$$N_{\Lambda_Y(U)^*} = \left\{N_x^{s,Y} \oplus N_x^{u,Y}\right\}_{x \in \Lambda_Y(U)^*}$$

denotes the splitting for the Linear Poincaré Flow P_t^Y of Y; see Theorem 2.33 in Sect. 2.6. For $x \in \Lambda_Y^*(U)$ we define the bundles $E_x^{cs,Y} = N_x^{s,Y} \oplus E_x^Y$ and $E_x^{cu,Y} = N_x^{u,Y} \oplus E_x^Y$.

Recall also that for Y near X and $p \in \mathrm{Per}_Y(\Lambda_Y(U))$ we denote by $E_p^{s,Y} \oplus E_p^{cu,Y}$ the splitting induced by the hyperbolic splitting along the periodic orbit as in Definition 5.36. In this case, we have $E_p^{cu,Y} = N_p^{u,Y} \oplus E_p^Y$ and $E_p^{s,Y} \subset E_p^{cs,Y} = N_p^{s,Y} \oplus E_p^Y$.

Using the property that a uniformly hyperbolic set has a unique locally defined continuation for flows close to the initial one, we obtain that, for every point whose orbit does not go away from Γ for any nearby flow, any tangent vector in E^{cs} close to the flow direction remains close to the flow direction under the action of the flow.

Lemma 5.46 *Let $X \in \mathscr{U}_U$ and Γ be a compact invariant set without singularities. Then there are neighborhoods $\mathscr{V}$ of X, V of Γ and $\gamma > 0$ such that for any $\varepsilon > 0$ there exists $T = T(\varepsilon) > 0$ satisfying: if $Y \in \mathscr{V}$, $y \in V \cap \Lambda_Y(U)$ with $Y^s(y) \in V$ for $0 \le s \le t$ and some $t \ge T$, and also $v \in N_y^{s,Y} \oplus E_y^Y$ with $\angle(v, Y(y)) < \gamma$, then $\angle(DY^t(y) \cdot v, Y(Y^t(y))) < \varepsilon$.*

The next result provides angle estimates for orbits passing near to a singularity. For a point y in $\Lambda_Y(U)$ and vectors v with angle bounded away from zero with the strong-stable bundle at the singularity, after passing near the singularity, $DY^t(v)$ lands in the direction of the central unstable bundle at $Y^t(y)$.

Given $\sigma_Y \in S(Y) \cap \Lambda_Y(U)$, $W^s_{loc}(\sigma_Y)$ ($W^u_{loc}(\sigma_Y)$ respectively) stands for the local stable (unstable) manifold at σ_Y. We set $\hat{W}^s_{loc}(\sigma_Y) = W^s_{loc}(\sigma_Y) \setminus \{\sigma_Y\}$ and $\hat{W}^u_{loc}(\sigma_Y) = W^u_{loc}(\sigma_Y) \setminus \{\sigma_Y\}$. Since σ_Y is Lorenz-like, there is a unique bundle $\hat{E}^{ss,Y}$ in $TW^s_{loc}(\sigma_Y)$ which is strongly contracted by the derivative of the flow. For each $y \in W^s_{loc}(\sigma_Y)$, $\hat{E}^{ss,Y}_y$ is the fiber of $\hat{E}^{ss,Y}$ at y.

In the following we use the notation from Sect. 5.4.4 for cross-sections near a δ-neighborhood of the singularities.

Definition 5.47 If $y \in B_\delta(\sigma_Y)$ we write y_* for the point in $\hat{W}^s_{loc}(\sigma_Y)$ satisfying $\operatorname{dist}(y, \hat{W}^s_{loc}(\sigma_Y)) = \operatorname{dist}(y, y_*)$.

Now we can state the result precisely.

Lemma 5.48 *Let $X \in \mathscr{U}_U$, $\sigma \in S(X) \cap \Lambda_X(U)$ and $\delta > 0$. There exists a neighborhood $\mathscr{V}$ of X such that given $\gamma > 0$ and $\varepsilon > 0$ there exists $r = r(\varepsilon, \gamma) > 0$ such that, for $Y \in \mathscr{V}$, $y \in B_\delta(\sigma) \cap \Lambda_Y(U)$ satisfying $\operatorname{dist}(y, \hat{W}^s_{loc}(\sigma_Y)) < r$ and for $v \in T_yM$ with $\angle(v, \hat{E}^{ss,Y}_{y_*}) > \gamma$, then $\angle(DY^{s_y}(y)(v), E^{cu,Y}_{Y^{s_y}(y)}) < \varepsilon$, where s_y is the smallest positive time such that $Y^{s_y}(y) \in \Sigma^{o,\pm}$.*

Given $\delta' \in (0, \delta)$ we define a neighborhood of the local stable manifold of σ in $\Sigma^{i,\pm}$ by

$$\Sigma^{i,\pm}_{\delta,\delta'} = \{x \in \Sigma^{i,\pm} : \operatorname{dist}(x, \hat{W}^s_{loc}(\sigma) \cap \Sigma^{i,\pm}) \le \delta'\}. \tag{5.10}$$

Finally, next result provides estimates for the angles after passing near a singularity: if a vector v in the central direction has angle bounded away from zero with the flow direction then, after passing near σ, $DX^t(v)$ becomes closer to the direction of the flow.

Lemma 5.49 *Let $X \in \mathscr{U}_U$, $\sigma \in S(X) \cap \Lambda_X(U)$ and $\delta > 0$. There is a neighborhood $\mathscr{V}$ of X such that, given $\varepsilon > 0$, $\kappa > 0$, $\delta > 0$ and cross-sections $\Sigma^{i,\pm}$, $\Sigma^{o,\pm}$ as above, there exists $\delta' > 0$ satisfying: for all $Y \in \mathscr{V}$, $p \in \Sigma^{i,\pm}_{\delta,\delta'}$ and $v \in N^{u,Y}_p \oplus E^Y_p$, if $\angle(v, Y(p)) > \kappa$, then $\angle(DY^{s_p}(p) \cdot v, Y(Y^{s_p}(p))) < \varepsilon$, where s_p is the first positive time such that $Y^{s_p}(p) \in \Sigma^{o,\pm}$.*

We postpone the proof of Lemmas 5.46, 5.48 and 5.49 to the end of this section, and continue with the proof of Propositions 5.41 and 5.42 assuming these results.

Since we have only a finite number of singularities, we can assume that the estimates given by the previous lemmas are simultaneously valid for all singularities of Y in $\Lambda_Y(U)$ and for all $Y \in \mathscr{V}$.

Proof of Proposition 5.41(a) We argue by contradiction. Using the property that hyperbolic singularities depend continuously on the vector field, we have that, if item (a) of Proposition 5.41 fails, then there are a singularity σ of X, $\gamma > 0$, a sequence of vector fields Y_n converging to X and a sequence of periodic points $p_n \in \mathrm{Per}_{Y_n}(\Lambda_{Y_n}(U))$ with $p_n \to \sigma$ such that

$$\angle\big(E^{s,Y_n}_{p_n}, \hat{E}^{ss,Y_n}_{\sigma_{Y_n}}\big) > \gamma. \tag{5.11}$$

We prove, using (5.11), that after a first passage through a neighborhood of a singularity, the stable direction and the flow direction become closer. This property persists up to the next return to that neighborhood. After a second passage through it, we show that the stable direction and the flow direction are close, and that the unstable direction and the flow direction are also close. This implies that the stable and the unstable direction are close to each other, leading to a contradiction to Theorem 5.38(b).

Fix a neighborhood $B_\delta(\sigma)$ and cross-sections $\Sigma^{o,\pm}$, $\Sigma^{i,\pm}$ contained in $B_\delta(\sigma)$ as in Sect. 5.4.4. Since $p_n \to \sigma$, we have that for all sufficiently large n there exists the smallest $t_n > 0$ such that $q_n = Y_n^{t_n}(p_n) \in \Sigma^{o,\pm}$.

Note that there exists $q \in \hat{W}^u_{loc}(\sigma) \cap \Lambda_X(U)$ such that $q_n \to q$.

Lemma 5.50 *The bound* (5.11) *implies that* $\angle(E^{s,Y_n}_{q_n}, Y_n(q_n)) \xrightarrow[n\to+\infty]{} 0$.

Proof We prove first that, as a consequence of (5.11), the stable direction at q_n is close to the central-unstable direction at q_n. Using some properties of the splitting given by the Poincaré flow, we deduce that the stable direction at q_n is necessarily close to the flow direction at q_n, and from this we deduce the statement of the claim.

By (5.11) and since $p_n \to \sigma$, by Lemma 5.48 we get

$$\angle\big(E^{s,Y_n}_{q_n}, N^{u,Y_n}_{q_n} \oplus E^{Y_n}_{q_n}\big) \xrightarrow[n\to+\infty]{} 0. \tag{5.12}$$

Now we deduce from (5.12) that $E^{s,Y_n}_{q_n}$ is leaning in the direction of the flow. Indeed, since $q_n \to q \in \Lambda^*_X(U)$, Theorem 2.33 for the Linear Poincaré Flow ensures that $\angle(N^{s,Y_n}_{q_n}, N^{u,Y_n}_{q_n}) > \frac{9}{10} \cdot \angle(N^{s,X}_q, N^{u,X}_q)$ for every n big enough. Because $N^{s(u)}_{q_n}$ is orthogonal to $Y_n(q_n)$, we deduce that

$$\angle(N^{s,Y_n}_{q_n} \oplus E^{Y^n}_{q_n}, N^{u,Y_n}_{q_n} \oplus E^{Y^n}_{q_n}) = \angle\big(N^{s,Y_n}_{q_n}, N^{u,Y_n}_{q_n}\big).$$

Hence $\angle(N^{s,Y_n}_{q_n} \oplus E^{Y^n}_{q_n}, N^{u,Y_n}_{q_n} \oplus E^{Y^n}_{q_n})$ is uniformly bounded away from zero. Since $E^{s,Y_n}_{q_n} \subset N^{s,Y_n}_{q_n} \oplus E^{Y^n}_{q_n}$ and $Y_n(q_n) = (N^{s,Y_n}_{q_n} \oplus E^{Y^n}_{q_n}) \cap (N^{u,Y_n}_{q_n} \oplus E^{Y^n}_{q_n})$, by (5.12) we obtain

$$\angle\big(E^{s,Y_n}_{q_n}, Y_n(q_n)\big) \xrightarrow[n\to\infty]{} 0. \tag{5.13}$$

This completes the proof of the lemma. □

Now we apply Lemma 5.49. For this, let δ be as above, $\kappa = c$ with c given by Theorem 5.38(b) and $\varepsilon < c/2$. Let δ' be given by Lemma 5.49.

Fix $\delta^* < \max\{\delta, \delta'\}$ and consider $U_{\delta^*} = \overline{U \setminus C_{\delta^*}}$. Since the singularities of $Y \in \mathscr{V}$ are continuations of the singularities of X, we can assume that $U_{\delta^*} \cap S(Y) \cap \Lambda_Y(U) = \emptyset$ for all $Y \in \mathscr{V}$.

Since σ is an accumulation point of $\{\mathscr{O}_{Y_n}(q_n)\}_{n\geq 1}$ we have that, for n large enough, there is a first positive time s_n such that $\tilde{q}_n = Y_n^{s_n}(q_n)$ belongs to C_{δ^*}. We can take s_n in such a way that $\tilde{q}_n \in \Sigma^{i,\pm}_{\delta,\delta'}$ (defined in (5.10)).

We assume, without loss of generality, that every $\tilde{q}_n$ belongs to the same cross-section $\Sigma^{i,\pm}_{\delta,\delta'}$ associated to the same singularity of Y_n and of X. Note that from the choice of δ^* we have $Y_n^s(q_n) \in U_{\delta^*}$ for all $0 \leq s \leq s_n$.

Next we prove that (5.13) is also true replacing q_n by $\tilde{q}_n$, that is,

$$\angle\big(E^{s,Y_n}_{\tilde{q}_n}, Y_n(\tilde{q}_n)\big) \xrightarrow[n\to+\infty]{} 0. \tag{5.14}$$

Indeed, if there exists $S > 0$ such that $s_n < S$ for infinitely many n, then (5.13) immediately implies (5.14).

Otherwise, let q be such that $Y_n^{s_n/2}(q_n) \xrightarrow[s_n\to\infty]{} q$. Then $\overline{\mathscr{O}_X(q)} \subset U_{\delta^*}$ which implies that $\omega_X(q) \subset \Omega_X(U_{\delta^*})$. By Corollary 5.45 we know that $\Omega_X(U_{\delta^*})$ is uniformly hyperbolic. Let V be a neighborhood of $\Omega_X(U_{\delta^*})$ given by Lemma 5.46. Now we establish that the time spent by the Y_n-orbit segment $\{Y_n^t(q_n), 0 \leq t \leq s_n\}$ outside V is uniformly bounded.

Lemma 5.51 *There exists $S > 0$ such that for all n there are $0 \leq s_n^1 < s_n^2 \leq s_n$ with $s_n^1 < S$ and $s_n - s_n^2 < S$ satisfying $Y_n^s(q_n) \in V$ for all $s_n^1 \leq s \leq s_n^2$.*

Proof It is enough to prove that there exists S' such that, given q_n and $0 < s_n' < s_n$ with $Y_n^{s_n'}(q_n) \notin V$, then either $s_n' < S'$ or $s_n - s_n' < S'$.

If this were not the case, there would exist s_n' such that $Y_n^{s_n'}(q_n) \notin V$ and both $s_n - s_n' \to +\infty$ and $s_n' \to +\infty$. Then we can take a sequence $Y_n^{s_n'}(q_n) \to q'$ with $q' \notin V$. This implies that $\overline{\mathscr{O}_X(q')} \subset U_{\delta^*}$. So $\omega_X(q') \subset \Omega_X(U_{\delta^*})$ and hence $\omega_X(q') \subset V$. Thus for large n we would get $Y_n^{s_n'}(q_n) \in V$, contradicting the assumption. This finishes the proof. □

Returning to the proof of (5.14), recall that $\angle(E^{s,Y_n}_{q_n}, Y_n(q_n))$ is arbitrarily small for n large enough, by relation (5.13). Now Lemma 5.46, together with Lemma 5.51, implies (5.14), since we know that the time spent by $Y_n^s(q_n)$ in V for $s \in [0, s_n]$ is arbitrarily big.

Now since $\tilde{q}_n \in \Sigma^{i,\pm}_{\delta,\delta'}$ there is a first time $r_n > 0$ such that $\hat{q}_n = Y_n^{r_n}(\tilde{q}_n) \in \Sigma^{o,\pm}$, by the choice of the cross-sections near the singularities. We prove that also $\angle(E^{s,Y_n}_{\hat{q}_n}, Y_n(\hat{q}_n)) \xrightarrow[n\to\infty]{} 0$.

If there exists $S > 0$ such that $0 < r_n < S$ for infinitely many n, then taking a subsequence we obtain the desired conclusion. Otherwise, taking a subsequence if necessary, we have $\tilde{q}_n \to \hat{W}^s_{loc}(\sigma) \cap \Sigma^{i,\pm}_{\delta,\delta'}$ and there exists $\hat{q} \in \hat{W}^u_{loc}(\sigma) \cap \Sigma^{o,\pm}$ such that $\hat{q}_n \to q$. Observe that there exists $d > 0$ satisfying: for any $y \in \hat{W}^s_{loc}(\sigma) \cap \Sigma^{i,\pm}_{\delta,\delta'}$ we have $\angle(X(y), \hat{E}^{ss}_y) > d$. So provided n is large enough we obtain

$$\angle\big(Y_n(\tilde{q}_n), \hat{E}^{ss,Y_n}_{\tilde{q}_n}\big) > d. \tag{5.15}$$

Combining (5.14) and (5.15) we obtain $\angle(E^{s,Y_n}_{\tilde{q}_n}, \hat{E}^{ss,Y_n}_{\tilde{q}_n}) > d$ for n large. Arguing as in the proof of Lemma 5.51, replacing q_n by $\tilde{q}_n$ for $n \geq 0$, we obtain

$$\lim_{n\to\infty} \angle\big(E^{s,Y_n}_{\tilde{\tilde{q}}_n}, Y_n(\widetilde{\tilde{q}}_n)\big) = 0. \tag{5.16}$$

Moreover from (5.14) Theorem 5.38(b) ensures that

$$\angle\big(E^{u,Y_n}_{\tilde{q}_n}, Y_n(\tilde{q}_n)\big) > c \text{ for } n \text{ big enough.} \tag{5.17}$$

Since $E^{u,Y_n}_{\tilde{q}_n} \subset N^{u,Y_n}_{\tilde{q}_n} \oplus E^{Y_n}_{\tilde{q}_n}$ from (5.17), Lemma 5.49 implies that

$$\angle\big(DY^{r_n}_n(E^{u,Y_n}_{\tilde{q}_n}), Y_n(\hat{q}_n)\big) < \varepsilon < c/2 \tag{5.18}$$

by the choice of ε.

Finally (5.16) and (5.18) combined with $E^{u,Y_n}_{\hat{q}_n} = DY^{r_n}_n(E^{u,Y_n}_{\tilde{q}_n})$ give

$$\angle\big(E^{u,Y_n}_{\hat{q}_n}, E^{s,Y_n}_{\hat{q}_n}\big) < c/2 \text{ for } n \text{ big enough.}$$

This contradicts Theorem 5.38(b). This contradiction completes the proof of Proposition 5.41(a). □

Proof of Proposition 5.41(b) We show that given Y near to X and a periodic point p of Y close to σ_Y, then $E^{cu,Y}_p$ is close to $\hat{E}^{cu,Y}_{\sigma_Y}$. We split the argument into the following claims.

Given $\delta, \delta' > 0$ we consider the cross-sections $\Sigma^{i,\pm}$ and $\Sigma^{i,\pm}_{\delta,\delta'}$ as in Sect. 5.4.4 and definition (5.10).

Lemma 5.52 *Let $X \in \mathscr{U}_U$, $\sigma \in S(X) \cap \Lambda_X(U)$ and $\delta > 0$. There are a neighborhood $\mathscr{V}$ of X such that, given $\gamma > 0$ and $\varepsilon > 0$, there exists $r = r(\varepsilon, \gamma) > 0$ such that, if $y \in \Sigma^{i,\pm}$ and $L_y \subset T_yM$ is a plane with $\angle(L_y, \hat{E}^{ss}_y) > \gamma$, then $\angle(DY^{s_y}(y) \cdot L_y, \hat{E}^{cu}_{\sigma_Y}) < \varepsilon$, where s_y is such that $Y^{s_y}(y) \in B_r(\sigma_Y)$ and $Y^s(y) \in B_\delta(\sigma_Y)$ for all $0 \leq s \leq s_y$.*

The proof of this claim is analogous to the proof of Lemma 5.48, which is presented at the end of this section.

Given $y \in \Sigma^{i,\pm}_{\delta,\delta'}$ let y_* be as in Definition 5.47.

Lemma 5.53 *Let $X \in \mathscr{U}_U$, $\sigma \in S(X) \cap \Lambda_X(U)$ and $\delta > 0$. There are a neighborhood $\mathscr{V}$ of X, $\gamma > 0$ and $\delta' > 0$ such that for all $Y \in \mathscr{V}$ and all $y \in \Lambda_Y(U) \cap \Sigma^{i,\pm}_{\delta,\delta'}$ we have $\angle(E^{cu,Y}_y, \hat{E}^{ss,Y}_{y_*}) > \gamma$.*

Assuming the lemmas, let us complete the proof of the proposition.

Observe that for p close to σ_Y there is $s_p > 0$ such that $\tilde{p} = Y_{-s_p}(p) \in \Sigma^{i,\pm}_{\delta,\delta'}$, where δ and δ' are as in Lemma 5.53. Let $\tilde{p}_*$ be as in Definition 5.47. By Lemma 5.53 we have $\angle(E^{cu,Y}_{\tilde{p}}, \hat{E}^{ss,Y}_{\tilde{p}_*}) > \gamma$. Hence, by Lemma 5.52, we get $\angle(DY^t(\tilde{p})(E^{cu,Y}_{\tilde{p}}), \hat{E}^{cu,Y}_{\sigma_Y})$ arbitrarily small, provided that p is close enough to σ_Y, concluding the proof of Proposition 5.41(b). □

Proof of Lemma 5.53 First we consider points $q \in \Sigma^{i,\pm}_{\delta,\delta'} \cap \Lambda_X(U) \cap \hat{W}^s_{loc}(\sigma)$.

In this case, observe that $\angle(E^{cu,X}_q, \hat{E}^{ss}_q) \geq \angle(E^{cu,X}_q, T_q W^s_{loc}(\sigma))$. By item 3 of Theorem 2.33 we have $N^{s,X}_q = T_q W^s_{loc}(\sigma) \cap N_q$ and since $X(q) \in T_q W^s_{loc}(\sigma)$ we have $T_q W^s_{loc}(\sigma) = N^{s,X}_q \oplus E^X_q$. We conclude that

$$\angle\big(E^{cu,X}_q, T_q W^s_{loc}(\sigma)\big) = \angle\big(E^{cu,X}_q, N^{s,X}_q \oplus E^X_q\big) = \angle\big(N^{u,X}_q, N^{s,X}_q\big). \tag{5.19}$$

Since $\Sigma^{i,\pm}_{\delta,\delta'}$ is compact and does not contain singularities by construction, Theorem 2.33 ensures that there is $\gamma = \gamma(\delta, \delta')$ such that $\angle(N^{u,X}_q, N^{s,X}_q) > \gamma$ for all $q \in \Sigma^{i,\pm}_{\delta,\delta'}$. Replacing this inequality in (5.19) we conclude the proof of the lemma in this case.

For p close enough to σ_Y, we have $\operatorname{dist}(\tilde{p}, \Sigma^{i,\pm}_{\delta,\delta'} \cap \hat{W}^s_{loc}(\sigma_Y))$ arbitrarily small. Using the continuous dependence of the splitting $N^{s,X} \oplus N^{u,X}$ with the flow together with Theorem 2.33, we find that the estimate (5.19) above still holds replacing q by $\tilde{p}$ and X by Y, concluding the proof. □

Proof of Proposition 5.42 Assume, by contradiction, that there exists a sequence of periodic points $p_n \notin C_\delta$ for flows $Y_n \to X$ such that

$$\angle\big(E^{cu,Y_n}_{p_n}, E^{s,Y_n}_{p_n}\big) \xrightarrow[n\to+\infty]{} 0. \tag{5.20}$$

We claim that $\overline{\cup_n \mathcal{O}_{Y_n}(p_n)} \cap S(X) \cap \Lambda_X(U) \neq \emptyset$. Indeed, if this were not the case, we would get $\delta^* > 0$ such that $\overline{\cup_n \mathcal{O}_{Y_n}(p_n)} \subset \Omega_X(U_{\delta^*})$. By Corollary 5.45 the set $\Omega_X(U_{\delta^*})$ is hyperbolic and so there are neighborhoods V and $\mathcal{V}$ of $\Omega_X(U_{\delta^*})$ and Y, respectively, and $c > 0$ satisfying $\angle(E^{s,Y}_p, E^{cu,Y}_p) > c$ for all $p \in \mathrm{Per}_Y(\Lambda_Y(U))$ such that $\mathcal{O}_Y(p) \subset V$. Since $Y_n \xrightarrow[n\to+\infty]{} X$ we have $\mathcal{O}_{Y_n}(p_n) \subset V$ for n sufficiently large. We conclude that $\angle(E^{s,Y_n}_{p_n}, E^{cu,Y_n}_{p_n}) > c$, leading to a contradiction. Thus $\overline{\cup_n \mathcal{O}_{Y_n}(p_n)} \cap S(X) \cap \Lambda_X(U) \neq \emptyset$ as claimed.

Fix $\delta > 0$ and take cross-sections $\Sigma^{i,\pm}$ and $\Sigma^{o,\pm}$ as in Sect. 5.4.4.

Since $\overline{\cup_n \mathcal{O}_{Y_n}(p_n)} \cap S(X) \cap \Lambda_X(U) \neq \emptyset$, there exists a positive s_n such that $\tilde{p}_n = Y^{s_n}_n(p_n) \in \Sigma^{i,\pm}$ for each n.

Now take $\kappa = c$ with c given by Theorem 5.38(b), $\varepsilon < c/2$ and δ' as in Lemma 5.49. Fix $\delta^* < \min\{\delta, \delta'\}$ and consider $U_{\delta^*} = \overline{U \setminus C_{\delta^*}}$. By Corollary 5.45 the subset $\Omega_X(U_{\delta^*})$ is hyperbolic.

From the choice of δ^*, we obtain that $Y_n^s(p_n) \in U_{\delta^*}$ for any $0 \le s \le s_n$. We assume, without loss of generality, that every $\tilde{p}_n$ is in a neighborhood of the same singularity σ. Reasoning as in the proof of Proposition 5.41(a) we prove that (5.20) implies that

$$\angle\big(E_{\tilde{p}_n}^{s,Y_n}, Y_n(\tilde{p}_n)\big) \xrightarrow[n\to+\infty]{} 0. \tag{5.21}$$

Once (5.21) is settled, the proof follows analogously to that of the previous proposition. □

We finally present the proofs Lemmas 5.46, 5.48 and 5.49.

Proof of Lemma 5.46 Since Γ is hyperbolic, there are $0 < \lambda_\Gamma < 1$ and $c > 0$ such that $N_\Gamma^{s,X} = E_\Gamma^{s,X} \oplus E^X$ with $\|DX^t \mid E^{s,X}\| < c \cdot \lambda_\Gamma^t$, and $c^{-1} < \|X \mid \Gamma\| < c$. Changing the metric in a neighborhood of Γ, we can assume without loss of generality that $E_x^{s,X}$ is orthogonal to E_x^X and $\|X(x)\| = 1$ for all $x \in \Gamma$. In other words, in the new metric $E_\Gamma^{s,X}$ coincides with the stable bundle $N_\Gamma^{s,X}$ of the Linear Poincaré Flow restricted to Γ.

For each $x \in \Gamma$, let $n_x^{s,X} \in N_x^{s,X}$ be a unit vector and consider the orthogonal basis $\mathcal{B}_x = \{X(x), n_x^{s,X}\}$ of $E_x^X \oplus N_x^{s,X}$. In this basis the matrix of $DX^t(x)$ restricted to $E_x^X \oplus N_x^{s,X}$ is

$$DX^t \mid \big(E_x^X \oplus N_x^{s,X}\big) = \begin{bmatrix} 1 & 0 \\ 0 & n_{x,t}^{s,X} \end{bmatrix},$$

where $\|n_{x,t}^{s,X}\| < c \cdot \lambda_\Gamma^t$.

Fix $t_0 > 0$ such that $\|n_{x,t_0}^{s,X}\| < 1/2$ for all $x \in \Gamma$. There exists $c' > 0$ such that $\|n_{x,t_0}^{s,X}\| > c'$ for all $x \in \Gamma$ by continuity of the flow and compactness of Γ. Taking a neighborhood V of Γ and a neighborhood $\mathcal{V} \subset \mathcal{U}_U$ of X, both sufficiently small, and a change of metric varying continuously with the flow, we can make $\|Y(y)\| = 1$ for all $Y \in \mathcal{V}$ and all $y \in \Lambda_Y(U)$. Thus the matrix of $DY^{t_0}(y)$ restricted to $E_y^Y \oplus N_y^{s,Y}$ with respect to the basis $\mathcal{B}_y = \{Y(y), n_y^{s,Y}\}$ is

$$DY^{t_0} \mid \big(E_y^Y \oplus N_y^{s,Y}\big) = \begin{bmatrix} 1 & \delta_y^Y \\ 0 & n_{y,t_0}^{s,Y} \end{bmatrix},$$

where $\delta_y^Y < \delta_0$ and δ_0 is small for Y sufficiently close to X. Moreover $\|n_{y,t_0}^{s,Y}\| < 1/2$. Hence

$$DY_{n\cdot t_0} \mid \big(E_y^Y \oplus N_y^{s,Y}\big) = \begin{bmatrix} 1 & \delta_{y,n}^Y \\ 0 & n_{y,n\cdot t_0}^{s,Y} \end{bmatrix},$$

with $\delta_{y,n}^Y \ll 2\delta_0$. Let $\varepsilon > 0$ and n_0 be such that $2^{-n} < \varepsilon$ for all $n \ge n_0$. Given $v \in E_y^Y \oplus N_y^{s,Y}$ we can write $v = (1, \gamma_0')$ in the basis B_y. Then for any positive integer m we get

$$\angle\big(DY_{n_0\cdot m}(y)\cdot v,(1,0)\big)\le\frac{\gamma_0'\cdot n^{s,Y}_{y,n_0\cdot m}}{1-\delta^Y_{y,n_0}\cdot n^{s,Y}_{y,n_0\cdot m}}<\frac{(1/2)^{n_0\cdot m}}{1-2\delta_0}.$$

For $t>n_0$ we write $t=m\cdot n_0+s$ with $0\le s\le n_0$ and then

$$\angle\big(DY^t(y)\cdot v, Y(Y^t(y))\big)<K\cdot\varepsilon$$

for some positive constant K, concluding the proof of Lemma 5.46. □

Proof of Lemma 5.48 We prove the lemma by introducing linearizing coordinates in a normal neighborhood V of σ. For this we assume that there is a neighborhood V of σ where all Y sufficiently near to X are linearizable. This is no restriction since we can always get rid of resonances between the eigenvalues by small C^∞ perturbations of the flow. Fix $\delta>0$ small so that $B_\delta(\sigma)\subset V$. Assume also that $\sigma_Y=\sigma$ and the eigenvalues of $DY(\sigma_Y)$ are the same as the ones of $DX(\sigma)$. Let $\lambda_2<\lambda_3<0<-\lambda_3<\lambda_1$ be the eigenvalues of $DX(\sigma)$. So, in local coordinates $\bar{x},\bar{y},\bar{z}$, we have that $Y\mid V$ can be written as

$$Y(\bar{x},\bar{w},\bar{z})=\begin{cases}\dot{\bar{x}}=\lambda_1\bar{x}\\ \dot{\bar{y}}=\lambda_2\bar{y}\\ \dot{\bar{z}}=\lambda_3\bar{z}.\end{cases}\tag{5.22}$$

Note that in this case for $y\in W^s_{loc}(\sigma)$

$$W^s_{loc}(\sigma)=V\cap\big(\{0\}\times\mathbb{R}^2\big),\qquad W^u_{loc}(\sigma)=V\cap\big(\mathbb{R}\times\{(0,0)\},$$
$$W^{cu}_{loc}(\sigma)=V\cap\big(\mathbb{R}\times\{0\}\times\mathbb{R}\big),\qquad \hat{E}^{ss,Y}_y=V\cap\big(\{0\}\times\mathbb{R}\times\{0\}\big).$$

For $y\in W^u_{loc}(\sigma)$ we have $\hat{E}^{cu,Y}_y=\mathbb{R}\times\{0\}\times\mathbb{R}$ and $\Sigma^{o,\pm}\cap W^u_{loc}(\sigma)=\{(\pm1,0,0)\}$.

For $y\in V$ and for $v=(v_1,v_2,v_3)$, if $t>0$ is such that $Y^s(y)\in V$ for all $0\le s\le t$, then

$$DY^t(y)\cdot v=(e^{\lambda_1 t}v_1,e^{\lambda_2 t}v_2,e^{\lambda_3}v_3).\tag{5.23}$$

Given two vectors v and w we denote the slope between v and w by slope(v,w).

Let $r>0$ and let $y\in B_\delta(\sigma)$ be such that dist$(y,\hat{W}^s_{loc}(\sigma))<r$. Let $v=(v_1,v_2,v_3)\in T_yM$ and let $t>0$ be such that $Y^s(y)\in V$ for all $0\le s\le t$. Then

$$\text{slope}\left(DY^t(y)\cdot v,\hat{E}^{cu,X}_\sigma\right)=\frac{|e^{\lambda_2 t}\cdot v_2|}{\sqrt{(e^{\lambda_1 t}\cdot v_1)^2+(e^{\lambda_3 t}\cdot v_3)^2}}.$$

On the other hand, assuming that $\angle(\hat{E}^{ss,X}_\sigma,v)=\angle((0,1,0),v)>\gamma$ we find that there exists $0<\hat{\gamma}<1$ such that $0\le|v_2|<\hat{\gamma}$. Hence $v_1^2+v_3^2>1-\hat{\gamma}^2$. This implies that either $v_1 > \sqrt{(1-\hat{\gamma}^2)/2}$, or $v_3 > \sqrt{(1-\hat{\gamma}^2)/2}$. Thus

$$\operatorname{slope}\left(DY^t(y)\cdot v, \hat{E}^{cu,X}_\sigma\right) \leq \frac{|e^{\lambda_2 t}\cdot v_2|}{|e^{\lambda_i t}\cdot v_i|} \leq \frac{\hat{\gamma}}{\sqrt{(1-\hat{\gamma}^2)/2)}}\cdot e^{(\lambda_2-\lambda_i)t}, \tag{5.24}$$

where $i\in\{1,3\}$ is chosen so that v_i satisfies $v_i^2 > \sqrt{(1-\gamma^2)/2}$. As both $\lambda_2-\lambda_3$ and $\lambda_2-\lambda_1$ are strictly negative, there exists $T = T(\varepsilon,\gamma) > 0$ such the bound given by (5.24) is smaller than ε for all $t > T$.

Now taking r sufficiently small, for $y \in (B_\delta(\sigma)\setminus \hat{W}^s_{loc}(\sigma))$, we can ensure that if $Y^t(y)\in\Sigma^{o,\pm}$, then $t > T$. These last two facts combined complete the proof. □

Proof of Lemma 5.49 For the proof of this lemma we use local linearizable coordinates in a neighborhood of σ as in the proof of Lemma 5.48.

Let $\delta > 0$ be small enough so that $B_\delta(\sigma)\subset V$ and consider $\Sigma^{i,\pm}$, $\Sigma^{o,\pm}$ as in Sect. 5.4.4. Take $\delta' > 0$ and consider $\Sigma^{i,\pm}_{\delta,\delta'}$ as in (5.10). Let $p\in\Sigma^{i,\pm}_{\delta,\delta'}\cap\Lambda_Y(U)$ and $v\in N^{u,Y}_p\oplus E^Y_p$ with $\angle(v, Y(p)) > \kappa > 0$. Write $v = a\cdot(1,0,0)+b\cdot(0,1,0)+c\cdot(0,0,1)$ with $a^2+b^2+c^2=1$.

Claim There are $R > 0$ and δ' such that, if p and v are as above, then $|a| > R$.

Proof By the continuity of the flow direction and the normal bundle splitting far from singularities, it suffices to verify the claim for $p\in W^s_{loc}(\sigma)\setminus\{\sigma\}$. In this case $E^{cs,Y}_p = \{0\}\times\mathbb{R}^2$. Thus all we need to prove is that $\angle(v, E^{cs,Y}_p) > \kappa$ for some $\kappa > 0$. For this, observe that since $\operatorname{dist}(p,\sigma) > \delta$, by Theorem 2.33 there is $k' = k'(\delta)$ such that $\angle(N^{s,Y}_p, N^{u,Y}_p) > k'$. Since $\angle(E^{cu,Y}_p, E^{cs,Y}_p) = \angle(N^{s,Y}_p, N^{u,Y}_p)$, we conclude that

$$\angle\left(E^{cu,Y}_p, E^{cs,Y}_p\right) > k'. \tag{5.25}$$

On the other hand, $v\in E^Y_p\oplus N^{u,Y}_p = E^{cu,Y}_p$ and $\angle(v, Y(p)) = \angle(v, E^{cs,Y}_p\cap E^{cu,Y}_p) > \kappa$ by hypothesis. This fact combined with (5.25) give the proof of the claim. □

Returning to the proof of Lemma 5.49, let $t_p > 0$ be such that $Y^{t_p}(p)\in\Sigma^u_\delta$. Next we prove that for δ' small we have

1. $\angle\big(Y(Y^{t_p}(p)), (1,0,0)\big)$ is small, and
2. $\angle\big(DY^{t_p}(p)(v), (1,0,0)\big)$ is small.

Observe that, if $\delta'\to 0$, then $t_p\to\infty$ and $Y^{t_p}(p)$ converges to a point in $\hat{W}^u_{loc}(\sigma)$, where the flow direction is $(1,0,0)$. Hence the continuity of the flow direction implies the first item above.

To prove the second item, recall (5.23). Then by the previous claim

$$\left|\frac{b\cdot e^{\lambda_2\cdot t_p}}{a\cdot e^{\lambda_1\cdot t_p}}\right| < e^{(\lambda_2-\lambda_1)\cdot t_p}\cdot\frac{|b|}{R}.$$

Similarly, we have $\left|\frac{c\cdot e^{\lambda_3\cdot t_p}}{a\cdot e^{\lambda_1\cdot t_p}}\right| < e^{(\lambda_3-\lambda_1)\cdot t_p}\cdot |c|/R$. Since $t_p \to \infty$ as $\delta' \to 0$, $R > 0$ and both $\lambda_2 - \lambda_1$ and $\lambda_3 - \lambda_1$ are negative numbers, we deduce that the bounds on both inequalities above tend to 0 when $\delta' \to 0$, concluding the proof of Lemma 5.49. □

Chapter 6
Singular-Hyperbolicity and Robustness

Under the assumption of singular-hyperbolicity one can show that at each point there exists a strong stable manifold; more precisely, the attractor is a subset of a lamination by strong stable manifolds. It is also possible to show the existence of local central manifolds tangent to the central unstable direction, see [110] and Sect. 6.1. Although these central manifolds do not behave as unstable manifolds, in the sense that their points are not necessarily asymptotic in the past, the fact that the flow expands volume along the central unstable direction implies rather strong properties.

We list some of these properties that give us a nice description of the dynamics of robustly transitive sets with singularities and, in particular, for robust attractors, or of singular-hyperbolic attracting sets with a dense subset of periodic orbits.

The first two properties do not depend on the fact that the set is either robustly transitive or an attractor, but only on the fact that the set is partially hyperbolic and that the flow expands volume in the central-unstable direction.

Lemma 6.1 *For a non-trivial (different from a finite set of singularities) partially hyperbolic set Λ with hyperbolic singularities of saddle type of a C^1 flow X, the flow direction is contained in the central bundle.*

Proposition 6.2 *Let Λ be a singular-hyperbolic compact set of $X^t \in \mathfrak{X}^1(M)$. Then any invariant compact set $\Gamma \subset \Lambda$ without singularities is a uniformly hyperbolic set.*

It is clear that a singular-hyperbolic attractor has only finitely many equilibria, since these are hyperbolic equilibria (thus isolated) contained in a compact set. Morales in [165] shows that there are smooth vector fields exhibiting singular-hyperbolic attractors with any given (finite) number of (Lorenz-like) equilibria in compact 3-manifolds. Moreover, in [164], Morales also obtains an example of a singular-hyperbolic *attracting* set *without singularities*.

For singular hyperbolic attracting sets having only one singularity we can obtain a partial converse to the results in Chap. 5.

V. Araújo, M.J. Pacifico, *Three-Dimensional Flows*, Ergebnisse der Mathematik und ihrer Grenzgebiete. 3. Folge / A Series of Modern Surveys in Mathematics 53,
DOI 10.1007/978-3-642-11414-4_6, © Springer-Verlag Berlin Heidelberg 2010

We start by stating a corollary of the arguments used to prove Theorems 5.10 and 5.11 (see Remarks 5.25 and 5.31 in Chap. 5). Observe that we assume partial hyperbolicity with volume expanding central direction but do not assume transitivity in the following statement.

Theorem 6.3 *Let Λ be a nonempty compact invariant isolated set for a three-dimensional flow $X^t \in \mathfrak{X}^1$. Assume that Λ is partially hyperbolic with volume expanding central direction. If σ is a singularity properly accumulated by regular orbits in Λ, then*

- *either σ is Lorenz-like for X and $W^{ss}_X(\sigma) \cap \Lambda = \{\sigma\}$;*
- *or σ is Lorenz-like for $-X$ and $W^{uu}_X(\sigma) \cap \Lambda = \{\sigma\}$.*

The conditions on the orbits accumulating the singularity allow us to use the Connecting Lemma to produce a saddle connection for a nearby C^1 flow which will contradict the existence of dominated splitting if the singularity is not Lorenz-like.

Definition 6.4 Let Λ be an isolated set of $X^t \in \mathfrak{X}^r(M)$. We say that Λ is C^r *robustly periodic* if there are an isolating block U of Λ and a neighborhood $\mathscr{U}$ of X in $\mathfrak{X}^r(M)$ such that $\Lambda_Y(U) = \overline{\mathrm{Per}(Y) \cap \Lambda_Y(U)}$ for all $Y \in \mathscr{U}$.

Examples of C^1 robustly periodic sets are the hyperbolic attractors and the geometric Lorenz attractor (see Sects. 2.3 and 3.3). These examples are also C^1 robustly transitive. On the other hand, the singular horseshoe (from Sect. 3.1) and the example by Morales and Pujals in [171] are neither C^1 robust transitive nor C^1 robustly periodic. These examples motivate the question whether *all* C^1 robust transitive sets for vector fields are C^1 robustly periodic. Arroyo and Pujals have recently obtained a positive answer to this question in [27]:

Theorem 6.5 *Every singular-hyperbolic attractor has a dense subset of periodic orbits.*

Combining this with Theorem 5.13 and Proposition 6.2 one deduces the following.

Corollary 6.6 *Every C^1 robustly transitive set of a three-dimensional manifold is C^1 robustly periodic.*

Moreover Theorem 6.5 together with Proposition 2.10 implies that singular-hyperbolic attractors are chaotic (as defined in Sect. 2.4) and, in particular, have sensitive dependence on initial conditions.

Corollary 6.7 *Every singular-hyperbolic attractor is chaotic.*

The existence of a periodic orbit for singular-hyperbolic *attracting sets* was first obtained by Bautista-Morales in [37]. However it is possible to construct *a singular-hyperbolic isolated set with a transitive orbit but without periodic orbits*, see [162].

All known singular-hyperbolic attractors are homoclinic classes. It has been conjectured by Morales in [164] that every singular-hyperbolic attractor is in fact a homoclinic class, which, from the Birkhoff-Smale Theorem 2.17, would imply Theorem 6.5.

Using the constructions of adapted cross-sections and of global Poincaré maps, explained in Sect. 6.1 and first presented in [21], we give a proof of this property (and of Theorem 6.5 as corollary through the Birkhoff-Smale Theorem) for singular-hyperbolic attractors. We observe that this property was proved for the geometric Lorenz attractors in Chap. 3.

Theorem 6.8 *Every singular-hyperbolic attractor for a C^1 flow X^t is a homoclinic class, that is, there exists a periodic orbit $\mathcal{O}_X(p)$ of X in Λ such that $\Lambda = H_X(p)$.*

In general, singular-hyperbolic attractors with only one singularity may not be C^r robustly transitive as already observed by Morales-Pujals in [171] and explained in Example 5.7 in Chap. 5.

Nevertheless, on compact 3-manifolds, C^r robustly periodic sets are C^r robustly transitive *among singular-hyperbolic attractors with only one singularity.*

Theorem 6.9 *A C^r robustly periodic singular-hyperbolic attractor, with only one singularity, on a compact 3-manifold is C^r robust.*

This result, first proved in [169], gives explicit sufficient conditions for robustness of attractors *depending on the perturbed flow*. One should aim to obtain sufficient conditions *depending only on the unperturbed flow*. This is still an open question.

We prove Theorem 6.8 in Sect. 6.2 (and Theorem 6.5 is a corollary of this as already observed). We present a proof of Theorem 6.9 following [169] in Sect. 6.3.

The next two results show that important features of hyperbolic attractors and of the geometric Lorenz attractor are present for singular-hyperbolic attractors, and so also for robust attractors with singularities.

Proposition 6.10 *A singular-hyperbolic attractor Λ of $X \in \mathfrak{X}^1(M)$ has positive Lyapunov exponent uniformly bounded away from zero at every orbit.*

The following generic property in the space $\mathfrak{X}^1(M)$ can also be deduced from Theorems 5.3 and 5.10.

Proposition 6.11 *For X in a residual subset (a set containing an intersection of an enumerable family of open dense subsets) of $\mathfrak{X}^1(M)$, each robust transitive set with singularities is the closure of the stable or unstable manifold of one of its hyperbolic periodic points.*

Now we present proofs of the propositions stated above using the results from Chap. 5.

Proof of Lemma 6.1 It is enough to show that the flow direction $X(x)$ at $x \in \Lambda \setminus S(X)$ is not contained in the uniformly contracting direction of the dominated splitting $E^s \oplus E^c$ and that the angle between E^X and E^s is bounded away from zero.

Indeed, the one-dimensional sub-bundle E^X generated by the flow direction at regular points $x \in \Lambda$ is DX^t-invariant. If E^X is not contained in E^s and the angle between the two is bounded away from zero, then at each $x \in \Lambda \setminus S(X)$ the sub-bundle E^X is contained in the graph of a linear map $\ell_x : E^c_x \to E^s_x$ whose norm is bounded from above uniformly on $x \in \Lambda \setminus S(X)$. The domination condition on the splitting now implies that the ratio

$$\frac{\|\ell_x(u)\|}{\|u\|} = \frac{\|DX^t_x \cdot \big(DX^{-t}_x \cdot \ell_x(u)\big)\|}{\|DX^t_x \cdot \big(DX^{-t}_x \cdot u\big)\|} = \frac{\|DX^t_x \cdot \ell_{X^{-t}(x)}(v)\|}{\|DX^t_x \cdot v\|}$$

for any given $u \in E^c_x \setminus \{0\}$, goes to zero exponentially fast with t going to $+\infty$, where $v = DX^{-t}_x \cdot u$ is such that $v + \ell_{X^{-t}(x)}(v) = DX^{-t}_x \cdot X(x) = X(X^{-t}(x))$ by the invariance of E^X. This means that in fact $\ell_x(u) = 0$ for all $u \in E^c_x$ and so $E^X \subset E^c$.

If we assume by contradiction that $E^X_x \in E^s_x$, then we get for all $t > 0$

$$\|X(x)\| = \|DX^t DX^{-t} X(x)\| \le K e^{-\lambda t} \|DX^{-t} X(x)\|$$

and so we obtain $\|DX^{-t} X(x)\| \ge \hat{K} e^{\lambda t}$ which grows exponentially fast with t. Moreover since both E^X and E^s are DX^t-invariant, we see that $X(z) \in E^s_z$ for all z in the alpha-limit set $\alpha_X(x)$ of the orbit of x. Therefore $\|X(X^t(z))\| = \|DX^t X(z)\| \to 0$ as $t \to +\infty$ and so there exists $\sigma \in S(X) \cap \alpha_X(x)$ such that $z \in W^s(\sigma)$.

If $\alpha_X(x) = \{\sigma\}$, then $X(X^{-t}(x)) = DX^{-t} X(x)$ should grow exponentially fast, because $X(x) \in E^s_x$, and also tend to zero, because $X^{-t}(x) \to \sigma \in S(X)$. This contradiction shows that $\alpha_X(x) \setminus \{\sigma\} \neq \emptyset$. In addition, since σ is hyperbolic of saddle type and accumulated by the backward orbit of the regular point x, then there must be a point $z \in \alpha_X(x) \cap W^u(\sigma) \setminus \{\sigma\}$ by the linearization of the flow near σ. We again deduce a contradiction because $X(X^{-t}(z))$ should grow exponentially fast and tend to zero. This concludes the proof that $E^X \not\subset E^s$.

Near the singularities of Λ, the flow direction is clearly away from the uniformly contracting direction at the singularities. The continuity of a dominated splitting ensures that this holds on a neighborhood U of the singularities in Λ. Away from the singularities the set $\Lambda \setminus U$ is compact, and so the continuity of the splittings $E^s \oplus E^c$ and the bundle E^X and the fact that $E^X \not\subseteq E^s$ ensure that there exists a positive lower bound for the angle between E^s and E^X on $\Lambda \setminus U$. This is enough to conclude the statement of the lemma. □

Proof of Proposition 6.2 The argument relies on the fact that the intersection of the dominated splitting $E^s \oplus E^{cu}$ with the normal bundle N_Γ over Γ induces a hyperbolic splitting for the Linear Poincaré Flow defined over Γ (recall the definition of the Linear Poincaré Flow in Sect. 2.6). Thus by Theorem 2.27 we conclude that Γ is uniformly hyperbolic, finishing the proof.

From the fact that Γ does not contain singularities, there exists $K > 0$ such that $1/K < \|X(x)\| < K$ for every $x \in \Gamma$. Consider the following splitting on the normal bundle N_Γ: define $N^u_x = E^{cu}_x \cap N_x$ and $N^s_x = E^{cs}_x \cap N_x$ for $x \in \Gamma$, where $E^{cs}_x = E^X_x \oplus E^s_x$.

Now we show that this splitting is hyperbolic for the Linear Poincaré Flow P_t over Γ. Note that for any $t \in \mathbb{R}$ the Jacobian of DX^t along the sub-bundle E^{cu}_x can be given by

$$\sin\angle\big(DX^t(x)\cdot n^u_x, X(X^t(x))\big)\cdot\|DX^t(x)\cdot n^u_x\|\cdot\frac{\|X(X^t(x))\|}{\|X(x)\|},$$

where $n^u_x \in N^u_x$ is any choice of a unit vector. The last expression is the same as

$$\big\|O_{X^t(x)}\big(DX^t(x)\cdot n^u_x\big)\big\|\cdot\frac{\|X(X^t(x))\|}{\|X(x)\|},$$

where $O_{X^t(x)}$ denotes the orthogonal projection from $T_{X^t(x)}M$ onto $N_{X^t(x)}$; recall Sect. 2.6. Thus

$$\big|\det(DX^t \mid E^{cu}_x)\big| = \big\|O_{X^t(x)}(DX^t(x)\cdot n^u_x)\big\|\cdot\frac{\|X(X^t(x))\|}{\|X(x)\|}. \tag{6.1}$$

Since the central direction is (c,λ)-volume expanding, we know that the value of the expression in (6.1) is bigger than $c\cdot e^{\lambda t}$. Hence we get

$$\big\|O_{X^t(x)}(DX^t(x)\cdot n^u_x)\big\| > \frac{c}{K^2}\cdot e^{\lambda t}\quad\text{for all } t\ge 0.$$

This proves that N^u is uniformly expanded by P^t.

To see that N^s is uniformly contracted by the Linear Poincaré Flow, first note that the splitting $E^s\oplus E^{cu}$ is partially hyperbolic along Γ. Thus there exists $A > 0$ such that $\angle(E^s_x, X(x)) > A$ for every $x \in \Gamma$. Hence we can find a_0 such that, for all $x \in \Gamma$ and $v \in N^s_x$ with $\|v\| = 1$, there exists $w \in E^s_x$ with $\|w\| = 1$ such that $v = aw + b\cdot\frac{X(x)}{\|X(x)\|}$ with $|a| < a_0$. Therefore we have

$$\begin{aligned}\big\|O_{X^t(x)}(DX^t(x)\cdot v)\big\| &= \left\|O_{X^t(x)}\left(DX^t(x)\cdot\left(aw + b\cdot\frac{X(x)}{\|X(x)\|}\right)\right)\right\|\\ &= \big\|O_{X^t(x)}(DX^t(x)\cdot(aw))\big\|\\ &\le \big\|DX^t(x)\cdot(aw)\big\| \le a_0\cdot K\cdot e^{-\lambda t}\end{aligned}$$

for some $K,\lambda > 0$ (recall that E^s is (K,λ) contracting). Thus N^s is uniformly contracted by P_t. Proposition 6.2 is proved. □

Proof of Proposition 6.10 Let Λ be as in the statement of Proposition 6.10. Given $x \in \Lambda$, if x is a singularity then the result follows from the fact that x is Lorenz-like for X. Now assume that $X(x) \neq 0$ and take $v \in E^{cu}_x$ with $\|v\| = 1$ and orthogonal to $X(x)$. We have for some $c, \lambda > 0$

$$c \cdot e^{\lambda t} \leq \left|\det(DX^t \mid E^{cu}_x)\right| \leq \left\|DX^t(x) \cdot v\right\| \cdot \frac{\|DX^t(x) \cdot X(x)\|}{\|X(x)\|}$$
$$= \left\|DX^t(x) \cdot v\right\| \cdot \frac{\|X(X^t(x))\|}{\|X(x)\|}$$

and then for $t > 0$ we get

$$\frac{1}{t} \log \left\|DX^t(x) \cdot v\right\| \geq \lambda + \frac{1}{t} \log c - \frac{1}{t} \log \frac{\|X(X^t(x))\|}{\|X(x)\|}.$$

Since $\|X(X^t(x))\|$ is uniformly bounded for all $t > 0$ by compactness of Λ, we see that $\limsup_{t \to +\infty} t^{-1} \log \|DX^t(x) \cdot v\| > 0$. □

Proof of Proposition 6.11 Let $\Lambda = \Lambda_X(U)$ be a robustly transitive set with singularities for $X \in \mathfrak{X}^1(M)$ with isolating neighborhood U. By Theorems 5.3 and 5.13 we can assume that Λ is a partially hyperbolic attractor for X. On the other hand, since the Kupka-Smale property is generic, we deduce that, for a generic subset $\mathscr{G}$ in a C^1 neighborhood $\mathscr{V}$ of X, $\Lambda = \Lambda_Y(U)$ has a hyperbolic periodic orbit p for all $Y \in \mathscr{G}$.

As Λ is an attractor, the unstable manifold $W^u(p)$ of any periodic point p of Λ is contained in Λ. In particular its closure $\overline{W^u(p)}$ is contained in Λ. We show that Λ is contained in $\overline{W^u(p)}$.

Let $q \in \Lambda$ be such that $\Lambda = \omega_Y(q)$ (recall that an attractor is transitive by definition). Let V be a small neighborhood of p. On the one hand, by transitivity, we can assume without loss of generality that $q \in V$. On the other hand, since Λ is partially hyperbolic, projecting q into $W^u(p)$ through the stable manifold of q, we can assume that q is actually contained in $W^u(p)$. Indeed, being in the same stable manifold, q and its projection have the same ω-limit sets.

Finally observe that $\omega_Y(q) \subset \overline{W^u(p)}$ because $W^u(p)$ is invariant by the flow. Thus $\Lambda = \omega_Y(q) \subset \overline{W^u(p)}$ finishing the proof. □

6.1 Cross-Sections and Poincaré Maps

For future reference we give here a few properties of *Poincaré maps*, that is, continuous maps $R : \Sigma \to \Sigma'$ of the form $R(x) = X^{t(x)}(x)$ between cross-sections Σ and Σ' of the flow near a singular-hyperbolic set. We always assume that the Poincaré time $t(\cdot)$ is large enough as explained in what follows.

These properties will be often used in the following chapters to obtain many dynamical and ergodic consequences of singular-hyperbolicity. In particular they will be used in Sect. 6.3.1 to prove Theorem 6.36.

We assume that Λ is a compact invariant subset for a flow $X \in \mathfrak{X}^1(M)$ such that

- either Λ is a singular-hyperbolic attractor,
- or Λ is a singular-hyperbolic attracting set with a dense subset of periodic orbits.

In both cases it has already been proved that every singularity in Λ is Lorenz-like. Indeed since in both cases every singularity is properly accumulated by regular orbits in Λ and attracting sets contain the unstable manifold of every hyperbolic critical element, the only possible type of singularity allowed by Theorem 6.3 is a Lorenz-like singularity.

We start by observing that cross-sections have co-dimension one foliations which are dynamically defined: the leaves $W^s(x, \Sigma) = W^s_{loc}(x) \cap \Sigma$ correspond to the intersections with the stable manifolds of the flow. We shall prove that these leaves are uniformly contracted and, assuming the cross-section is *adapted*, that the foliation is invariant:

$$R(W^s(x, \Sigma)) \subset W^s(R(x), \Sigma') \quad \text{for all } x \in \Sigma.$$

Moreover, we will show that R is uniformly expanding in the transverse direction. Then we analyze the flow close to singularities, again by means of cross-sections.

6.1.1 Stable Foliations on Cross-Sections

We recall classical facts about partially hyperbolic systems, especially the existence of strong-stable and center-unstable foliations. The standard reference is [110].

We have that Λ is a singular-hyperbolic isolated set of $X \in \mathfrak{X}^1(M)$ with invariant splitting $T_\Lambda M = E^s \oplus E^{cu}$ with $\dim E^{cu} = 2$. Let $\tilde{E}^s \oplus \tilde{E}^{cu}$ be a continuous extension of this splitting to a small neighborhood U_0 of Λ. For convenience we take U_0 to be forward invariant. Then $\tilde{E}^s$ may be chosen invariant under the derivative: just consider at each point the direction formed by those vectors which are strongly contracted by DX^t for positive t. In general $\tilde{E}^{cu}$ is not invariant. However we can consider a cone field around it on U_0

$$C^{cu}_a(x) = \{v = v^s + v^{cu} : v^s \in \tilde{E}^s_x \text{ and } v^{cu} \in \tilde{E}^{cu}_x \text{ with } \|v^s\| \le a \cdot \|v^{cu}\|\}$$

which is forward invariant for $a > 0$:

$$DX^t(C^{cu}_a(x)) \subset C^{cu}_a(X^t(x)) \quad \text{for all large } t > 0. \tag{6.2}$$

Moreover we may take $a > 0$ arbitrarily small, reducing U_0 if necessary. For notational simplicity we write E^s and E^{cu} for $\tilde{E}^s$ and $\tilde{E}^{cu}$ in all that follows.

The next result says that there are locally strong-stable and center-unstable manifolds, defined at every regular point $x \in U_0$ and which are embedded disks tangent to $E^s(x)$ and $E^{cu}(x)$, respectively. The strong-stable manifolds are locally invariant. Given any $x \in U_0$ define the strong-stable manifold $W^{ss}(x)$ and the stable-manifold $W^s(x)$ as in Sect. 2.3.

Given $\varepsilon > 0$, denote $I_\varepsilon = (-\varepsilon, \varepsilon)$ and let $\mathscr{E}^1(I_1, M)$ be the set of C^1 embedding maps $f : I_1 \to M$ endowed with the C^1 topology.

Proposition 6.12 (Stable and center-unstable manifolds) *There are continuous maps $\phi^{ss} : U_0 \to \mathscr{E}^1(I_1, M)$ and $\phi^{cu} : U_0 \to \mathscr{E}^1(I_1 \times I_1, M)$ such that, given any $0 < \varepsilon < 1$ and $x \in U_0$, if we denote $W^{ss}_\varepsilon(x) = \phi^{ss}(x)(I_\varepsilon)$ and $W^{cu}_\varepsilon(x) = \phi^{cu}(x)(I_\varepsilon \times I_\varepsilon)$, then*

(a) $T_x W^{ss}_\varepsilon(x) = E^s(x)$;
(b) $T_x W^{cu}_\varepsilon(x) = E^{cu}(x)$;
(c) *$W^{ss}_\varepsilon(x)$ is a neighborhood of x inside $W^{ss}(x)$;*
(d) *if $y \in W^{ss}(x)$ then there is $T \geq 0$ such that $X^T(y) \in W^{ss}_\varepsilon(X^T(x))$ (local invariance);*
(e) *$d(X^t(x), X^t(y)) \leq K \cdot e^{-\lambda t} \cdot d(x, y)$ for all $t > 0$ and all $y \in W^{ss}_\varepsilon(x)$.*

The constants $K, \lambda > 0$ are taken as in the definition of (K, λ)-splitting at the beginning of Chap. 5, and the distance $d(x, y)$ is the intrinsic distance between two points on the manifold $W^{ss}_\varepsilon(x)$, given by the length of the shortest smooth curve contained in $W^{ss}_\varepsilon(x)$ connecting x to y.

Denoting $E^{cs}_x = E^s_x \oplus E^X_x$, where E^X_x is the direction of the flow at x, it follows that

$$T_x W^{ss}(x) = E^s_x \quad \text{and} \quad T_x W^s(x) = E^{cs}_x.$$

We fix ε once and for all. Then we call $W^{ss}_\varepsilon(x)$ the local *strong-stable manifold* and $W^{cu}_\varepsilon(x)$ the local *center-unstable manifold* of x.

Now let Σ be a *cross-section* to the flow, that is, a C^2 embedded compact disk transverse to X: at every point $z \in \Sigma$ we have $T_z\Sigma \oplus E^X_z = T_z M$ (recall that E^X_z is the one-dimensional subspace $\{s \cdot X(z) : s \in \mathbb{R}\}$). For every $x \in \Sigma$ we define $W^s(x, \Sigma)$ to be the connected component of $W^s(x) \cap \Sigma$ that contains x. This defines a foliation $\mathscr{F}^s_\Sigma$ of Σ into co-dimension 1 sub-manifolds of class C^1.

Remark 6.13 Given any cross-section Σ and a point x in its interior, we may always find a smaller cross-section also with x in its interior and which is the image of the square $[0, 1] \times [0, 1]$ by a C^2 diffeomorphism h that sends horizontal lines inside leaves of $\mathscr{F}^s_\Sigma$. In what follows we always assume that cross-sections are of this kind; see Fig. 6.1. We denote by $\text{int}(\Sigma)$ the image of $(0, 1) \times (0, 1)$ under the above-mentioned diffeomorphism, which we call the *interior* of Σ.

We also assume that each cross-section Σ is contained in U_0, so that every $x \in \Sigma$ is such that $\omega(x) \subset \Lambda$.

Remark 6.14 In general, we can not choose the cross-section such that $W^s(x, \Sigma) \subset W^{ss}_\varepsilon(x)$. The reason is that we want cross-sections to be C^2. Cross-sections of class C^1 are enough for the proof of expansiveness in Sect. 7.2.1 but C^2 is needed for the construction of the physical measure in Sect. 7.3.3 and for the absolute continuity results in Sect. 7.3.10. See Sect. 2.7.2 for the technical definitions.

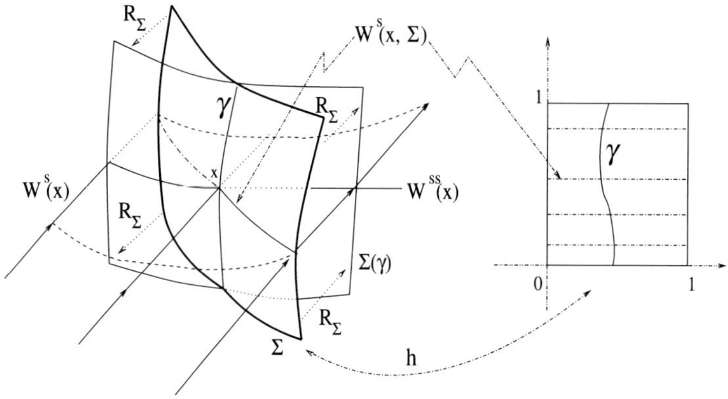

Fig. 6.1 The sections Σ, $\Sigma(\gamma)$, the manifolds $W^s(x)$, $W^{ss}(x)$, $W^s(x, \Sigma)$ and the projection R_Σ, on the right. On the left, the square $[0, 1]^2$ is identified with Σ through the map h, where $\mathscr{F}^s_\Sigma$ becomes the horizontal foliation and γ is a transverse curve, and solid lines with arrows indicate the flow direction

On the one hand $x \mapsto W^{ss}_\varepsilon(x)$ is usually not differentiable if we assume that X is only of class C^1; see e.g. [193]. On the other hand, assuming that the cross-section is small with respect to ε, and choosing any curve $\gamma \subset \Sigma$ crossing transversely every leaf of $\mathscr{F}^s_\Sigma$, we may consider a Poincaré map

$$R_\Sigma : \Sigma \to \Sigma(\gamma) = \bigcup_{z \in \gamma} W^{ss}_\varepsilon(z)$$

with Poincaré time close to zero; see Fig. 6.1. This is a homeomorphism onto its image, close to the identity, such that $R_\Sigma(W^s(x, \Sigma)) \subset W^{ss}_\varepsilon(R_\Sigma(x))$. So, identifying the points of Σ with their images under this homeomorphism, we may pretend that indeed $W^s(x, \Sigma) \subset W^{ss}_\varepsilon(x)$. We shall often do this in the sequel, to avoid cumbersome technicalities.

6.1.2 Hyperbolicity of Poincaré Maps

Let Σ be a small cross-section to X and let $R : \Sigma \to \Sigma'$ be a Poincaré map $R(y) = X^{t(y)}(y)$ to another cross-section Σ' (possibly $\Sigma = \Sigma'$). Note that R need not correspond to the first time that the orbits of Σ encounter Σ'.

The splitting $E^s \oplus E^{cu}$ over U_0 induces a continuous splitting $E^s_\Sigma \oplus E^{cu}_\Sigma$ of the tangent bundle $T\Sigma$ to Σ (and analogously for Σ'), defined by

$$E^s_\Sigma(y) = E^{cs}_y \cap T_y\Sigma \quad \text{and} \quad E^{cu}_\Sigma(y) = E^{cu}_y \cap T_y\Sigma. \tag{6.3}$$

We now show that if the Poincaré time $t(x)$ is sufficiently large then (6.3) defines a hyperbolic splitting for the transformation R on the cross-sections, at least restricted to Λ, in the following sense.

Proposition 6.15 *Let $R : \Sigma \to \Sigma'$ be a Poincaré map as before with Poincaré time $t(\cdot)$. Then $DR_x(E^s_\Sigma(x)) = E^s_{\Sigma'}(R(x))$ at every $x \in \Sigma$ and $DR_x(E^{cu}_\Sigma(x)) = E^{cu}_{\Sigma'}(R(x))$ at every $x \in \Lambda \cap \Sigma$.*

Moreover for every given $0 < \lambda < 1$ there exists $T_1 = T_1(\Sigma, \Sigma', \lambda) > 0$ such that, if $t(\cdot) > T_1$ at every point, then

$$\|DR \mid E^s_\Sigma(x)\| < \lambda \quad \textit{and} \quad \|DR \mid E^{cu}_\Sigma(x)\| > 1/\lambda \quad \textit{at every } x \in \Sigma.$$

Remark 6.16 In what follows we use K as a generic notation for large constants depending only on a lower bound for the angles between the cross-sections and the flow direction, and on upper and lower bounds for the norm of the vector field on the cross-sections. The conditions on T_1 in the proof of the proposition depend only on these bounds as well. In all our applications, all these angles and norms will be uniformly bounded from zero and infinity, and so both K and T_1 may be chosen uniformly.

Proof The differential of the Poincaré map at any point $x \in \Sigma$ is given by

$$DR(x) = P_{R(x)} \circ DX^{t(x)} \mid T_x\Sigma,$$

where $P_{R(x)}$ is the projection onto $T_{R(x)}\Sigma'$ along the direction of $X(R(x))$. Note that $E^s_\Sigma(x)$ is tangent to $\Sigma \cap W^s(x) \supset W^s(x, \Sigma)$. Since the stable manifold $W^s(x)$ is invariant, we have invariance of the stable bundle:

$$DR(x)\big(E^s_\Sigma(x)\big) = E^s_{\Sigma'}\big(R(x)\big).$$

Moreover for all $x \in \Lambda$ we have

$$DX^{t(x)}\big(E^{cu}_\Sigma(x)\big) \subset DX^{t(x)}\big(E^{cu}_x\big) = E^{cu}_{R(x)}.$$

As $P_{R(x)}$ is the projection along the vector field, it sends $E^{cu}_{R(x)}$ to $E^{cu}_{\Sigma'}(R(x))$. This proves that the center-unstable bundle is invariant restricted to Λ, that is, $DR(x)(E^{cu}_\Sigma(x)) = E^{cu}_{\Sigma'}(R(x))$.

Next we prove the expansion and contraction statements. We start by noting that $\|P_{R(x)}\| \le K$. Then we consider the basis $\{\frac{X(x)}{\|X(x)\|}, e^u_x\}$ of E^{cu}_x, where e^u_x is a unit vector in the direction of $E^{cu}_\Sigma(x)$. Since the flow direction is invariant, the matrix of $DX^t \mid E^{cu}_x$ relative to this basis is upper triangular:

$$DX^{t(x)} \mid E^{cu}_x = \begin{bmatrix} \frac{\|X(R(x))\|}{\|X(x)\|} & \star \\ 0 & a \end{bmatrix}.$$

Moreover

$$\frac{1}{K} \cdot \det\big(DX^{t(x)} \mid E^{cu}_x\big) \le \frac{\|X(R(x))\|}{\|X(x)\|} a \le K \cdot \det\big(DX^{t(x)} \mid E^{cu}_x\big).$$

Then

$$\begin{aligned}\|DR(x)\,e_x^u\| &= \big\|P_{R(x)}\big(DX^{t(x)}(x)\cdot e_x^u\big)\big\| = \|a\cdot e_{R(x)}^u\| = |a| \\ &\geq K^{-3}\,\big|\det\big(DX^{t(x)}\mid E_x^{cu}\big)\big| \geq K^{-3}\lambda^{-t(x)} \geq K^{-3}\,\lambda^{-t_1}.\end{aligned}$$

Taking t_1 large enough we ensure that the latter expression is larger than $1/\lambda$.

To prove that $\|DR \mid E^s_\Sigma(x)\| < \lambda$, let us consider unit vectors $e_x^s \in E_x^s$ and $\hat{e}_x^s \in E^s_\Sigma(x)$, and write

$$e_x^s = a_x\cdot\hat{e}_x^s + b_x\cdot\frac{X(x)}{\|X(x)\|}.$$

Since $\angle(E_x^s, X(x)) \geq \angle(E_x^s, E_x^{cu})$ and the latter is uniformly bounded from zero, we have $|a_x| \geq \kappa$ for some $\kappa > 0$ which depends only on the flow. Then

$$\begin{aligned}\|DR(x)\,e_x^s\| &= \big\|P_{R(x)}\circ\big(DX^{t(x)}(x)\cdot e_x^s\big)\big\| \\ &= \frac{1}{|a_x|}\left\|P_{R(x)}\circ\left(DX^{t(x)}(x)\left(\hat{e}_x^s - b_x\frac{X(x)}{\|X(x)\|}\right)\right)\right\| \\ &= \frac{1}{|a_x|}\,\big\|P_{R(x)}\circ\big(DX^{t(x)}(x)\cdot\hat{e}_x^s\big)\big\| \leq \frac{K}{\kappa}\lambda^{t(x)} \leq \frac{K}{\kappa}\lambda^{t_1}.\end{aligned} \tag{6.4}$$

Once more it suffices to take t_1 large to ensure that the right hand side is less than λ. □

Given a cross-section Σ, a positive number ρ, and a point $x\in\Sigma$, we define the unstable cone of width ρ at x by

$$C_\rho^u(x) = \{v = v^s + v^u : v^s\in E^s_\Sigma(x),\ v^u\in E^{cu}_\Sigma(x) \text{ and } \|v^s\|\leq\rho\|v^u\|\} \tag{6.5}$$

(we omit the dependence on the cross-section in our notations).

Let $\rho > 0$ be any small constant. In the following consequence of Proposition 6.15 we assume the neighborhood U_0 has been chosen sufficiently small, depending on ρ and on a bound on the angles between the flow and the cross-sections.

Corollary 6.17 *For any* $R:\Sigma\to\Sigma'$ *as in Proposition* 6.15, *with* $t(\cdot) > T_1$, *and any* $x\in\Sigma$, *we have* $DR(x)(C_\rho^u(x))\subset C^u_{\rho/2}(R(x))$ *and*

$$\|DR_x(v)\| \geq \frac{5}{6}\lambda^{-1}\cdot\|v\| \quad \textit{for all } v\in C_\rho^u(x).$$

Proof Proposition 6.15 immediately implies that $DR_x(C_\rho^u(x))$ is contained in the cone of width $\rho/4$ around $DR(x)(E^{cu}_\Sigma(x))$ relative to the splitting

$$T_{R(x)}\Sigma' = E^s_{\Sigma'}(R(x))\oplus DR(x)\big(E^{cu}_\Sigma(x)\big).$$

(We recall that E^s_Σ is always mapped to $E^s_{\Sigma'}$.) The same is true for E^{cu}_Σ and $E^{cu}_{\Sigma'}$, restricted to Λ. So the previous observation already gives the conclusion of the first part of the corollary in the special case of points in the attractor. Moreover to prove the general case we only have to show that $DR(x)(E^{cu}_{\Sigma}(x))$ belongs to a cone of width less than $\rho/4$ around $E^{cu}_{\Sigma'}(R(x))$. This is easily done with the aid of the flow invariant cone field C^{cu}_a in (6.2), as follows. On the one hand,

$$DX^{t(x)}\big(E^{cu}_{\Sigma}(x)\big) \subset DX^{t(x)}\big(E^{cu}_{x}\big) \subset DX^{t(x)}\big(C^{cu}_{a}(x)\big) \subset C^{cu}_{a}\big(R(x)\big).$$

We note that $DR(x)(E^{cu}_{\Sigma}(x)) = P_{R(x)} \circ DX^{t(x)}(E^{cu}_{\Sigma}(x))$. On the other hand, since $P_{R(x)}$ maps $E^{cu}_{R(x)}$ to $E^{cu}_{\Sigma'}(R(x))$ and the norms of both $P_{R(x)}$ and its inverse are bounded by some constant K (see Remark 6.16), we conclude that $DR(x)(E^{cu}_{\Sigma}(x))$ is contained in a cone of width b around $E^{cu}_{\Sigma'}(R(x))$, where $b = b(a, K)$ can be made arbitrarily small by reducing a. We keep K bounded, by assuming that the angles between the cross-sections and the flow are bounded from zero and then, reducing U_0 if necessary, we can make a small so that $b < \rho/4$. This concludes the proof since the expansion estimate is a trivial consequence of Proposition 6.15. □

As usual a *curve* is the image of a compact interval $[a, b]$ by a C^1 map. We use $\ell(\gamma)$ to denote its length. By a *cu-curve* in Σ we mean a curve contained in the cross-section Σ and whose tangent direction is contained in the unstable cone $T_z\gamma \subset C^u_{\rho}(z)$ for all $z \in \gamma$.

The previous lemma ensures that the image of a cu-curve by Poincaré maps between cross-sections is another cu-curve. The next lemma says that *the length of cu-curves linking the stable leaves of nearby points x, y must be bounded by the distance between x and y.*

Lemma 6.18 *Let us assume that ρ has been fixed, sufficiently small. Then there exists a constant κ such that, for any pair of points $x, y \in \Sigma$, and any cu-curve γ joining x to some point of $W^s(y, \Sigma)$, we have $\ell(\gamma) \le \kappa \cdot d(x, y)$.*

Here d is the intrinsic distance in the C^2 surface Σ, that is, the length of the shortest smooth curve inside Σ connecting two given points in Σ.

Proof We consider coordinates on Σ for which x corresponds to the origin, $E^{cu}_{\Sigma}(x)$ corresponds to the vertical axis, and $E^{s}_{\Sigma}(x)$ corresponds to the horizontal axis; through these coordinates we identify Σ with a subset of its tangent space at x, endowed with the Euclidean metric. In general this identification is not an isometry, but the distortion is uniformly bounded, which is taken care of by the constants C_1 to C_4 in what follows.

The hypothesis that γ is a cu-curve implies that its velocity vector $\dot\gamma(s)$ is contained in the cone of width $C_1 \cdot \rho$ centered at $\gamma(s)$ for all values of the parameter s. In the coordinates described above this means that we may write $\gamma(s) = (\xi(s), s)$ for some C^1 function $\xi : [0, s_0] \to [0, +\infty)$ with $\xi(0) = 0$, $\xi(s) > 0$ for all $s > 0$ and $|\dot\xi| \le C_1\rho$.

On the other hand, stable leaves are close to being horizontal, that is, fixing some stable leaf through $y \in \Sigma$ we may write it as a graph $(u, \eta(u))$ for a C^1 function $\eta : (-u_0, u_0) \to \mathbb{R}$ with $\eta(0) = d > 0$ and $|\dot\eta| \le C_2\rho$ (see Fig. 6.2).

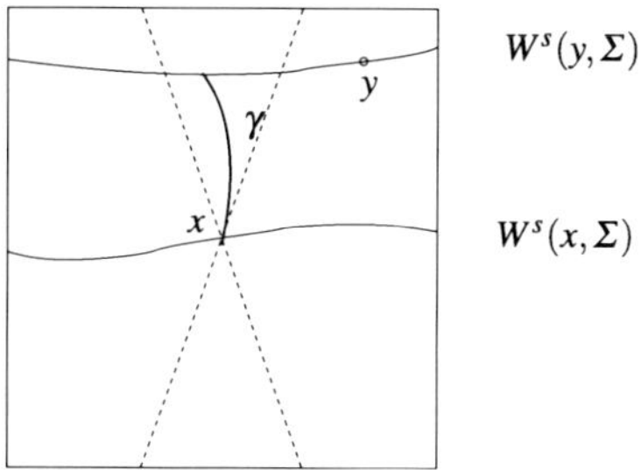

Fig. 6.2 The stable manifolds on the cross-section and the cu-curve γ connecting them

Observe now that $h = \eta \circ \xi$ satisfies $|h'| \le \delta = C_1 C_2 \rho^2$ and $h(0) = d$. Thus $|h(s) - d| \le \delta \cdot s$ and hence $h(s_*) = 0$ for some $0 < s_* < d/(1-\delta) < d(1+\delta)$. But this means that

$$\begin{cases} u = \xi(s)) \\ s = \eta(\xi(s)) = \eta(u) \end{cases} \quad \text{or} \quad \gamma(s) = (\xi(s), s) = (u, \eta(u)),$$

that is, we have an intersection between γ and the stable leaf at a distance from x along γ bounded by $d(1+\delta)\sqrt{1+(C_1\rho)^2} < d(1 + C_3\rho)$, where C_3 is a constant depending on C_1 and C_2 only.

Finally y has coordinates $(\eta(u_1), u_1)$ for some $|u_1| < u_0$ and since $u_0 < \rho$ we obtain $\eta(u_1) \ge d - \delta u_1 > d - \delta\rho$. Hence in Euclidean coordinates $\|x - y\| > d - \delta\rho = d(1 - \delta\rho/d)$ and then $d(x, y) > C_4 d$ for some $C_4 > 0$ depending on all the previous constants (remember that $d < \rho$ also) including the distortion due to the change of metric. It follows that the length of γ is bounded by $\kappa \cdot d(x, y)$ where we write $\kappa = (1+\delta)\sqrt{1+(C_1\rho)^2}/C_4$. □

In what follows we take T_1 in Proposition 6.15 for $\lambda = 1/3$. From Sect. 7.3.3 onwards we will need to decrease λ once by taking a bigger T_1.

6.1.3 Adapted Cross-Sections

Now we exhibit stable manifolds for Poincaré transformations $R : \Sigma \to \Sigma'$. The natural candidates are the intersections $W^s(x, \Sigma) = W^s_\varepsilon(x) \cap \Sigma$ which we introduced previously. These intersections are tangent to the corresponding sub-bundle E^s_Σ and so, by Proposition 6.15, they are contracted by the transformation. For our purposes it is also important that the stable foliation be invariant:

$$R(W^s(x, \Sigma)) \subset W^s(R(x), \Sigma') \quad \text{for every } x \in \Sigma. \tag{6.6}$$

In order to have this we restrict our class of cross-sections whose center-unstable boundary is disjoint from Λ. Recall (Remark 6.13) that we are considering cross-sections Σ that are diffeomorphic to the square $[0, 1] \times [0, 1]$, with the horizontal lines $[0, 1] \times \{\eta\}$ being mapped to stable sets $W^s(y, \Sigma)$. The *stable boundary* $\partial^s \Sigma$

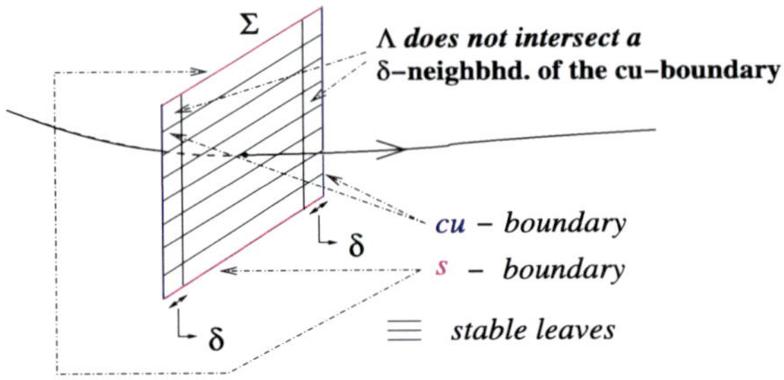

Fig. 6.3 An adapted cross-section for Λ

is the image of $[0,1]\times\{0,1\}$. The *center-unstable boundary* $\partial^{cu}\Sigma$ is the image of $\{0,1\}\times[0,1]$. The cross-section is δ-*adapted* if

$$d(\Lambda\cap\Sigma,\partial^{cu}\Sigma)>\delta,$$

where d is the intrinsic distance in Σ; see Fig. 6.3. We define a *horizontal strip* of Σ as the image $h([0,1]\times I)$ for any compact subinterval I, where $h:[0,1]\times[0,1]\to\Sigma$ is the coordinate system on Σ as in Remark 6.13. Notice that every horizontal strip is a δ-adapted cross-section.

In order to prove that adapted cross-sections do exist, we need the following result.

Lemma 6.19 *Let Λ be either a transitive singular-hyperbolic Lyapunov stable set, or a connected singular-hyperbolic attracting set admitting a dense subset of periodic orbits. Then every point $x\in\Lambda$ is in the closure of $W^{ss}(x)\setminus\Lambda$.*

Note that a singular-hyperbolic attractor satisfies the first condition of the statement of Lemma 6.19. We need the following simple result.

Lemma 6.20 *Let $X\in\mathfrak{X}^1(M)$ and let Λ be a compact invariant partially hyperbolic subset containing a strong-stable disk γ, that is γ is a neighborhood of some point of $W^{ss}(x)\cap\Lambda$ with the relative topology of $W^{ss}(x)$, for some $x\in\Lambda$. Then $L=\alpha_X(\gamma)=\{\alpha(z):z\in\gamma\}$ contains all stable disks through its points.*

Proof The partial hyperbolic assumption on Λ ensures that every one of its points has a strong-stable manifold. Moreover $W^{ss}(z)\subset\Lambda$ for every $z\in\alpha(\gamma)$, since any compact part of the strong-stable manifold of z is accumulated by backward iterates of any small neighborhood of $x\in\gamma$ inside $W^{ss}(x)$. Here we are using the fact that the contraction along the strong-stable manifold, which becomes an expansion for negative time, is uniform. □

Proof of Lemma 6.19 The proof is by contradiction. Let us suppose that there exists $x\in\Lambda$ such that x is in the interior of $W^{ss}(x)\cap\Lambda$. By Lemma 6.20 the set $\alpha(x)$ contains all stable manifolds through its points.

It follows that $\alpha(x)$ does not contain any singularity: indeed, Theorems 5.10 or 6.3 prove that the strong-stable manifold of each singularity meets Λ only at the singularity (observe that every singularity of Λ is properly accumulated by regular orbits in Λ).

Therefore by Proposition 6.2 the invariant set $\alpha(x) \subset \Lambda$ is hyperbolic. It also follows from the previous remarks that the set

$$H = \overline{\cup\{W^{ss}(y) : y \in \alpha(x) \cap \Lambda\}}$$

is contained in Λ. Also by the same argument as before, this set contains the strong-stable manifolds of all its points. Hence H does not contain any singularity, that is, H is uniformly hyperbolic.

We claim that $\overline{W^u(H)}$, the closure of the union of the unstable manifolds of the points of H, is an open set (it is clearly a closed set).

First we show that $W^u(H)$ is open. Note that H contains the whole stable manifold $W^s(z)$ of every $z \in H$: this is because H is invariant and contains the strong-stable manifold of z. Note that the union of the strong-unstable manifolds through the points of $W^s(z)$ contains a neighborhood of z. This proves that $W^u(H)$ is a neighborhood of H. Thus the backward orbit of any point in $W^u(H)$ must enter the interior of $W^u(H)$. Since the interior is an invariant set, this proves that $W^u(H)$ is open, as claimed.

Now observe that, because $W^u(H)$ is open and invariant, the strong-stable manifold of any $z \in W^u(H)$ is contained in $W^u(H)$, which is contained in Λ since we are assuming that Λ is either Lyapunov stable or attracting. Therefore taking limits we see that $W^{ss}(w) \subset \overline{W^u(H)}$ for all $w \in \overline{W^u(H)}$. This implies that $\overline{W^u(H)}$ does not contain singularities and is hyperbolic. Finally the unstable manifolds of points in $\overline{W^u(H)}$ are well defined by hyperbolicity and are contained in $\overline{W^u(H)}$, just by taking limits of points in $W^u(H)$. Hence $\overline{W^u(H)}$ contains its stable and unstable manifolds, and so it is an open set inside Λ.

Since Λ is also a connected set (which is always the case if Λ is transitive) we obtain $\Lambda = \overline{W^u(H)}$. This means that any singularity $\sigma \in \Lambda$ must be in $\overline{W^u(H)}$, a contradiction. The proof of the lemma is complete. □

Corollary 6.21 *For any $x \in \Lambda$ there exist points $x^+ \notin \Lambda$ and $x^- \notin \Lambda$ in distinct connected components of $W^{ss}(x) \setminus \{x\}$.*

Proof Otherwise there would exist a whole segment of the strong-stable manifold entirely contained in Λ. Considering any point in the interior of this segment, we would get a contradiction to Lemma 6.19. □

Lemma 6.22 *Let $x \in \Lambda$ be a regular point, that is, such that $X(x) \neq 0$. Then there exist $\delta > 0$ and a δ-adapted cross-section Σ at x.*

Proof Fix $\varepsilon > 0$ as in the stable manifold theorem. Any cross-section Σ_0 at x sufficiently small with respect to $\varepsilon > 0$ is foliated by the intersections $W^s_\varepsilon(x) \cap \Sigma_0$. By Corollary 6.21, we may find points $x^+ \notin \Lambda$ and $x^- \notin \Lambda$ in each of the connected components of $W^s_\varepsilon(x) \cap \Sigma_0$. Since Λ is closed, there are neighborhoods $V^\pm$ of $x^\pm$ disjoint from Λ. Let $\gamma \subset \Sigma_0$ be some small curve through x, transverse to

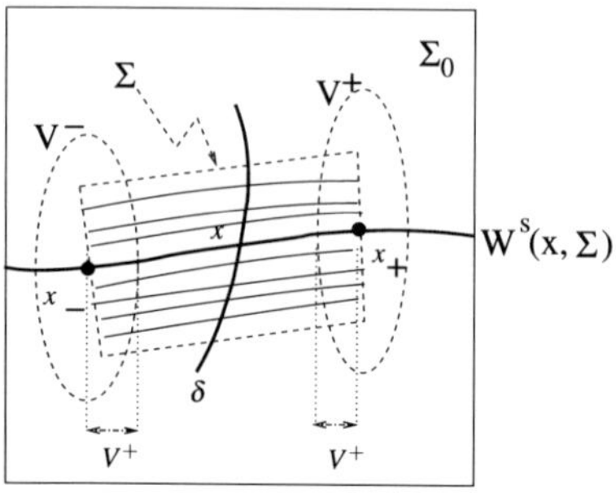

Fig. 6.4 The construction of a δ-adapted cross-section for a regular $x \in \Lambda$

$W^s_\varepsilon(x) \cap \Sigma_0$. Then we may find a continuous family of segments inside $W^s_\varepsilon(y) \cap \Sigma_0$, for $y \in \gamma$ with endpoints contained in $V^\pm$. The union Σ of these segments is a δ-adapted cross-section, for some $\delta > 0$; see Fig. 6.4. □

We are going to show that, if the cross-sections are adapted, then we have the invariance property (6.6). Given $\Sigma, \Sigma' \in \Xi$ we set $\Sigma(\Sigma') = \{x \in \Sigma : R(x) \in \Sigma'\}$, the domain of the return map from Σ to Σ'.

Lemma 6.23 *Given $\delta > 0$ and δ-adapted cross-sections Σ and Σ', there exists $T_2 = T_2(\Sigma, \Sigma') > 0$ such that, if $R : \Sigma(\Sigma') \to \Sigma'$ defined by $R(z) = R_{t(z)}(z)$ is a Poincaré map with time $t(\cdot) > T_2$, then*

1. $R(W^s(x, \Sigma)) \subset W^s(R(x), \Sigma')$ *for every $x \in \Sigma(\Sigma')$, and also*
2. $d(R(y), R(z)) \le \frac{1}{2} d(y, z)$ *for every $y, z \in W^s(x, \Sigma)$ and $x \in \Sigma(\Sigma')$.*

Proof This is a simple consequence of the relation (6.4) from the proof of Proposition 6.15: the tangent direction to each $W^s(x, \Sigma)$ is contracted at an exponential rate $\|DR(x)\, e^s_x\| \le Ce^{-\lambda t(x)}$. Choosing T_2 sufficiently large we ensure that

$$Ce^{-\lambda T_2} \cdot \sup\{\ell(W^s(x, \Sigma)) : x \in \Sigma\} < \delta.$$

In view of the definition of δ-adapted cross-section this gives part (1) of the lemma. Part (2) is entirely analogous: it suffices that $Ce^{-\lambda T_2} < 1/2$. □

Remark 6.24 Clearly we may choose $T_2 > T_1$. Remark 6.16 applies to T_2 as well.

The following is a technical consequence of the uniform contraction and the way cross-sections were chosen near real stable leaves.

Lemma 6.25 *Let Σ be a δ-adapted cross-section. Then given any $r > 0$ there exists ρ such that* $\operatorname{dist}(X^s(y), X^s(z)) < r$ *if $d(y, z) < \rho$ for all $s > 0$, every y, $z \in W^s(x, \Sigma)$, and every $x \in \Sigma$.*

Proof Let y and z be as in the statement. As in Remark 6.14, we may find $z' = X^\tau(z)$ in the intersection of the orbit of z with the strong-stable manifold of y satisfying

$$\frac{1}{K} \le \frac{\operatorname{dist}(y, z')}{d(y, z)} \le K \quad \text{and} \quad |\tau| \le K \cdot d(y, z).$$

Then, given any $s > 0$,

$$\begin{aligned}\operatorname{dist}(X^s(y), X^s(z)) &\le \operatorname{dist}(X^s(y), X^s(z')) + \operatorname{dist}(X^s(z'), X^s(z)) \\ &\le C \cdot e^{-\lambda s} \cdot \operatorname{dist}(y, z') + \operatorname{dist}(X^{s+\tau}(z), X^s(z)) \\ &\le KC \cdot e^{\gamma s} \cdot d(y, z) + K|\tau| \le (KC + K^2) \cdot d(y, z).\end{aligned}$$

Taking $\rho < r/(KC + K^2)$ we get the statement of the lemma. □

A very useful consequence of the hyperbolicity of Poincaré maps is the following criterion for the existence of a periodic orbit.

Lemma 6.26 *Let $x \in \Lambda$ be a regular point and suppose that there exist another regular point $z \in W^{ss}(x)$ and a sequence $t_n \to +\infty$ satisfying $X^{t_n}(z) \to x$ when $n \to \infty$. Then x belongs to a periodic orbit.*

Proof Take an adapted cross-section Σ through x. The conditions on z imply that there exists a Poincaré return map R defined on some substrip $\Sigma(\Sigma)$ containing $W = W^s(x, \Sigma)$, and that this line W is forward invariant $R(W) \subset W$. The contracting property given by Lemma 6.23 ensures there exists a periodic point p for R. Therefore p belongs to a periodic orbit for the flow and to the line W. Hence $z \in W^s(p)$ and so $\omega(z) = \mathcal{O}(p)$. Thus $x = p$ since there can be only one intersection $\mathcal{O}(p)$ with Σ on the same stable manifold. □

From Proposition 6.2 any compact invariant subset H of a singular-hyperbolic set Λ is uniformly hyperbolic, and of saddle type. Using adapted cross-sections we can say a bit more.

Lemma 6.27 *Let Λ be a singular-hyperbolic set. Suppose that one of the following conditions holds*

1. *Λ is Lyapunov stable and transitive, and H is a compact proper invariant subset of Λ;*
2. *Λ is an attracting set with a dense subset of periodic orbits, H is the set of accumulation points of a branch of the unstable manifold of some singularity σ of Λ, and H does not contain σ.*

Then either $H \subset S(X)$ or, for any adapted cross-section Σ through some regular point of H, the intersection $H \cap \Sigma$ is totally disconnected.

Note that the compact invariant set H is covered by a finite number of tubular flow-boxes or flow-boxes near singularities $U_{\Sigma_i} = X^{(-\varepsilon,\varepsilon)}(\Sigma_i)$, for $\varepsilon > 0$ small and

$i = 1, \dots, k$. From Lemma 6.27 we conclude that each $U_{\Sigma_i} \cap H$ has topological dimension one. Hence H in each case of the statement above is a one-dimensional set. For the definition and main properties of topological dimensions see e.g. [117].

Proof We follow the arguments in Morales [161]. If H is not contained in the set of singularities, fix a regular point $x \in H \cap \Sigma$. From Lemma 6.19 together with Remark 6.14 we have that the connected component C of $H \cap \Sigma$ containing x cannot contain intervals inside $W^s(x, \Sigma)$. Then either $C = \{x\}$ or C contains some point y in $\Sigma \setminus W^s(x, \Sigma)$. We show that the latter cannot happen in each case according to the assumption in the statement.

Observe first that since Σ is adapted there are no points of $H \cap \Sigma$ near the center-unstable boundary $\partial^{cu}\Sigma$. Hence there must be some point $h_0 \in H$ in the interior of the substrip Σ' of Σ formed by the points of Σ between the two horizontal lines $W^s(x, \Sigma)$ and $W^s(y, \Sigma)$. For otherwise $y \in C$ would be disconnected from x.

1. If Λ is transitive, then there exists $w \in \Sigma'$ close to h_0 with $\omega(w) = \Lambda$. Arguing as above, there must exist a point $\zeta \in H \cap W^s(w, \Sigma)$, for otherwise y and x would be in different connected components of $H \cap \Sigma \setminus W^s(w, \Sigma)$. Then $\Lambda = \omega(w) = \omega(\zeta) \subset H$. This is not possible because H is a proper subset of Λ.
2. Let $H = \omega(z)$ for some $z \in W^u(\sigma) \setminus \{\sigma\}$ and some singularity σ, as in item 2 of the statement, and suppose that H is not a singularity. Let Σ be some-cross section through some regular point h of H. Since $\mathrm{Per}(X)$ is dense in Λ, we can find a sequence p_n of points in periodic orbits such that $p_n \xrightarrow[n\to+\infty]{} \sigma$. By assumption we can find a point w in the positive orbit of z such that $w \in \Sigma$ close to h.

   Observe that, since $W^u(\sigma)$ is one-dimensional, we can assume without loss of generality that $z \in \Sigma_\sigma^{o,\pm}$ for some outgoing cross-section near σ. Then there are points $p_n' \in \mathcal{O}(p_n)$ satisfying $p_n' \xrightarrow[n\to+\infty]{} z$. So we can also find points $\tilde{p}_n \in \mathcal{O}(p_n)$ such that $\tilde{p}_n \xrightarrow[n\to+\infty]{} w$.

   As before, there exists a point $\zeta_n \in H \cap W^s(\tilde{p}_n, \Sigma)$. Hence we can find a sequence ζ_n' in the positive orbit of ζ_n, and thus in H, arbitrarily close to p_n'. Hence we can also find a sequence $\tilde{\zeta}_n \in H$ tending to σ. But then $\sigma \in H$, which is a contradiction.

We conclude that either $H \subset S(X)$ (and H is a singularity different from σ in the scenario of item 2), or the connected component of $H \cap \Sigma$ containing x is formed by x itself. □

6.1.4 Global Poincaré Return Map

Here we construct a global Poincaré map for the flow near the singular-hyperbolic attractor Λ. We then use the hyperbolicity properties of this map to reduce the dynamics to a one-dimensional piecewise expanding map through a quotient map over the stable leaves. This in turn enables us to use the rich known features of such low dimensional maps to deduce many properties of singular-hyperbolic attractors.

6.1.4.1 Cross-Sections and Invariant Foliations

We observe first that by Lemma 6.22 we can take a δ-adapted cross-section at each non-singular point $x \in \Lambda$. We know also that near each singularity σ_k there is a flow-box U_{σ_k} as in Sect. 5.4.4; see Fig. 5.3

Using a tubular neighborhood construction near any given δ-adapted cross-section Σ, we linearize the flow in an open set $U_\Sigma = X^{(-\varepsilon,\varepsilon)}(\mathrm{int}(\Sigma))$ for a small $\varepsilon > 0$, containing the interior of the cross-section. This provides an open cover of the compact set Λ by flow-boxes near the singularities and tubular neighborhoods around regular points.

We let $\{U_{\Sigma_i}, U_{\sigma_k} : i = 1, \dots, l;\ k = 1, \dots, s\}$ be a finite cover of Λ, where $s \geq 1$ is the number of singularities in Λ, and we set $T_3 > 0$ to be an upper bound for the time it takes any point $z \in U_{\Sigma_i}$ to leave this tubular neighborhood under the flow, for any $i = 1, \dots, l$. We assume without loss that $T_2 > T_3$.

To define the Poincaré map R, for any point z in one of the cross-sections in

$$\Xi = \{\Sigma_j, \Sigma^{i,\pm}_{\sigma_k}, \Sigma^{o,\pm}_{\sigma_k} : j = 1, \dots, l; k = 1, \dots, s\},$$

we consider $\hat{z} = X^{T_2}(z)$ and wait for the next time $t(z)$ when the orbit of $\hat{z}$ hits again one of the cross-sections. Then we define $R(z) = X^{T_2 + t(z)}(z)$ and say that $\tau(z) = T_2 + t(z)$ is the *Poincaré time* of z. If the point z never returns to one of the cross-sections, then the map R is not defined at z (e.g. at the lines $\ell^\pm$ in the flow-boxes near a singularity). Moreover by Lemma 6.23, if R is defined for $x \in \Sigma$ on some $\Sigma \subset \Xi$, then R is defined for every point in $W^s(x, \Sigma)$. Hence *the domain of $R \mid \Sigma$ consists of strips of Σ*. The smoothness of $(t, x) \mapsto X^t(x)$ ensures that the strips

$$\Sigma(\Sigma') = \{x \in \Sigma : R(x) \in \Sigma'\} \tag{6.7}$$

have non-empty interior in Σ for every $\Sigma, \Sigma' \in \Xi$. When R maps to an outgoing strip near a singularity σ_k, there might be a boundary of the strip corresponding to the line $\ell^\pm_k$ of points which fall in the stable manifold of σ_k.

Remark 6.28 Consider the Poincaré map given by the *first return map* $R_0 : \Xi \to \Xi$ defined simply as $R_0(z) = X^{T(z)}(z)$, where

$$T(z) = \inf\{t > 0 : X^t(z) \in \Xi\}$$

is the time the X-orbit of $z \in \Xi$ takes to arrive again at Ξ. This map R_0 is not defined on those points z which do not return and, moreover, R_0 might not satisfy the lemmas of Sect. 6.1.2, since we do not know whether the flow from z to $R_0(z)$ has enough time to gain expansion. However the stable manifolds are still well defined.

By the definitions of R_0 and of R we see that R *is induced by* R_0, i.e., *if* R *is defined for* $z \in \Xi$, *then there exists an integer* $r(x)$ *such that*

$$R(z) = R_0^{r(z)}(z).$$

We note that, since the number of cross-sections in Ξ is finite and the time t_2 is a constant, the function $r : \Xi \to \mathbb{N}$ is bounded: there exists $r_0 \in \mathbb{N}$ such that $r(x) \leq r_0$ for all $x \in \Xi$.

6.1.4.2 Finite Number of Strips

We show that fixing a cross-section $\Sigma \in \Xi$ the family of all possible strips as in (6.7) covers Σ except for finitely many stable leaves $W^s(x_i, \Sigma), i = 1, \ldots, m = m(\Sigma)$. Moreover we also show that each strip given by (6.7) has finitely many connected components. Thus the number of strips in each cross-section is finite.

We first recall that each $\Sigma \in \Xi$ is contained in the forward invariant open neighborhood U_0 which is an isolating neighborhood for Λ; see Sect. 6.1.1. So $x \in \Sigma$ is such that $\omega(x) \subset \Lambda$. Note that R is locally smooth for all points $x \in \text{int}(\Sigma)$ such that $R(x) \in \text{int}(\Xi)$ by the Tubular Flow Theorem and the smoothness of the flow, where $\text{int}(\Xi)$ is the union of the interiors of each cross-section of Ξ. Let $\partial^s \Xi$ denote the union of all the leaves forming the stable boundary of every cross-section in Ξ.

Lemma 6.29 *The set of discontinuities of* R *in* $\Xi \setminus \partial^s \Xi$ *is contained in the set of points* $x \in \Xi \setminus \partial^s \Xi$ *such that*:

1. *either* $R(x)$ *is defined and belongs to* $\partial^s \Xi$;
2. *or there is some time* $0 < t \leq T_2$ *such that* $X^t(x) \in W^s_{loc}(\sigma)$ *for some singularity* σ *of* Λ.

Moreover this set is contained in a finite number of stable leaves of the cross-sections $\Sigma \in \Xi$.

Proof We divide the proof into several steps.

Step 1 Cases (1) and (2) in the statement of the lemma correspond to all possible discontinuities of R in $\Xi \setminus \partial^s \Xi$.

Let x be a point in $\Sigma \setminus \partial^s \Sigma$ for some $\Sigma \in \Xi$, *not* satisfying any of the conditions in items (1) and (2). Then $R(x)$ is defined and $R(x)$ belongs to the interior of some cross-section Σ'. By the smoothness of the flow and by the flow-box theorem we have that R is smooth in a neighborhood of x in Σ. Hence any discontinuity point for R must be in one the situations (1) or (2).

Step 2 Points satisfying item (2) are contained in finitely many stable leaves in each $\Sigma \in \Xi$.

Indeed if we set $W = X^{[-T_2,0]}(\cup_\sigma W^s_{loc}(\sigma))$, where the union above is taken over all singularities σ of Λ, then W is a compact sub-manifold of M with boundary, tangent to the center-stable sub-bundle $E^s \oplus E^X$. This means that W is transverse to any cross-section of Ξ.

Hence the intersection of W with any $\Sigma \in \Xi$ is a one-dimensional sub-manifold of Σ. Thus the number of connected components of the intersection is finite in each Σ. This means that there are finitely many points $x_1, \dots, x_k \in \Sigma$ such that

$$W \cap \Sigma \subset W^s(x_1, \Sigma) \cup \cdots \cup W^s(x_k, \Sigma).$$

Step 3 Points satisfying item (1) are contained in a finite number of stable leaves of each $\Sigma \in \Xi$.

We argue by contradiction. Assume that the set of points D of Σ sent by R into stable boundary points of some cross-section of Ξ is such that

$$L = \{W^s(x, \Sigma) : x \in D\}$$

has *infinitely many lines*. Note that D in fact equals L by Lemma 6.23. Then there exists an accumulation line $W^s(x_0, \Sigma)$. Since the number of cross-sections in Ξ is finite we may assume that $W^s(x_0, \Sigma)$ is accumulated by *distinct* $W^s(x_i, \Sigma)$ with $x_i \in D$ satisfying $R(x_i) \in W^s(z, \Sigma') \subset \partial^s \Sigma'$ for a fixed $\Sigma' \in \Xi$, $i \geq 1$. We may assume that x_i tends to x_0 when $i \to \infty$, that x_0 is in the interior of $W^s(x_0, \Sigma)$ and that the x_i are all distinct—in particular the points x_i do not belong to any periodic orbit of the flow since we can choose the x_i anywhere in the stable set $W^s(x_i, \Sigma)$.

As a preliminary result we show that $R(x_i) - X^{s_i}(x_i)$ is such that s_i is a bounded sequence in the real line. For otherwise $s_i \to \infty$ and this means, by definition of R, that the orbit of $X^{T_2}(x_i)$ is very close to the local stable manifold of some singularity σ of Λ and that $R(x_i)$ belongs to the outgoing cross-section near this singularity: $R(x_i) \in \Sigma_\sigma^{o,\pm}$. Hence it must be that $X^{s_i}(x_i)$ tends to the stable manifold of σ when $i \to \infty$ and that $R(x_i)$ tends to the stable boundary of $\Sigma_\sigma^{o,+}$. Since no point in any cross-section in Ξ is sent by R into this boundary line, we get a contradiction.

Now the smoothness of the flow and the fact that $W^s(z, \Sigma')$ is closed imply that $R(x_0) \in W^s(z, \Sigma')$ also since we have the following

$$R(x_0) = \lim_{i\to\infty} R(x_i) = \lim_{i\to\infty} X^{s_i}(x_i) = X^{s_0}(x_0) \quad \text{and} \quad \lim_{i\to\infty} s_i = s_0.$$

Moreover $R(W^s(x_0, \Sigma)) \subset W^s(z, \Sigma')$ and $R(x_0)$ is in the interior of the image $R(W^s(x_0, \Sigma))$, and then $R(x_i) \in R(W^s(x_0, \Sigma))$ for all i big enough. This means that there exists a sequence $y_i \in W^s(x_0, \Sigma)$ and a sequence of real numbers τ_i such that $X^{\tau_i}(y_i) = R(y_i) = R(x_i)$ for all sufficiently big integers i. By construction we have $x_i \neq y_i$ and both belong to the same orbit. Since x_i, y_i are in the same cross-section we get $x_i = X^{\alpha_i}(y_i)$ with $|\alpha_i| \geq T_3$ for all big i.

However we also have $\tau_i \to s_0$ because $R(y_i) = R(x_i) \to R(x_0)$, $y_i \in W^s(x_0, \Sigma)$ and $R \mid W^s(x_0, \Sigma)$ is smooth. Thus $|s_i - \tau_i| \to 0$. But $|s_i - \tau_i| = |\alpha_i| \geq T_3 > 0$. This is a contradiction.

This proves that D is contained in finitely many stable leaves.

Combining the three steps above we conclude the proof of the lemma. □

Let Γ be the finite set of stable leaves of Ξ provided by Lemma 6.29 together with $\partial^s \Xi$. Then the complement $\Xi \setminus \Gamma$ of this set is formed by finitely many open strips where R is smooth. Each of these strips is then a connected component of the sets $\Sigma(\Sigma')$ for $\Sigma, \Sigma' \in \Xi$.

6.1.5 The One-Dimensional Piecewise Expanding Map

We choose a C^2 cu-curve γ_Σ transverse to $\mathscr{F}^s_\Sigma$ in each $\Sigma \in \Xi$. Then the projection p_Σ along leaves of $\mathscr{F}^s_\Sigma$ onto γ_Σ is a C^1 map, since the stable leaves $W^s(x, \Sigma)$ are defined through every point of $\Sigma \in \Xi$ and each depends C^1 smoothly on the base point, by the Stable Manifold Theorem. We set

$$I = \bigcup_{\Sigma, \Sigma' \in \Xi} \operatorname{int}\big(\Sigma(\Sigma')\big) \cap \gamma_\Sigma$$

and observe that, by the properties of $\Sigma(\Sigma')$ obtained earlier at the beginning of the previous Sect. 6.1.4.1, the set I is diffeomorphic to a *finite union of non-degenerate open intervals* $I_1, \ldots, I_m$ by a C^2 diffeomorphism, and $p_\Sigma \mid p_\Sigma^{-1}(I)$ becomes a C^1 submersion. Note that since Ξ is finite we can choose γ_Σ so that p_Σ has bounded derivative: there exists $\beta_0 > 1$ such that

$$\frac{1}{\beta_0} \le \big|Dp_\Sigma \mid \gamma\big| \le \beta_0 \textit{ for every cu-curve } \gamma \text{ inside any } \Sigma \in \Xi.$$

According to Lemma 6.23, Proposition 6.15 and Corollary 6.17, the Poincaré map $R : \Xi \to \Xi$ takes stable leaves of $\mathscr{F}^s_\Sigma$ inside stable leaves of the same foliation and is hyperbolic. In addition a cu-curve $\gamma \subset \Sigma$ is taken by R into a cu-curve $R(\gamma)$ in the image cross-section. Hence the map

$$f : I \to I \quad \text{given by } I \ni z \mapsto p_{\Sigma'}\Big(R\big(W^s(z, \Sigma) \cap \Sigma(\Sigma')\big)\Big)$$

for $\Sigma, \Sigma' \in \Xi$ is C^1 for points in the interior of I_i, $i = 1, \ldots, m$, and we have

$$|Df| = \big|D\big(p_{\Sigma'} \circ R \circ \gamma_\Sigma\big)\big| \ge \frac{1}{\beta_0} \cdot \sigma. \tag{6.8}$$

Thus choosing t_1 (and consequently t_2) big enough so that $\sigma/\beta_0 > 2$ in Proposition 6.15, we deduce that f *is piecewise expanding with finitely many branches.*

6.1.6 Denseness of Periodic Orbits and the One-Dimensional Map

We fix in every $\Sigma \in \Xi$ a cu-curve γ_Σ which is transversal to every stable leaf of Σ and take the projection $p_\Sigma : \Sigma \to \gamma_\Sigma$ by $x \in \Sigma \mapsto W^s(x, \Sigma) \cap \gamma_\Sigma$ and the associated one-dimensional piecewise expanding map $f : \cup_j I_j \to I := \overline{\cup_j I_j}$, as defined in Sect. 6.1.5. It is important to recall that each I_j is taken to be an *open* interval. As usual in one-dimensional dynamics, we denote in what follows $|J|$ for the length or Lebesgue measure of an interval $J \subset I$.

The following is the first step towards the proof that a singular-hyperbolic attractor is a homoclinic class.

Lemma 6.30 *Let $f : \cup_j I_j \to I$ be a piecewise C^1 expanding map with finitely many branches $I_1, \dots, I_l$ such that each I_j is a non-empty open interval, $|Df \mid I_j| \ge \sigma > 2$ and $I \setminus (\cup_j I_j)$ is finite.*

Then for each small $\delta > 0$ there exists $n = n(\delta)$ such that, for every non-empty open interval $J \subset \cup_j I_j$ with $|J| \ge \delta$, we can find $0 \le k \le n$, a sub-interval $\hat{J}$ of J and $1 \le j \le l$ satisfying

$$f^k \mid \hat{J} : \hat{J} \to I_j \quad \textit{is a diffeomorphism.}$$

In addition, f has finitely many periodic orbits $\mathcal{O}(p_1), \dots, \mathcal{O}(p_k)$ contained in $\cup_j I_j$, and every non-empty open interval J admits an open sub-interval $\hat{J}$, a periodic point p_j and an iterate n such that $f^n \mid \hat{J}$ is a diffeomorphism onto a neighborhood of p_j.

Proof We start by setting $J_0 = J$, which is contained in some I_j by assumption, and define a sequence J_k of intervals by induction, as follows.

For each $k \ge 1$, assuming that J_k is defined and *strictly contained* in some element of the family $I_1, \dots, I_l$, we consider $f(J_k)$. If the interval $f(J_k)$ contains some of the intervals $I_1, \dots, I_l$, we are done. Otherwise $f(J_k)$ is an interval not containing any $I_1, \dots, I_l$.

Let J_{k+1} be the largest connected component of

$$f(J_k) \cap (\cup_j I_j) = f(J_k) \setminus (\cup_j \partial I_j)$$

(we recall that each I_j is taken to be an open interval). Then there exists a sub-interval $\tilde{J}_k$ of J_k such that $f \mid \tilde{J}_k : \tilde{J}_k \to J_{k+1}$ is a diffeomorphism. Moreover by assumption we have $|J_{k+1}| \ge \sigma \cdot |J_k|/2$, with $\sigma > 2$.

Therefore there must be some integer k such that J_k covers some element of the family $I_1, \dots, I_k$ and the process stops, for otherwise the length of $J_k \subset I$ would grow without bound. Moreover the number of steps k satisfies $(\sigma/2)^k \cdot |J| \le |I|$, and thus k is bounded from above depending only on $|J|$.

Now assuming that $J_k \supset I_j$ for some $k \ge 1$ and $j \in \{1, \dots, l\}$, consider the diffeomorphism g we obtain by composing the maps in the previous induction process

$$\tilde{J}_0 \to \tilde{J}_1 \to \cdots \to \tilde{J}_k \supset I_j.$$

Then $g^{-1}(I_j)$ is the required interval $\hat{J} \subset \tilde{J}_0 \subset J$ and $g = f^k$ as in the statement.

This is enough to deduce that every such map f has a periodic orbit, since the number of smoothness domains $I_1, \dots, I_l$ is finite. Indeed, the map g is a diffeomorphism between a sub-interval $\tilde{J}_0$ of J_0, contained in some interval I_i, and some interval $\tilde{J}_k$ covering an interval I_j. We write this relation as $i \to j$.

We can now start with $J_0 = I_j$ and iterate. We obtain a map from the finite set $\{1, \dots, l\}$ to itself which must have a loop: there exists a sequence

$$i_1 \to i_2 \to \cdots \to i_k = i_j$$

with $1 \le j < k$ and $i_1, i_2, \dots, i_k \in \{1, \dots, l\}$. This means that there exists a sub-interval $\tilde{J}$ of I_j which is sent diffeomorphically by some power g of f into an interval $\hat{J}$ covering I_j, that is, $g : \tilde{J} \subset I_j \to \hat{J} \supset I_j$, so that there is a fixed point for g, which is a periodic point p of f.

Now we note that starting with any given I_{i_1} as above implies that the periodic point p has some pre-image q in I_{i_1}. Moreover the iterates $q, f(q), \dots, f^m(q) = p$ all belong to $\cup_j I_j$ by construction. Since the starting interval I_{i_1} is arbitrary, the proof is complete. □

Since I is a Baire space we obtain the second step towards the proof that a singular-hyperbolic attractor is a homoclinic class.

Proposition 6.31 *In the same setting of Lemma* 6.30 *there exists a periodic point $p \in L$ and a neighborhood W of p satisfying*:

- *for each interval $J \subset W$ there are $\ell \in \mathbb{N}$ and a sub-interval $J_0 \subset J$ such that*

$$f^\ell \mid J_0 : J_0 \to W \quad \textit{is a diffeomorphism.}$$

In addition, W contains a dense subset of periodic orbits of f.

Proof Let $L_i = \cup_{n\in\mathbb{Z}} f^n(p_i) \setminus (\cup_{n\ge 0} f^{-n}(\cup_j \partial I_j))$ for each periodic orbit given by Lemma 6.30. The closure of these sets are f-invariant sets by construction and satisfy

$$\overline{L}_1 \cup \cdots \cup \overline{L}_k = I$$

by Lemma 6.30 again. Since I is a Baire space, one of these sets, $L = \overline{L}_1$ say, must have non-empty interior. Moreover, by the construction of L, given a point $x \in L$ and a neighborhood V_0 of x, there exists $y \in L \cap V_0$ and $n \in \mathbb{N}$ with $f^n(y) = p_1$, and a neighborhood $V_1 \subset V_0$ of y such that $f^n \mid V_1$ is a diffeomorphism onto a neighborhood of p_1. Then we can assume that p_1 belongs to the interior of L.

This shows that every non-empty open interval J satisfying $J \cap L \ne \emptyset$ admits $n \in \mathbb{N}$ such that $f^n(J)$ is a neighborhood of p_1. Thus, given $0 < \delta < \min\{|I_i|, i = 1, \dots, l\}$ we may find $n_0 \in \mathbb{N}$ such that $1/n_0 < \delta/2$ and partition the interval I into n_0 equally sized sub-intervals $J_k := [k/n_0, (k+1)/n_0]$, $k = 0, \dots, n_0 - 1$. For each such interval J_k intersecting L there exists n_k such that $f^{n_k}(J_k)$ contains a non-empty open interval W_k which contains p_1. Let W be the intersection of all these finitely many intervals, which is also an interval containing p_1.

Now any interval $J \subset W \subset L$ admits a sub-interval $\hat{J}$ and $n \in \mathbb{N}$ such that $f^n \mid \hat{J}$ is a diffeomorphism from $\hat{J}$ to some I_i. By the choice of δ there exists some $J_k \subset I_i$ and by construction $I_i \subset L$. Hence there exist $j \in \mathbb{N}$ and a sub-interval $\tilde{J}$ of J_k such that $f^j \mid J_k$ is a diffeomorphism between $\tilde{J}$ and W. Therefore composing the maps above we get $\ell = n + j$ and a sub-interval J_0 of J as in the statement of the lemma.

Observe now that W contains a dense subset of periodic orbits of f, since each sub-interval W_0 of $W \subset L$ admits a smaller sub-interval $W_1 \subset W_0$ and an iterate k such that $f^k \mid W_1 : W_1 \to W \supset W_1$ is a diffeomorphism, so that there exists a periodic point in $W_1 \subset W_0$. The proof is complete. □

6.1.7 Crossing Strips and the One-Dimensional Map

The family $\mathscr{C}(\Xi)$ of connected components of the sub-strips $\{\Sigma(\Sigma') : \Sigma' \in \Xi\}$, for any given $\Sigma \in \Xi$, is the set of maximal connected sub-strips of Σ where R is smooth; see Sect. 6.1.4.

Definition 6.32 Given strips $S_1 = [0, 1] \times J_1 \subset \Sigma_1$ and $S_2 = [0, 1] \times J_2 \subset \Sigma_2$, for some $\Sigma_1, \Sigma_2 \in \Xi$, where J_1, J_2 are compact sub-intervals of $[0, 1]$, we say that $R^k(S_1)$ *crosses* S_2 for some $k \geq 1$, if there exists a sub-strip $S = [0, 1] \times J$ of S_1 such that

$$p_{\Sigma_2}\big(\overline{R^k(S)}\big) \supseteq p_{\Sigma_2}(\overline{S_2}) \quad \text{and} \quad R^k(S) \quad \text{is diffeomorphic to } S.$$

Given two strips S_1, S_2 we write $S_1 \xrightarrow{k} S_2$ if $R^k(S_1)$ crosses S_2.

The usefulness of the previous results on the dynamics of the one-dimensional map is given by the following simple criterion for two given strips to cross.

Lemma 6.33 *If there exists an interval $J \subset I$ and an integer n such that $f^n \mid J$ is a diffeomorphism between J and its image $f^n(J)$, then the strip $S = [0, 1] \times J$ crosses the strip $\hat{S} = [0, 1] \times f^n(J)$.*

Proof Indeed since $f \circ p = p \circ R$ by the definition of f though the projection p, we have

$$f^n\big(p(S)\big) = p\big(R^n(S)\big) = f^n(J)$$

and thus $R^n(S)$ projects on $p(\hat{S})$. Hence $S \xrightarrow{n} \hat{S}$. □

Moreover this has the following important consequence for *cu*-curves.

Lemma 6.34 *In the same setting as Lemma* 6.33, *if γ is a cu-curve which crosses S, that is, $p \mid (\gamma \cap S)$ is a diffeomorphism between γ and J, then $R^n(\gamma)$ also crosses $\hat{S}$. In particular, $R^n(\gamma) \pitchfork W^s(x, \Sigma) \neq \emptyset$ for all $x \in \hat{S}$, where $\Sigma \in \Xi$ is such that $S \subset \Sigma$.*

Proof Since $f^n \mid J$ is a diffeomorphism with its image and J is precisely the image of $p \mid \gamma$, the smooth conjugation between R and f ensures that $p \circ R^n \mid \gamma$ is also a diffeomorphism with $f^n(J)$. Therefore for each $x = (t, y) \in \hat{S}$ the intersection between $R^n(\gamma)$ and $W^s(x, \Sigma)$ is given by $R^n \circ (p \circ R^n \mid \gamma)^{-1}(y)$. The transversality of the intersection is clear from the fact that cu-curves are sent to cu-curves by the Poincaré map R. □

6.2 Singular-Hyperbolic Attractors are Homoclinic Classes

Now we assume that Λ is a singular-hyperbolic attractor for a C^1 flow X^t. Recall that given a hyperbolic periodic orbit p for X the associated *homoclinic class* $H(p)$ is given by the closure of the set of transversal intersections of the stable and unstable manifolds of the orbit of p; see Chap. 2.

We consider the global Poincaré return map R for a C^1 flow X^t associated to a singular-hyperbolic attractor Λ. Let f be the one-dimensional map obtained in Sect. 6.1.5.

The main step in the argument is the following.

Proposition 6.35 *There exists a strip S_0 such that every sub-strip S of S_0 eventually crosses S_0, that is, there exists $k \in \mathbb{N}$ satisfying $S \xrightarrow{k} S_0$.*

We prove this result arguing with the one-dimensional map f associated with Ξ and X through the projection p.

Proof Consider the strip $S_0 := p^{-1}(W)$ in the adapted cross-section Σ_i corresponding to the interval I_j which contains W, where W is the interval whose existence is assured by Proposition 6.31. Then, given any sub-strip S of S_0 and considering $J = \pi(S)$, we know from Proposition 6.31 that there exists a sub-interval J_0 of J such that $f^n \mid J_0 : J_0 \to W$ is a diffeomorphism.

Lemma 6.33 ensures that the strip $S' = p^{-1}(J_0)$ eventually crosses S_0, as claimed. □

It is well known that every Lyapunov stable set contains the unstable manifold of its periodic orbits. Hence, given any point P of a periodic orbit $\mathcal{O}(P)$ of X in a singular-hyperbolic attractor Λ, we have $\overline{W^u(P)} \subset \Lambda$ and also $H(P) \subset \Lambda$ (we recall that each periodic orbit inside a singular-hyperbolic attractor is hyperbolic).

In addition, for the local unstable manifold $W^u_\varepsilon(P)$ for $P \in \Sigma$, with some $\Sigma \in \Xi$, there exists a strip S contained in some connected component of $\mathcal{C}(\Xi)$ such that $S \cap W^u_\varepsilon(P)$ is a neighborhood of p inside $W^u_\varepsilon(P) \cap \Sigma$. That is, $p_\Sigma(W^u_\varepsilon(P) \cap \Sigma)$ contains a neighborhood of $p_\Sigma(P)$.

Let w be a point of a transitive orbit inside Λ. Then there exists $t > 0$ such that $X^t(w) \in S$, and so there exists $y \in W^u_\varepsilon(P) \cap W^s(X^t(w), \Sigma)$. This implies that $\omega_X(y) = \omega_X(w) = \Lambda$ and so $\overline{W^u(P)} \supset \Lambda$.

Next we show that there exists a periodic orbit $\mathcal{O}(P)$ in Λ such that $\overline{W^u(P)} = H(P)$. Since $\overline{W^u(P)} = \Lambda$ this proves that Λ is a homoclinic class.

We observe that from Lemma 6.30 and Proposition 6.31 we can find a periodic stable leaf $W^s(y, \Sigma)$ for some $\Sigma \in \Xi$. The uniform contraction of stable leaves on Ξ ensures that there exists a unique periodic point P of R on this leaf, which corresponds to a periodic orbit of X in Λ. This shows that *every C^1 singular-hyperbolic attractor admits some periodic orbit.*

Let S_0 be the strip contained in some connected component of $\mathcal{C}(\Xi)$, such that $S_0 \cap W^u_\varepsilon(P)$ is a neighborhood of P inside $W^u_\varepsilon(P) \cap \Sigma$, given by Proposition 6.35. We consider D a small connected bounded disk inside $W^u(P)$. Then we have that $X^{-t}(D)$ converges uniformly to $\mathcal{O}(P)$ when $t \to +\infty$ by the definition of unstable manifold of a hyperbolic periodic orbit. Therefore there exists $t_0 > 0$ such that $X^{-t_0}(D) \cap S_0$ contains a *cu*-curve γ. In particular γ is not tangent to the center-stable direction at any point.

According to Proposition 6.35, by the choice of S_0, the strip $S = p^{-1}(p(\gamma))$ admits $n \geq 1$ such that $S \xrightarrow{n} S_0$. This implies that $\gamma_n = f^n(\gamma_0)$ eventually crosses S_0. But $\gamma_0 \subset \gamma$ is a piece of $W^u(P)$ which is invariant under the flow. Hence $\gamma_n \subset W^u(P)$ is a *cu*-curve which crosses S_0. Lemma 6.34 implies that $\gamma_n \pitchfork W^s(P, \Sigma) \neq \emptyset$.

Again the invariance of the center-stable and center-unstable manifolds imply that $\gamma \pitchfork W^s(P) \neq \emptyset$ first, and then also that $D \pitchfork W^s(P) \neq \emptyset$. Since D is an arbitrary small disk inside $W^u(P)$, this shows that a dense subset of $W^u(P)$ consists of homoclinic points. This is enough to conclude that $\overline{W^u(p)} \subset H(p)$. This completes the proof that a singular-hyperbolic attractor is a homoclinic class.

6.3 Sufficient Conditions for Singular-Hyperbolic Attractors to Be Robust

Here we present a proof of Theorem 6.9. This is based on the following result whose proof we postpone to Sect. 6.3.1.

Theorem 6.36 *Let Λ be a singular-hyperbolic attracting set of $X \in \mathfrak{X}^r$ for some $r \geq 1$. Suppose that Λ is connected and contains a dense subset of periodic orbits. Moreover assume that Λ contains only one singularity and is not transitive.*

Then for every neighborhood U of Λ there exists a flow Y close to X in the C^r topology such that $\Lambda_Y(U) \not\subset \Omega(Y)$.

Let Λ be a singular-hyperbolic attractor of a C^r flow X on a compact 3-manifold M. Assume that Λ is C^r robustly periodic and has a unique singularity σ.

Denote by U a neighborhood of Λ such that $\Lambda_Y(U) \cap \mathrm{Per}(Y)$ is dense in $\Lambda_Y(U)$ for every flow Y which is C^r close to X. Clearly $\Lambda_Y(U)$ is a singular-hyperbolic set of Y for all Y close to X.

Because Λ has a unique singularity, which is Lorenz-like, then $\Lambda_Y(U)$ has a unique singularity as well. Indeed, by Theorem 6.3, every singularity of $\Lambda_Y(U)$

must be either singular-hyperbolic for Y or for $-Y$. In both cases the singularities are hyperbolic and bifurcations are not allowed for any Y close to X. Hence if Λ_Y had more than one singularity there would exist at least two distinct singularities in the original set Λ, by the property of analytic continuation of any hyperbolic critical element.

Recalling that Λ is an attractor by assumption, and thus transitive in particular, we see that Λ is connected, and so we can assume that the neighborhood U above is connected. Then $\Lambda_Y(U)$ is connected as well.

Summarizing: $\Lambda_Y(U)$ is a connected singular-hyperbolic attracting set of Y containing only one singularity.

Were Λ not C^r robust, there would exist Y close to X such that $\Lambda_Y(U)$ is not transitive. In this case $\Lambda_Y(U)$ would satisfy all the conditions of Theorem 6.36. Hence there would exist Z close to Y satisfying $\Lambda_Z(U) \not\subset \Omega(Z)$. But we are assuming that $\Lambda_Z(U) \cap C(Z)$ is dense in $\Lambda_Z(U)$ and $C(Z)$ is always contained in $\Omega(Z)$.

This contradiction completes the proof of Theorem 6.9, assuming Theorem 6.36.

6.3.1 Denseness of Periodic Orbits and Transitivity with a Unique Singularity

Here we start the proof of Theorem 6.36 We present the proof as a sequence of several simpler results which will be proved in the sequel.

Let $X \in \mathfrak{X}^r$ and let Λ be a singular-hyperbolic set of X satisfying the conditions in the statement of Theorem 6.36: thus Λ contains a unique singularity σ, it has a dense subset of periodic orbits and it is a singular hyperbolic *non-connected* attracting set. The singularity is Lorenz-like by Theorem 6.3. Then $W^{ss}(\sigma)$ divides $W^s(\sigma)$ into two connected components, which we denote by $W^{s,+}$ and $W^{s,-}$.

Note that $\Lambda \neq \{\sigma\}$, for otherwise we would get an attracting set consisting of a singularity with an expanding eigenvalue, which is impossible. Therefore the set of periodic orbits in Λ is non-empty.

A crucial result in this setting is that *the unstable manifold of every periodic orbit in Λ crosses the stable manifold of the singularity transversely.* We present a proof in Sect. 6.3.2 following the arguments in [168].

Theorem 6.37 *Let Λ be either a singular-hyperbolic attractor, or a connected singular-hyperbolic attracting set with a dense subset of periodic orbits. Then for every $p \in \mathrm{Per}(X) \cap \Lambda$ there exists a singularity σ of Λ such that $W^u(p)$ and $W^s(\sigma)$ intersect transversely.*

The intersections provided by this results together with the uniqueness assumption on $S(X)$ enables us to use the two connected components $W^{s,+}$ and $W^{s,-}$ of $W^s(\sigma)$ to relate two distinct periodic orbits of Λ or to deduce non-trivial consequences if Λ is not transitive or a disconnected set.

We do this by considering the following invariant subsets of Λ:

$$P^{\pm} = \{p \in \mathrm{Per}(X) \cap \Lambda : W^u_X(p) \pitchfork W^{s,\pm}(\sigma) \neq \emptyset\} \quad \text{and} \quad H^{\pm} = \overline{P^{\pm}}.$$

The rest of this section is devoted to proving the following result, which we then use to prove Theorem 6.36.

Theorem 6.38 *Let Λ be a connected singular-hyperbolic attracting set of a flow $X \in \mathfrak{X}^r, r \geq 1$, on a closed three-manifold M. Suppose that Λ contains a dense subset of periodic orbits and a unique singularity. Moreover assume that Λ is not transitive. Then H^+ and H^- are homoclinic classes of X.*

From Theorem 6.37 and the assumption that $S(X) \cap \Lambda$ is a singleton together with denseness of periodic orbits in Λ, we easily deduce that $P^{\pm}$ cover the whole attractor.

Lemma 6.39 *Let Λ be a connected singular-hyperbolic attracting set with dense periodic orbits and only one singularity σ. Then $\Lambda = H^+ \cup H^-$.*

In this setting we can state Theorem 6.38 in the following useful way: *a singular hyperbolic attracting set having dense periodic orbits with only one singularity is either transitive or the union of two homoclinic classes.*

Since each element $\mathscr{O} \in \mathrm{Per}(X) \cap \Lambda$ is hyperbolic of saddle-type, then $W^u(\mathscr{O}) \setminus \{\mathscr{O}\}$ has two connected components. For $\mathscr{O} \in P^{\pm}$ one of these components intersects $W^{s,\pm}(\sigma)$. We write that component $W^{u,\pm}(\mathscr{O})$.

Now we show that both H^+ and H^- are transitive sets.

Lemma 6.40 *Let Λ be a singular-hyperbolic attracting set with dense periodic orbits and only one singularity σ. Then H^+ and H^- are transitive. Moreover $H^{\pm} \subset \overline{W^{u,\pm}(q)} = \overline{W^{u,\pm}(q) \pitchfork W^{s,\pm}(\sigma)}$ for all $q \in P^{\pm}$.*

Proof Let p, q be two points in distinct orbits inside H^+ (the argument for H^- is analogous). Then their unstable manifolds intersect transversely on the same side of the stable manifold of the unique singularity. Hence through the local behavior of the flow near a singularity, flowing to an incoming cross-section $\Sigma = \Sigma^{i,+}$, we obtain two small curves $\gamma \subset W^u(p) \cap \Sigma$ and $\zeta \subset W^u(q) \cap \Sigma$ crossing ℓ^+ transversely. See Fig. 5.3.

Fix neighborhoods U of p and V of q. Since periodic orbits are dense in Λ and $\gamma \subset \Lambda$ (because Λ is an attracting set), then we can find a periodic orbit r so close to p that

- $W^s(r, \Sigma)$ intersects both γ and ζ transversely;
- the orbit of r intersects U.

Hence taking $z \in \zeta \pitchfork W^s(r, \Sigma) \subset W^u(q) \cap W^s(r)$ we see that the positive orbit of z visits U infinitely many times, and the negative orbit of z converges to $\mathscr{O}(q)$, and thus visits V infinitely many times. This means that there exists some $t > 0$ such that $X^t(V \cap \Lambda) \cap (U \cap \Lambda) \neq \emptyset$. Since U and V where arbitrarily chosen, this proves transitivity.

Recall the convention $W^{u,+}(q)$ for the branch of $W^u(q) \setminus \mathcal{O}(q)$ which intersects $W^{s,+}(\sigma)$. The above argument also shows that the $W^{u,+}(q)$ is arbitrarily near p, that is, $P^+ \subset \overline{W^{u,+}(q)}$ for every $q \in P^+$, and thus $H^+ \subset \overline{W^{u,+}(q)}$.

Since Λ is attracting we have $W^{u,+}(q) \subset \Lambda$. Therefore given any $y \in W^{u,+}(q)$ there is a sequence $p_n \in \mathrm{Per}(X) \cap \Lambda$ such that $p_n \xrightarrow[n\to+\infty]{} y$. By Theorem 6.37 together with the inclination lemma, we deduce that $W^s(\sigma)$ crosses $W^{u,+}(q)$ very near y. This shows that $\overline{W^s(\sigma) \cap W^{u,+}(q)} \supset W^{u,+}(q)$.

Analogously with $-$ instead of $+$. Note that the intersections above are always transverse. The lemma is proved. □

From this we deduce the following condition for transitivity.

Lemma 6.41 *In the same setting as the previous lemma, suppose that there exists a sequence $\{p_n\}_{n\geq 1} \subset P^-$ converging to some point in $W^{s,+}(\sigma)$ (or similarly interchanging $+$ with $-$). Then Λ is transitive.*

Proof Fix $p \in P^+$ and let $p_n \in P^-$ be as in the statement. From the construction of flow-boxes near singularities in Sect. 5.4.4 we can fix an adapted cross-section $\Sigma = \Sigma^{i,+}$ through $W^{s,+}(\sigma)$ and an open arc $J \subset \Sigma \cap W^u(p)$ intersecting $W^{s,+}(\sigma)$ transversely.

Again by the behavior of the flow near singularities we can assume that $p_n \in \Sigma$ for every n. By the choice of adapted cross-sections, we know that the local stable manifolds $W^s(p_n, \Sigma)$ of p_n inside Σ intersect J transversely, for every big enough n.

The inclination lemma 2.15 applied to the positive orbit of $J \subset W^u_X(p)$ together with the assumption $p_n \in P^-$ imply that $W^u(p) \cap W^{s,-}(\sigma) \neq \emptyset$. Hence $p \in P^-$.

This shows that $H^+ \subset H^-$. Thus $\Lambda = H^-$ by Lemma 6.39, and from Lemma 6.40 we conclude that Λ is transitive. □

Proposition 6.42 *In the same setting as above, if there is $z \in W^u(\sigma) \setminus \{\sigma\}$ such that $\sigma \in \omega(z)$, then Λ is transitive.*

Proof Let z be as in the statement. By the local dynamics in flow-boxes near σ we can assume that there are points $z_n \in \Sigma^{i,+}$ in the positive orbit of z such that $z_n \to z_0 \in \ell$, where $\ell = W^{s,+}(\sigma) \cap \Sigma$. (The argument for the $-$ case above is analogous.)

If $\mathrm{Per}(X) \cap P^- = \emptyset$, then we would have $\mathrm{Per}(X) \cap \Lambda \subset P^+$ by Theorem 6.37. In this case Λ would be transitive by Lemma 6.40. Hence we can assume that there exists $q \in P^-$.

This allows us to choose a sequence of points $w_n \in W^u(q)$ in the same side as z is, such that $w_n \xrightarrow[n\to+\infty]{} w \in W^{s,-}(\sigma)$. Since Λ is attracting, it contains the unstable manifolds of its points and so $w \in \Lambda$. Thus we can find a sequence $p_n \in \mathrm{Per}(X) \cap \Lambda$ tending to w, whose orbit passes very close to z. Consequently there are $p_n' \in \mathcal{O}(p_n)$ converging to z_0.

We have found a sequence of periodic orbits accumulating simultaneously on $W^{s,+}(\sigma)$ and on $W^{s,-}(\sigma)$. Arguing by contradiction, suppose that Λ is not transitive. Then Lemma 6.41 would imply that $p_n' \notin P^+$ and $p_n' \notin P^-$ for all n large enough. This contradicts Theorem 6.37 and concludes the proof. □

We assume that Λ is not transitive and use the previous results to disconnect Λ.

Lemma 6.43 *If Λ is not transitive, then for all $q \in P^\pm$ we have*

$$H^\pm = \overline{W^{u,\pm}(q) \pitchfork W^{s,\pm}(\sigma)} = \overline{W^{u,\pm}(q)}.$$

Proof Fix $q \in P^+$ (for P^- the argument is the same). From Lemma 6.40 it is enough to show that every point $y \in W^{u,+}(q)$ is an accumulation point of elements of P^+. This implies that y is accumulated by points in $W^{u,+}(q) \cap W^{s,+}(\sigma)$ by the inclination lemma and, in addition, also ensures that $y \in H^+$.

By denseness of periodic orbits there exists a sequence $p_n \in \mathrm{Per}(X) \cap \Lambda$ such that $p_n \xrightarrow[n\to+\infty]{} y$. Then $p_n \in P^+$ for all n big enough, for otherwise we would get $y \in H^-$ and thus $H^+ \subset \overline{W^{u,+}(q)} = \overline{\mathcal{O}(y)} \subset H^-$, since H^- is invariant. Hence $\Lambda = H^+$. This contradicts the assumption that Λ is not transitive. □

Theorem 6.44 *If Λ is not transitive, then for all $a \in W^u(\sigma) \setminus \{\sigma\}$ there exists a periodic orbit $O \subset \Lambda$ such that $a \in W^s(O)$, that is, $\omega(a) = O$.*

Note that by Theorem 6.37 the periodic orbits given by Theorem 6.44 are homoclinically related to σ.

Proof Fix $a \in W^u(\sigma) \setminus \{\sigma\}$ and assume that $\omega(a)$ is *not* a periodic orbit.

Since Λ is not transitive and periodic orbits are dense by assumption, we have $P^+ \neq P^-$ and both are non-empty. Take $p \in P^+$ and $q \in P^-$.

Using the flow we can assume that a belongs to some outgoing cross-section $\Sigma = \Sigma^{o,\pm}$ of a flow box near σ. Since the unstable manifolds of p and q cross $W^s(\sigma)$ on sides opposite to $W^{ss}(\sigma)$, both their intersections with $W^s(\sigma)$ contain a curve having σ as an accumulation point and tangent to the eigendirection corresponding to the weak contracting eigenvalue of σ; see Fig. 6.5. Using the flow-box near σ we can find a curve $I = I_a$ contained in Σ through a such that $I \setminus \{a\}$ is formed by two arcs $I^+ \subset W^u(p)$ and $I^- \subset W^u(q)$; see Fig. 6.5. Observe that the tangent space of I is contained in $E^c \cap T\Sigma$ by construction.

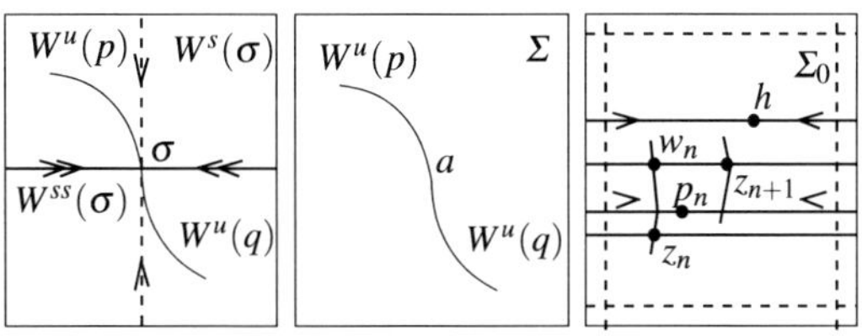

Fig. 6.5 The stable manifold of σ, the unstable manifolds of p, q and the points in Σ_0

Proposition 6.42 ensures that $\sigma \notin \omega(a)$ since Λ is not transitive. Therefore from Proposition 6.2 we see that $H = \omega(a)$ is a uniformly hyperbolic saddle-type set. Moreover $I \subset \Lambda$ because Λ is a closed attracting set.

Consider an adapted cross-section Σ_0 through some point of H. Then by item 2 of Lemma 6.27 and shrinking Σ_0 if necessary, we can assume that *the stable boundary* $\partial^s \Sigma_0$ *of* Σ_0 *does not touch* H. Moreover since $\Sigma_0 \setminus H$ is open we can in addition assume that $d(\Sigma_0 \cap H, \partial^s \Sigma) > \delta$ for some $\delta > 0$, just as in the definition of δ-adapted cross-section, but now in the center-unstable direction.

Using a tubular flow construction we can linearize X in an *open* tube-like set $U_{\Sigma_0} = X^{(-\varepsilon,\varepsilon)}(\text{int}(\Sigma_0))$ for a small $\varepsilon > 0$. We can cover H by a finite number $\Xi = \{U_{\Sigma_0}, \dots, U_{\Sigma_l}\}$ of this type of *open tubular flow-boxes,* since $H \cap S(X) = \emptyset$, H is compact and H satisfies item 2 of Lemma 6.27.

Consider the Poincaré map $R : \Xi \cap H \to \Xi$ defined by $z \in \Xi \cap H \mapsto X^{T_2+t(z)}(z)$ where T_2 is defined in Sect. 6.1 and $\tau(z)$ is the first return time of $X^{T_2}(z)$ to Ξ. The map is defined on entire strips of Ξ by the construction of adapted cross-sections.

Fix now $h_0 \in H$ and let z_n be points in the positive orbit of z such that

$$d\big(z_n, R^n(h_0)\big) \xrightarrow[n\to+\infty]{} 0.$$

Note that $h_n = R^n(h_0)$ always belongs to the interior of Ξ and the same is true of z_n. Observe that there exists a corresponding sequence of images I_n of I such that $z_n \in I_n \subset \Xi$. Since I is transverse to the flow direction, we have that z_n belongs to the interior of I_n. In addition, the expansion of R in the central-unstable direction and the fact that z_n is δ-away from the boundary of Ξ ensures that there is an arc J_n with length bounded away from zero such that $z_n \in J_n \subset I_n$.

Let h be a limit point of h_n. Hence J_n converges in the C^1 topology to an interval in $W^u(h)$ (recall that $h \in H$ and H is uniformly hyperbolic). Since Ξ has finitely many components, we can assume that h_n, J_n and h all belong to the same component Σ_0 of Ξ.

Notice that we cannot have $z_n \in W^s(h, \Sigma_0)$ for infinitely many n, for otherwise by Lemma 6.26 we conclude that h is periodic and $z \in W^s(h)$, and thus $H = \omega(z) = \mathcal{O}(h)$ contradicting the assumption. Hence $z_n \notin W^s(h, \Sigma_0)$ for all big enough n.

Therefore the intersection of $J_n \setminus \{z_n\}$ with $W^s(z_{n+1}, \Xi)$ is non-empty for big enough n. If w_n belongs to this intersection, then it is either in the image of I^+ or in the image of I^- inside J_n. We write $J_n^{\pm}$ for the corresponding components.

Now we use the fact that periodic orbits are dense. Assume that $w_n \in J_n^+$ and take $p_n \in \text{Per}(X) \cap \Sigma_0$ close to a point in J_n^- near z_{n+1}; see the rightmost rectangle in Fig. 6.5. Then we ensure that

$$W^s(p_n) \pitchfork J_n^+ \neq \emptyset \neq J_{n+1}^- \pitchfork W^s(p_n)$$

which implies that

$$W^s(p_n) \pitchfork W^u(p) \neq \emptyset \neq W^u(q) \pitchfork W^s(p_n).$$

By the choice of p_n we know that $\mathcal{O}(p_n)$ goes very close to $W^{s,-}(\sigma)$. We can find a sequence of such orbits converging to a point in $W^{s,-}(\sigma)$. Since Λ is not transitive,

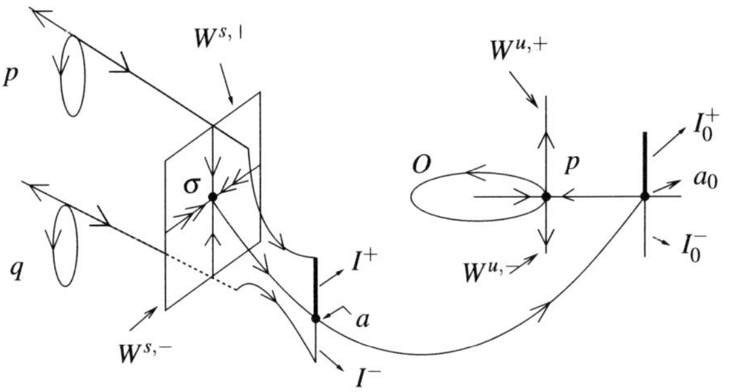

Fig. 6.6 Definition of $W^{u,+}$ and $W^{u,-}$

by Lemma 6.41 we must have $p_n \in P^-$. But then p must be in P^- by the inclination lemma 2.15. Since p was an arbitrary point in P^+, we conclude that $P^+ \subset P^-$ and so $\Lambda = H^-$ is transitive, a contradiction.

Otherwise we have $w_n \in J_n^-$ and by the same arguments we deduce that $q \in P^+$, implying that $\Lambda = H^+$ is transitive as before.

Hence $\omega(z)$ must be a periodic orbit, as claimed. □

The orbit $\mathcal{O}$ provided by Theorem 6.44 is hyperbolic of saddle-type (because it carries a dominated splitting with volume expanding central direction). Hence there are two connected components $W^{u,\pm}$ of the unstable manifold of O such that $W^{u,+} \cup W^{u,-} = W^u(O) \setminus O$. The labels $\pm$ on each component are chosen according to whether the corresponding component is accumulated by the unstable manifold of a periodic point in P^+ or P^-, as in the proof of Theorem 6.44; see Fig. 6.6. The above convention does not depend on $p \in P^+, q \in P^-$ nor on I^+, I^- (this is easily proved using the Inclination Lemma).

The next results show that the choice of signs for the branches of $W^u(\mathcal{O})$ coincides with the previous convention for the unstable manifolds of periodic orbits in Λ.

Lemma 6.45 *We have $W^{u,+} \cap W^{s,-}(\sigma) = \emptyset$ and $W^{u,+} \cap W^{s,+}(\sigma) \neq \emptyset$, and the similar facts interchanging $+$ and $-$. In particular $\mathcal{O} \in P^+ \cap P^-$.*

Proof Arguing by contradiction, note that if $W^{u,+} \cap W^{s,-}(\sigma) \neq \emptyset$, then because this intersection is transverse and every $p \in P^+$ has an unstable manifold accumulating on $W^{u,+}$, we deduce that $p \in P^-$, and again $P^+ \subset P^-$. Thus $\Lambda = H^-$ is transitive, a contradiction. Similarly exchanging $+$ with $-$ in the above argument.

For the other part, if $W^{u,+} \cap W^{s,+}(\sigma) = \emptyset$, then $W^u(\mathcal{O}) \cap W^s(\sigma) = \emptyset$ since $\Lambda \cap W^{ss}(\sigma) = \{\sigma\}$ by Theorem 5.10, contradicting Theorem 6.37. □

Lemma 6.46 *Assume that $W^s(p) \pitchfork W^{u,+} \neq \emptyset$ for some $p \in \mathrm{Per}(X) \cap \Lambda$. Then $\overline{W^s(p) \cap W^{u,+}} = \overline{W^{u,+}}$. Similarly replacing $+$ by $-$.*

Proof Choose a neighborhood U of $x \in W^{u,+}$. By Lemma 6.43 we have in particular $\overline{W^{s,+}(\sigma) \cap W^{u,+}} = \overline{W^{u,+}}$. Then we can find a point $y \in W^{s,+}(\sigma) \cap W^{u,+} \cap U$.

Let γ be a curve through y inside $U \cap W^{u,+}$ transverse to $W^{s,+}(\sigma)$. Then the positive orbit of γ contains open arcs which converge in the C^1 topology to any compact neighborhood of $\mathcal{O}$ inside $W^{u,+}$, by the inclination lemma. Hence the positive orbit of γ intersects $W^s(p)$ by the assumption on p. Therefore there exists a point of $W^s(p)$ in U, proving that $W^{u,+} \subset \overline{W^s(p) \cap W^{u,+}}$. □

Now we are ready to consider homoclinic classes inside Λ (see Sect. 2.5.4 for the definition and basic properties).

Lemma 6.47 *For $p \in P^{\pm}$ such that $W^s(p) \pitchfork W^{u,\pm} \neq \emptyset$, the homoclinic class $H(p)$ equals $\overline{W^{u,\pm}}$.*

Observe that, since periodic orbits are dense we can choose $p \in \mathrm{Per}(X) \cap \Lambda$ very close to $W^{u,+}$ to obtain the condition on p in Lemma 6.47. Then by Lemma 6.43 we see that $H^+ = \overline{W^{u,\pm}} = H(p)$ is a homoclinic class. This completes the proof of Theorem 6.38.

Proof of Lemma 6.47 Fix $z \in P^+$, $y \in W^{u,\pm}$ and a neighborhood U of y. By definition there exists an arc $I \subset W^u(p)$ whose forward orbit crosses $W^{s,+}(\sigma)$. Lemma 6.46 ensures that we can find a disk D transverse to $W^{u,\pm}$ inside $W^s(p) \cap U$.

The inclination lemma implies that the positive orbit of a sub-arc $J \subset I$ accumulates $W^{u,+}$. Then there exists $t > 0$ such that $X^t(J) \pitchfork D \neq \emptyset$. This means that $H(p) \cap U \neq \emptyset$. Since U was arbitrarily chosen and $H(p)$ is closed by definition, we find that $y \in H(p)$. Hence $W^{u,+} \subset H(p)$ and $\overline{W^{u,\pm}} \subset H(p)$.

For the opposite inclusion note that by the assumption $W^s(p) \pitchfork W^{u,\pm} \neq \emptyset$ and the inclination lemma, we have $\overline{W^{u,\pm}} \supset W^u(p) \supset H(p)$. □

Proof of Theorem 6.36 First, by Lemma 6.47, we must have $\overline{W^{u,\pm}} \cap W^{s,-}(\sigma) = \emptyset$. For otherwise we can find a sequence $p_n \in P^+$ converging to a point in $W^{s,-}(\sigma)$. By Lemma 6.41 this implies that Λ is transitive, a contradiction.

Therefore there exists a neighborhood B of $\overline{W^{u,\pm}}$ disjoint from $W^{s,-}(\sigma)$. Let $J = [a, b]$ be a fundamental neighborhood of $W^{ss}(p_0)$ for some $p_0 \in \mathcal{O}$, where $\mathcal{O}$ is the periodic orbit given by Theorem 6.44. That is, J is an arc with $b = X^t(a)$ for some $t > 0$ such that $X^s(a) \notin W^{ss}(p_0)$ for all $0 < s < t$. Take $V \subset B$ a small neighborhood of J such that every point of V belongs to a stable manifold of a point in $V \cap W^{u,\pm}$. The forward orbits of points in V never leave B, since $\overline{W^{u,\pm}}$ is invariant.

We are going to describe a perturbation of the flow X close to the point $a \in W^u(\sigma) \setminus \{\sigma\}$ (which defines the orbit $\mathcal{O} = \omega(a)$). Consider the following cross-sections of X (recall the definition of flow-box near a singularity in Sect. 5.4.4):

- $\Sigma^{o,+}$ containing a in its interior and $\Sigma' = X^1(\Sigma^{o,+})$.
- Σ_0 intersecting O in a single point in the center-unstable boundary.
- Σ^- a substrip of $\Sigma^{i,-}$ which is a one-sided neighborhood of ℓ^- not touching B on the same side of a.
- Σ^+ a substrip of $\Sigma^{i,+}$ which is a one-sided neighborhood of ℓ^+ also on the same side of a.

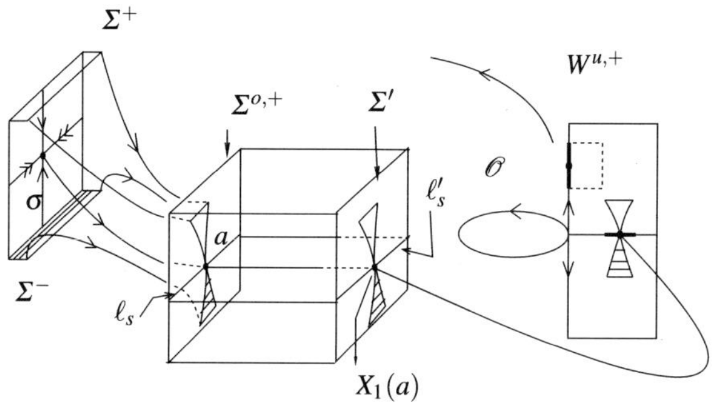

Fig. 6.7 The unperturbed flow X

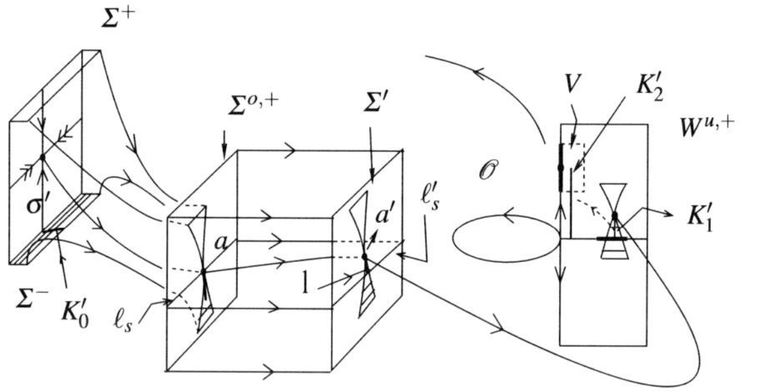

Fig. 6.8 The perturbed flow Y

Observe that the positive orbit of any point in $\Sigma^+ \cup \Sigma^-$ by X will cross $\Sigma^{o,+}$. Define $W = X^{[0,1]}(\Sigma^{o,+})$. The support of the perturbation from X to Y sketched in Figs. 6.7 and 6.8 is contained in W. This perturbation is standard, see e.g. [190], amounting to "push a' upwards so that its image under the flow of Y lands in Σ_O above the stable manifold of $\mathcal{O}$".

Recall that $\Lambda = \Lambda_X(U) = \cap_{t>0} X^t(U)$. Since Λ is not transitive there exists $q \in P^-$ and so there is an interval K_0 in $\Sigma^- \cap W^u(q)$ crossing Σ^- as in Fig. 6.8.

Denote by q', $W^u(q')$, σ', K_0' the continuation of these objects for the perturbed flow Y. The Y-flow carries K_0' to an interval K_1' as in Fig. 6.8.

Note that $K_0' \subset \Lambda_Y(U)$ since $\Lambda_Y(U)$ is an attracting set, $q' \in \Lambda_Y(U)$ and $K_0' \subset W^u(q')$.

We claim that $K_0' \not\subset \Omega(Y)$.

Arguing by contradiction, assume that $K_0' \subset \Omega(Y)$ and choose $x \in \mathrm{int}(K_0')$.

On the one hand, the flow of Y carries points nearby x to V as sketched in Fig. 6.8, close to the line K_2'. By assumption on K_0', x is non-wandering for Y. In particular there exists $x' \in K_0'$ close to x such that *the positive Y-orbit of x' returns to Σ^-*.

On the other hand, by construction, the positive orbit of every point in V by the flow of X does not intersect Σ^-.

Since $Y = X$ outside W we conclude that the positive orbit of x' by Y intersects Σ^+ by the definition of $W^{u,+}$. The positive orbit of such an intersection passes through the flow-box W and arrives at V again. Then we conclude that *the positive*

Y-orbit of x' never returns to Σ^-. This contradiction proves that $K_0' \not\subset \Omega(Y)$, as claimed.

This implies that $\Lambda_Y(U) \not\subset \Omega(Y)$ and finishes the proof of Theorem 6.36. □

6.3.2 Unstable Manifolds of Periodic Orbits Inside Singular-Hyperbolic Attractors

Here we present a proof of Theorem 6.37 following the proof presented in [166].

Let Λ be either a singular-hyperbolic attractor, or a connected singular-hyperbolic attracting set having a dense subset of periodic orbits.

We start by showing that the closure of the unstable manifold of any periodic orbit in Λ must contain some singularity of the flow.

Lemma 6.48 *Let Λ be a connected singular-hyperbolic attracting set containing either a dense subset of periodic orbits, or a dense regular orbit. Fix a periodic point $p_0 \in \mathrm{Per}(X) \cap \Lambda$ (necessarily hyperbolic of saddle-type). Let $J = [a, b]$ be an arc on a connected component of $W^{uu}(p_0) \setminus \{p_0\}$ with $a \neq b$. Then $H = \overline{\cup_{t>0} X^t(J)}$ contains some singularity of Λ.*

Proof Observe that $H = \overline{W_0^u(p_0)} \subset \Lambda$ by construction, where $W_0^u(p_0)$ is the connected component of $W^u(p_0) \setminus \mathcal{O}(p_0)$ containing J. In addition H contains the unstable manifolds through any of its points, since every point in H is accumulated by forward iterates of the arc J.

Consider the set $W^{ss}(H) = \cup\{W^{ss}(y) : y \in H\}$. Note that $W^u(y) \subset H$ for $y \in H$ and the family $\{W^{ss}(z)\}_{y \in W^u(y)}$ covers an open neighborhood of y, and so $W^{ss}(H)$ is a neighborhood of H in M.

Let x be a point in $W^{ss}(H)$. Then by forward iteration this point is sent close to H. This shows that x is in the interior of $W^{ss}(H)$ and hence $W^{ss}(H)$ is open in M. Thus $H^s = W^{ss}(H) \cap \Lambda$ is an open neighborhood of H in Λ. If Λ is transitive, we can take $z \in H^s$ such that $\omega(z) = \Lambda$ and, since $\omega(z) = \omega(x)$ for some $x \in H^s$, we conclude that $\Lambda \subset H$ and so $H \cap S(X) \neq \emptyset$.

If Λ is not transitive, we claim that either $H^s \cap S(X) \neq \emptyset$, or the closure of H^s is an open subset of Λ (besides being clearly a closed set).

First note that, if $\sigma \in H^s \cap S(X)$, then $\sigma \in W^{ss}(y)$ for some $y \in H$ implying that $\sigma \in H$. For otherwise we would get $y \in W^{ss}(\sigma) \cap \Lambda \setminus \{\sigma\}$, a contradiction to Theorem 6.3.

Suppose that $H^s \cap S(X) = \emptyset$. From Proposition 6.2 we know that H^s is a uniformly hyperbolic compact subset of Λ. Then every $w \in H^s$ has a well defined strong-unstable manifold. Moreover $W_\varepsilon^{uu}(w) \subset \Lambda \cap H^s$ for some $\varepsilon > 0$, because Λ is attracting and H^s is open. We conclude that H^s contains the unstable manifold of all its points. Hence taking limits we obtain that the closure $\overline{H^s}$ also contains every unstable manifold. Analogously we see that $\overline{H^s}$ contains the strong-stable manifold $W^{ss}(z) \cap \Lambda$ relative to Λ for all $z \in \overline{H^s}$. The union of the unstable manifolds through all points in the strong-stable manifolds provides a neighborhood of $\overline{H^s}$ in Λ.

Since Λ is connected we obtain $\overline{H^s} = \Lambda$. Hence there exists some singularity σ of Λ in the closure of the stable manifolds of H inside Λ. Let h_n be points in H^s converging to σ. Hence the orbits of h_n contain points h_n' very close to $W^u(p_0)$ by definition of H^s. Using the assumption of dense periodic orbits, consider a periodic orbit p_n very close to h_n. Then the orbit of p_n will be close to $W_0^u(p_0)$ and so $W^s(p_n) \pitchfork W_0^u(p_0) \neq \emptyset$ (to see this, consider an adapted cross-section Σ through h_n', a small tubular flow-box through Σ and recall that stable manifolds cross Σ horizontally). The inclination lemma now ensures that $\overline{W_0^u(p_0)} = H$ contains p_n. Thus H is arbitrarily close to σ. Therefore the closed set H contains some singularity of $S(X) \cap \Lambda$. □

Fix p_0 and $\sigma \in S(X) \cap H$ as in the statement of Lemma 6.48. We can assume that J is a fundamental domain for $W^u(p_0)$, that is, $b = X^T(a)$ with $T > 0$ the first return time of the orbit of a to $W^{uu}(p_0)$, i.e., $X^t(a) \notin W^{uu}(p_0)$ for all $0 < t < T$.

Fix ingoing adapted cross-sections $\hat{\Sigma}_\sigma^{i,\pm}$ of every $\sigma \in S(X) \cap \Lambda$ and horizontal substrips $\Sigma_\sigma^{i,\pm}$ around $\ell_\sigma^\pm$ of small width so that $\mathcal{O}(p_0)$ does not touch $\Sigma_\sigma^{i,\pm}$. We assume that $\hat{\Sigma}_\sigma^{i,\pm} \setminus \Sigma_\sigma^{i,\pm}$ have nonempty interiors.

Consider also a cross-section Σ_p containing p_0. We can then take $J = [a, b]$ so close to p_0 that $J \subset \operatorname{int}(\Sigma_p)$ and $X^{-t}(J)$ never intersects $\Sigma_\sigma^{i,\pm}$ for all $t > 0$ and every $\sigma \in S(X) \cap \Lambda$.

Since $S(X) \cap \overline{W^u(p_0)} \neq \emptyset$ there exists a Poincaré map R from a subset D of Σ_p to $\operatorname{int}(\cup_\sigma \Sigma_\sigma^{i,\pm})$ given by the first return time $\tau(x)$ of $x \in D$. Without loss of generality assume that $R(b) \in \operatorname{int}(\Sigma_\sigma^{i,+})$ for some singularity σ fixed from now on. We drop the σ from the notation of the cross-sections in what follows.

Note that $R(a)$ must equal $R(b)$. Using this with some tubular flow-boxes together with the fact that $\Sigma^{i,+}$ is an adapted cross-section, we show that the image of J under R must cross $\Sigma^{i,+}$ from one stable boundary to the other, thus intersecting ℓ^+. Since $\ell^+ = \Sigma^{i,+} \cap W_{loc}^s(\sigma)$, this argument proves Theorem 6.37.

Observe that, because both J and $R(a) = R(b)$ belong to the interior of the respective cross-sections to X, then there exists a tubular flow-box, given by Theorem 2.13, and open neighborhoods $V \subset \Sigma_p$ of b and $W \subset \Sigma^{i,+}$ of $R(b)$, such that $V \subset D$, that is $R \mid V : V \to W$ is well defined and a diffeomorphism. Moreover, since J is transverse to the stable foliation in Σ_p, then the image $R(V \cap J)$ is also transverse to the stable foliation of $\Sigma^{i,+}$. In addition, since Λ is attracting, we see that J and $R(J \cap V)$ are contained in Λ. Because $\Sigma^{i,+}$ is adapted, the image of J is δ-away the center-unstable boundary. Identifying the arc $[a, b]$ with some interval $[a, b] \subset \mathbb{R}$ we define (see Fig. 6.9)

$$q = \sup\{s \in [a, b] : R([a, s]) \subset \operatorname{int}(\Sigma^{i,+})\}.$$

By the existence of the pair V, W we have $q > a$. Moreover given $s \in (a, q)$ and covering the compact arcs $[a, s]$ and $R([a, s])$ by a finite number of open tubular flow-boxes $U_1, \dots, U_k$ we easily see that $R([a, s])$ is connected. Indeed, $R([a, s])$ is the union of a sequence $R([s_i, s_{i+1}])$ of arcs inside $U_i \cap \operatorname{int}(\Sigma^{i,+})$, where $a = s_0 < s_1 < \cdots < s_k = s$ and $R \mid U_i \cap \operatorname{int}(\Sigma_p) : U_i \cap \operatorname{int}(\Sigma_p) \to U_i \cap \operatorname{int}(\Sigma^{i,+})$ is a diffeomorphism, $i = 1, \dots, k$.

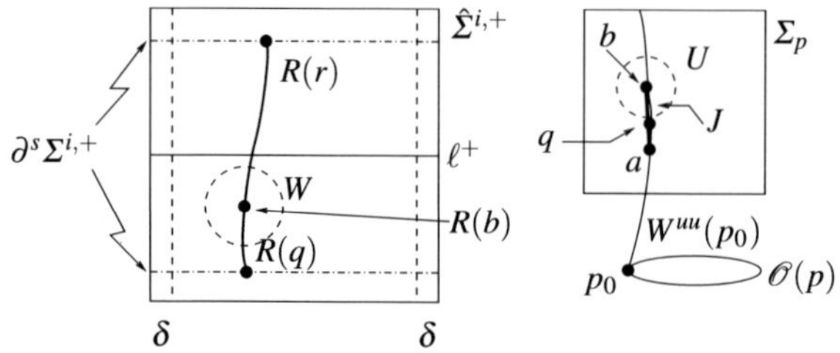

Fig. 6.9 The arc J and cross-sections Σ_p, $\Sigma_0^{i,+}$

Note that, by the choice of $\Sigma^{i,+}$ strictly inside $\hat{\Sigma}^{i,+}$, *if q belongs to the domain D of R*, then there exists a tubular flow-box U_0 taking q to $R(q)$, so that $R(q)$ is well defined. Hence $R(q) = \lim_{s \nearrow q} R(s)$ is not on the center-unstable boundary $\partial^{cu} \Sigma^{i,+}$ by construction. Moreover using the tubular flow-box U_0 we see that $R(q) \in \partial^s \Sigma^{i,+}$. For otherwise, in case $R(q) \in \text{int}(\Sigma^{i,+})$, we would be able to extend the definition of R along J through the flow-box U_0.

Now apply the same arguments to

$$r = \inf\{s \in [a,b] : R([s,b]) \subset \text{int}(\Sigma^{i,+})\}.$$

We obtain $R(a) = R(b)$ and $R(q), R(r) \in \partial^s \Sigma^{i,+}$ *if r belongs to the domain of D*. We obtain in this way $\gamma = R([a,q] \cup [r,b])$, a connected smooth arc joining two points in the stable boundary.

If $R(q), R(r)$ belong to the same stable-manifold on $\partial^s \Sigma^{i,+}$, then by smoothness and connectedness there must be a tangency between γ and the stable foliation on $\Sigma^{i,+}$. This is a contradiction.

Hence $R(q), R(r)$ are on different stable leaves on the boundary of $\Sigma^{i,+}$, and thus γ crosses ℓ^+ transversely. This means that $W^u(p_0) \pitchfork W^s(\sigma) \neq \emptyset$. The proof of Theorem 6.37 now rests on the claim that *both q and r belong to the domain of R*. To prove this claim we need the following result, whose proof we postpone.

Lemma 6.49 *Let $\tilde{\Sigma}$ be a cross-section of X containing a compact cu-curve ζ, which is the image of a regular parametrization $\zeta : [0,1] \to \tilde{\Sigma}$, and assume that ζ is contained in Λ. Let Σ be another cross-section of X. Suppose that ζ* falls off Σ, *that is*

- *the positive orbit of $\zeta(t)$ visits* $\text{int}(\Sigma)$ *for all $t \in [0,1)$;*
- *and the ω-limit of $\zeta(1)$ is disjoint from Σ.*

Then $\zeta(1)$ belongs either to the stable manifold of some periodic orbit p in Λ, or to the stable manifold of some singularity.

Observe that $[a,q]$ (and $[r,b]$) *fall off* $\Sigma^{i,+}$, *if q (and r) does not belong to D*. Then $\omega(q)$ (and $\omega(r)$) is either a periodic orbit in Λ, or a singularity. In the first case the arc $J \supset [a,q]$ is transverse at q to the stable manifold of a periodic orbit p. The inclination lemma ensures that there exists a fundamental domain L of $W^{uu}(p)$ accumulated by iterates of the open arc (a,q); see Fig. 6.10. Hence the flow takes *every point* of L through $\Sigma^{i,+}$. As before the image of L by the corresponding first

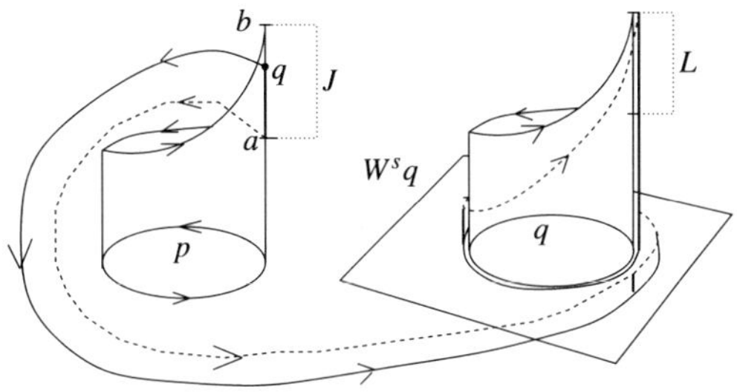

Fig. 6.10 How I accumulates $D^u(x^*)$

return map must be a *cu*-curve C in $\Sigma^{i,+}$. Moreover since the endpoints of L are on the same orbit of the flow, C must be a *closed cu*-curve. This is impossible.

This contradiction shows that either q (and r) does not fall off $\Sigma^{i,+}$, so that q (and r) is in the domain of R, or q is in the stable manifold of some singularity. In the former case, we are done. In the latter case, since the stable manifold is transverse to $W^{uu}(p_0)$ by the assumption of singular-hyperbolicity, we obtain the statement of the theorem as well.

Now to finish the proof of Theorem 6.37 we prove the remaining lemma.

Proof of Lemma 6.49 Define $H = \omega(\zeta(1))$ and suppose that H is not a singularity. By an argument similar to the proof of Lemma 6.27 we find that H has totally disconnected intersection with any cross-section.

Indeed, consider an adapted cross-section Σ_x of X through $x \in H$ and consider the connected component C of $H \cap \Sigma_x$ containing x. As in the proof of Lemma 6.27, we have $C \cap W^s(x, \Sigma_x) = \{x\}$.

If there exists $y \in C \setminus W^s(x, \Sigma_x)$ consider the horizontal strip S of Σ_x between the stable leaves $W^s(x, \Sigma_x)$ and $W^s(y, \Sigma_x)$. Then there exists a point w of $H \cap \operatorname{int}(S)$, for otherwise $W^s(w, \Sigma_x)$ would disconnect y from x. From this we find ξ in the positive orbit of $\zeta(1)$ inside $\operatorname{int}(S)$ and close to w. But ζ is a *cu*-curve. Hence, considering the tubular flow on a neighborhood around the piece of orbit from $\zeta(1)$ to ξ, we find in the image of ζ under the tubular flow a *cu*-curve ζ', a connected image of a neighborhood of $\zeta(1)$ in ζ, with ξ as a boundary point. (Here we use the hyperbolicity of the Poincaré maps between cross-sections assuming that the time from $\zeta(1)$ to ξ is big enough.)

So we have a positive iterate of a point $\zeta(s)$ in $\operatorname{int}(S)$ for some $s \in [0, 1)$. Now we use the density of periodic orbits to find a point of a periodic orbit p' very close to $\zeta(s)$ in $\operatorname{int}(S)$. Then the orbit of p' crosses $\operatorname{int}(\Sigma)$ by the assumption on the curve ζ. Again there exists $h \in H \cap W^s(p', S)$. This means that the orbit of h will cross $\operatorname{int}(\Sigma)$. Since $h \in \omega(\zeta(1))$, then the orbit of $\zeta(1)$ must cross $\operatorname{int}(\Sigma)$ also. We have reached a contradiction.

We conclude that $\Sigma_x \cap H$ is totally disconnected.

Hence we can cover the set H with a finite number of flow-boxes around the singularities contained in H together with finitely many tubular neighborhoods through adapted cross-sections, i.e., sets of the form $X^{(-\varepsilon,\varepsilon)}(\Sigma_x)$. Let Ξ be the collection of adapted cross-sections used in this cover, some of them ingoing or outgoing cross-sections around singularities.

Since $\Sigma_x \cap H$ is totally disconnected (if the intersection is non-empty), then H is contained in the interior of these flow-boxes. Thus $\Sigma_x \cap H$ is not only δ-away from the center-unstable boundary of Σ, but δ-away from the stable boundary of Σ as well, for some uniform $\delta > 0$ valid for every cross-section of Ξ.

The definition of H ensures that $\zeta^t(1) = X^t(\zeta(1))$ is, for big enough $t > 0$, contained in a small closed neighborhood W around H, which can be taken disjoint from the reference section Σ.

Let $t_n \to +\infty$ be such that $\zeta_n(1) = \zeta^{t_n}(1) \in \text{int}(\Xi)$ for all $n \geq 1$. Since Ξ is a finite collection of sections, we can assume without loss of generality that $\zeta_n(1)$ always belongs to the same section $S \in \Xi$.

Observe that the positive orbit of $\zeta(s)$, with $s < 1$ and close to 1, enters W by continuity of the flow, but does not stay in W, since it must cross $\text{int}(\Sigma)$. Then the first return of $\zeta(s)$ to S, which we write $\zeta_n(s)$, is well defined for $s < 1$ and close to 1.

For infinitely many values of n there exists some $s_n \in [0, 1)$ such that $\zeta_n([s_n, 1])$ is contained in S, the orbit segment from $\zeta(s)$ to $\zeta_n(s)$ is disjoint from Σ for all $s_n \leq s \leq 1$, and $\zeta_n(s_n)$ is in the boundary of S. For otherwise we would get $\zeta_n([0, 1]) \subset \text{int}(S) \subset W$ and so $\zeta(s)$ would never reach Σ.

This means that the cu-curve $\gamma_n = \zeta_n([s_n, 1])$ has length at least δ inside S and

- either the end point $\zeta_n(1)$ of γ_n has a subsequence contained in the same stable manifold inside S, which by Lemma 6.26 implies that $\zeta_n(1)$ is in the stable manifold of a periodic orbit, and thus H is a periodic orbit;
- or γ_n has an accumulation curve inside S in the C^1 topology (using the Arzelá-Ascoli Theorem, since γ_n have bounded derivative by definition of cu-curve and length bounded away from aero, and S is compact), so that we can find a point $\zeta_n(s)$ in the stable manifold of $\zeta_m(1)$, for m, n very big. This is impossible because the positive orbit of $\zeta_n(s)$ would stay forever close to the orbit of $\zeta_m(1)$, inside W, and would never reach Σ.

We conclude that H is a periodic orbit if it is not a singularity. The proof of Lemma 6.49 is complete. □

Chapter 7
Expansiveness and Physical Measure

Here we obtain further consequences of singular-hyperbolicity. A singular-hyperbolic attractor is chaotic in two senses: *it is robustly expansive and so has sensitive dependence on initial conditions; and it supports a unique physical measure with non-zero Lyapunov exponent.*

7.1 Statements of the Results and Overview of the Arguments

Theorem 7.1 *Let Λ be a singular-hyperbolic attractor of $X \in \mathfrak{X}^1(M)$. Then Λ is expansive.*

The proof of Theorem 7.1 is the content of Sect. 7.2 based on the arguments in [21].

The reasoning is based on analyzing Poincaré return maps of the flow to a convenient (δ-adapted) cross-section. We use the family of adapted cross-sections and corresponding Poincaré maps R, whose Poincaré time $t(\cdot)$ is large enough, obtained assuming that the attractor Λ is singular-hyperbolic in Sect. 6.1. These cross-sections have co-dimension 1 foliations, which are dynamically defined, whose leaves are uniformly contracted and invariant under the Poincaré maps. In addition R is uniformly expanding in the transverse direction and this also holds near the singularities.

From here we argue by contradiction: if the flow is not expansive on Λ, then we can find a pair of orbits hitting the cross-sections infinitely often on pairs of points uniformly close. We derive a contradiction by showing that the uniform expansion in the transverse direction to the stable foliation must take the pairs of points apart, unless one orbit is on the stable manifold of the other.

This argument only depends on the existence of, firstly, finitely many Lorenz-like singularities on a compact partially hyperbolic invariant attracting subset $\Lambda = \Lambda_X(U)$, with volume expanding central direction and, secondly, a family of adapted cross-sections with Poincaré maps between them, whose derivative is hyperbolic. It is straightforward that, if these conditions are satisfied for a flow X^t of $X \in \mathfrak{X}^1(M^3)$,

V. Araújo, M.J. Pacifico, *Three-Dimensional Flows*, Ergebnisse der Mathematik und ihrer Grenzgebiete. 3. Folge / A Series of Modern Surveys in Mathematics 53,
DOI 10.1007/978-3-642-11414-4_7,

then the same conditions hold for all C^1 nearby flows Y^t and for the maximal invariant subset $\Lambda_Y(U)$ *with the same family of cross-sections* which are also adapted to $\Lambda_Y(U)$ (as long as Y is C^1-close enough to X). We recall that the notation $\Lambda_Y(U)$ denotes $\cap_{t>0} Y^t(U)$ for some trapping region U. To obtain the adapted cross-sections to start with, however, we need to assume that we have a singular-hyperbolic attractor, that is, we need to assume transitivity.

Corollary 7.2 *A singular-hyperbolic attractor $\Lambda = \Lambda_X(U)$ is* robustly expansive, *that is, there exists a neighborhood $\mathcal{U}$ of X in $\mathfrak{X}^1(M)$ such that $\Lambda_Y(U)$ is expansive for each $Y \in \mathcal{U}$, where U is an isolating neighborhood of Λ.*

Indeed, since transversality, partial hyperbolicity and volume expanding central direction are robust properties, and also the hyperbolicity of the Poincaré maps depends on the central volume expansion, all we need to do is to check that a given δ-adapted cross-section Σ to X is also adapted to $Y \in \mathfrak{X}^1$ for every Y sufficiently C^1 close to X. But $\Lambda_X(U)$ and $\Lambda_Y(U)$ are close in the Hausdorff distance if X and Y are close in the C^0 distance, by Lemma 2.3. Thus, if Σ is δ-adapted we can find a C^1-neighborhood $\mathcal{U}$ of X in $\mathfrak{X}^1$ such that Σ is $(\delta/10)$-adapted to every flow Y^t generated by a vector field in $\mathcal{U}$.

7.1.1 Robust Sensitiveness

We already know that expansiveness implies sensitive dependence on initial conditions or, in other words, that the flow is chaotic. An argument with the same flavor as the proof of expansiveness provides the following.

Theorem 7.3 *Every singular-hyperbolic isolated set $\Lambda = \cap_{t\in\mathbb{R}} \overline{X^t(U)}$ is* robustly chaotic, *i.e.*, *there exists a neighborhood $\mathcal{U}$ of X in $\mathfrak{X}^1(M)$ such that $\cap_{t\in\mathbb{R}} \overline{Y^t(U)}$ is chaotic for each $Y \in \mathcal{U}$, where U is an isolating neighborhood of Λ.*

We present the proof of this result after the proof of expansiveness, in Sect. 7.2.8.

It is natural to consider the converse: are all robustly chaotic isolated sets singular-hyperbolic? Straightforward adaptations of the arguments of Chap. 5 enable us to prove the converse for attractors under a mild condition on the singularities of the vector field.

Theorem 7.4 *Let Λ be an attractor for $X \in \mathfrak{X}^1(M^3)$ such that every singularity in its trapping region is hyperbolic with no resonances. Then Λ is singular-hyperbolic if, and only if, Λ is robustly chaotic.*

This means that *if we can show that arbitrarily close orbits, in an isolating neighborhood of an attractor, are driven apart, for the future as well as for the past, by the evolution of the system, and this behavior persists for all C^1 nearby vector fields, then the attractor is singular-hyperbolic.*

The condition on the singularities amounts to restricting the possible three-dimensional vector fields in the above statement to an open a dense subset of all C^1 vector fields. Indeed, the hyperbolic and no-resonance condition on a singularity σ means that:

- either $\lambda \neq \Re(\omega)$ if the eigenvalues of $DX(\sigma)$ are $\lambda \in \mathbb{R}$ and $\omega, \overline{\omega} \in \mathbb{C}$;
- or σ has only real eigenvalues with different norms.

As we will see, these conditions precisely allow us to deduce the same conclusions of Lemma 5.22, so that the only possible singularities in the setting of Theorem 7.4 are Lorenz-like singularities.

We can also deduce the same conclusion of Theorem 7.4 for an isolated set, not necessarily transitive, if we assume that its singularities are Lorenz-like, as follows.

Theorem 7.5 *Let Λ be a compact isolated invariant proper subset of M with respect to a C^1 vector field X whose singularities are Lorenz-like. Then Λ is singular-hyperbolic if, and only if, Λ is robustly chaotic.*

The conditions on the singularities in both Theorems 7.4 and 7.5 are in principle easy to check numerically for any given system.

We remark that the invariant set in the above statements may have no singularities. This is the case of a hyperbolic set of saddle-type for X. We also note that the singular-horseshoe is in the setting of Theorem 7.5.

One interpretation of these results is that, among three-dimensional flows, *robustly chaotic behavior is always associated either to (uniform) hyperbolicity* (if there are no singularities properly accumulated by regular orbits in the invariant set) *or to singular-hyperbolicity*. Since both these types of invariant sets are by now well known, this may have relevant consequences for the study of all sorts of mathematical models involving flows in three-dimensional manifolds.

In recent ongoing PhD thesis work of Laura Senos, at Universidade Federal do Rio de Janeiro, a generalized version of this result is obtained, removing the non-resonance conditions and the assumptions of future and past expansiveness (future and past chaotic), showing that any robustly expansive attractor for a 3-flow is necessarily a singular-hyperbolic attractor.

Proof of Theorems 7.4 and 7.5 As mentioned in Sect. 2.4 a chaotic system, in our terminology, cannot have either attracting or repelling critical elements. Hence, a robustly chaotic attractor Λ for a vector field X has no attracting or repelling critical elements in a neighborhood for all vector fields Y which are C^1 close to X. Now Theorem 2.33 implies that the Linear Poincaré Flow over $\Lambda_Y^*(U) = \Lambda_Y(U) \setminus S(Y)$ admits a dominated splitting, for each vector field Y close enough to X in the C^1 topology. As in the proofs in Chap. 5, we can now classify the possible singularities appearing in the attractor Λ.

Since each singularity $\sigma \in \Lambda$ is assumed to be hyperbolic, then $DX(\sigma)$ cannot admit complex eigenvalues, for otherwise we can use the transitive orbit and the Connecting Lemma to produce a saddle-focus connection as in the proof of

Lemma 5.22. The non-resonance condition ensures that this saddle-focus connection would imply the existence of an attracting or repelling periodic orbit for an arbitrarily C^1 close vector field. Having only real eigenvalues, σ must be Lorenz-like, either for X or for $-X$. For otherwise we would produce a saddle-connection, again by the Connecting Lemma through the existence of a dense orbit in Λ, exhibiting an inclination-flip type of connecting orbit for an arbitrarily C^1 close vector field. This contradicts the domination of the splitting for the Linear Poincaré Flow. Finally σ cannot be a Lorenz-like singularity for $-X$ inside an attractor in our setting, according to Lemma 5.32.

At this point we have all the assumptions of Theorem 7.5: every singularity σ in Λ is Lorenz-like for the vector field X, there are no attracting or repelling critical elements for every C^1 close vector field Y. Hence, since an attractor is in particular isolated, we can apply Theorem 5.34 and conclude that Λ is singular-hyperbolic.

This concludes the proof that robustly chaotic isolated sets with Lorenz-like singularities, or robustly chaotic attractors under open and dense conditions on the singularities, are singular-hyperbolic. This is a part of the statement of Theorems 7.4 and 7.5. The other part of the statements follows from Theorem 7.3. □

7.1.2 Existence and Uniqueness of a Physical Measure

It was proved by Colmenarez in [72] that, if Λ is a singular-hyperbolic attractor of a C^2 flow X with a dense set of periodic orbits, then the central direction $E^{cu}_{\widetilde{\Lambda}}$ can be continuously decomposed into $E^u \oplus E^X$ along each orbit of $\widetilde{\Lambda}$, where the E^u direction is non-uniformly hyperbolic, that is, has a positive Lyapunov exponent, and $\widetilde{\Lambda} = \Lambda \setminus \cup_{\sigma \in S(X) \cap \Lambda} W^u(\sigma)$. In [71], again under the assumption of a dense set of periodic orbits, Colmenarez showed that every C^2 singular-hyperbolic attractor supports a physical probability measure—see Sect. 2.7.1 of Chap. 1 for the relevant definitions.

However, in another recent work, Arroyo and Pujals [27] show that every singular-hyperbolic attractor has a dense set of periodic orbits. Indeed, as shown in Chap. 6, every singular-hyperbolic attractor is a homoclinic class. So the denseness assumption is no restriction.

Here we give an independent proof of the existence of physical measures which does not use denseness of periodic orbits, and this enables us to obtain the hyperbolicity of the physical measure and, also, to show that it is a Gibbs state. We follow the proof presented in [21].

Theorem 7.6 *Let Λ be a singular-hyperbolic attractor of a flow $X \in \mathfrak{X}^2(M)$ on a three-dimensional manifold. Then Λ supports a unique physical probability measure μ which is ergodic, hyperbolic and whose ergodic basin covers a full Lebesgue measure subset of the topological basin of attraction, i.e., $B(\mu) = W^s(\Lambda)$ Lebesgue mod 0. Moreover the support of μ is the whole attractor Λ.*

Here we need to assume that $(X^t)_{t\in\mathbb{R}}$ is a flow of class C^2 since for the construction of physical measures a bounded distortion property for one-dimensional maps is needed. These maps are naturally obtained as quotient maps over the set of stable leaves, which form a $C^{1+\alpha}$ foliation of a finite number of cross-sections associated to the flow if the flow is C^2; see Sect. 2.7.2. This will be detailed in Sect. 7.3.

We recall from Sect. 2.7 of Chap. 1 that hyperbolicity of a measure means *non-uniform hyperbolicity*: the tangent bundle over Λ splits into a sum $T_zM = E^s_z \oplus E^X_z \oplus F_z$ of three one-dimensional invariant subspaces defined for μ-a.e. $z \in \Lambda$ and depending measurably on the base point z, where μ is the physical measure in the statement of Theorem 7.6, E^X_z is the flow direction (with zero Lyapunov exponent) and F_z is the direction with positive Lyapunov exponent.

Theorem 7.6 is another statement of sensitiveness, this time applying to the whole essentially open set $B(\Lambda)$. Indeed, since non-zero Lyapunov exponents express the idea that the orbits of infinitesimally close-by points tend to move apart from each other, this theorem means that most orbits in the basin of attraction separate under forward iteration. See Kifer [122], and Metzger [153], and references therein, for previous results about invariant measures and stochastic stability of the geometric Lorenz models.

In the uniformly hyperbolic setting, it is well known that physical measures for hyperbolic attractors admit a disintegration into conditional measures along the unstable manifolds of almost every point which are absolutely continuous with respect to the induced Lebesgue measure on these sub-manifolds; see [60, 62, 201, 266].

Here the existence of unstable manifolds is guaranteed by the hyperbolicity of the physical measure: the strong-unstable manifolds $W^{uu}(z)$ are the "integral manifolds" in the direction of the one-dimensional sub-bundle F, tangent to F_z at almost every $z \in \Lambda$. The tools developed to prove Theorem 7.6 enable us to prove that the physical measure obtained there has absolutely continuous disintegration along the center-unstable direction; see Sect. 2.7 of Chap. 1 for the definition of conditional measures and the notion of adapted foliated neighborhoods of a point.

Theorem 7.7 *Let Λ be a singular-hyperbolic attractor for a C^2 three-dimensional flow. Then the physical measure μ supported in Λ has a disintegration into absolutely continuous conditional measures μ_γ along center-unstable surfaces $\gamma \in \Pi_\delta(x)$ such that $\frac{d\mu_\gamma}{dm_\gamma}$ is uniformly bounded from above, for all δ-adapted foliated neighborhoods $\Pi_\delta(x)$ and every $\delta > 0$. Moreover* $\operatorname{supp}(\mu) = \Lambda$.

Remark 7.8 The proof that $\operatorname{supp}(\mu) = \Lambda$ which we present depends on the absolutely continuous disintegration property of the physical measure μ and the transitivity of X on Λ. However, most singular-hyperbolic attractors are topologically mixing in the C^1 topology [166] and the Lorenz geometric models are always topologically mixing [142], so we should expect a more general argument proving $\operatorname{supp}(\mu) = \Lambda$ without the need to obtain *first* that μ is a *cu*-Gibbs measure or *SRB*-measure.

Remark 7.9 It follows from the proof that the densities of the conditional measures μ_γ are bounded from below away from zero on $\Lambda \setminus B$, where B is any neighborhood of the singularities $\sigma(X \mid \Lambda)$. In particular the densities tend to zero as we get closer to the singularities of Λ.

The absolute continuity property along the center-unstable sub-bundle given by Theorem 7.7 ensures that

$$h_\mu(X^1) = \int \log \left| \det(DX^1 \mid E^{cu}) \right| d\mu,$$

by the characterization of probability measures satisfying the Entropy Formula, obtained in [129]. The above integral is the sum of the positive Lyapunov exponents along the sub-bundle E^{cu} by Oseledets' Theorem [147, 269]. Since in the direction E^{cu} there is only one positive Lyapunov exponent along the one-dimensional direction F_z, for μ-a.e. z, the ergodicity of μ then shows that the following is true.

Corollary 7.10 *If Λ is a singular-hyperbolic attractor for a C^2 three-dimensional flow X^t, then the physical measure μ supported in Λ satisfies the Entropy Formula*

$$h_\mu(X^1) = \int \log \| DX^1 \mid F_z \| \, d\mu(z).$$

Again by the characterization of measures satisfying the Entropy Formula we deduce that μ *has absolutely continuous disintegration along the strong-unstable direction*, along which the Lyapunov exponent is positive, and thus μ *is a u-Gibbs state* [201]. This also shows that μ *is an equilibrium state for the potential* $-\log \| DX^1 \mid F_z \|$ with respect to the diffeomorphism X^1. We note that the entropy $h_\mu(X^1)$ of X^1 is the entropy of the flow X^t with respect to the measure μ [269].

Hence we are able to extend most of the basic results on the ergodic theory of hyperbolic attractors to the setting of singular-hyperbolic attractors.

7.2 Expansiveness

For the proof of Theorem 7.1 we need the construction of cross-sections and Poincaré return maps, which is the subject of Sect. 6.1. We use the construction and notations defined there in what follows.

7.2.1 Proof of Expansiveness

Here we prove Theorem 7.1. The proof is by contradiction: let us suppose that there exist $\varepsilon > 0$, a sequence $\delta_n \to 0$, a sequence of functions $h_n \in \mathcal{K}$ (see Sect. 2.4 of

Chap. 1 for the definition of expansiveness), and sequences of points x_n, $y_n \in \Lambda$ such that

$$d\big(X^t(x_n), X^{h_n(t)}(y_n)\big) \le \delta_n \quad \text{for all } t \in \mathbb{R}, \tag{7.1}$$

but

$$X^{h_n(t)}(y_n) \notin X^{[t-\varepsilon,t+\varepsilon]}(x_n) \quad \text{for all } t \in \mathbb{R}. \tag{7.2}$$

The main step in the proof is a reduction to a forward expansiveness statement about Poincaré maps which we state in Theorem 7.11 below.

We are going to use the following observation: there exists some regular (i.e. non-equilibrium) point $z \in \Lambda$ which is accumulated by the sequence of ω-limit sets $\omega(x_n)$. To see that this is so, we start by observing that accumulation points do exist, since Λ is compact. Moreover, if the ω-limit sets accumulate on a singularity then they also accumulate on at least one of the corresponding unstable branches which, of course, consist of regular points. We fix such a z once and for all. Replacing our sequences by subsequences, if necessary, we may suppose that for every n there exists $z_n \in \omega(x_n)$ such that $z_n \to z$.

Let Σ be a δ-adapted cross-section at z, for some small δ. Reducing δ (but keeping the same cross-section) we may ensure that z is in the interior of the subset

$$\Sigma_\delta = \{y \in \Sigma : d(y, \partial\Sigma) > \delta\}.$$

By definition, x_n returns infinitely often to the neighborhood of z_n which, in its turn, is close to z. Thus, dropping a finite number of terms in our sequences if necessary, we see that the orbit of x_n intersects Σ_δ infinitely many times. Let t_n be the time corresponding to the nth intersection.

Replacing x_n, y_n, t, and h_n by $x^{(n)} = X^{t_n}(x_n)$, $y^{(n)} = X^{h_n(t_n)}(y_n)$, $t' = t - t_n$, and $h'_n(t') = h_n(t' + t_n) - h_n(t_n)$, we may suppose that $x^{(n)} \in \Sigma_\delta$, while preserving both relations (7.1) and (7.2). Moreover there exists a sequence $\tau_{n,j}$, $j \ge 0$ with $\tau_{n,0} = 0$ such that

$$x^{(n)}(j) = X^{\tau_{n,j}}(x^{(n)}) \in \Sigma_\delta \quad \text{and} \quad \tau_{n,j} - \tau_{n,j-1} > \max\{t_1, t_2\} \tag{7.3}$$

for all $j \ge 1$, where t_1 is given by Proposition 6.15 and t_2 is given by Lemma 6.23.

Theorem 7.11 *Given $\varepsilon_0 > 0$ there exists $\delta_0 > 0$ such that, if $x \in \Sigma_\delta$ and $y \in \Lambda$ satisfy*

(a) *there exist τ_j such that*

$$x_j = X^{\tau_j}(x) \in \Sigma_\delta \quad \textit{and} \quad \tau_j - \tau_{j-1} > \max\{t_1, t_2\} \quad \textit{for all } j \ge 1;$$

(b) $\operatorname{dist}(X^t(x), X^{h(t)}(y)) < \delta_0$, *for all $t > 0$ and some $h \in \mathcal{K}$,*

then there exists $j \ge 1$ such that $X^{h(\tau_j)}(y) \in W^{ss}_{\varepsilon_0}(X^{[\tau_j-\varepsilon_0,\tau_j+\varepsilon_0]}(x))$.

We postpone the proof of Theorem 7.11 until the next section and explain first why it implies Theorem 7.1. We are going to use the following observation.

Fig. 7.1 Distances near a point in the stable-manifold

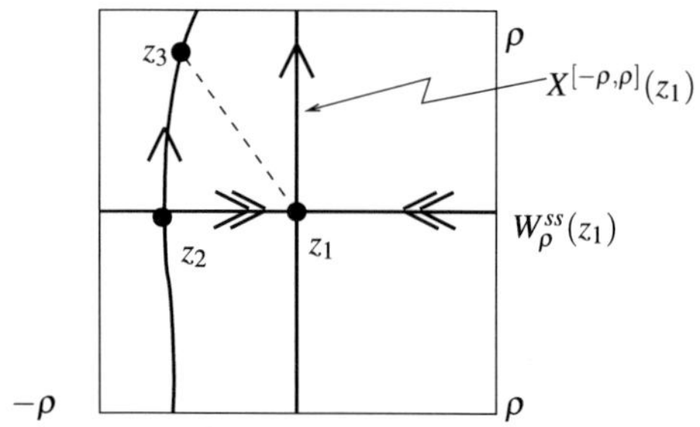

Lemma 7.12 *There exist $\rho > 0$ small and $c > 0$, depending only on the flow, such that, if z_1, z_2, z_3 are points in Λ satisfying $z_3 \in X^{[-\rho,\rho]}(z_2)$ and $z_2 \in W^{ss}_\rho(z_1)$, with z_1 away from any equilibria of X, then*

$$\operatorname{dist}(z_1, z_3) \geq c \cdot \max\{\operatorname{dist}(z_1, z_2), \operatorname{dist}(z_2, z_3)\}.$$

Proof This is a direct consequence of the fact that the angle between E^{ss} and the flow direction is bounded from zero which, in its turn, follows from the fact that the latter is contained in the center-unstable sub-bundle E^{cu}. Indeed consider the C^1 surface $X^{[-\rho,\rho]}(W^{ss}_\rho(z_1))$ for small enough $\rho > 0$. The Riemannian metric here is uniformly close to the Euclidean one and we may choose coordinates on $[-\rho, \rho]^2$ putting z_1 at the origin, sending $W^{ss}_\rho(z_1)$ to the segment $[-\rho, \rho] \times \{0\}$ and $X^{[-\rho,\rho]}(z_1)$ to $\{0\} \times [-\rho, \rho]$; see Fig. 7.1. Then the angle α between $X^{[-\rho,\rho]}(z_2)$ and the horizontal is bounded from below away from zero and the existence of c follows by standard arguments using the Euclidean metric. □

We fix $\varepsilon_0 = \varepsilon$ as in (7.2) and then consider δ_0 as given by Theorem 7.11. Next, we fix n such that $\delta_n < \delta_0$ and $\delta_n < (c + \sup_{z\in\Lambda} \|X(z)\|)\rho$, and apply Theorem 7.11 to $x = x^{(n)}$, $y = y^{(n)}$ and $h = h_n$. Hypothesis (a) in the theorem corresponds to (7.3) and, with these choices, hypothesis (b) follows from (7.1). Therefore we obtain $X^{h(s)}(y) \in W^{ss}_\varepsilon(X^{[s-\varepsilon,s+\varepsilon]}(x))$. In other words, there exists $|\tau| \leq \varepsilon$ such that $X^{h(s)}(y) \in W^{ss}_\varepsilon(X^{s+\tau}(x))$. Hypothesis (7.2) implies that $X^{h(s)}(y) \neq X^{s+\tau}(x)$. Hence, since strong-stable manifolds are expanded under backward iteration, there exists a maximum $\theta > 0$ such that

$$X^{h(s)-t}(y) \in W^{ss}_\rho(X^{s+\tau-t}(x)) \quad \text{and} \quad X^{h(s+\tau-t)}(y) \in X^{[-\rho,\rho]}(X^{h(s)-t}(y))$$

for all $0 \leq t \leq \theta$; see Fig. 7.2. Moreover $s = \tau_j$ for some $j \geq 1$ so that x is close to cross-section of the flow which we can assume is uniformly bounded away from the equilibria, and then we can assume that $\|X(X^t(x))\| \geq c$ for $0 \leq t \leq \theta$. Since θ is maximum

$$\text{either } \operatorname{dist}\left(X^{h(s)-t}(y), X^{s+\tau-t}(x)\right) \geq \rho$$
$$\text{or } \operatorname{dist}\left(X^{h(s+\tau-t)}(y), X^{h(s)-t}(y)\right) \geq c\rho \text{ for } t = \theta.$$

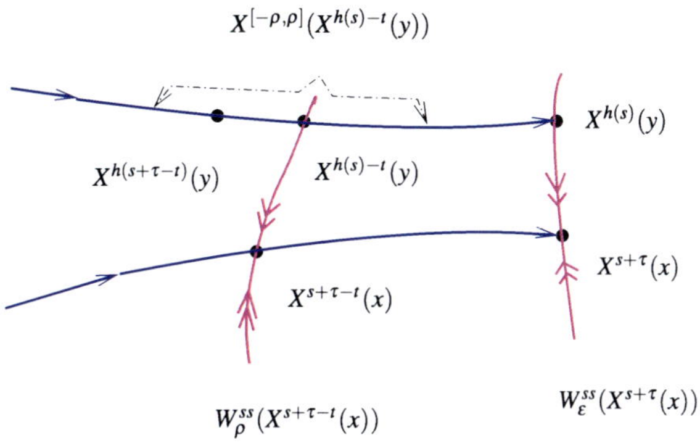

Fig. 7.2 Relative positions of the strong-stable manifolds and orbits

Using Lemma 7.12, we conclude that $\mathrm{dist}(X^{s+\tau-t}(x), X^{h(s+\tau-t)}(y)) \geq c\rho > \delta_n$ which contradicts (7.1). This contradiction reduces the proof of Theorem 7.1 to that of Theorem 7.11.

7.2.2 Infinitely Many Coupled Returns

We start by outlining the proof of Theorem 7.11. There are three steps.

- The first one, which we carry out in the present section, is to show that to each return x_j of the orbit of x to Σ there corresponds a nearby return y_j of the orbit of y to Σ. The precise statement is in Lemma 7.13 below.
- The second, and most crucial step, is to show that there exists a smooth Poincaré map, with large return time, defined on the whole strip of Σ in between the stable manifolds of x_j and y_j. This is done in Sects. 7.2.3 to 7.2.6.
- The last step, Sect. 7.2.7, is to show that these Poincaré maps are uniformly hyperbolic, in particular, they expand cu-curves uniformly (recall the definition of cu-curve in Sect. 6.1.2).

The theorem is then easily deduced: to prove that $X^{h(s)}(y)$ is in the orbit of $W^{ss}_{\varepsilon}(x)$ it suffices to show that $y_j \in W^s(x_j, \Sigma)$, by Remark 6.14. The latter must be true, for otherwise, by hyperbolicity of the Poincaré maps, the stable manifolds of x_j and y_j would move apart as $j \to \infty$, and this would contradict condition (b) of Theorem 7.11. See Sect. 7.2.7 for more details.

Lemma 7.13 *There exists $K > 0$ such that, in the setting of Theorem 7.11, there exists a sequence $(\upsilon_j)_{j\geq 0}$ such that*

1. $y_j = X^{\upsilon_j}(y)$ *is in Σ for all $j \geq 0$.*
2. $|\upsilon_j - h(\tau_j)| < K \cdot \delta_0$ *and $d(x_j, y_j) < K \cdot \delta_0$.*

Proof By assumption $d(x_j, X^{h(\tau_j)}(y)) < K \cdot \delta_0$ for all $j \geq 0$. In particular $y'_j = X^{h(\tau_j)}(y)$ is close to Σ. Using a flow-box in a neighborhood of Σ we obtain

$X^{\varepsilon_j}(y'_j) \in \Sigma$ for some $\varepsilon_j \in (-K \cdot \delta_0, K \cdot \delta_0)$. The constant K depends only on the vector field X and the cross-section Σ (more precisely, on the angle between Σ and the flow direction). Taking $\upsilon_j = h(\tau_j) + \varepsilon_j$ we get the first two claims in the lemma. The third one follows from the triangle inequality; it may be necessary to replace K by a larger constant, still depending on X and Σ only. □

7.2.3 Semi-global Poincaré Map

Since we took the cross-section Σ to be adapted, we may use Lemma 6.23 to conclude that there exist Poincaré maps R_j with $R_j(x_j) = x_{j+1}$ and $R_j(y_j) = y_{j+1}$ and sending $W^s_\varepsilon(x_j, \Sigma)$ and $W^s_\varepsilon(y_j, \Sigma)$ inside the lines $W^s_\varepsilon(x_{j+1}, \Sigma)$ and $W^s_\varepsilon(y_{j+1}, \Sigma)$, respectively. The goal of this section is to prove that R_j *extends to a smooth Poincaré map on the whole strip* Σ_j *of* Σ *bounded by the stable manifolds of* x_j *and* y_j.

We first outline the proof. For each j we choose a curve γ_j transverse to the stable foliation of Σ, connecting x_j to y_j and such that γ_j is disjoint from the orbit segments $[x_j, x_{j+1}]$ and $[y_j, y_{j+1}]$. Using Lemma 6.23 in the same way as in the last paragraph, we see that it suffices to prove that R_j extends smoothly to γ_j. For this purpose we consider a tube-like domain $\mathscr{T}_j$ consisting of local stable manifolds through an *immersed surface* S_j whose boundary is formed by γ_j and γ_{j+1} and the orbit segments $[x_j, x_{j+1}]$ and $[y_j, y_{j+1}]$; see Fig. 7.3. We will prove that the orbit of any point in γ_j must leave the tube through γ_{j+1} in finite time. We begin by showing that the tube contains no singularities. This uses hypothesis (b) together with the local dynamics near Lorenz-like singularities. Next, using hypothesis (b) together with a Poincaré-Bendixson argument on S_j, we conclude that the forward orbit of any point in $\mathscr{T}_j$ must leave the tube. Another argument, using hyperbolicity properties of the Poincaré map, shows that orbits through γ_j must leave $\mathscr{T}_j$ through γ_{j+1}. In the sequel we detail these arguments.

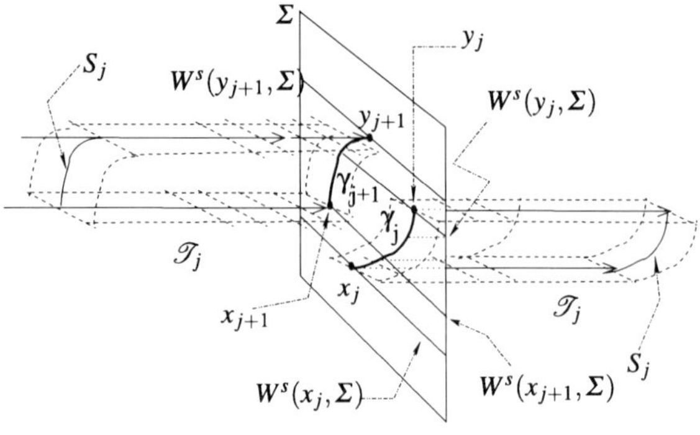

Fig. 7.3 A tube-like domain

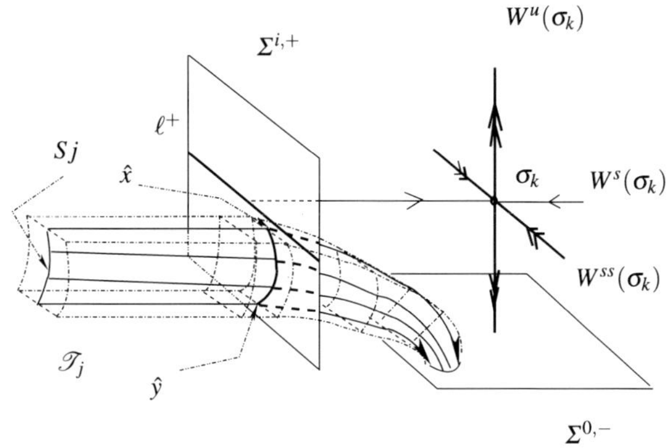

Fig. 7.4 Entering the flow-box of a singularity

7.2.4 A Tube-Like Domain Without Singularities

Since we took γ_j and γ_{j+1} disjoint from the orbit segments $[x_j, x_{j+1}]$ and $[y_j, y_{j+1}]$, the union of these four curves is an embedded circle. We recall that the two orbit segments are close to each other: by hypothesis (b)

$$d(X^t(x), X^{h(t)}(y)) < \delta_0 \quad \text{for all } t \in [t_j, t_{j+1}].$$

Assuming that δ_0 is smaller than the radius of injectiveness of the exponential map of the ambient manifold (i.e., $\exp_x : T_x M \to M$ is locally invertible in a δ_0-neighborhood of x in M for any $x \in M$), there exists a unique geodesic linking each $X^t(x)$ to $X^{h(t)}(y)$, and it varies continuously (even smoothly) with t. Using these geodesics we easily see that the union of $[y_j, y_{j+1}]$ with γ_j and γ_{j+1} is homotopic to a curve inside the orbit of x, with endpoints x_j and x_{j+1}, and so it is also homotopic to the segment $[x_j, x_{j+1}]$. This means that the previously mentioned embedded circle is homotopic to zero. It follows that there is a *smooth immersion* $\phi : [0, 1] \times [0, 1] \to M$ such that

- $\phi(\{0\} \times [0, 1]) = \gamma_j$ and $\phi(\{1\} \times [0, 1]) = \gamma_{j+1}$;
- $\phi([0, 1] \times \{0\}) = [y_j, y_{j+1}]$ and $\phi([0, 1] \times \{1\}) = [x_j, x_{j+1}]$.

Moreover $S_j = \phi([0, 1] \times [0, 1])$ may be chosen such that (see Fig. 7.4)

- all the points of S_j are at distance less than δ_1 from the orbit segment $[x_j, x_{j+1}]$, for some uniform constant $\delta_1 > \delta_0$ which can be taken arbitrarily close to zero, reducing δ_0 if necessary;
- the intersection of S_j with an incoming cross-section of any singularity (Sect. 7.3.3.1) is transverse to the corresponding stable foliation.

Then we define $\mathscr{T}_j$ to be the union of the local stable manifolds through the points of that disk.

Proposition 7.14 *The domain $\mathscr{T}_j$ contains no singularities of the flow.*

Proof By construction, every point of $\mathscr{T}_j$ is at distance $\leq \varepsilon$ from S_j and, consequently, at distance $\leq \varepsilon + \delta_1$ from $[x_j, x_{j+1}]$. So, taking ε and δ_0 much smaller than the sizes of the cross-sections associated to the singularities (Sect. 7.3.3.1), we immediately get the conclusion of the proposition in the case when $[x_j, x_{j+1}]$ is disjoint from the incoming cross-sections of all singularities. In the general case we must analyze the intersections of the tube with the flow-boxes at the singularities. The key observation is in the following statement whose proof we postpone.

Lemma 7.15 *Suppose that* $[x_j, x_{j+1}]$ *intersects an incoming cross-section* Σ_k^i *of some singularity* σ_k *at some point* $\hat{x}$ *with* $d(\hat{x}, \partial \Sigma_k^i) > \delta$. *Then* $[y_j, y_{j+1}]$ *intersects* Σ_k^i *at some point* $\hat{y}$ *with* $d(\hat{x}, \hat{y}) < K \cdot \delta_0$ *and, moreover,* $\hat{x}$ *and* $\hat{y}$ *are in the same connected component of* $\Sigma_k^i \setminus W_{loc}^s(\sigma_k)$.

Let us recall by construction that the intersection of S_j with the incoming cross-section Σ_k^i is transverse to the corresponding stable foliation; see Fig. 7.4. By the previous lemma this intersection is entirely contained in one of the connected components of $\Sigma_k^i \setminus W_{loc}^s(\sigma_k)$. Since $\mathscr{T}_j$ consists of local stable manifolds through the points of S_j, its intersection with Σ_k^i is contained in the region bounded by the stable manifolds $W^s(\hat{x}, \Sigma_k^i)$ and $W^s(\hat{y}, \Sigma_k^i)$, and so it is entirely contained in a connected component of $\Sigma_k^i \setminus W_{loc}^s(\sigma_k)$. In other words, the crossing of the tube $\mathscr{T}_j$ through the flow-box is disjoint from $W_{loc}^s(\sigma_k)$; in particular, it does not contain the singularity. Repeating this argument for every intersection of the tube with a neighborhood of some singularity, we get the conclusion of the proposition. □

Proof of Lemma 7.15 The first part of the lemma is proved in exactly the same way as Lemma 7.13. We have

$$\hat{x} = X^{r_0}(x) \quad \text{and} \quad \hat{y} = X^{s_0}(y)$$

with $|s_0 - h(r_0)| < K\delta_0$. The proof of the second part is by contradiction and relies, fundamentally, on the local description of the dynamics near the singularity. Associated to $\hat{x}$ and $\hat{y}$ we have the points $\tilde{x} = X^{r_1}(x)$ and $\tilde{y} = X^{s_1}(y)$, where the two orbits leave the flow-box associated to the singularity. If $\hat{x}$ and $\hat{y}$ are in opposite sides of the local stable manifold of σ_k, then $\tilde{x}$ and $\tilde{y}$ belong to different outgoing cross-sections of σ_k. Our goal is to find some $t \in \mathbb{R}$ such that

$$\operatorname{dist}\left(X^t(x), X^{h(t)}(y)\right) > \delta_0,$$

thus contradicting hypothesis (b).

We assume by contradiction that $\hat{x}$, $\hat{y}$ are in different connected components of $\Sigma_k^{i,\pm} \setminus \ell^{\pm}$. There are two cases to consider. We suppose first that $h(r_1) > s_1$ and note that $s_1 \gg s_0 \approx h(r_0)$, so that $s_1 > h(r_0)$. It follows that there exists $t \in (r_0, r_1)$ such that $h(t) = s_1$ since h is non-decreasing and continuous. Then $X^t(x)$ is on one side of the flow-box of σ_k, whereas $X^{h(t)}(y)$ belongs to the outgoing cross-section at the other side of the flow-box. Thus $\operatorname{dist}(X^t(x), X^{h(t)}(y))$ has the order of magnitude of the diameter of the flow-box, which we may assume to be much larger than δ_0.

Now we suppose that $s_1 \geq h(r_1)$ and observe that $h(r_1) > h(r_0)$, since h is increasing. We recall also that $X^{h(r_0)}(y)$ is close to $\hat{y}$, near the incoming cross-section, so that the whole orbit segment from $X^{h(r_0)}(y)$ to $X^{s_1}(y)$ is contained in (a small neighborhood of) the flow-box, to one side of the local stable manifold of σ_j. The previous observation means that this orbit segment contains $X^{h(r_1)}(y)$. However, $X^{r_1}(x)$ belongs to the outgoing cross-section at the opposite side of the flow-box, and so $\operatorname{dist}(X^{r_1}(x), X^{h(r_1)}(y))$ has the order of magnitude of the diameter of the flow-box, which is much larger than δ_0. □

7.2.5 *Every Orbit Leaves the Tube*

Our goal in this subsection is to show that the forward orbit of every point $z \in \mathcal{T}_j$ leaves the tube in finite time. The proof is based on a Poincaré-Bendixson argument applied to the flow induced by X^t on the disk S_j.

We begin by defining this induced flow. For the time being, we make the following simplifying assumption:

(H) $S_j = \phi([0,1] \times [0,1])$ is an embedded disk and the stable manifolds $W^s_\varepsilon(\xi)$ through the points $\xi \in S_j$ are pairwise disjoint.

This condition provides a well-defined continuous projection $\pi : \mathcal{T}_j \to S_j$ by assigning to each point $z \in \mathcal{T}_j$ the unique $\xi \in S_j$ whose local stable manifold contains z. The (not necessarily complete) flow Y^t induced by X^t on S_j is given by $Y^t(\xi) = \pi(X^t(\xi))$ for the largest interval of values of t for which this is defined. It is clear, just by continuity, that given any subset E of S_j at a positive distance from ∂S_j, there exists $\varepsilon > 0$ such that $Y^t(\xi)$ is defined for all $\xi \in E$ and $t \in [0, \varepsilon]$. In fact this remains true even if E approaches the curve γ_j (since Σ is a cross-section for X^t, the flow at γ_j points inward S_j) or the X^t-orbit segments $[x_j, x_{j+1}]$ and $[y_j, y_{j+1}]$ on the boundary of S_j (because they are also Y^t-orbit segments). Thus we only have to worry about the distance to the remaining boundary segment:

($\star$) given any subset E of S_j at positive distance from γ_{j+1}, there exists $\varepsilon > 0$ such that $Y^t(\xi)$ is defined for all $\xi \in E$ and $t \in [0, \varepsilon]$.

We observe also that for points ξ close to γ_{j+1} the flow $Y^t(\xi)$ must intersect γ_{j+1}, after which it is no longer defined.

Now we explain how to remove condition (H). In this case, the induced flow is naturally defined on $[0,1] \times [0,1]$ rather than S_j, as we now explain. *Recall that $\phi : [0,1] \times [0,1] \to M$ is an immersion.* So given any $w \in [0,1] \times [0,1]$ there exist neighborhoods U of w and V of $\phi(w)$ such that $\phi : U \to V$ is a diffeomorphism. Moreover, just by continuity of the stable foliation, choosing V sufficiently small we may ensure that each stable manifold $W^s_\varepsilon(\xi)$, $\xi \in V$, intersects V only at the point ξ. This means that we have a well-defined projection π from $\cup_{\xi \in V} W^s_\varepsilon(\xi)$ to V associating to each point z in the domain the unique element of V whose stable manifold contains z. Then we may define $Y^t(w)$ for small t by

$$Y^t(w) = \phi^{-1}(\pi(X^t(\phi(w)))).$$

As before, we extend Y^t to a maximal domain. This defines a (partial) flow on the square $[0,1] \times [0,1]$ such that both $[0,1] \times \{i\}$, $i \in \{0,1\}$ are trajectories.

Remark 7.16 An equilibrium ζ for the flow Y^t corresponds to an equilibrium of X in the local strong-stable manifold of ζ in M by the definition of Y^t through the projection π.

Notice also that forward trajectories of points in $\{0\} \times [0,1]$ enter the square. Hence, the only way trajectories may exit is through $\{1\} \times [0,1]$. So we have the following reformulation of property ($\star$):

($\star$) given any subset E of $[0,1] \times [0,1]$ at positive distance from $\{1\} \times [0,1]$, there exists $\varepsilon > 0$ such that $Y^t(w)$ is defined for all $w \in E$ and $t \in [0,\varepsilon]$.

Moreover for points w close to $\{1\} \times [0,1]$ the flow $Y^t(\xi)$ must intersect $\{1\} \times [0,1]$, after which it is no longer defined.

Proposition 7.17 *Given any point $z \in \mathcal{T}_j$ there exists $t > 0$ such that $X^t(z) \notin \mathcal{T}_j$.*

Proof The proof is by contradiction. First, we assume condition (H). Suppose there exists $z \in \mathcal{T}_j$ whose forward orbit remains in the tube for all times. Let $z_0 = \pi(z)$. Then $Y^t(z_0)$ is defined for all $t > 0$, and so it makes sense to speak of the ω-limit set $\omega(z_0)$. The orbit $Y^t(z_0)$ can not accumulate on γ_{j+1} for otherwise it would leave S_j. Therefore $\omega(z_0)$ is a compact subset of S_j at positive distance from γ_{j+1}. Using property (U) we can find a uniform constant $\varepsilon > 0$ such that $Y^t(w)$ is defined for every $t \in [0,\varepsilon]$ and every $w \in \omega(z_0)$. Since $\omega(z_0)$ is an invariant set, we can extend Y^t to a complete flow on it.

In particular we may fix $w_0 \in \omega(z_0)$ and $w \in \omega(w_0)$, and apply the arguments in the proof of the Poincaré-Bendixson Theorem. On the one hand, if we consider a cross-section S to the flow at w, the forward orbits of z_0 and w_0 must intersects it on monotone sequences; on the other hand, every intersection of the orbit of w_0 with S is accumulated by points in the orbit of z_0. This implies that w is in the orbit of w_0 and, in fact, that the later is periodic.

We consider the disk $D \subset S_j$ bounded by the orbit of w_0. The flow Y^t is complete restricted to D and so we may apply the Poincaré-Bendixson theorem (see [190]) once more, and conclude that Y^t has some singularity ζ inside D. This implies by Remark 7.16 that X^t has a singularity in the local stable manifold of ζ, which contradicts Proposition 7.14. This contradiction completes the proof of the proposition, under assumption (H). The general case is treated in the same way, just dealing with the flow induced on $[0,1] \times [0,1]$ instead of on S_j. □

7.2.6 The Poincaré Map Is Well-Defined on Σ_j

We have shown that for the induced flow Y^t on S_j (or, more generally, on $[0,1] \times [0,1]$) every orbit must eventually cross γ_{j+1} (respectively, $\{1\} \times [0,1]$). Hence

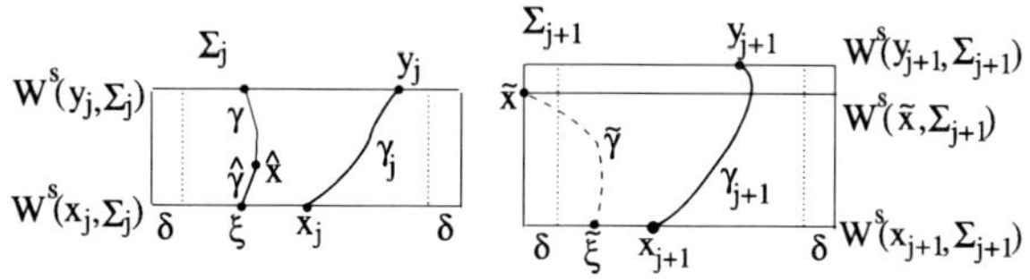

Fig. 7.5 Exiting the tube at Σ_{j+1} flowing from Σ_j

there exists a continuous Poincaré map

$$r : \gamma_j \to \gamma_{j+1}, \quad r(\xi) = Y^{\theta(\xi)}(\xi).$$

By compactness the Poincaré time $\theta(\cdot)$ is bounded. We are going to deduce that every forward X^t-orbit eventually leaves the tube $\mathscr{T}_j$ through Σ_{j+1}, which proves that R_j is defined on the whole strip of Σ_j between the manifolds $W^s(x_j, \Sigma_j)$ and $W^s(y_j, \Sigma_j)$, as claimed in Sect. 7.2.2.

To this end, let γ be a *central-unstable curve in Σ_δ connecting the stable manifolds $W^s(x_j, \Sigma)$ and $W^s(y_j, \Sigma)$*. Observe that γ is inside $\mathscr{T}_j$. For each $z \in \gamma$, let $t(z)$ be the smallest positive time for which $X^{t(z)}(z)$ is on the boundary of $\mathscr{T}_j$.

The crucial observation is that, in view of the construction of Y^t, *each $X^{t(z)}(z)$ belongs to the (global) stable manifold of $Y^{t(z)}(\pi(z))$*. We observe also that for $\{\xi\} = \gamma \cap W^s(x_j, \Sigma)$ we have $Y^t(\xi) = X^t(\xi)$ and so $t(\xi) = \theta(\xi)$.

Now we take $z \in \gamma$ close to ξ. Just by continuity, the X^t-trajectory of ξ and z remain close, and by the forward contraction along stable manifolds, the X^t-trajectory of ξ remains close to the segment $[x_j, x_{j+1}]$. Moreover the orbit of z cannot leave the tube through the union of the local strong stable manifolds passing through $[x_j, x_{j+1}]$, for otherwise it would contradict the definition of Y^t. Hence the trajectory of z must leave the tube through Σ_{j+1}. In other words $X^{t(z)}(z)$ is a point of Σ_{j+1}, close to $\tilde{\xi} = X^{t(\xi)}(\xi)$.

Let $\hat{\gamma} \subset \gamma_j$ be the *largest connected subset* which contains x_j such that $X^{t(z)}(z) \in \Sigma_{j+1}$ for all $z \in \hat{\gamma}$. We want to prove that $\hat{\gamma} = \gamma$ since this implies that R_j extends to the whole γ and so, using Lemma 6.23, to the whole Σ_j.

The proof is by contradiction. We assume $\hat{\gamma}$ is not the whole of γ, and let $\hat{x}$ be the endpoint different from ξ. Then by definition of $\mathscr{F}^s_\Sigma$ and of Y^t (from Sect. 7.2.5) $\tilde{x} = X^{t(\hat{x})}(\hat{x})$ is on the center-unstable boundary $\partial^{cu}\Sigma_{j+1}$ of the cross-section Σ_{j+1}, between the stable manifolds $W^s(x_{j+1}, \Sigma_{j+1})$ and $W^s(y_{j+1}, \Sigma_{j+1})$; see Fig. 7.5. By the choice of γ and by Corollary 6.17, $\tilde{\gamma} = \{X^{t(z)}(z) : z \in \hat{\gamma}\}$ is a cu-curve.

On the one hand, by Lemma 6.18, the distance between $\tilde{x}$ and $\tilde{\xi} = X_{t(\xi)}(\xi)$ dominates the distance between their stable manifolds and $\ell(\tilde{\gamma})$:

$$\ell(\tilde{\gamma}) \le \kappa \cdot d(\tilde{\xi}, \tilde{x}) \le \kappa \cdot d\big(W^s(x_{j+1}, \Sigma), W^s(\tilde{x}, \Sigma)\big).$$

We note that $\ell(\tilde{\gamma})$ is larger than δ, since $\tilde{\xi}$ is in Λ and the section Σ_{j+1} is adapted. On the other hand, the distance between the two stable manifolds is smaller than the distance between the stable manifold of x_{j+1} and the stable manifold of y_{j+1}, and this is smaller than $K \cdot \delta_0$. Since δ_0 is much smaller than δ, this is a contradiction. This proves the claim that $X^{t(z)}(z) \in \Sigma$ for all $z \in \gamma$.

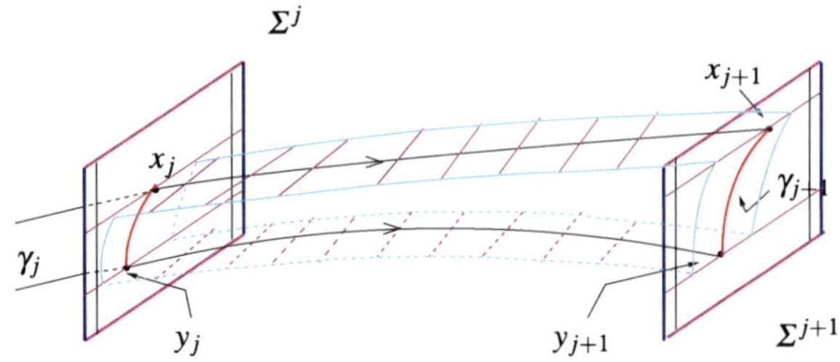

Fig. 7.6 Expansion within the tube

7.2.7 Expansiveness of the Poincaré Map

We have shown that there exists a well defined Poincaré return map R_j on the whole strip between the stable manifolds of x_j and y_j inside Σ. By Proposition 6.15 and Corollary 6.17 we know that the map R_j is hyperbolic where defined and, moreover, that the length of each cu-curve is expanded by a factor of 3 by R_j (since we chose $\lambda = 1/3$ in Sect. 6.1.2). Hence the distance between the stable manifolds $R_j(W^s(x_j, \Sigma))$ and $R_j(W^s(y_j, \Sigma))$ is increased by a factor strictly larger than one; see Fig. 7.6. This contradicts item (2) of Lemma 7.13 since this distance will eventually become larger than $K \cdot \delta_0$. Thus y_j must be in the stable manifold $W^s(x_j, \Sigma)$. Since the strong-stable manifold is locally flow-invariant and $X^{h(\tau_j)}(y)$ is in the orbit of $y_j = X^{\upsilon_j}(y)$, then $X^{h(\tau_j)}(y) \in W^s(x_j) = W^s(X^{\tau_j}(x))$; see Lemma 7.13.

According to Lemma 7.13 we have $|\upsilon_j - h(\tau_j)| < K \cdot \delta_0$ and, by Remark 6.14, there exists a small $\varepsilon_1 > 0$ such that

$$R_\Sigma(y_j) = X^t(y_j) \in W^{ss}_\varepsilon(x_j) \quad \text{with } |t| < \varepsilon_1.$$

Therefore the piece of the orbit $\mathcal{O}_y = X^{[\upsilon_j - K\cdot\delta_0 - \varepsilon_1, \upsilon_j + K\cdot\delta_0 + \varepsilon_1]}(y)$ must contain $X^{h(\tau_j)}(y)$. We note that this holds for all sufficiently small values of $\delta_0 > 0$ fixed from the beginning.

Let $\varepsilon_0 > 0$ be given and let us consider the piece of the orbit $\mathcal{O}_x := X^{[\tau_j - \varepsilon_0, \tau_j + \varepsilon_0]}(x)$ and the piece of the orbit of x whose strong-stable manifolds intersect $\mathcal{O}_y$, i.e.,

$$\mathcal{O}_{xy} = \left\{X^s(x) : \exists \tau \in [\upsilon_j - K \cdot \delta_0 - \varepsilon_1, \upsilon_j + K \cdot \delta_0 + \varepsilon_1] \text{ s. t. } X^\tau(y) \in W^{ss}_\varepsilon\left(X^s(x)\right)\right\}.$$

Since $y_j \in W^s(x_j)$ we conclude that $\mathcal{O}_{xy}$ is a neighborhood of $x_j = X^{\tau_j}(x)$ which can be made as small as we want by taking δ_0 and ε_1 small enough. In particular this ensures that $\mathcal{O}_{xy} \subset \mathcal{O}_x$ and so $X^{h(\tau_j)}(y) \in W^{ss}_\varepsilon(X^{[\tau_j - \varepsilon_0, \tau_j + \varepsilon_0]}(x))$. This finishes the proof of Theorem 7.11.

7.2.8 Singular-Hyperbolicity and Chaotic Behavior

Now we can present a proof that singular-hyperbolicity implies chaotic behavior for the past and the future.

Proof of Theorem 7.3 The assumption of singular-hyperbolicity on an isolated proper subset Λ with isolating neighborhood U ensures that the maximal invariant subsets $\cap_{t\in\mathbb{R}}\overline{Y^t(U)}$ for all C^1 nearby flows Y are also singular-hyperbolic. Therefore to deduce robust chaotic behavior in this setting it is enough to show that a proper isolated invariant compact singular-hyperbolic subset is chaotic.

Let Λ be a singular-hyperbolic isolated proper subset for a C^1 flow. Then there exists a strong-stable manifold $W^{ss}(x)$ through each of its points x. We claim that this implies that Λ is past chaotic. Indeed, assume by contradiction that we can find $y \in W^{ss}(x)$ such that $y \neq x$ and $\operatorname{dist}(X^{-t}(y), X^{-t}(x)) < \varepsilon$ for every $t > 0$, for some small $\varepsilon > 0$. Then, because $W^{ss}(x)$ is uniformly contracted by the flow in positive time, there exists $\lambda > 0$ such that

$$\operatorname{dist}(y, x) \leq Const \cdot e^{-\lambda t} \operatorname{dist}\left(X^{-t}(y), X^{-t}(x)\right) \leq Const \cdot \varepsilon e^{-\lambda t}$$

for all $t > 0$, a contradiction since $y \neq x$. Hence for any given small $\varepsilon > 0$ we can always find a point y arbitrarily close to x (it is enough to choose y is the strong-stable manifold of x) such that its past orbit separates from the orbit of x.

To obtain future chaotic behavior, we argue by contradiction: we assume that Λ is not future chaotic. Then for every $\varepsilon > 0$ we can find a point $x \in \Lambda$ and an open neighborhood V of x such that the future orbit of each $y \in V$ is ε-close to the future orbit of x, that is, $\operatorname{dist}(X^t(y), X^t(x)) \leq \varepsilon$ for all $t > 0$.

First, x is not a singularity, because all the possible singularities inside a singular-hyperbolic set are hyperbolic saddles and so each singularity has a unstable manifold. Likewise, x cannot be in the stable manifold of a singularity. Therefore $\omega(x)$ contains some regular point z. Let Σ be a transversal section to the flow X^t at z.

Hence there are infinitely many times $t_n \to +\infty$ such that $x_n := X^{t_n}(x) \in \Sigma$ and $x_n \to z$ when $n \to +\infty$. Taking Σ sufficiently small looking only to very large times, the assumption on V ensures that each $y \in V$ admits also an infinite sequence $t_n(y) \xrightarrow[n\to+\infty]{} +\infty$ satisfying

$$y_n := X^{t_n(y)}(y) \in \Sigma \quad \text{and} \quad \operatorname{dist}(y_n, x_n) \leq 10\varepsilon.$$

We can assume that $y \in V$ does not belong to $W^s(x)$, since $W^s(x)$ is a C^1 immersed sub-manifold of M. Hence we consider the connected components $\gamma_n := W^s(x_n, \Sigma)$ and $\xi_n := W^s(y_n, \Sigma)$ of $W^s(x) \cap \Sigma$ and $W^s(y) \cap \Sigma$, respectively. We recall that we can assume that every y in a small neighborhood of Λ admits an invariant stable manifold because we can extend the invariant stable cone fields from Λ to a small neighborhood of Λ. We can also extend the invariant center-unstable cone fields from Λ to this same neighborhood, so that we can also define the notion of *cu*-curve in Σ in this setting.

The assumption on V ensures that there exists a *cu*-curve ζ_n in Σ connecting γ_n to ξ_n, because $X^{t_n}(V) \cap \Sigma$ is an open neighborhood of x_n containing y_n. But we can assume without loss of generality that $t_{n+1} - t_n > \max\{t_1, t_2\}$, forgetting some returns to Σ in between if necessary and relabeling the times t_n. Thus Proposition 6.15 applies and the Poincaré return maps associated to the returns to Σ considered above are hyperbolic.

The same argument as in the proof of expansiveness, in Sect. 7.2, guarantees that there exists a flow-box connecting $\{x_n, y_n\}$ to $\{x_{n+1}, y_{n+1}\}$ and sending ζ_n into a cu-curve $R(\zeta_n)$ connecting γ_{n+1} and ξ_{n+1}, for every $n \geq 1$.

The hyperbolicity of the Poincaré return maps ensures that the length of $R(\zeta_n)$ grows by a factor greater than one. Therefore, since y_n, x_n are uniformly close, this implies that the length of ζ_1 and the distance between γ_1 and ξ_1 must be zero. This contradicts the choice of $y \neq W^s(x)$.

This contradiction shows that Λ is future chaotic, and concludes the proof. □

7.3 Singular-Hyperbolic Attractors Are Non-uniformly Hyperbolic

Here we start the proof of Theorem 7.6.

7.3.1 The Starting Point

We show in Sect. 7.3.3 that, choosing a *global Poincaré section* Ξ (with several connected components) for X on Λ, we can reduce the transformation R to the quotient over the stable leaves. We can do this using Lemma 6.23 with the exception of finitely many leaves Γ, corresponding to the points whose orbit falls into the local stable manifold of some singularity or are sent into the stable boundary $\partial^s \Sigma$ of some $\Sigma \in \Xi$ by R, where the return time function τ is discontinuous.

As explained in Sect. 6.1.4 the global Poincaré map $R : \Xi \to \Xi$ induces in this way a map $f : \mathscr{F} \setminus \Gamma \to \mathscr{F}$ on the leaf space, diffeomorphic to a finite union of open intervals I, which is piecewise expanding and admits finitely many $\upsilon_1, \dots, \upsilon_l$ ergodic absolutely continuous (with respect to Lebesgue measure on I) invariant probability measures whose basins cover Lebesgue almost all points of I. In what follows we will simply say "absolutely continuous invariant measures" and use the shorthand acim's when referring to this kind of measures.

Moreover the Radon-Nikodym derivatives (densities) $\frac{d\upsilon_k}{d\lambda}$ are *bounded from above* and *the support of each* υ_k *contains nonempty open intervals*, and so the basin $B(\upsilon_k)$ contains nonempty open intervals Lebesgue modulo zero, $k = 1, \dots, l$.

7.3.1.1 Description of the Construction

Later we unwind the reductions made in Sect. 7.3.3 and obtain a physical measure for the original flow at the end of the proof.

We divide the construction of the physical measure for Λ into the following steps.

1. The compact metric space Ξ is endowed with a partition $\mathscr{F}$ and map $R : \Xi \setminus \Gamma \to \Xi$, where Γ is a finite set of elements of $\mathscr{F}$ (see Sect. 6.1.4.1). The map R preserves the partition $\mathscr{F}$ and contracts its elements by Lemma 6.23. We have a

finite family $\upsilon_1, \ldots, \upsilon_l$ of absolutely continuous invariant probability measures for the induced quotient map $f : \mathcal{F} \setminus \Gamma \to \mathcal{F}$.

We show in Sect. 6.1.5 that each υ_i defines a R-invariant ergodic probability measure η_i. In Sect. 7.3.5 we show that the basin $B(\eta_i)$ is a union of strips of Ξ, and η_i are therefore physical measures for R. Moreover these basins cover Ξ:

$$\lambda^2\big(\Xi \setminus (B(\eta_1) \cup \cdots \cup B(\eta_l))\big) = 0,$$

where λ^2 is the area measure on Ξ.

2. We then pass from R-invariant physical measures $\eta_1, \ldots, \eta_l$ to invariant probability measures $\nu_1, \ldots, \nu_l$ for the suspension semiflow over R with roof function τ. In the process we keep the ergodicity (Sect. 7.3.7) and the basin property (Sect. 7.3.7) of the measures: the whole space $\Xi \times [0, +\infty)/\sim$ where the semiflow is defined equals the union of the ergodic basins of the ν_i Lebesgue modulo zero.
3. Finally in Sect. 7.3.8 we convert each physical measure ν_i for the semiflow into a physical measure μ_i for the original flow. We use the fact that the semiflow is semi-conjugated to X_t on a neighborhood of Λ by a local diffeomorphism. Uniqueness of the physical measure μ is then deduced in Sect. 7.3.8.1 through the existence of a dense regular orbit in Λ (recall that our definition of attractor *demands* transitivity) and by the observation that the basin of μ contains open sets Lebesgue modulo zero. In Sect. 7.3.9 we show that μ is (non-uniformly) hyperbolic.

The details are expounded in the following sections.

7.3.2 The Hölder Property of the Projection

Recall the construction of the Global Poincaré Map R near a singular-hyperbolic attractor Λ for a C^1 flow in Chap. 6 and the one-dimensional reduction. Assuming that X is of class C^2 then the projections along stable leaves on cross-sections are Hölder-C^1. Hence the one-dimensional map f we obtain through reduction to the quotient leaf space has the property that $1/|Df|$ is Hölder continuous. This is essential to construct an absolutely continuous invariant probability measure for f, as we now explain.

We assume now that the flow $(X^t)_{t\in\mathbb{R}}$ is C^2. Under this condition it is well known [147, 193] that the stable leaf $W^s(x, \Sigma)$ is a C^2 embedded disk for every $x \in \Sigma \in \Xi$ and these leaves define a C^1 foliation $\mathcal{F}^s_\Sigma$ of each $\Sigma \in \Xi$ with a Hölder-C^1 holonomy (since the leaves are one-dimensional).

From Sect. 2.7.2 we know that in this setting the holonomy (projection) along transverse curves to $\mathcal{F}^s_\Sigma$ are $C^{1+\alpha}$ for some $0 < \alpha < 1$ which depends on X only, since they can be seen as maps between subsets of the real line.

Recall also Remark 6.14: the projections we are dealing with consist really of the composition of two projections: the first along the strong-stable leaves and the second along the flow to Σ. Since the flow is assumed to be C^2, the end result is a holonomy map in Σ which is Hölder-C^1.

This shows that the map f obtained here is in fact a C^1 piecewise expanding map such that $1/|f'|$ is Hölder restricted to each I_j. Indeed

- for I_j corresponding to a flow-box between cross-sections of the flow away from singularities, $Df \mid I_j$ is α-Hölder for some $0 < \alpha < 1$, and so for all $x, y \in I_j$ we have

$$\frac{1}{|f'(x)|} - \frac{1}{|f'(y)|} \le \frac{|f'(x) - f'(y)|}{|f'(x)f'(y)|} \le \frac{C}{(3/2)^2} \cdot |x - y|^\alpha;$$

- otherwise, $f \mid I_j$ corresponds to the flow near a singularity. Hence $1/|Df|$ on such I_j is α-Hölder continuous according to Remark 7.18 on the estimation of Poincaré times near singularities.

7.3.2.1 Existence and Finiteness of Acim's

It is well known [112, 266, 275] that C^1 piecewise expanding maps f of the interval such that $1/|Df|$ has bounded variation have finitely many absolutely continuous invariant probability measures whose basins cover Lebesgue almost all points of I.

Using an extension of the notion of bounded variation (defined below) it was shown in [120] that the results of existence and finiteness of absolutely continuous ergodic invariant measures can be extended to C^1 piecewise expanding maps f such that $g = 1/|f'|$ is α-Hölder for some $\alpha \in (0, 1)$. These functions are of *universally bounded variation*, i.e.,

$$\sup_{a=a_0<a_1<\cdots<a_n=b} \left(\sum_{j=1}^{n} \left| g(a_i) - g(a_{i-1}) \right|^{1/\alpha} \right)^\alpha < \infty,$$

where the supremum is taken over all finite partitions of the interval $I = [a, b]$. Moreover from [120, Theorem 3.2] the densities φ of the absolutely continuous invariant probability measures for f satisfy the following: there exists constants $A, C > 0$ such that

$$\int \operatorname{osc}(\varphi, \varepsilon, x)\, dx \le C \cdot \varepsilon^\alpha \quad \text{for all } 0 < \varepsilon \le A,$$

where $\operatorname{osc}(\varphi, \varepsilon, x) = \operatorname{ess\,sup}_{y,z \in B(x,\varepsilon)} |\varphi(y) - \varphi(z)|$ and the essential supremum is taken with respect to Lebesgue measure. From this we can find a sequence $\varepsilon_n \to 0$ such that $\operatorname{osc}(\varphi, \varepsilon_n, \cdot) \xrightarrow[n\to\infty]{} 0$ (with respect to Lebesgue measure). This implies that supp(φ) *contains non-empty open intervals*.

Indeed, for a given small $\delta > 0$ let $\rho > 0$ be so small and n so big that $W = \{\varphi > \rho\}$ and $V = \{\operatorname{osc}(\varphi, \varepsilon_n, \cdot) > \rho/2\}$ satisfy $\lambda(I \setminus W) < \delta$ and $\lambda(V) < \delta$. Then $\lambda(W \cap I \setminus V) > 1 - 2\delta > 0$. Let x be a Lebesgue density point of $W \cap I \setminus V$. Then there exists a positive Lebesgue measure subset of $B(x, \varepsilon_n)$ where $\varphi > \rho$. By definition of $\operatorname{osc}(\varphi, \varepsilon_n, x)$ this implies that $\varphi(y) > \rho/2 > 0$ for Lebesgue almost every $y \in B(x, \varepsilon_n)$, and thus $B(x, \varepsilon_n) \subset \operatorname{supp}(\varphi)$.

In addition from [120, Theorem 3.3] there are finitely many ergodic absolutely continuous invariant probability measures $\upsilon_1, \ldots, \upsilon_l$ of f and every absolutely continuous invariant probability measure υ decomposes into a convex linear combination $\upsilon = \sum_{i=1}^{l} a_i \upsilon_i$. From [120, Theorem 3.2] considering any subinterval $J \subset I$ and the normalized Lebesgue measure $\lambda_J = (\lambda \mid J)/\lambda(J)$ on J, every weak* accumulation point of $n^{-1} \sum_{j=0}^{n-1} f_*^j(\lambda_J)$ is an absolutely continuous invariant probability measure υ for f (since the indicator function of J is of generalized $1/\alpha$-bounded variation). Hence the basin of the $\upsilon_1, \ldots, \upsilon_l$ cover I Lebesgue modulo zero: $\lambda(I \setminus (B(\upsilon_1) \cup \cdots \cup B(\upsilon_l)) = 0$.

Note that from [120, Lemma 1.4] we also know that *the density φ of any absolutely continuous f-invariant probability measure is bounded from above.*

7.3.3 Integrability of the Global Return Time

Here we study the integrability of the Poincaré time τ with respect to the Lebesgue area measure on the adapted cross-sections Ξ associated to the flow-boxes which cover Λ.

7.3.3.1 Poincaré Times Near Singularities

Recall that, since singularities are Lorenz-like, the unstable manifold $W^u(\sigma_k)$ is one-dimensional, and there is a one-dimensional strong-stable manifold $W^{ss}(\sigma_k)$ contained in the two-dimensional stable manifold $W^s(\sigma_k)$. Most important for what follows, the attractor intersects the strong-stable manifold at the singularity only, by Theorem 5.10.

Hence for some $\delta > 0$ we may take δ-adapted cross-sections containing $\Sigma^{o,\pm}$ and $\Sigma^{i,\pm}$, in U_0, as in Sect. 5.4.4. Reducing the cross-sections if necessary, i.e., taking $\delta > 0$ small enough, we ensure that the Poincaré times are larger than T_2, so that the same conclusions as in the previous subsections apply here. Indeed using linearizing coordinates it is easy to see that for points $z = (x_1, x_2, \pm 1) \in \Sigma^{i,\pm}$ the time $\tau^{\pm}$ taken by the flow starting at z to reach one of $\Sigma^{o,\pm}$ depends on x_1 only and is given by

$$\tau^{\pm}(x_1) = -\frac{\log x_1}{\lambda_1}.$$

We then fix these cross-sections once and for all and define for small $\varepsilon > 0$ the *flow-box*

$$U_{\sigma_k} = \bigcup_{x \in \Sigma^{i,\pm} \setminus \ell^{\pm}} X_{(-\varepsilon, \tau^{\pm}(x)+\varepsilon)}(x) \cup (-\delta, \delta) \times (-\delta, \delta) \times (-1, 1)$$

which is an open neighborhood of σ_k with σ_k the unique zero of $X \mid U_{\sigma_k}$. We note that the function $\tau^\pm : \Sigma^{i,\pm} \to \mathbb{R}$ is integrable with respect to the Lebesgue (area) measure over $\Sigma^{i,\pm}$: we say that *the exit time function in a flow-box near each singularity is Lebesgue integrable.*

More precisely, we can determine the expression of the Poincaré maps between ingoing and outgoing cross-sections easily though linearized coordinates

$$\Sigma^{i,+} \cap \{x_1 > 0\} \to \Sigma^{0,+}, \quad (x_1, x_2, 1) \mapsto \left(1, x_2 \cdot x_1^{-\lambda_2/\lambda_1}, x_1^{-\lambda_3/\lambda_1}\right). \tag{7.4}$$

This shows that the map obtained by identifying points with the same x_2 coordinate, i.e., points in the same stable leaf, is simply $x_1 \mapsto x_1^\beta$ where $\beta = -\lambda_3/\lambda_1 \in (0, 1)$. For the other possible combinations of ingoing and outgoing cross-sections the Poincaré maps have a similar expression. This will be useful to construct physical measures for the flow in Chap. 7.

Remark 7.18 Note that for the map $f(x) = x^\beta$ we have $1/Df(x) = x^{1-\beta}/\beta$ which is Hölder continuous.

7.3.3.2 The Global Poincaré Return Time

We claim that *the Poincaré time τ is integrable with respect to the Lebesgue area measure on Ξ*. Indeed, given $z \in \Xi$, the point $\hat{z} = X^{t_2}(z)$ is either inside a flow-box U_{σ_k} of a singularity σ_k, or not. In the former case, the time $\hat{z}$ takes to reach an outgoing cross-section $\Sigma^{o,\pm}_{\sigma_k}$ is bounded by the exit time function $\tau^\pm_{\sigma_k}$ of the corresponding flow-box, which is integrable. In the latter case, $\hat{z}$ takes a time of at most $2T_3$ to reach another cross-section, by definition of T_3. Thus the Poincaré time on Ξ is bounded by $t_2 + 2T_3$ plus a sum of finitely many integrable functions, one for each flow-box near a singularity, by finiteness of the number of singularities, of the number of cross-sections in Ξ and of the number of strips at each cross-section. This proves the claim.

Remark 7.19 Given $z \in \Sigma \in \Xi$ we write $\tau^k(z) = \tau(R^{k-1}(z)) + \cdots + \tau(z)$ for $k \geq 1$ and so $\tau = \tau^1$. Then $R^k(W^s(z, \Sigma)) \subset \Sigma'$ for some $\Sigma' \in \Xi$, since

$$R^k\big(W^s(z, \Sigma)\big) \subset X^{\tau^k(z)}\big(W^s(z, \Sigma)\big) \subset X^{\tau^k(z)}(U),$$

and the length $\ell(R^k(W^s(z, \Sigma)))$ is uniformly contracted as $\tau^k(z) \xrightarrow[k \to +\infty]{} +\infty$ Also

$$d\Big(R^k\big(W^s(z, \Sigma)\big), \partial^{cu}\Sigma'\Big) > \delta/2$$

for all big enough k, by the definitions of U and of δ-adapted cross-section. (The distance $d(A, B)$ between two sets A, B is $\inf\{d(a, b) : a \in A, b \in B\}$.) We may assume that this property holds for all stable leaves $W^s(z, \Sigma)$, all $z \in \Sigma$ and every $\Sigma \in \Xi$ for all $k \geq k_0$, for some fixed large $k_0 \in \mathbb{N}$, by the uniform contraction property of R in the stable direction.

Remark 7.20 By Lemma 6.23 the Poincaré time τ is constant on stable leaves $W^s(x, \Sigma)$ for all $x \in \Sigma \in \Xi$. Thus there exists a return time function τ_I on I such that $\tau = \tau_I \circ p$, where $p : \Xi \to \gamma_\Xi$ is the joining of all p_Σ, $\Sigma \in \Xi$ and $\gamma_\Xi = \{\gamma_\Sigma : \Sigma \in \Xi\}$. The integrability of τ with respect to λ^2 (see Sect. 6.1.4.1) implies the λ-integrability of τ_I naturally since $(p_\Sigma)_*\lambda^2 \ll \lambda$ and $\tau_I \circ p = \tau$.

7.3.4 Suspending Invariant Measures

Here we show how to construct an invariant measure for a transformation from an invariant measure for the quotient map obtained from a partition of the space. We show also that, if the measure is ergodic on the quotient, then we also obtain ergodicity on the starting space.

In Sect. 7.3.5 we apply these results to the global Poincaré map R of a singular-hyperbolic attractor and its corresponding one-dimensional quotient map f. We use the properties of the one-dimensional map given in Sect. 6.1.4 to build physical measures for f first.

Later we extend the transformation to a semi-flow through a suspension construction and show that each invariant and ergodic measure for the transformation corresponds to a unique measure for the semi-flow with the same properties.

In Sect. 7.3.7 we again apply these results to the transformation R to obtain physical measures for the suspension semiflow over R with roof function τ.

7.3.4.1 Reduction to the Quotient Map

Let Ξ be a compact metric space, $\Gamma \subset \Xi$ and let $F : (\Xi \setminus \Gamma) \to \Xi$ be a measurable map. We assume that there exists a partition $\mathcal{F}$ of Ξ into measurable subsets, having Γ as an element, which is

- *invariant*: the image of any $\xi \in \mathcal{F}$ distinct from Γ is contained in some element η of $\mathcal{F}$;
- *contracting*: the diameter of $F^n(\xi)$ goes to zero when $n \to \infty$, uniformly over all the $\xi \in \mathcal{F}$ for which $F^n(\xi)$ is defined.

We denote $p : \Xi \to \mathcal{F}$ the canonical projection, i.e., p assigns to each point $x \in \Xi$ the atom $\xi \in \mathcal{F}$ that contains it. By definition, $A \subset \mathcal{F}$ is measurable if and only if $p^{-1}(A)$ is a measurable subset of Ξ and likewise A is open if, and only if, $p^{-1}(A)$ is open in Ξ. The invariance condition means that there is a uniquely defined map

$$f : (\mathcal{F} \setminus \{\Gamma\}) \to \mathcal{F} \quad \text{such that } f \circ p = p \circ F.$$

Clearly, f is measurable with respect to the measurable structure we introduced in $\mathcal{F}$. We assume from now on that the leaves are sufficiently regular so that $\Xi/\mathcal{F}$ is a metric space with the topology induced by p.

Let μ_f be any probability measure on $\mathscr{F}$ invariant under the transformation f. For any bounded function $\psi : \Xi \to \mathbb{R}$, let $\psi_- : \mathscr{F} \to \mathbb{R}$ and $\psi_+ : \mathscr{F} \to \mathbb{R}$ be defined by

$$\psi_-(\xi) = \inf_{x\in\xi} \psi(x) \quad \text{and} \quad \psi_+(\xi) = \sup_{x\in\xi} \psi(x).$$

Lemma 7.21 *Given any continuous function $\psi : \Xi \to \mathbb{R}$, both limits*

$$\lim_n \int (\psi \circ F^n)_- \, d\mu_f \quad \textit{and} \quad \lim_n \int (\psi \circ F^n)_+ \, d\mu_f \tag{7.5}$$

exist, and they coincide.

Proof Let ψ be fixed as in the statement. Given $\varepsilon > 0$, let $\delta > 0$ be such that $|\psi(x_1) - \psi(x_2)| \le \varepsilon$ for all x_1, x_2 with $d(x_1, x_2) \le \delta$. Since the partition $\mathscr{F}$ is assumed to be contracting, there exists $n_0 \ge 0$ such that $\operatorname{diam}(F^n(\xi)) \le \delta$ for every $\xi \in \mathscr{F}$ and any $n \ge n_0$. Let $n + k \ge n \ge n_0$. By definition,

$$(\psi \circ F^{n+k})_-(\xi) - (\psi \circ F^n)_-(f^k(\xi)) = \inf\bigl(\psi \mid F^{n+k}(\xi)\bigr) - \inf\bigl(\psi \mid F^n(f^k(\xi))\bigr).$$

Observe that $F^{n+k}(\xi) \subset F^n(f^k(\xi))$. So the difference on the right-hand side is bounded by

$$\sup\bigl(\psi \mid F^n(f^k(\xi))\bigr) - \inf\bigl(\psi \mid F^n(f^k(\xi))\bigr) \le \varepsilon.$$

Therefore

$$\left| \int (\psi \circ F^{n+k})_- \, d\mu_f - \int (\psi \circ F^n)_- \circ f^k \, d\mu_f \right| \le \varepsilon.$$

Moreover, one may replace the second integral by $\int (\psi \circ F^n)_- \, d\mu_f$, because μ_f is f-invariant.

At this point we have shown that $\{\int (\psi \circ F^n)_- \, d\mu_f\}_{n\ge 1}$ is a Cauchy sequence in $\mathbb{R}$. In particular, it converges. The same argument proves that $\{\int (\psi \circ F^n)_+ \, d\mu_f\}_{n\ge 1}$ is also convergent. Moreover, keeping the previous notations,

$$0 \le (\psi \circ F^n)_+(\xi) - (\psi \circ F^n)_-(\xi) = \sup\bigl(\psi \mid F^n(\xi)\bigr) - \inf\bigl(\psi \mid F^n(\xi)\bigr) \le \varepsilon$$

for every $n \ge n_0$. So the two sequences in (7.5) must have the same limit. The lemma is proved. □

Corollary 7.22 *There exists a unique probability measure μ_F on Ξ such that*

$$\int \psi \, d\mu_F = \lim \int (\psi \circ F^n)_- \, d\mu_f = \lim \int (\psi \circ F^n)_+ \, d\mu_f$$

for every continuous function $\psi : \Xi \to \mathbb{R}$. Besides, μ_F is invariant under F. Moreover the correspondence $\mu_f \mapsto \mu_F$ is injective.

Proof Let $\hat{\mu}(\psi)$ denote the value of the two limits. Using the expression for $\hat{\mu}(\psi)$ in terms of $(\psi \circ F^n)_-$ we immediately obtain

$$\hat{\mu}(\psi_1 + \psi_2) \geq \hat{\mu}(\psi_1) + \hat{\mu}(\psi_2).$$

Analogously, the expression of $\hat{\mu}(\psi)$ in terms of $(\psi \circ F^n)_+$ gives the opposite inequality. The function $\hat{\mu}(\cdot)$ is therefore additive. Moreover, $\hat{\mu}(c\psi) = c\hat{\mu}(\psi)$ for every $c \in \mathbb{R}$ and every continuous function ψ. Therefore $\hat{\mu}(\cdot)$ is a linear real operator in the space of continuous functions $\psi : \Xi \to \mathbb{R}$.

Clearly, $\hat{\mu}(1) = 1$ and the operator $\hat{\mu}$ is non-negative: $\hat{\mu}(\psi) \geq 0$ if $\psi \geq 0$. By the Riesz-Markov theorem, there exists a unique measure μ_F on Ξ such that $\hat{\mu}(\psi) = \int \psi \, d\mu_F$ for every continuous ψ. To conclude that μ_F is invariant under F it suffices to note that

$$\hat{\mu}(\psi \circ F) = \lim_n \int (\psi \circ F^{n+1})_- \, d\mu_f = \hat{\mu}(\psi)$$

for every ψ.

To prove that the map $\mu_f \mapsto \mu_F$ is injective, let $\mu_F = \mu'_F$ be obtained from μ_f and μ'_f respectively. For any continuous function $\varphi : \mathscr{F} \to \mathbb{R}$ we have that $\psi = \varphi \circ p : \Xi \to \mathbb{R}$ is continuous. But

$$\mu_f\big((\psi \circ F^n)_\pm\big) = \mu_f\big((\varphi \circ p \circ F^n)_\pm\big) = \mu_f\big((\varphi \circ f^n \circ p)_\pm\big) = \mu_f(\varphi \circ f^n) = \mu_f(\varphi)$$

for all $n \geq 1$ by the f-invariance of μ_f. Hence by definition

$$\mu_f(\varphi) = \mu_F(\psi) = \mu'_F(\psi) = \mu'_f(\varphi)$$

and so $\mu_f = \mu'_f$. This finishes the proof of the corollary. □

Remark 7.23 We note that $\int \psi \, d\mu_F = \lim_n \int (\psi \circ F^n)_\# \, d\mu_f$ for every continuous $\psi : \Xi \to \mathbb{R}$ and any choice of a sequence $(\psi \circ F^n)_\# : \mathscr{F} \to \mathbb{R}$ with

$$\inf(\psi \mid F^n(\xi)) \leq (\psi \circ F^n)_\#(\xi) \leq \sup(\psi \mid F^n(\xi)).$$

Moreover we can define $\int \psi \, d\mu_F$ for any measurable $\psi : \Xi \to \mathbb{R}$ such that

$$\lim_{n \to +\infty} \big(\sup(\psi \mid F^n(\xi)) - \inf(\psi \mid F^n(\xi))\big) = 0$$

uniformly in $n \in \mathbb{N}$ and in $\xi \in \mathscr{F}$. This will be useful in what follows.

Lemma 7.24 *Let $\psi : \Xi \to \mathbb{R}$ be a continuous function and $\xi \in \mathscr{F}$ be such that*

$$\lim_n \frac{1}{n} \sum_{j=0}^{n-1} (\psi \circ F^k)_\pm(f^j(\xi)) = \int (\psi \circ F^k)_\pm \, d\mu_f$$

for every $k \geq 1$. Then $\lim_n \frac{1}{n} \sum_{j=0}^{n-1} \psi(F^j(x)) = \int \psi \, d\mu_F$ for every $x \in p^{-1}(\xi)$.

Proof Let us fix ψ and ξ as in the statement. Then by definition of $(\psi \circ F^k)_\pm$ and by the properties of $\mathscr{F}$ we have

$$(\psi \circ F^k)_-\big(f^j(\xi)\big) \le (\psi \circ F^k)\big(F^j(x)\big) \le (\psi \circ F^k)_+\big(f^j(\xi)\big)$$

for all $x \in \xi$ and $j, k \ge 1$. Given $\varepsilon > 0$, by Corollary 7.22 there exists $k_0 \in \mathbb{N}$ such that

$$\mu_F(\psi) - \frac{\varepsilon}{2} \le \mu_f\big((\psi \circ F^k)_-\big) = \mu_f\big((\psi \circ F^k)_+\big) \le \mu_F(\psi) + \frac{\varepsilon}{2}$$

for all $k \ge k_0$ and there exists $n_0 \in \mathbb{N}$ such that

$$\left| \frac{1}{n} \sum_{j=0}^{n-1} (\psi \circ F^k)_\pm\big(f^j(\xi)\big) - \mu_f\big((\psi \circ F^k)_\pm\big) \right| < \frac{\varepsilon}{2}$$

for all $n \ge n_0 = n_0(k)$. Hence we have

$$\mu_F(\psi) - \varepsilon \le \frac{1}{n} \sum_{j=0}^{n-1} (\psi \circ F^k)(F^j(x)) \le \mu_F(\psi) + \varepsilon,$$

for all $n \ge n_0(k)$. Since n can be made arbitrarily big and $\varepsilon > 0$ can be taken arbitrarily small we have concluded the proof of the lemma. □

Corollary 7.25 *If μ_f is f-ergodic, then μ_F is ergodic for F.*

Proof Since $\Xi/\mathscr{F}$ is a metric space with the topology induced by p, $C^0(\mathscr{F}, \mathbb{R})$ is dense in $L^1(\mathscr{F}, \mathbb{R})$ for the L^1-topology and $p : \Xi \to \mathscr{F}$ is continuous. Hence there exists a subset $\mathscr{E}$ of $\mathscr{F}$ with $\mu_f(\mathscr{E}) = 1$ such that the conclusion of Lemma 7.24 holds for a subset $E = p^{-1}(\mathscr{E})$ of Ξ. To prove the corollary it is enough to show that $\mu_F(E) = 1$.

Let $\chi_E = \chi_{\mathscr{E}} \circ p$ and take $\psi_n : \mathscr{F} \to \mathbb{R}$ a sequence of continuous functions such that $\psi_n \to \chi_{\mathscr{E}}$ when $n \to +\infty$ in the L^1 topology with respect to μ_f. Then $\varphi_n = \psi_n \circ p$ is a sequence of continuous functions on Ξ such that $\varphi_n \to \chi_E$ when $n \to +\infty$ in the L^1 norm with respect to μ_F.

Then it is straightforward to check that

$$\mu_F(\varphi_n) = \lim_{k \to +\infty} \mu_f\big((\varphi_n \circ F^k)_-\big) = \lim_{k \to +\infty} \mu_f(\psi_n \circ f^k) = \mu_f(\psi_n)$$

which converges to $\mu_f(\mathscr{E}) = 1$. Since $\mu_F(\varphi_n)$ tends to $\mu_F(E)$ when $n \to +\infty$, we conclude that $\mu_F(E) = 1$, as we wanted. □

7.3.5 Physical Measure for the Global Poincaré Map

Let us now apply these results (with R replacing F) to the case of the global Poincaré map for a singular-hyperbolic attractor.

From the previous results in Sects. 7.3.3 and 7.3.4.1, the finitely many acim's $\upsilon_1, \ldots, \upsilon_l$ for the one-dimensional quotient map f uniquely induce R-invariant ergodic probability measures $\eta_1, \ldots, \eta_l$ on Ξ.

We claim that the basins of each $\eta_1, \ldots, \eta_l$ have positive Lebesgue area λ^2 on Ξ and cover λ^2 almost every point of $p^{-1}(I)$. Indeed the uniform contraction of the leaves $\mathscr{F}^s_\Sigma \setminus \Gamma$ provided by Lemma 6.23 implies that the forward time averages of any pair x, y of points in $\xi \in \mathscr{F} \setminus p(\Gamma)$ on continuous functions $\varphi : \Xi \to \mathbb{R}$ are equal, i.e.,

$$\lim_{n\to+\infty}\left[\frac{1}{n}\sum_{j=0}^{n-1}\varphi\big(R^j(x)\big) - \frac{1}{n}\sum_{j=0}^{n-1}\varphi\big(R^j(y)\big)\right] = 0.$$

Hence $B(\eta_i) \supset p^{-1}(B(\upsilon_i)), i = 1, \ldots, l$. This shows that $B(\eta_i)$ contains an entire strip except for a subset of λ^2-null measure, because $B(\upsilon_i)$ contains some open interval λ modulo zero. Since $p_*(\lambda^2) \ll \lambda$ we get in particular

$$\lambda^2\big(B(\eta_i)\big) > 0 \quad \text{and} \quad \lambda^2\left(p^{-1}(I) \setminus \bigcup_{i=1}^{l} B(\eta_i)\right) = p_*(\lambda^2)\left(I \setminus \bigcup_{i=1}^{l} B(\upsilon_i)\right) = 0,$$

showing that $\eta_1, \ldots, \eta_l$ are physical measures whose basins cover $p^{-1}(I)$ Lebesgue almost everywhere. We observe that $p^{-1}(I) \subset \Xi$ is forward invariant under R, and thus it contains $\Lambda \cap \Xi$.

7.3.6 Suspension Flow from the Poincaré Map

Let Ξ be a measurable space, Γ some measurable subset of Ξ, and $F : (\Xi \setminus \Gamma) \to \Xi$ a measurable map. Let $\tau : \Xi \to (0, +\infty]$ be a measurable function such that $\inf \tau > 0$ and $\tau \equiv +\infty$ on Γ.

Let $\sim$ be the equivalence relation on $\Xi \times [0, +\infty)$ generated by $(x, \tau(x)) \sim (F(x), 0)$, that is, $(x, s) \sim (\tilde{x}, \tilde{s})$ if and only if there exist

$$(x, s) = (x_0, s_0),\ (x_1, s_1), \ldots, (x_N, s_N) = (\tilde{x}, \tilde{s})$$

in $\Xi \times (0, +\infty)$ such that, for every $1 \le i \le N$,

$$\begin{aligned} &\text{either} \quad x_i = F(x_{i-1}) \quad \text{and} \quad s_i = s_{i-1} - \tau(x_{i-1}); \\ &\text{or} \quad x_{i-1} = F(x_i) \quad \text{and} \quad s_{i-1} = s_i - \tau(x_i). \end{aligned}$$

We denote by $V = \Xi \times [0, +\infty)/\sim$ the corresponding quotient space and by $\pi : \Xi \times [0, +\infty) \to V$ the canonical projection which induces on V a topology and a Borel σ-algebra of measurable subsets of V.

Definition 7.26 The *suspension of F with roof function (or return-time) τ* is the semi-flow $(X_\tau^t)_{t\geq 0}$ defined on V by

$$X_\tau^t(\pi(x,s)) = \pi(x, s+t) \quad \text{for every } (x,s) \in \Xi \times [0,+\infty) \text{ and } t > 0.$$

It is easy to see that this is indeed well defined as in Sect. 2.3.2.2. In what follows we write X^t instead of X_τ^t since τ is fixed and no ambiguity can arise.

Remark 7.27 If F is injective then we can also define

$$X^{-t}\big(\pi(x,s)\big) = \pi\big(F^{-n}(x), s + \tau(F^{-n}(x)) + \cdots + \tau(F^{-1}(x)) - t\big)$$

for every $x \in F^n(\Xi)$ and $0 < t \leq s + \tau(F^{-n}(x)) + \cdots + \tau(F^{-1}(x))$. The expression on the right does not depend on the choice of $n \geq 1$. In particular, the restriction of the semi-flow $(X^t)_{t\geq 0}$ to the maximal invariant set

$$\Lambda = \left\{(x,t) : x \in \bigcap_{n\geq 0} F^n(\Xi) \text{ and } t \geq 0\right\}$$

extends, in this way, to a flow $(X^t)_{t\in\mathbb{R}}$ on Λ.

Let μ_F be any probability measure on Ξ that is invariant under F. Then the product $\mu_F \times dt$ of μ_F by Lebesgue measure on $[0,+\infty)$ is an infinite measure, invariant under the trivial flow $(x,s) \mapsto (x, s+t)$ in $\Xi \times [0,+\infty)$. In what follows we assume that the return time is integrable with respect to μ_F, i.e.,

$$\mu_F(\tau) = \int \tau \, d\mu_F < \infty. \tag{7.6}$$

In particular $\mu_F(\Gamma) = 0$. Then we introduce the probability measure μ_X on V defined by

$$\int \varphi \, d\mu_X = \frac{1}{\mu_F(\tau)} \int \int_0^{\tau(x)} \varphi(\pi(x,t)) \, dt \, d\mu_F(x)$$

for each bounded measurable $\varphi : V \to \mathbb{R}$.

We observe that the correspondence $\mu_F \mapsto \mu_X$ defined above is injective. Indeed for any bounded measurable $\psi : \Xi \to \mathbb{R}$, defining φ on $\{x\} \times [0, \tau(x))$ to equal $\mu_F(\tau) \cdot \psi(x)/\tau(x)$ gives a bounded measurable map $\varphi : V \to \mathbb{R}$ (since $\inf \tau > 0$) such that $\mu_X(\varphi) = \mu_F(\psi)$. Hence if $\mu_X = \mu_X'$ then $\mu_F = \mu_F'$.

Lemma 7.28 *The measure μ_X is invariant under the semi-flow $(X^t)_{t\geq 0}$.*

Proof It is enough to show that $\mu_X((X^t)^{-1}(B)) = \mu_X(B)$ for every measurable set $B \subset V$ and any $0 < t < \inf \tau$. Moreover, we may suppose that B is of the form $B = \pi(A \times J)$ for some $A \subset \Xi$ and J a bounded interval in $[0, \inf(\tau \mid A))$. This is because these sets form a basis for the σ-algebra of measurable subsets of V.

Let B be of this form and let (x, s) be any point in Ξ with $0 \le s < \tau(x)$. Then $X^t\pi(x, s) \in B$ if and only if $\pi(x, s+t) = \pi(\tilde{x}, \tilde{s})$ for some $(\tilde{x}, \tilde{s}) \in A \times J$. In other words, $\pi(x, s) \in (X^t)^{-1}(B)$ if and only if there exists some $n \ge 0$ such that

$$\tilde{x} = F^n(x) \quad \text{and} \quad \tilde{s} = s + t - \tau(x) - \cdots - \tau(F^{n-1}(x)).$$

Since $s < \tau(x)$, $t < \inf \tau$, and $\tilde{s} \ge 0$, it is impossible to have $n \ge 2$. So,

- either $\tilde{x} = x$ and $\tilde{s} = s + t$ (corresponding to $n = 0$),
- or $\tilde{x} = F(x)$ and $\tilde{s} = s + t - \tau(x)$ (corresponding to $n = 1$).

The two possibilities are mutually exclusive: for the first one (x, s) must be such that $s + t < \tau(x)$, whereas in the second case $s + t \ge \tau(x)$. This shows that we can write $(X^t)^-(B)$ as a disjoint union $(X^t)^-(B) = B_1 \cup B_2$ with

$$B_1 = \pi\big\{(x, s) : x \in A \text{ and } s \in (J - t) \cap [0, \tau(x))\big\}$$
$$B_2 = \pi\big\{(x, s) : F(x) \in A \text{ and } s \in (J + \tau(x) - t) \cap [0, \tau(x))\big\}.$$

Since $t > 0$ and $\sup J < \tau(x)$, we have $(J - t) \cap [0, \tau(x)) = (J - t) \cap [0, +\infty)$ for every $x \in A$. So, by definition, $\mu_X(B_1)$ equals

$$\frac{1}{\mu_F(\tau)} \int_A \ell\Big((J - t) \cap [0, \tau(x))\Big)\, d\mu_F(x) = \frac{1}{\mu_F(\tau)} \mu_F(A) \cdot \ell\Big((J - t) \cap [0, +\infty)\Big).$$

Similarly $\inf J \ge 0$ and $t < \tau(x)$ imply that

$$(J + \tau(x) - t) \cap [0, \tau(x)) = \tau(x) + (J - t) \cap (-\infty, 0).$$

Hence $\mu_X(B_2)$ is given by

$$\frac{1}{\mu_F(\tau)} \int_{F^{-1}(A)} \ell\Big((J - t) \cap (-\infty, 0)\Big)\, d\mu_F(x) = \frac{\mu_F(F^{-1}(A))}{\mu_F(\tau)} \ell\Big((J - t) \cap (-\infty, 0)\Big).$$

Since μ_F is invariant under F, we may replace $\mu_F(F^{-1}(A))$ by $\mu_F(A)$ in the last expression. It follows that

$$\mu_X\Big((X^t)^{-1}(B)\Big) = \mu_X(B_1) + \mu_X(B_2) = \frac{1}{\mu_F(\tau)} \mu_F(A) \cdot \ell\big((J - t)\big).$$

Clearly, the last term may be written as $\mu_F(\tau)^{-1} \mu_F(A) \cdot \ell(J)$ which, by definition, is the same as $\mu_X(B)$. This proves that μ_X is invariant under the semi-flow and ends the proof. □

Given a bounded measurable function $\varphi : V \to \mathbb{R}$, let $\hat{\varphi} : \Xi \to \mathbb{R}$ be defined by

$$\hat{\varphi}(x) = \int_0^{\tau(x)} \varphi(\pi(x, t))\, dt. \qquad (7.7)$$

Observe that $\hat{\varphi}$ is integrable with respect to μ_F and

$$\int \hat{\varphi}\, d\mu_F = \mu_F(\tau) \cdot \int \varphi\, d\mu_X,$$

by the definition of μ_X.

Lemma 7.29 *Let $\varphi : V \to \mathbb{R}$ be a bounded function and let $\hat{\varphi}$ be as above. We assume that $x \in \Xi$ is such that $\tau(F^j(x))$ and $\hat{\varphi}(F^j(x))$ are finite for every $j \geq 0$, and also*

(a) $\lim_n \frac{1}{n}\sum_{j=0}^{n-1} \tau(F^j(x)) = \int \tau\, d\mu_F$, *and*
(b) $\lim_n \frac{1}{n}\sum_{j=0}^{n-1} \hat{\varphi}(F^j(x)) = \int \hat{\varphi}\, d\mu_F$.

Then $\lim_{T\to+\infty} \frac{1}{T}\int_0^T \varphi(\pi(x, s+t))\, dt = \int \varphi\, d\mu_X$ *for every* $\pi(x, s) \in V$.

Proof Let x be fixed, satisfying (a) and (b). Given any $T > 0$ we define $n = n(T)$ by

$$T_{n-1} \leq T < T_n \quad \text{where } T_j = \tau(x) + \cdots + \tau(F^j(x)) \text{ for } j \geq 0.$$

Then using $(y, \tau(y)) \sim (F(y), 0)$ we get

$$\begin{aligned} \frac{1}{T}\int_0^T \varphi(\pi(x, s+t))\, dt = \frac{1}{T}\Bigg[\sum_{j=0}^{n-1} \int_0^{\tau(F^j(x))} &\varphi(\pi(F^j(x), t))\, dt \\ + \int_0^{T-T_{n-1}} \varphi(\pi(F^n(x), t))\, dt &- \int_0^s \varphi(\pi(x, t))\, dt\Bigg]. \end{aligned} \tag{7.8}$$

Using the definition of $\hat{\varphi}$, we may rewrite the first term on the right-hand side as

$$\frac{n}{T} \cdot \frac{1}{n}\sum_{j=0}^{n-1} \hat{\varphi}(F^j(x)). \tag{7.9}$$

Now we fix $\varepsilon > 0$. Assumption (a) and the definition of n imply that

$$n \cdot \left(\int \tau\, d\mu_F - \varepsilon\right) \leq T_{n-1} \leq T \leq T_n \leq (n+1) \cdot \left(\int \tau\, d\mu_F + \varepsilon\right)$$

for large enough n. Observe also that n goes to infinity as $T \to +\infty$, since $\tau(F^j(x)) < \infty$ for every j. So, for large T,

$$\mu_F(\tau) - \varepsilon \leq \frac{T}{n} \leq \frac{n+1}{n}\big(\mu_F(\tau) + \varepsilon\big) \leq \mu_F(\tau) + 2\varepsilon.$$

This proves that T/n converges to $\mu_F(\tau)$ when $T \to +\infty$. Consequently, assumption (b) implies that (7.9) converges to

$$\frac{1}{\mu_F(\tau)} \int \hat{\varphi}\, d\mu_F = \int \varphi\, d\mu_X.$$

Now we prove that the remaining terms in (7.8) converge to zero when T goes to infinity. Since φ is bounded,

$$\left| \frac{1}{T} \int_0^{T-T_{n-1}} \varphi(\pi(F^n(x),t))\, dt \right| \le \frac{T-T_{n-1}}{T} \sup |\varphi|. \tag{7.10}$$

Using the definition of n once more,

$$T - T_{n-1} \le T_n - T_{n-1} \le (n+1)\Big(\int \tau\, d\mu_F + \varepsilon\Big) - n\Big(\int \tau\, d\mu_F - \varepsilon\Big)$$

whenever n is large enough. Then

$$\frac{T-T_{n-1}}{T} \le \frac{\int \tau\, d\mu_F + (2n+1)\varepsilon}{n(\int \tau\, d\mu_F - \varepsilon)} \le \frac{4\varepsilon}{\int \tau\, d\mu_F - \varepsilon}$$

for all large enough T. This proves that $(T - T_{n-1})/T$ converges to zero, and then so does the left-hand side of (7.10). Finally, it is clear that

$$\frac{1}{T} \int_0^s \varphi(\pi(x,t))\, dt \to 0 \quad \text{when } T \to +\infty.$$

This completes the proof of the lemma. □

Corollary 7.30 *If μ_F is ergodic then μ_X is ergodic.*

Proof Let $\varphi : V \to \mathbb{R}$ be any bounded measurable function, and let $\hat{\varphi}$ be as in (7.7). As already noted, $\hat{\varphi}$ is μ_F integrable. It follows that $\hat{\varphi}(F^j(x)) < \infty$ for every $j \ge 0$, at μ_F-almost every point $x \in \Xi$. Moreover, by the Ergodic Theorem, condition (b) in Lemma 7.29 holds μ_F-almost everywhere. For the same reasons, $\tau(F^j(x))$ is finite for all $j \ge 0$, and condition (a) in the lemma is satisfied for μ_F-almost all $x \in \Xi$.

This shows that Lemma 7.29 applies to every point x in a subset $A \subset \Xi$ with $\mu_F(A) = 1$. It follows that

$$\lim_{T \to +\infty} \frac{1}{T} \int_0^T \varphi(X^t(z))\, dt = \int \varphi\, d\mu_X$$

for every point z in $B = \pi(A \times [0, +\infty))$. Since the latter has $\mu_X(B) = 1$, we have shown that the Birkhoff average of φ is constant μ_X-almost everywhere. Then the same is true for any integrable function, as bounded functions are dense in $L^1(\mu_X)$. Thus μ_X is ergodic and the corollary is proved. □

7.3.7 Physical Measures for the Suspension

Using the results from Sects. 7.3.5 and 7.3.6, it is straightforward to obtain ergodic probability measures $\nu_1, \dots, \nu_l$ invariant under the suspension $(X^t_\tau)_{t\geq 0}$ of R with return time τ, corresponding to the R-physical probability measures $\eta_1, \dots, \eta_l$ respectively.

Now we use Lemma 7.29 to show that each ν_i is a physical measure for $(X^t_\tau)_{t\geq 0}$, $i = 1, \dots, l$. Let $x \in \Sigma \cap B(\nu_i)$ for a fixed $\Sigma \in \Xi$ and $i \in \{1, \dots, l\}$. According to Remark 7.20 the return time τ_I on I is Lebesgue integrable, and thus υ_i-integrable also since $\frac{d\upsilon_i}{d\lambda}$ is bounded. Hence τ is η_i-integrable by the construction of η_i from υ_i (see Sect. 7.3.4.1).

Lemma 7.29, together with the fact that η_i is physical for R, ensures that $B(\nu_i)$ contains the positive X^t_τ orbit of almost every point $(x, 0)$, $x \in B(\nu_i)$, with respect to λ^2 on $B(\eta_i)$. If we denote by $\lambda^3 = \pi_*(\lambda^2 \times dt)$ a natural volume measure on V, then we get $\lambda^3(B(\nu_i)) > 0$.

This also shows that the basins $B(\nu_1), \dots, B(\nu_l)$ cover λ^3-almost every point in $V_0 = \pi(p^{-1}(I) \times [0, +\infty))$. Notice that this subset is a neighborhood of the suspension $\pi((\Lambda \cap \Xi \setminus \Gamma) \times [0, +\infty))$ of $\Lambda \cap \Xi \setminus \Gamma$.

7.3.8 Physical Measure for the Flow

Here we extend the previous conclusions to the original flow, completing the proof of Theorem 7.6.

We relate the suspension $(X^t_\tau)_{t\geq 0}$ of R with return time τ to $(X^t)_{t\geq 0}$ in U as follows. We define

$$\Phi : \Xi \times [0, +\infty) \to U \quad \text{by } (x, t) \mapsto X^t(x)$$

and since $\Phi(x, \tau(x)) = (R(x), 0) \in \Xi \times \{0\}$, this map naturally defines a quotient map

$$\phi : V \to U \quad \text{such that} \quad \phi \circ X^t_\tau = X^t \circ \phi, \quad \text{for all } t \geq 0, \tag{7.11}$$

through the identification $\sim$ from Sect. 7.3.6.

Let $\Xi_\tau = \{(x, t) \in (\Xi \setminus \Gamma) \times [0, +\infty) : 0 < t < \tau(x)\}$. Note that Ξ_τ is an open set in V and that $\pi \mid \Xi_\tau : \Xi_\tau \to \Xi_\tau$ is a homeomorphism (the identity). Then the map $\phi \mid \Xi_\tau$ is a local diffeomorphism into $V_0 = \phi(\Xi \times [0, +\infty)) \subset U$ by the natural identification given by π and by the Tubular Flow Theorem, since points in Ξ_τ are not sent into singularities of X. Notice that Ξ_τ is a full Lebesgue (λ^3) measure subset of V. Thus ϕ is a semi-conjugation modulo zero. Note also that the number of pre-images of ϕ is globally bounded by r_0 from Remark 6.28.

Therefore the measures ν_i constructed for the semiflow X^t_τ in the previous Sect. 7.3.7 define physical measures $\mu_i = \phi_*(\nu_i)$, $i = 1, \dots, l$, whose basins cover a full Lebesgue (m) measure subset of V_0, which is a neighborhood of Λ. Indeed the semi-conjugacy (7.11) ensures that $\phi(B(\nu_i)) \subset B(\mu_i)$ and, since ϕ is a local diffeomorphisms on a full Lebesgue measure subset, then

$$m\Big(V_0 \setminus \phi\big(B(\nu_1) \cup \cdots \cup B(\nu_l)\big)\Big) = 0.$$

Since $V_0 \subset U$ we have

$$W^s(\Lambda) = \bigcup_{t<0} X^t(V_0).$$

Moreover X^t is a diffeomorphism for all $t \in \mathbb{R}$, and thus preserves subsets of zero m measure. Hence $\cup_{t<0} X^t(B(\mu_1) \cup \cdots \cup B(\mu_l))$ has full Lebesgue measure in $W^s(\Lambda)$. In other words, Lebesgue (m) almost every point x in the basin $W^s(\Lambda)$ of Λ is such that $X^t(x) \in B(\mu_i)$ for some $t > 0$ and $i = 1, \ldots, l$.

7.3.8.1 Uniqueness of the Physical Measure

The set Λ is an attractor. According to our definition of attractor there exists $z_0 \in \Lambda$ such that $\{X^t(z_0) : t > 0\}$ is a dense regular orbit in Λ.

We prove uniqueness of the physical measure by contradiction, assuming that the number l of distinct physical measures is bigger than one. Then we can take distinct physical measures η_1, η_2 for R on Ξ associated to distinct physical measures μ_1, μ_2 for $X \mid \Lambda$. Then there are open sets $U_1, U_2 \subset \Xi$ such that

$$U_1 \cap U_2 = \emptyset \quad \text{and} \quad \lambda^2\big(B(\eta_i) \setminus U_i\big) = 0, \quad i = 1, 2.$$

For a very small $\zeta > 0$ we consider the open subsets $V_i = X^{(-\zeta,\zeta)}(U_i)$, $i = 1, 2$, of U such that $V_1 \cap V_2 = \emptyset$. According to the construction of μ_i we have $m(B(\mu_i) \setminus V_i) = 0$, $i = 1, 2$.

The transitivity assumption ensures that there are positive times $T_1 < T_2$ (exchanging V_1 and V_2 if needed) such that $X^{T_i}(z_0) \in V_i$, $i = 1, 2$. Since V_1, V_2 are open sets and $g = X^{T_2 - T_1}$ is a diffeomorphism, there exists a small open set $W_1 \subset V_1$ such that $g \mid W_1 : W_1 \to V_2$ is a C^1 diffeomorphism into its image $W_2 = g(W_1) \subset V_2$.

Now the C^1 smoothness of $g \mid W_1$ ensures that a full Lebesgue (m) measure subset of W_1 is sent into a full Lebesgue measure subset of W_2. By the definition of g and the choice of V_1, V_2, there exists a point in $B(\mu_1) \cap W_1$ whose positive orbit contains a point in $B(\mu_2) \cap W_2$, and thus $\mu_1 = \mu_2$. Hence *singular-hyperbolic attractors have a unique physical probability measure* μ.

7.3.9 Hyperbolicity of the Physical Measure

For the hyperbolicity of the measure μ we note that

- the sub-bundle E^s is one-dimensional and uniformly contracting, and thus on the E^s-direction the Lyapunov exponent is negative for every point in U;

- the sub-bundle E^{cu} is two-dimensional, dominates E^s, contains the flow direction and is volume expanding, and thus by Oseledets' Theorem [147, 269] the sum of the Lyapunov exponents on the direction of E^{cu} is given by $\mu(\log|\det DX_1 \mid E^{cu}|) > 0$. Hence there is a positive Lyapunov exponent for μ-almost every point on the direction of E^{cu}.

We already know from Sect. 2.7 that an expanding direction in E^{cu} does not coincide with the flow direction $E_z^X = \{s \cdot X(z) : s \in \mathbb{R}\}$, $z \in \Lambda$, since E_z^X always has zero Lyapunov exponent for regular points for a smooth flow on a compact manifold.

This shows that at μ-almost every point z the Oseledets splitting of the tangent bundle has the form

$$T_z M = E_z^s \oplus E_z^X \oplus F_z,$$

where F_z is the one-dimensional measurable sub-bundle of vectors with positive Lyapunov exponent. The proof of Theorem 7.6 is complete.

7.3.10 *Absolutely Continuous Disintegration of the Physical Measure*

Here we prove Theorem 7.7. We let μ be a physical ergodic probability measure for a singular-hyperbolic attractor Λ of a C^2-flow in an open subset $U \subset M^3$, obtained through the sequence of reductions of the dynamics of the flow X^t to the suspension flow X_τ^t of the Poincaré map R and return time function τ, with corresponding X_τ^t-invariant measure ν obtained from the R-invariant measure η. In addition η is obtained through the ergodic invariant measure υ of the one-dimensional map $f : I \to I$. This is explained in Sect. 7.3.5. We know that μ is hyperbolic as explained in Sect. 7.3.8.

Let us fix $\delta_0 > 0$ small. Then by Pesin's non-uniformly hyperbolic theory [86, 196, 216] we know that there exists a compact subset $K \subset \Lambda$ such that $\mu(\Lambda \setminus K) < \delta_0$, and there exists $\delta_1 > 0$ for which every $z \in K$ admits a strong-unstable manifold $W_{\delta_1}^{uu}(z)$ with inner radius δ_1. We refer to this kind of set as a *Pesin set*. The *inner radius* of $W_{\delta_1}^{uu}(z)$ is defined as the length of the shortest smooth curve in this manifold from z to its boundary. Moreover $K \ni z \mapsto W_{\delta_1}^{uu}(z)$ is a continuous map $K \to \mathscr{E}^1(I_1, M)$ (recall the notations in Sect. 6.1.1).

The suspension flow X_τ^t defined on V in Sect. 7.3.6 is semi-conjugated to the X^t-flow on an open subset of U through a finite-to-1 local homeomorphism ϕ, defined in Sect. 7.3.8, which takes orbits to orbits and preserves time as in (7.11). Hence there exists a corresponding set $K' = \phi^{-1}(K)$ satisfying the same properties of K with respect to X_τ^t, where the constants δ_0, δ_1 are changed by at most a constant factor due to ϕ^{-1} by the compactness of K. In what follows we use the measure $\nu = (\phi^{-1})_*\mu$ instead of μ and write K for K'.

We fix a density point $x_0 \in K$ of $\nu \mid K$. We may assume that $x_0 \in \Sigma$ for some $\Sigma \in \Xi$. Otherwise if $x_0 \notin \Xi$, since $x_0 = (x, t)$ for some $x \in \Sigma$, $\Sigma \in \Xi$ and $0 < t <$

$T(x)$, then we use $(x,0)$ instead of x_0 in the following arguments, but we still write x_0. Clearly the length of the unstable manifold through $(x,0)$ is unchanged due to the form of the suspension flow, at least for small values of δ_1. Since ν is given as a product measure on the quotient space V (see Sect. 7.3.7), we may assume without loss of generality that x_0 is a density point of η on $\Sigma \cap K$.

We set $W^u(x,\Sigma)$ to be the connected component of $W^u(x) \cap \Sigma$, the unstable manifold of x that contains x, for $x \in K \cap \Sigma$. Recall that $W^u(x) \subset \Lambda$ because Λ is an attracting set. Then $W^u(x,\Sigma)$ has inner radius bigger than some positive value $\delta_2 > 0$ for $x \in K \cap \Sigma$, which depends only on δ_1 and the angle between $W^{uu}_{\delta_1}(x)$ and $T_x\Sigma$.

Let us define $\mathscr{F}^s(x_0,\delta_2) = \{W^s(x,\Sigma) : x \in W^u(x_0,\Sigma)\}$ and the corresponding horizontal strip $F^s(x_0,\delta_2) = \cup_{\gamma \in \mathscr{F}^s(x_0,\delta_2)} \gamma$. Points $z \in F^s(x_0,\delta_2)$ can be specified using coordinates $(x,y) \in W^u(x_0,\Sigma) \times \mathbb{R}$, where x is given by $W^u(x_0,\Sigma) \cap W^s(z,\Sigma)$ and y is the length of the shortest smooth curve connecting x to z in $W^s(z,\Sigma)$. Let us consider

$$\mathscr{F}^u(x_0,\delta_2) = \{W^u(z,\Sigma) : z \in \Sigma \quad \text{and} \quad W^u(z,\Sigma) \quad \text{crosses } F^s(x_0,\delta_2)\},$$

where we say that a curve γ *crosses* $F^s(x_0,\delta_2)$ if the trace of γ can be written as the graph of a map $W^u(x_0,\Sigma) \to W^s(x_0,\Sigma)$ using the coordinates outlined above. We stress that $\mathscr{F}^u(x_0,\delta_2)$ is not restricted to leaves through points of K.

We may assume that the set $F^u(x_0,\delta_2) = F^u(x_0,\delta_2) = \cup_{\gamma \in \mathscr{F}^u(x_0,\delta_2)} \gamma$ satisfies $\eta(F^u(x_0,\delta_2)) > 0$ up to taking a smaller $\delta_2 > 0$, since x_0 is a density point of $\eta \mid K \cap \Sigma$. Let $\hat{\eta}$ be the measure on $\mathscr{F}^u(x_0,\delta_2)$ given by

$$\hat{\eta}(A) = \eta\Big(\bigcup_{\gamma \in A} \gamma\Big) \quad \text{for every measurable set } A \subset \mathscr{F}^u(x_0,\delta_2).$$

Proposition 7.31 *The measure $\eta \mid F^u(x_0,\delta_2)$ admits a disintegration into conditional measures η_γ along $\hat{\eta}$-a.e. $\gamma \in \mathscr{F}^u(x_0,\delta_2)$ such that $\eta_\gamma \ll \lambda_\gamma$, where λ_γ is the measure (length) induced on γ by the natural Riemannian measure λ^2 (area) on Σ. Moreover there exists $D_0 > 0$ such that*

$$\frac{1}{D_0} \le \frac{d\eta_\gamma}{d\lambda_\gamma} \le D_0, \quad \eta_\gamma\text{-almost everywhere for } \hat{\eta}\text{-almost every } \gamma.$$

This is enough to conclude the proof of Theorem 7.7 since both δ_0 and δ_2 can be taken arbitrarily close to zero, so that all unstable leaves $W^u(x,\Sigma)$ through almost every point with respect to η will support a conditional measure of η.

Indeed, to obtain the disintegration of ν along the center-unstable leaves that cross any small ball around a density point x_0 of K, we project that neighborhood of x_0 along the flow in negative time onto a cross-section Σ. Then we obtain the family $\{\eta_\gamma\}$, the disintegration of η along the unstable leaves $\gamma \in \mathscr{F}^u$ on a strip F^s of Σ, and consider the family $\{\eta_\gamma \times dt\}$ of measures on $\mathscr{F}^u \times [0,T]$ to obtain a disintegration of ν, where $T > 0$ is a fixed time slightly smaller than the return time of the points in the strip F^s; see Fig. 7.7.

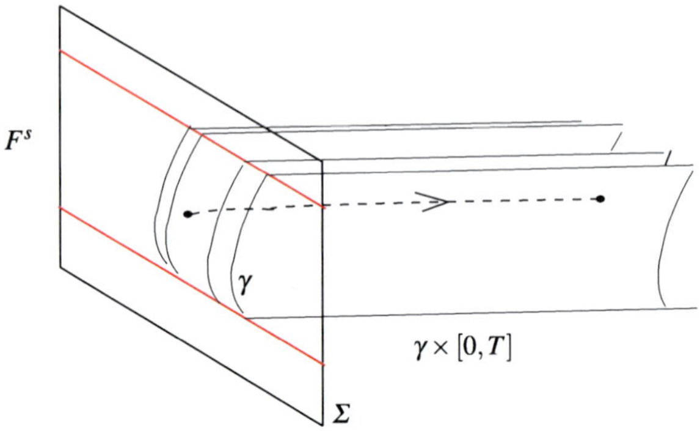

Fig. 7.7 Center-unstable leaves on the suspension flow

In fact, $\eta_\gamma \times dt \ll \lambda_\gamma \times dt$ and $\lambda_\gamma \times dt$ is the induced (area) measure on the center-unstable leaves by the volume measure λ^3 on V, and it can be given by restricting the volume form λ^3 to the surface $\gamma \times [0,T]$ which we write λ^3_γ, for $\gamma \in \mathscr{F}^u$. Thus, by Proposition 7.31 and by the definition of ν, we have

$$\nu_\gamma = \eta_\gamma \times dt = \frac{d\eta_\gamma}{d\lambda_\gamma}\cdot\lambda^3_\gamma, \quad \gamma\in\mathscr{F}^u,$$

and the densities of the conditional measures $\eta_\gamma \times dt$ with respect to λ^3_γ are also uniformly bounded from above and from below away from zero – we have left out the constant factor $1/\mu(\tau)$ to simplify the notation.

Since $\mu = \phi_*\nu$ and ϕ is a finite-to-1 local diffeomorphism when restricted to Ξ_τ, then μ also has an absolutely continuous disintegration along the center-unstable leaves. The densities on unstable leaves γ are related by the expression (where m_γ denotes the area measure on the center-unstable leaves induced by the volume form m)

$$\mu_\gamma = \phi_*(\nu_\gamma) = \phi_*\left(\frac{d\eta_\gamma}{\lambda_\gamma}\cdot\lambda^3_\gamma\right) = \left(\frac{1}{\det D(\phi \mid \gamma\times[0,T])}\cdot\frac{d\eta_\gamma}{\lambda_\gamma}\right)\circ\phi^{-1}\cdot m_\gamma,$$

for $\gamma\in\mathscr{F}^u$, which implies that the densities along the center-unstable leaves are uniformly bounded from above.

Indeed observe first that the number of pre-images of x under ϕ is uniformly bounded by r_0 from Remark 6.28, i.e., by the number of cross-sections of Ξ hit by the orbit of x from time 0 to time t_2. Moreover the tangent bundle of $\gamma\times[0,T]$ is sent by $D\phi$ into the bundle E^{cu} by construction. Then, recalling that $\phi(x,t) = X^t(x)$, if e_1 is a unit tangent vector at $x\in\gamma$, $\hat{e}_1$ is the unit tangent vector at $\phi(x,0)\in W^u(x,\Sigma)$ and e_2 is the flow direction at (x,t), we get

$$D\phi(x,t)(e_1) = DX_t\big(X^t(x)\big)(\hat{e}_1) \quad \text{and}$$
$$D\phi(x,t)(e_2) = DX^t\big(X^t(x)\big)\big(X(x,0)\big) = X\big(X^t(x)\big).$$

Hence $D(\phi \mid \gamma \times [0,T])(x,t) = DX^t \mid E^{cu}_{\phi(x,t)}$ for $(x,t) \in \gamma \times [0,T]$ and so

$$\left|\det D\big(\phi \mid \gamma \times [0,T]\big)(x,t)\right| = J^c_t(x).$$

Now the volume expanding property of X^t along the center-unstable sub-bundle, together with the fact that the return time function τ is not bounded from above near the singularities, show that the densities of μ_γ are uniformly bounded from above throughout Λ but not from below. In fact, this shows that these densities will tend to zero close to the singularities of X in Λ.

This finishes the proof of Theorem 7.7 except for the proof of Proposition 7.31 and of $\operatorname{supp}(\mu) = \Lambda$, which we present in what follows.

7.3.11 Constructing the Disintegration

Here we prove Proposition 7.31. We split the proof into several lemmas keeping the notations of the previous sections.

Let λ^2, $R : p^{-1}(I) \to \Xi$, $\mathcal{F}^u(x_0,\delta_2)$, $F^u(x_0,\delta_2)$ and η be as before, where $x_0 \in K \cap \Sigma$ is a density point of $\eta \mid K$ and K is a compact Pesin set.

7.3.11.1 Existence of Hyperbolic Times for f and Consequences for R

Now we show that a full measure subset of $\mathcal{F}^u(x_0,\delta_2)$ has absolutely continuous disintegrations, which is enough to conclude the proof of Proposition 7.31 except for the bounds on the densities.

We need the notion of *hyperbolic time* for the one-dimensional map f [9]. We know that $\log(\frac{1}{f'})$ is α-Hölder and the boundaries Γ_0 of the intervals $I_1,\dots,I_n$ can be taken as a *singular set* for f (where the map is not defined or is not differentiable) which behaves like a *power of the distance to* Γ_0, as follows. Denoting by d the usual distance on the intervals I, there exist $B > 0$ and $\beta > 0$ such that

- $\frac{1}{B} \cdot d(x,\Gamma_0)^\beta \le |f'| \le B \cdot d(x,\Gamma_0)^{-\beta}$;
- $\left|\log|f'(x)| - \log|f'(y)|\right| \le B \cdot d(x,y) \cdot d(x,\Gamma_0)^{-\beta}$,

for all $x, y \in I$ with $d(x,y) < d(x,\Gamma_0)/2$. This is true for f since it was shown in Sect. 6.1.5 that $f' \mid I_j$ either is bounded from above and below away from zero, or else is of the form x^β with $\beta \in (0,1)$.

Given $\delta > 0$ we define $d_\delta(x,\Gamma_0) = d(x,\Gamma_0)$ if $d(x,\Gamma_0) < \delta$ and 1 otherwise.

Definition 7.32 Given $b, c, \delta > 0$ we say that $n \ge 1$ is a (b,c,δ)-hyperbolic time for $x \in I$ if

$$\prod_{j=n-k}^{n-1} \left|f'\big(f^j(x)\big)\right|^{-1} \le e^{-ck} \quad \text{and} \quad \prod_{j=n-k}^{n-1} d_\delta\big(f^j(x),\Gamma_0\big) \ge e^{-bk} \tag{7.12}$$

for all $k = 0,\dots,n-1$.

Since f has positive Lyapunov exponent υ-almost everywhere, i.e.,

$$\lim_{n\to+\infty}\frac{1}{n}\log\left|(f^n)'(x)\right|>0\quad\text{for }\upsilon\text{-almost all }x\in I,$$

and since $\frac{d\upsilon}{d\lambda}$ is bounded from above (where λ is the Lebesgue length measure on I), it follows that $|\log d(x,\Gamma_0)|$ is υ-integrable and, for any given $\varepsilon>0$, we can find $\delta>0$ such that

$$\lim_{n\to\infty}\frac{1}{n}\sum_{j=0}^{n-1}-\log d_\delta\big(f^j(x),\Gamma_0\big)=\int-\log d_\delta(x,\Gamma_0)\,d\upsilon(x)<\varepsilon,$$

for υ-a.e. $x\in I$. This means that f is *non-uniformly expanding* and has *slow recurrence to the singular set*. Hence we are in the setting of the following result.

Theorem 7.33 (Existence of a positive frequency of hyperbolic times) *Let $f:I\to I$ be a map with $\log|f'|$ α-Hölder, behaving like a power of the distance to a singular set Γ_0, non-uniformly expanding and with slow recurrence to Γ_0 with respect to an absolutely continuous invariant probability measure υ. Then for $b,c,\delta>0$ small enough there exists $\theta=\theta(b,c,\delta)>0$ such that υ-a.e. $x\in I$ has infinitely many (b,c,δ)-hyperbolic times. Moreover if we write $0<n_1<n_2<n_2<\cdots$ for the hyperbolic times of x then their asymptotic frequency satisfies*

$$\liminf_{N\to\infty}\frac{\#\{k\ge 1:n_k\le N\}}{N}\ge\theta\quad\textit{for }\upsilon\textit{-a.e. }x\in I.$$

Proof A complete proof can be found in [9, Sect. 5] with weaker assumptions corresponding to Theorem C in that paper. □

From now on we fix values of (b,c,δ) so that the conclusions of Theorem 7.33 are true.

We now outline the properties of these special times. For detailed proofs see [9, Proposition 2.8] and [7, Proposition 2.6, Corollary 2.7, Proposition 5.2].

Proposition 7.34 *There are constants $\beta_1,\beta_2>0$ depending on (b,c,δ) and f only such that, if n is a (b,c,δ)-hyperbolic time for $x\in I$, then there are neighborhoods $W_k(x)\subset I$ of $f^{n-k}(x)$, $k=1,\dots,n$, such that*

1. *$f^k\mid W_k(x)$ maps $W_k(x)$ diffeomorphically to the ball of radius β_1 around $f^n(x)$;*
2. *for every $1\le k\le n$ and $y,z\in W_n(x)$*

$$d\big(f^{n-k}(y),f^{n-k}(z)\big)\le e^{-ck/2}\cdot d\big(f^n(y),f^n(z)\big);$$

3. *for $y,z\in W_n(x)$*

$$\frac{1}{\beta_2}\le\frac{\left|(f^n)'(y)\right|}{\left|(f^n)'(z)\right|}\le\beta_2.$$

The conjugacy $p \circ R = f \circ p$ between the actions of the Poincaré map and the one-dimensional map on the space of leaves, together with the bounds on the derivative (6.8), enables us to extend the properties given by Proposition 7.34 to any cu-curve inside $B(\eta)$, as follows.

Let $\gamma : J \to \Xi$ be a cu-curve in $\Xi \setminus \Gamma$ such that $\gamma(s) \in B(\eta)$ for Lebesgue almost every $s \in J$, where J is a non-empty interval—such a curve exists since the basin $B(\eta)$ contains entire strips of some section $\Sigma \in \Xi$ except for a subset of zero area. Note that we have the following limit in the weak* topology,

$$\lim_{n\to+\infty} \lambda_\gamma^n = \eta \quad \text{where } \lambda_\gamma^n = \frac{1}{n}\sum_{j=0}^{n-1} R_*^j(\lambda_\gamma),$$

by the choice of γ and by an easy application of the Dominated Convergence Theorem.

Proposition 7.35 *There are constants $\kappa_0, \kappa_1 > 0$ depending on (b, c, δ) and on R_0, β_0, β_1 and β_2 only such that, if $x \in \gamma$ and n a big enough (b, c, δ)-hyperbolic time for $p(x) \in I$, then there are neighborhoods $V_k(x)$ of $R^{n-k}(x)$ on $R^{n-k}(x)(\gamma)$, $k = 1, \dots, n$, such that*

1. *$R^k \mid V_k(x)$ maps $V_k(x)$ diffeomorphically to the ball of radius κ_0 around $R^n(x)$ on $R^n(\gamma)$;*
2. *for every $1 \le k \le n$ and $y, z \in V_n(x)$*
$$d_{R^{n-k}(\gamma)}\big(R^{n-k}(y), R^{n-k}(z)\big) \le \beta_0 \cdot e^{-ck/2} \cdot d_{R^n(\gamma)}\big(R^n(y), R^n(z)\big);$$
3. *for $y, z \in V_n(x)$*
$$\frac{1}{\kappa_1} \le \frac{\big|D(R^n \mid \gamma)(y)\big|}{\big|D(R^n \mid \gamma)(z)\big|} \le \kappa_1,$$
4. *the inducing time of R^k on $V_k(x)$ is constant, i.e., $r^{n-k} \mid V_k(x) \equiv const$.*

Here d_γ denotes the distance along γ given by the shortest smooth curve in γ joining two given points and λ_γ denotes the normalized Lebesgue length measure induced on γ by the area form λ^2 on Ξ.

Proof of Proposition 7.35 Let $x_0 = p(x)$ and $W_k(x_0)$ be given by Proposition 7.34, $k = 1, \dots, n$. Then $p(\gamma)$ is an interval in I and $p \mid \gamma : \gamma \to p(\gamma)$ is a diffeomorphism—we may take γ with smaller length if needed.

If n is big enough, then $W_n(x_0) \subset p(\gamma)$. Moreover the conjugacy implies that the following maps are all diffeomorphisms

$$\begin{array}{ccc} V_k(x) & \xrightarrow{R^k} & R^k(V_k(x)) \\ p\downarrow & & \downarrow p \\ W_k(x_0) & \xrightarrow{f^k} & B\big(f^k(x_0), \kappa_0\big) \end{array},$$

Fig. 7.8 Hyperbolic times and projections

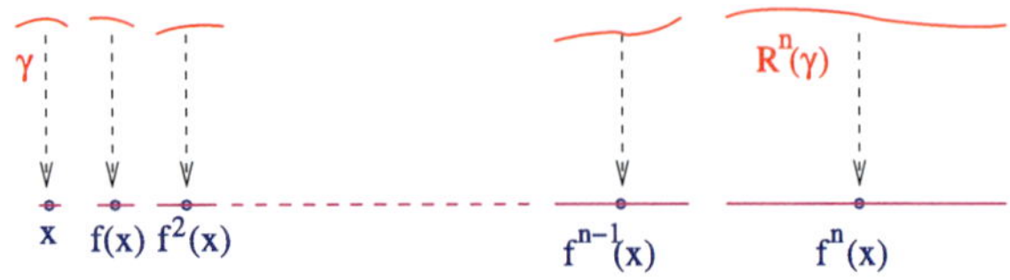

and the diagram commutes, where $V_k(x) = (p \mid R^{n-k}(\gamma))^{-1}(W_k(x_0)), k = 1, \dots, n$; see Fig. 7.8. Using the bounds (6.8) to compare derivatives we get $\kappa_0 = \beta_1/\beta_0$ and $\kappa_1 = \beta_0 \cdot \beta_2$.

To prove item (4) we just note that by definition of (b, c, δ)-hyperbolic time none of the sets $W_k(x_0)$ may intersect Γ_0. According to the definition of Γ_0, this means that orbits through $x, y \in V_k(x)$ cannot cut different cross-sections in Ξ before the next return in time $\tau(x), \tau(y)$ respectively. Hence every orbit through $W_k(x_0)$ cuts the same cross-sections on its way to the next return cross-section. In particular the number of cross-section cuts is the same, i.e., $r \mid V_k(x)$ is constant, $k = 1, \dots, n$. Hence by definition of r^k we obtain the statement of item (4) since $R(V_k(x)) = V_{k-1}(x)$ by definition. This completes the proof of the proposition. □

7.3.11.2 Approximating η by Push Forwards of Lebesgue Measure at Hyperbolic Times

We define for $n \geq 1$

$$H_n = \{x \in \gamma : n \text{ is a } (b, c/2, \delta)\text{-hyperbolic time for } p(x)\}.$$

As a consequence of items (1-2) of Proposition 7.35, H_n is an open subset of γ and for any $x \in \gamma \cap H_n$ we can find a connected component γ^n of $R^n(\gamma) \cap B(R^n(x), \kappa_0)$ containing x such that $R^n \mid V_n(x) : V_n(x) \to \gamma^n$ is a diffeomorphism. In addition, γ^n is a cu-curve according to Corollary 6.17, and by item (3) of Proposition 7.35 we deduce that

$$\frac{1}{\kappa_1} \leq \frac{d\big(R^n_*(\lambda_\gamma) \mid B(R^n(x), \kappa_0)\big)}{d\lambda_{\gamma^n}} \leq \kappa_1, \quad \lambda_{\gamma^n} - \text{a.e. on } \gamma^n, \tag{7.13}$$

where λ_{γ^n} is the Lebesgue induced measure on γ^n for any $n \geq 1$, and we normalize both measures so that $((R^n)_*(\lambda_\gamma) \mid B(R^n(x), \kappa_0))(\gamma^n) = \lambda_{\gamma^n}(\gamma^n)$, i.e., their masses on γ^n are the same.

Moreover the set $R^n(\gamma \cap H_n)$ has an at most countable number of connected components which are diffeomorphic to open intervals. Each of these components is a cu-curve with diameter bigger than κ_0 and hence we can find a *pairwise disjoint family* γ_i^n *of* κ_0*-neighborhoods around* $R^n(x_i)$ *in* $R^n(\gamma)$, *for some* $x_i \in H_n$, *with maximum cardinality*, such that

$$\Delta_n = \bigcup_i \gamma_i^n \subset R^n\big(\gamma \cap H_n\big) \quad \text{and} \quad \big(R^n_*(\lambda_\gamma) \mid \Delta_n\big)(\Delta_n) \geq \frac{1}{2\kappa_1} \cdot \lambda_\gamma(H_n). \tag{7.14}$$

Indeed, since $R^n(\gamma \cap H_n)$ *is one-dimensional*, for each connected component the family Δ_n may miss a set of points of length at most equal to the length of one γ_i^n, for otherwise we would manage to include an extra κ_0-neighborhood in Δ_n. Hence we have in the worst case (assuming that there is only one set γ_i^n for each connected component)

$$\lambda_{\gamma^n}\big(R^n(\gamma \cap H_n) \setminus \Delta_n\big) \le \lambda_{\gamma^n}\Big(\bigcup_i \gamma_i^n\Big) = \lambda_{\gamma^n}(\Delta_n)$$

so that

$$\lambda_{\gamma^n}(\Delta_n) \ge \frac{1}{2} \cdot \lambda_{\gamma^n}\big(R^n(\gamma \cap H_n)\big)$$

and the constant κ_1 comes from (7.13).

For a fixed small $\rho > 0$ we consider $\Delta_{n,\rho}$ given by the balls γ_i^n with the same center $x_{n,i}$ but a reduced radius of $\kappa_0 - \rho$. Then the same bound in (7.14) still holds with $2\kappa_1$ replaced by $3\kappa_1$.

We write D_n for the family of disks from $\cup_{j\ge 1}\Delta_j$ with the same expanding iterate (the disks with the same centers as those from $D_{n,\rho}$ but with their original size).

We define the following sequences of measures for each $n \ge 1$

$$\omega_\rho^n := \frac{1}{n}\sum_{j=0}^{n-1} R_*^j(\lambda_\gamma) \mid \Delta_{j,\rho}; \qquad \omega_0^n := \frac{1}{n}\sum_{j=0}^{n-1} R_*^j(\lambda_\gamma) \mid \Delta_j; \qquad \overline{\lambda}_\gamma^n = \lambda_\gamma^n - \omega_\rho^n;$$

$$\text{and also} \quad \omega_0^n := \frac{1}{n}\sum_{j=0}^{n-1} R_*^j(\lambda_\gamma) \mid \Delta_j, \quad \text{where } \Delta_j = \Delta_{j,0}.$$

Then any weak* limit point $\tilde{\eta} = \lim_k \omega_\rho^{n_k}$ for some subsequence $n_1 < n_2 < \cdots$, as well as

$$\overline{\eta} = \lim_k \overline{\lambda}_\gamma^{n'_k}$$

(where n'_k may be taken as a subsequence of n_k), are R-invariant measures which after normalization satisfy $\eta = a\tilde{\eta} + b\overline{\eta}$ for some $a, b \ge 0$ with $a + b = 1$.

We claim that $\tilde{\eta} \not\equiv 0$, and thus $\eta = \tilde{\eta}$ as a consequence of the ergodicity of η. In fact, we can bound the mass of ω_ρ^n from below using the density of hyperbolic times from Theorem 7.33 and the bound from (7.14) through the following Fubini-Toneli-type argument. Write $\#_n(J) = \#J/n$ for any $J \subset \{0, \dots, n-1\}$, the uniform discrete measure on the first n integers. Also set $\chi_i(x) = 1$ if $x \in H_i$ and zero otherwise, $i = 0, \dots, n-1$. Then

$$\omega_\rho^n(M) \ge \frac{1}{3\kappa_1 \cdot n}\sum_{j=0}^{n-1} \lambda_\gamma(H_j) = \frac{n}{3\kappa_1 n}\iint \chi_i(x)\, d\lambda_\gamma(x)\, d\#_n(i)$$

$$= \frac{1}{3\kappa_1} \iint \chi_i(x)\, d\#_n(i)\, d\lambda_\gamma(x) \geq \frac{\theta}{6\kappa_1} > 0,$$

for every n big enough by the choice of γ.

7.3.11.3 Approximating Unstable Curves by Images of Curves at Hyperbolic Times

We now observe that, since $\eta(F^u(x_0,\delta_2)) > 0$ and x_0 is a density point for the measure $\eta \mid F^u(x_0,\delta_2)$, then $\omega^n_\rho(F^u(x_0,\delta_2)) \geq c$ for some constant $c > 0$ for all big enough n. If we assume that $\delta_2 < \rho$, which poses no restriction, then we see that the cu-curves from $D_{j,\rho}$ intersecting $F^u(x_0,\delta_2)$ will cross this horizontal strip when we restore their original size. Thus the leaves $\cup_{j=0}^{n-1} D_j$ in the support of ω^n_0 which intersect $F^s(x_0,\delta_2)$ cross this strip. Given any sequence γ^{n_k} of leaves in D_{n_k} crossing $F^s(x_0,\delta_2)$ with $n_1 < n_2 < n_3 < \cdots$, then there exists a C^1 limit leaf γ^∞ also crossing $F^u(x_0,\delta_2)$, by the Ascoli-Arzela Theorem. We claim that this leaf coincides with the unstable manifold of its points, i.e., $\gamma^\infty = W^u(x,\Sigma)$ for all $x \in \gamma^\infty$. This shows that the accumulation curves γ^∞ are defined independently of the chosen sequence γ^{n_k} of curves in Σ.

To prove the claim let us fix $l > 0$ and take a large k so that $n_k \gg l$. We note that for any distinct $x, y \in \gamma^\infty$ there are $x_k, y_k \in \gamma^{n_k}$ such that $(x_k, y_k) \to (x, y)$ when $k \to \infty$. Then for x_k, y_k there exists a neighborhood V_{n_k} in γ such that $\gamma^{n_k} = R^{n_k}(V_{n_k})$.

We take $j = n_k - l$. We can now write for some $w_k, z_k \in V_{n_k}$

$$\begin{aligned} d(x_k, y_k) &= d\Big(R^{n_k - j}\big(R^j(w_k)\big), R^{n_k - j}\big(R^j(z_k)\big)\Big) \\ &\geq \frac{e^{lc/4}}{\beta_0} \cdot d\big(R^{n_k - l}(w_k), R^{n_k - l}(z_k)\big). \end{aligned}$$

Note that each pair $R^{n_k-l}(w_k), R^{n_k-l}(z_k)$ belongs to a section $\Sigma_l \in \Xi$ and that $R^l(R^{n_k-l}(w_k)) = x_k$ and $R^l(R^{n_k-l}(z_k)) = y_k$. Letting $k \to \infty$ we obtain limit points $(R^{n_k-l}(w_k), R^{n_k-l}(z_k)) \to (w_l, z_l)$ in some section $\Sigma \in \Xi$ (recall that Ξ is a finite family of compact adapted cross-sections) satisfying

$$R^l(w_l) = x, \quad R^l(z_l) = y \quad \text{and} \quad d(w_l, z_l) \leq \beta_0 e^{-lc/4} \cdot d(x, y).$$

Since this is true for any $l > 0$ we conclude that y is in the unstable manifold of x with respect to R, i.e., $y \in W^u_R(x)$, and thus $y \in W^u(x,\Sigma)$ by the following lemma. This proves the claim.

Lemma 7.36 *In the same setting as above, we have* $W^u_R(x) = W^u(x,\Sigma)$.

Proof On the one hand, let $y_0 \in W^u_R(x)$. Then for each $l \geq 1$ there exist y_l, x_l in some section $\Sigma_l \in \Xi$ such that

1. $R^l(y_l) = y_0$, $R^l(x_l) = x$;
2. $d(y_l, x_l) \to 0$ when $l \to +\infty$.

This means that there are $s_l, t_l \xrightarrow[l\to+\infty]{} +\infty$ such that

$$X^{s_l}(y_l) = y_0, \quad X^{t_l}(x_l) = x \quad \text{and} \quad \operatorname{dist}\big(X^{-s_l}(y_0), X^{-t_l}(x)\big) \xrightarrow[l\to+\infty]{} 0.$$

But this ensures that the points y_l, x_l belong to the same smoothness domain of R in each Σ_l, for otherwise $x_{l-1} = R(x_l)$ and $y_{l-1} = R(y_l)$ would be in distinct sections of Ξ and the distance between them would be larger than some constant bounded away from zero. Since there are only finitely many such smoothness domains on Ξ, the flow orbits between the pair x_l, y_l and x_{l-1}, y_{l-1} are contained in a tubular neighborhood for the flow of X. Hence condition 2 above implies that the distance between the orbits of x_l and y_l under the flow of X up until the points x_{l-1}, y_{l-1} are uniformly close and tend to zero. Moreover, as a consequence of the construction of the cross-sections in Ξ, as stated in Remark 6.14, we also have $y_0 = X^{t_l+\varepsilon_l}(y_l)$ with $\varepsilon_l \to 0$ as $l \to +\infty$. Altogether we get

$$\sup\Big\{\operatorname{dist}\big(X^{-t+\varepsilon_l}(y_0), X^{-t}(x)\big) : t_l < t < t_{l+1}\Big\} \xrightarrow[l\to+\infty]{} 0.$$

Thus $y_0 \in W^u(x) \cap \Sigma = W^u(x, \Sigma)$.

On the other hand, let us take $y_0 \in W^u(x, \Sigma)$. Then, by Remark 6.14, there exists ε so that $z_0 = X^{\varepsilon}(y_0) \in W^{uu}(x)$, with $|\varepsilon|$ small and tending to 0 when we let $y_0 \to x$. Let $t_l > 0$ be such that $X^{-t_l}(x) = w_l \in \Sigma$ for $l \geq 1$. Then we have

$$\operatorname{dist}\big(X^{-t_l}(z_0), X^{-t_l}(x)\big) \xrightarrow[l\to\infty]{} 0 \tag{7.15}$$

and so there exists ε_l such that $X^{\varepsilon_l - t_l}(z_0) = z_l = X^{\varepsilon_l+\varepsilon-t_l}(y_0) \in \Sigma$ with $|\varepsilon_l|$ small. Notice that (7.15) ensures that $|\varepsilon_l| \to 0$ also.

Hence there exists $\delta = \delta(\varepsilon, \varepsilon_l)$ satisfying $\delta \to 0$ when $(\varepsilon + \varepsilon_l) \to 0$ and also $d(z_l, w_l) < \delta$ for all $l \geq 1$. Since $R^l(z_l) = y_0$ we conclude that $y_0 \in W^u_R(x)$, finishing the proof. □

7.3.11.4 Upper and Lower Bounds for Densities Through Approximation

We define $\mathscr{F}^u_\infty$ to be the family of all leaves γ^∞ obtained as C^1 accumulation points of leaves in

$$\mathscr{F}^u_n = \big\{\xi \in \cup_{j=0}^{n-1} D_j : \xi \quad \text{crosses } F^s(x_0, \delta_2)\big\}.$$

We note that $\mathscr{F}^u_\infty \subset \mathscr{F}^u(x_0, \delta_2)$. Since for all n we have $\omega^n_0 \geq \omega^n_\rho$, then $\omega^n_0(\cup \mathscr{F}^u_n) > c$ and we obtain $\eta(\cup \mathscr{F}^u_\infty) \geq c$.

Lemma 7.37 *In our setting $\mathscr{F}^u_\infty$ equals $\mathscr{F}^u(x, \delta_2)$ except possibly in a η-measure zero subset.*

Proof Let $\mathscr{A}$ be the family of leaves of $\mathscr{F}^u(x,\delta_2)$ which are not accumulated by leaves of $\mathscr{F}_n^u$, and let us assume that $\eta(\mathscr{A})>0$. Since the family of leaves provides a measurable partition of $\mathscr{A}$, we can consider the quotient measure $\hat{\eta}$ of $\eta \mid \mathscr{A}$ on the space $\tilde{\mathscr{A}}$ of leaves which, by the dimension of the objects considered, can be identified with a subset of an interval. Therefore $\tilde{\eta}$-almost every leaf $\zeta \in \mathscr{A}$ is a density point of $\tilde{\mathscr{A}}$. We fix a density point ζ in this sense.

For any given $\varepsilon>0$ we can always find a δ-neighborhood V_δ of ζ such that $0<\delta<\varepsilon$ and $\eta(\partial V_\delta)=0$, so that $\omega_\rho^n(V_\delta)\to\eta(V_\delta)>0$ when $n\to+\infty$. Therefore ζ will be δ-close to a leaf ξ_n of $\mathscr{F}_n^u$ for every big enough n. This contradicts the definition of $\mathscr{A}$ so that we must have $\eta(\mathscr{A})=0$, as claimed. □

By definition of $\mathscr{F}_n^u$ and by (7.13) we see that $\omega_0^n \mid F_n^u$ disintegrates along the partition $\mathscr{F}_n^u$ of $F_n^u=\cup\mathscr{F}_n^u$ into measures ω_ξ^n having density with respect to λ_ξ uniformly bounded from above and below, for almost every $\xi\in\mathscr{F}_n^u$.

To take advantage of this in order to prove Proposition 7.31 we consider a sequence of increasing partitions $(\mathscr{V}_k)_{k\ge 1}$ of $W^s(x_0,\Sigma)$ whose diameter tends to zero. This defines a sequence $\mathscr{P}_k$ of partitions of $\tilde{\mathscr{F}}=\cup_{0\le n\le\infty}\mathscr{F}_n^u$ as follows: we fix $k\ge 1$ and say that two elements $\xi\in\mathscr{F}_i^u, \xi'\in\mathscr{F}_j^u, 0\le i,j\le\infty$ are in the same atom of $\mathscr{P}_k$ when both intersect $W^s(x,\Sigma)$ in the same atom of $\mathscr{V}_k$ and either $i,j\ge k$ or $i=j<k$.

If q is the projection $q:\tilde{\mathscr{F}}\to W^s(x_0,\Sigma)$ given by the transverse intersection $\xi\cap W^s(x_0,\Sigma)$ for all $\xi\in\tilde{\mathscr{F}}$, then $\tilde{\mathscr{F}}$ can be identified with a subset of the real line. Thus we may assume without loss that the union $\partial\mathscr{P}_k$ of the boundaries of $\mathscr{P}_k$ satisfies $\eta(\partial\mathscr{P}_k)=\hat{\eta}(\partial\mathscr{P}_k)=0$ for all $k\ge 1$, by suitably choosing the sequence $\mathscr{V}_k$.

Upper and Lower Bounds for Densities Given $\zeta\in\tilde{\mathscr{F}}$ we write $p:F^u(x_0,\delta_2)\to\zeta$ as the projection along stable leaves and ω for ω_0. We also write $\mathscr{P}_k(\zeta)$ for the atom of $\mathscr{P}_k$ which contains ζ. Then, since $\mathscr{P}_k(\zeta)$ is a union of leaves, for any given Borel set $B\subset\zeta$ and $n\ge 1$, we have

$$\omega^n\big(\mathscr{P}_k(\zeta)\cap p^{-1}(B)\big)=\int\omega_\xi^n\big(\mathscr{P}_k(\zeta)\cap p^{-1}(B)\big)\,d\hat{\omega}^n(\xi) \tag{7.16}$$

through disintegration, where $\hat{\omega}^n$ is the measure on $\tilde{\mathscr{F}}$ induced by ω^n. Moreover by (7.13) and because each curve in $\tilde{\mathscr{F}}$ crosses $F^u(x_0,\delta_2)$

$$\frac{1}{\kappa_1\kappa_2}\cdot\lambda_\zeta(B)\le\frac{1}{\kappa_1}\cdot\lambda_\xi\big(p^{-1}(B)\big)\le\omega_\xi^n\big(\mathscr{P}_k(\zeta)\cap p^{-1}(B)\big) \tag{7.17}$$

$$\omega_\xi^n\big(\mathscr{P}_k(\zeta)\cap p^{-1}(B)\big)\le\kappa_1\cdot\lambda_\xi\big(p^{-1}(B)\big)\le\kappa_1\kappa_2\cdot\lambda_\zeta(B) \tag{7.18}$$

for all $n,k\ge 1$ and $\hat{\omega}^n$-a.e. $\xi\in\tilde{\mathscr{F}}$, where $\kappa_2>0$ is a constant such that

$$\frac{1}{\kappa_2}\cdot\lambda_\zeta\le\lambda_\xi\le\kappa_2\cdot\lambda_\zeta \quad \text{for all } \xi\in\tilde{\mathscr{F}}.$$

This constant exists because the angle between the stable leaves in any $\Sigma\in\Xi$ and any *cu*-curve is bounded from below; see Fig. 7.9.

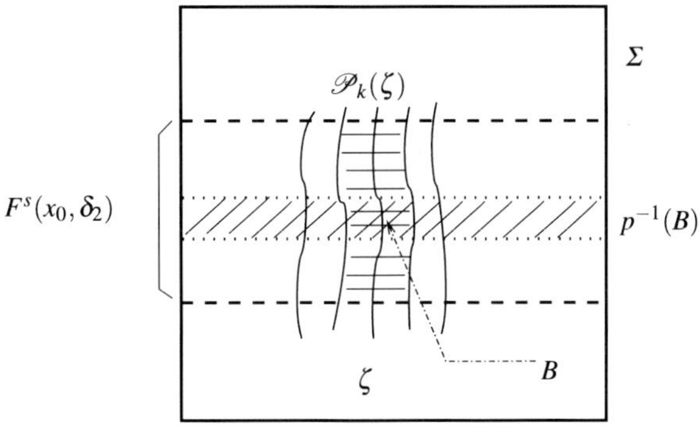

Fig. 7.9 Leaves crossing $F^s(x_0, \delta_2)$ and the projection p

Finally letting $\zeta \in \mathscr{F}^u_\infty$ and choosing B such that $\eta(\partial p^{-1}(B)) = 0$ (which poses no restriction) and assuming that $\eta(\partial(\mathscr{P}_k(\zeta) \cap p^{-1}(B))) = 0$, we get from (7.16), (7.17) and (7.18) for all $k \geq 1$

$$\frac{1}{\kappa_1\kappa_2} \cdot \lambda_\zeta(B) \cdot \hat{\eta}\big(\mathscr{P}_k(\zeta)\big) \leq \eta\big(\mathscr{P}_k(\zeta) \cap p^{-1}(B)\big) \leq \kappa_1\kappa_2 \cdot \lambda_\zeta(B) \cdot \hat{\eta}\big(\mathscr{P}_k(\zeta)\big) \quad (7.19)$$

by the weak* convergence of ω^n to η. Thus to conclude the proof we are left to check that $\eta(\partial(\mathscr{P}_k(\zeta) \cap p^{-1}(B))) = 0$. To do this we observe that $\mathscr{P}_k(\zeta) \cap p^{-1}(B)$ can be written as the product $q(\mathscr{P}_k(\zeta)) \times B$. Hence the boundary is equal to

$$\big(\partial q(\mathscr{P}_k(\zeta)) \times B\big) \cup \big(q(\mathscr{P}_k(\zeta)) \times \partial B\big) \subset q^{-1}\big(\partial q(\mathscr{P}_k(\zeta))\big) \cup p^{-1}(B)$$

and the right-hand side has η-zero measure by construction.

This completes the proof of Proposition 7.31 since we have $\{\zeta\} = \cap_{k\geq 1} \mathscr{P}_k(\zeta)$ for all $\zeta \in \tilde{\mathscr{F}}$ and, by the Theorem of Radon-Nikodym, the bounds in (7.19) imply that the disintegration of $\eta \mid \cup \mathscr{F}^u_\infty$ along the curves $\zeta \in \mathscr{F}^u_\infty$ is absolutely continuous with respect to Lebesgue measure along these curves and with uniformly bounded densities from above and from below.

7.3.12 The Support Covers the Whole Attractor

Finally to conclude that supp$(\mu) = \Lambda$ it is enough to show that supp(μ) contains some cu-curve $\gamma : (a,b) \to \Sigma$ in some subsection $\Sigma \in \Xi$. Indeed, see Fig. 7.10, letting $x_0 \in \Lambda \cap \Sigma$ be a point of a forward dense regular X-orbit and fixing $c \in (a,b)$ and $\varepsilon > 0$ such that $a < c - \varepsilon < c + \varepsilon < b$, then there exists $t > 0$ satisfying dist$(\gamma(c), X^t(x_0)) < \rho$ for any $\rho > 0$. Since $W^s(X^t(x_0), \Sigma) \pitchfork (\gamma \mid (c-\varepsilon, c+\varepsilon)) = \{z\}$ (because γ is a cu-curve in Σ and $\rho > 0$ can be made arbitrarily small, where $\pitchfork$ means transverse intersection), then, by the construction of the adapted cross-section Σ (see Sect. 6.1), this means that $z \in W^s(X^t(x_0))$. Hence the ω-limit sets of z and x_0 are equal to Λ. Thus supp$(\mu) \supseteq \Lambda$ because supp(μ) is X-invariant and closed, and $\Lambda \supseteq$ supp(μ) because Λ is an attracting set.

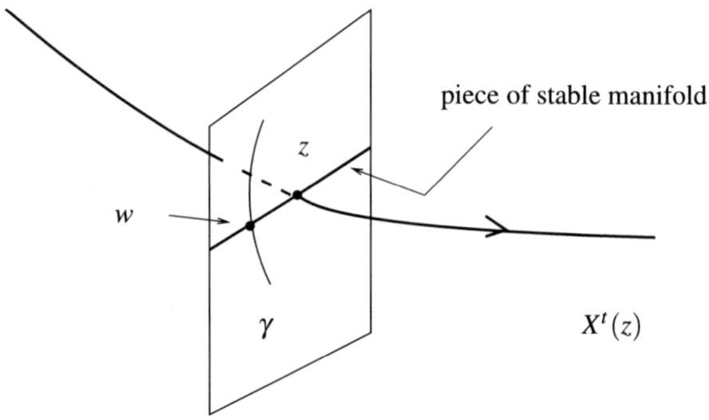

Fig. 7.10 Transitiveness and support of the physical measure

We now use (7.19) to show that $\hat{\eta}$-almost every $\gamma \in \tilde{\mathscr{F}}$ is contained in $\mathrm{supp}(\eta)$, which is contained in $\mathrm{supp}(\mu)$ by the construction of μ from η in Sect. 7.3.5. In fact, $\hat{\eta}$-almost every $\zeta \in \tilde{\mathscr{F}}$ is a density point of $\hat{\eta} \mid \tilde{\mathscr{F}}$ and so, for any one ζ of these curves, we have $\hat{\eta}(\mathscr{P}_k(\zeta)) > 0$ for all $k \geq 1$. Fixing $z \in \zeta$ and choosing $\varepsilon > 0$ we may find $k \geq 1$ big enough and a small enough open neighborhood B of z in ζ such that

$$\mathscr{P}_k(\zeta) \cap p^{-1}(B) \subset B(z, \varepsilon) \cap \Sigma \quad \text{and} \quad \eta\big(\mathscr{P}_k(\zeta) \cap p^{-1}(B)\big) > 0,$$

by the left-hand inequality in (7.19). Since $\varepsilon > 0$ and $z \in \zeta$ were arbitrarily chosen, this shows that $\zeta \subset \mathrm{supp}(\eta) \subset \mathrm{supp}(\mu)$ and completes the proof of Theorem 7.7.

Chapter 8
Singular-Hyperbolicity and Volume

It is well known that a C^2 dynamical system admitting a *hyperbolic basic set* with positive measure must be globally hyperbolic: see e.g. Bowen-Ruelle [62] and Bochi-Viana [52]. The construction of the geometric Lorenz models, presented in Chap. 3, forces the divergence of the vector field to be strictly negative in a isolating neighborhood of the attractor. This feature is also present in the Lorenz system of equations (2.2) for the classical parameters. It is then trivial to show that the corresponding attractor has zero volume.

Using only the partial hyperbolic character of singular-hyperbolicity we show in Sect. 8.2, following [8] for flows which are Hölder-C^1, that either singular-hyperbolic attractors have zero volume or else the flow is globally hyperbolic, that is, an Anosov flow (without singularities).

Using the fact that C^∞ flows are dense among C^1 flows in the C^1 topology, we extend this result in Sect. 8.2.3 to a locally generic subset of all C^1 flows exhibiting robust attractors. This extends the results of Bowen cited above.

A similar result can be obtained for invariant sets satisfying the weaker condition of dominated splitting for the Linear Poincaré Flow among incompressible Hölder-C^1 flows in 3-manifolds, which we present first in Sect. 8.1 following [19]. This extends the results of Bochi-Viana mentioned earlier and will be very useful to complete the proof of the C^1 generic dichotomy for conservative flows on 3-manifolds in Chap. 9.

8.1 Dominated Decomposition, Incompressibility and Zero Volume

Given a C^1 flow X^t say that an invariant subset Λ is *regular* if $\Lambda \cap S(X) = \emptyset$, that is, Λ does not contain zeroes of the vector field X. In what follows we write $\mathfrak{X}^r_\mu(M)$ for the family of C^r vector fields which preserve the volume form (Lebesgue measure) μ on the manifold M.

V. Araújo, M.J. Pacifico, *Three-Dimensional Flows*, Ergebnisse der Mathematik und ihrer Grenzgebiete. 3. Folge / A Series of Modern Surveys in Mathematics 53, DOI 10.1007/978-3-642-11414-4_8, © Springer-Verlag Berlin Heidelberg 2010

Theorem 8.1 *Given a compact 3-manifold M, there exists an open and dense subset $\mathcal{G} \subset \mathfrak{X}^2_\mu(M)$ such that for every $X \in \mathcal{G}$ with a regular invariant set Λ (not necessarily closed) satisfying:*

- *the Linear Poincaré Flow over Λ has a dominated decomposition; and*
- *Λ has positive volume: $\mu(\Lambda) > 0$;*

then X is Anosov and the closure of Λ is the whole of M.

Before giving the proof of the theorem, we can deduce the following useful consequence.

Corollary 8.2 *Given a compact 3-manifold M, there exists $D \subseteq \mathfrak{X}^1_\mu(M)^*$ such that D is C^1-dense and, for $X \in D$, X^t is aperiodic (that is, $\mu(Per(X^t)) = 0$), X is of class C^s ($s \geq 2$) and all its sets with dominated splitting for the Linear Poincaré Flow have zero or full measure.*

Proof We take the C^s-residual given by Robinson's version of Kupka-Smale Theorem for incompressible flows; see [223] and Sect. 2.5.10.

This residual set of vector fields is of class C^s and the associated flows have countable periodic points. Since $\mathfrak{X}^s_\mu(M)$ is a Baire space (with respect to C^s topology), it follows that we have a C^s-dense set D and therefore a C^1-dense set, of vector fields with countable periodic orbits on $\mathfrak{X}^s_\mu(M)$. We know that $\mathfrak{X}^s_\mu(M)$ is C^1-dense on $\mathfrak{X}^1_\mu(M)$ (by Zuppa [277]), and so D is C^1-dense on $\mathfrak{X}^1_\mu(M)$ and all vector fields in D are C^s. Finally, we use Theorem 8.1 and the corollary is proved. □

8.1.1 Dominated Splitting and Regularity

Here we prove that positive volume regular invariant subsets with dominated splitting for a 3-dimensional Hölder-C^1 flow cannot admit singularities in its closure and thus are essentially uniformly hyperbolic sets. This result will be used to prove Theorem 8.1.

We denote by $\mathfrak{X}^{1+}(M)$ the set of all C^1 vector fields X whose derivative DX is Hölder continuous with respect to the given Riemannian norm, and we say that $X \in \mathfrak{X}^{1+}(M)$ is of *class* C^{1+} or *Hölder*-C^1. We clearly have

$$\mathfrak{X}^1(M) \supset \mathfrak{X}^{1+}(M) \supset \mathfrak{X}^r(M), \quad \text{for every } r \geq 2.$$

Proposition 8.3 *Let $X \in \mathfrak{X}^{1+}_\mu(M)$ be given, where M is a 3-manifold. Assume that Λ is a regular X^t-invariant subset of M with positive volume and admitting a dominated splitting. Then the closure A of the set of Lebesgue density points of Λ does not contain singularities.*

According to Proposition 2.34, a compact invariant set without singularities of a C^1 three-dimensional vector field admitting a dominated splitting for the Linear Poincaré Flow is a uniformly hyperbolic set. Then we obtain the following.

Corollary 8.4 *Let M be a compact 3-manifold, let $X \in \mathfrak{X}_\mu^{1+}(M)$ and let Λ be a regular X^t-invariant subset of M with positive volume and admitting a dominated splitting. Then the closure A of the set of Lebesgue density points of Λ is a hyperbolic set.*

This implies in particular that there are neither singular-hyperbolic sets nor partially hyperbolic sets with positive volume for Hölder-C^1 incompressible flows on three-dimensional manifolds. A similar conclusion for singular-hyperbolic sets was obtained by Arbieto-Matheus in [22] but assuming that the invariant compact subset is robustly transitive.

The proof of Proposition 8.3 is divided into several steps, which we state and prove as a sequence of lemmas in the rest of this subsection.

8.1.1.1 Bounded Angles, Eigenvalues and Lorenz-Like Singularities

Denote by $D(\Lambda)$ the subset of the Lebesgue density points of Λ, that is, $x \in D(\Lambda)$ if $x \in \Lambda$ and

$$\lim_{r\to 0^+} \frac{\mu(\Lambda \cap B(x,r))}{\mu(B(x,r))} = 1.$$

is well known (see e.g. [233] or [178]) that almost all points of a measurable set are Lebesgue density points, that is, $\mu(\Lambda \setminus D(\Lambda)) = 0$. Moreover, since every nonempty open subset of M has positive μ-measure, we see that $D(\Lambda)$ is contained in the closure of Λ.

Assume that Λ is a X^t-invariant set without singularities such that $\mu(\Lambda) > 0$ and write A for the closure of $D(\Lambda)$ in what follows. Note that A is contained in the closure of Λ.

Lemma 8.5 *Suppose that the Linear Poincaré Flow over Λ has a dominated splitting for X. Then there exist a neighborhood V of Λ, a neighborhood $\mathfrak{U}$ of X in $\mathfrak{X}^1(M)$ (not necessarily contained in the space of conservative flows) and $\eta > 0$ such that, for every $Y \in \mathfrak{U}$, every periodic orbit contained in U is hyperbolic of saddle type and its eigenvalues λ_1 and λ_2 satisfy $\lambda_1 < -\eta$ and $\lambda_2 > \eta$. Moreover the angle between the unstable and stable directions of these periodic orbits is greater than η.*

Proof The Dominated Splitting for the Linear Poincaré Flow extends by continuity to every *regular* orbit $\mathcal{O}$ which remains close to Λ for a C^1 nearby flow Y; this is Lemma 2.29. The domination implies that the eigenvalues $\lambda_1 \le \lambda_2$ of $\mathcal{O}$ satisfy $\lambda_1 + 2\kappa \le \lambda_2$ for some $\kappa > 0$ which only depends on the domination constant of Λ.

Since the flow Y is close to being conservative, we have $|\lambda_1 + \lambda_2| \le \varepsilon$, where we can take $\varepsilon < \kappa/2$ just by letting Y be in a small C^1-neighborhood of X.

Thus we have $-\lambda_2 - \varepsilon \le \lambda_1$ which implies that $-\lambda_2 - \varepsilon + 2\kappa \le \lambda_1 + 2\kappa \le \lambda_2$, and so $2\lambda_2 \ge 2\kappa - \varepsilon > 0$ on the one hand. On the other hand $\lambda_1 \le \varepsilon - \lambda_2$ implies that $\lambda_1 \le \varepsilon - (\kappa - \varepsilon/2) = 3\varepsilon/2 - \kappa < 0$.

Hence there exists $\eta > 0$, independent of Y in a C^1 neighborhood of X, and independent of the periodic orbit $\mathcal{O}$ of Y in a neighborhood of Λ, such that $\lambda_1 < -\eta$ and $\lambda_2 > \eta$, as stated.

For the angle bound we argue by contradiction as in the proof of Theorem 5.37: assume there exists a sequence of flows $Y_n \xrightarrow[n\to+\infty]{C^1} X$ and of periodic orbits $\mathcal{O}_n$ of Y_n contained in the neighborhood V of Λ such that the angle α_n between the unstable subspace and the stable direction satisfies $\alpha_n \xrightarrow[n\to+\infty]{} 0$.

Then as in the proof of Theorem 5.37 we can find (through the flow version of Frank's Lemma, Theorem 2.24) an arbitrarily small C^1 perturbation Z_n of Y_n, for all big enough $n \ge 1$, sending the stable direction close to the unstable direction along the periodic orbit, such that the orbit of $\mathcal{O}_n$ becomes a sink or a source for Z_n. This contradicts the first part of the statement of the lemma. □

In the next lemma, recall the definition of Lorenz-like singularity from Chap. 2.

Lemma 8.6 *Assume that $X \in \mathfrak{X}^1_\mu(M)$ is such that all singularities are hyperbolic with no resonances (real eigenvalues are all distinct). Then the singularities $S(X) \cap \Lambda$ are all Lorenz-like for X or for $-X$.*

Remark 8.7 The assumptions of the lemma above hold true for an open and dense subset of all C^r vector fields, whether volume preserving or not.

Proof Fix σ in $S(X) \cap \Lambda$ if this set is nonempty (otherwise there is nothing to prove). By assumption on X we known that σ is hyperbolic. As in Chap. 5 we show first that σ has only real eigenvalues. For otherwise we would get a conjugate pair of complex eigenvalues $\omega, \overline{\omega}$ and a real one λ and, by reversing time if needed, we can assume that $\lambda < 0 < \mathrm{Re}(\omega)$. Since $\mu(\Lambda) > 0$ there are infinitely many distinct orbits of Λ passing through every given neighborhood of σ, because each regular orbit of a flow is a regular curve, and so does not fill volume in a three-dimensional manifold.

Using the Connecting Lemma of Hayashi adapted to conservative flows (see e.g. [271]) we can find a C^1-close flow Y preserving the same measure μ with a saddle-focus connection associated to the continuation σ_Y of the singularity σ. By a small perturbation of the vector field we can assume that Y is of class C^∞ and still C^1-close to X (see e.g. [277]).

We can now unfold the saddle-focus connection to obtain a Shil'nikov bifurcation as in [47]; see also Sect. 3.2.3. In this way we produce a periodic orbit with all Lyapunov exponents equal to zero (an elliptic closed orbit) for a C^1-close flow and near Λ. This contradicts Lemma 8.5, since such an orbit will be contained in a neighborhood of Λ. This shows that complex eigenvalues are not allowed for any singularity in Λ.

Let then $\lambda_2 \le \lambda_3 \le \lambda_1$ be the eigenvalues of σ. We have $\lambda_2 < 0 < \lambda_1$ because σ is hyperbolic. The preservation of volume implies that $\lambda_2 = -(\lambda_1 + \lambda_3) < 0$, so that $-\lambda_3 < \lambda_1$. We have now two cases:

$\lambda_3 < 0$: this implies that $\lambda_2 < \lambda_3 < 0 < -\lambda_3 < \lambda_1$ by the non-resonance assumption, and σ is Lorenz-like for X;

$\lambda_3 > 0$: since $\lambda_1 = -(\lambda_2 + \lambda_3) > 0$ the non-resonance assumption ensures that $\lambda_2 < -\lambda_3 < 0 < \lambda_3 < \lambda_1$, and so σ is Lorenz-like for $-X$.

The proof is complete. □

8.1.1.2 Invariant Manifolds of a Positive Volume Set with Dominated Splitting for the Linear Poincaré Flow

Now we show that there can be no equilibria in Λ. To do this we show that Λ admits generalized stable and unstable manifolds which are fully contained in Λ. This in turn implies that the existence of equilibria in Λ leads to a contradiction to the assumption of dominated decomposition.

Invariant Manifolds and (Non-uniform) Hyperbolicity An embedded disk $\gamma \subset M$ is a (local) *strong-unstable manifold*, or a *strong-unstable disk*, if $\mathrm{dist}(X^{-t}(x), X^{-t}(y))$ tends to zero exponentially fast as $t \to +\infty$, for every $x, y \in \gamma$. In the same way γ is called a (local) *strong-stable manifold*, or a *strong-stable disk*, if $\mathrm{dist}(X^t(x), X^t(y)) \to 0$ exponentially fast as $n \to +\infty$, for every $x, y \in \gamma$.

We observe that since Λ has positive volume, by Lemma 8.5 and its proof, the Lebesgue measure μ_Λ normalized and restricted to Λ is a (non-uniformly) *hyperbolic invariant probability measure*; see e.g. [33]: indeed every Lyapunov exponent of μ_Λ is non-zero, except along the direction of the flow.

Assuming from now on that $X \in \mathfrak{X}^{1+}(M)$ we know that, according to the non-uniform hyperbolic theory (see [33, 196, 197]), there are smooth strong-stable and strong-unstable disks tangent to the directions corresponding to negative and positive Lyapunov exponents, respectively, at μ_Λ almost every point. The sizes of these disks depend measurably on the point as well as the rates of exponential contraction and expansion. We can define as before the strong-stable, strong-unstable, stable and unstable manifolds at μ_Λ almost all points.

In addition, since μ is a smooth invariant measure, we can use [34, Theorem 11.3] and conclude that there are at most countably many ergodic components of μ_Λ. *Therefore we assume from now on that μ_Λ is ergodic without loss of generality.*

In addition, hyperbolic smooth ergodic invariant probability measures for a C^{1+} dynamics are in the setting of Katok's Closing Lemma; see [119] or [34, Sect. 15]. In particular, the support of μ_Λ is contained in the closure of the closed orbits inside Λ, that is,

$$\operatorname{supp}(\mu_A) \subset \overline{\operatorname{Per}(X) \cap A}, \tag{8.1}$$

where the periodic points in our setting are all hyperbolic by Lemma 8.5.

Almost All Invariant Manifolds Are Contained in *A* Now we adapt the arguments of Bochi-Viana in [52] to our setting to deduce the following. Let μ_u and μ_s denote the measures induced on (strong-)unstable and (strong-)stable manifolds by the Lebesgue volume form μ.

Lemma 8.8 *For μ_A almost every x the corresponding invariant manifolds satisfy*

$$\mu_s\big(W^{ss}(x) \setminus A\big) = 0 \quad \textit{and} \quad \mu_u\big(W^{uu}(x) \setminus A\big) = 0,$$

that is, the invariant manifolds are $\mu_{u,s}$ $\bmod 0$ contained in A.

In addition, since A is closed and every open subset of either $W^{ss}(x)$ or $W^{uu}(x)$ has positive μ_s or μ_u measure, respectively, then we see that in fact

$$W^{ss}(x) \subset A \quad \text{and} \quad W^{uu}(x) \subset A \quad \text{for } \mu - \text{almost every } x. \tag{8.2}$$

To prove Lemma 8.8 we need a bounded distortion property along invariant manifolds which is provided by [34, Theorems 11.1 & 11.2] and stated in Theorem 2.36.

Recurrent and Lebesgue Density Points We are now ready to start the proof of Lemma 8.8.

Let us take a strong-unstable disk $W^{uu}(x)$ satisfying simultaneously

- $x \in \mathcal{H}(\kappa)$,
- $\mu_u(W^{uu}(x) \cap A) > 0$ and
- x is a μ_u density point of $W^{uu}(x) \cap A$.

For this it is enough to take κ big enough since, by the absolute continuity of the foliation of strong-unstable disks, a positive volume subset, such as $\mathcal{H}(\kappa)$, must intersect almost all strong-stable disks on a subset of μ_u positive measure, see e.g. [216].

Using the Recurrence Theorem we can also assume without loss of generality that x is recurrent inside $\mathcal{H}(\kappa)$, that is, there exists a strictly increasing sequence of integers $n_1 < n_2 < \cdots$ such that

$$x_k := f^{n_k}(x) \in \mathcal{H}(\kappa) \quad \text{for all } k \in \mathbb{N} \text{ and } x_k \xrightarrow[k\to\infty]{} x.$$

Therefore we can consider the disk $W_k = f^{-n_k}(W^{uu}_{loc}(x_k))$. Observe that $W_k \subset W^{uu}_{loc}(x)$ is a neighborhood of x and, since the sizes of the strong-unstable disks on $\mathcal{H}(\kappa)$ are uniformly bounded, we see that $\operatorname{diam}(W_k) \to 0$ exponentially fast as $k \to +\infty$.

Now $W^{uu}_{loc}(x)$ is one-dimensional in our setting and thus the shrinking of W_k to x together with the f-invariance of A are enough to ensure

$$\frac{\mu_u\big(f^{-n_k}(W^{uu}_{loc}(x_k)\setminus A)\big)}{\mu_u\big(f^{-n_k}(W^{uu}_{loc}(x_k))\big)}=\frac{\mu_u(W_k\setminus A)}{\mu_u(W_k)}\xrightarrow[k\to\infty]{}0.$$

Finally the bounded distortion given by Theorem 2.36 implies that

$$\begin{aligned}\frac{\mu_u\big(f^{-n_k}(W^{uu}_{loc}(x_k)\setminus A)\big)}{\mu_u\big(f^{-n_k}(W^{uu}_{loc}(x_k))\big)}&=\frac{\int_{W^{uu}_{loc}(x_k)\setminus A}|\det Df^{-n_k}\mid E^u(z)|\,d\mu_u(z)}{\int_{W^{uu}_{loc}(x_k)}|\det Df^{-n_k}\mid E^u(z)|\,d\mu_u(z)}\\&\geq\frac{1}{h^u_\kappa}\cdot\frac{\mu_u\big(W^{uu}_{loc}(x_k)\setminus A\big)}{\mu_u(W^{uu}_{loc}(x_k))},\end{aligned}$$

which means that

$$\frac{\mu_u\big(W^{uu}_{loc}(x_k)\setminus A\big)}{\mu_u(W^{uu}_{loc}(x_k))}\leq h_\kappa\cdot\frac{\mu_u(W_k\setminus A)}{\mu_u(W_k)}$$

for all $k\geq 1$. Hence we get $\mu_u(W^{uu}_{loc}(x)\setminus A)=0$ by the choice of x_k and the continuous dependence of the strong-unstable disks on the points of the hyperbolic block $\mathcal{H}(\kappa)$. The argument for the stable direction is the same. Since the points of a full μ_A measure subset have all the properties we used, this concludes the proof of Lemma 8.8 and of the property (8.2).

Dense Invariant Manifolds of a Periodic Orbit Now we use the density of periodic points in A (property (8.1)). Consider again a hyperbolic block $\mathcal{H}(\kappa)$ with a big enough $\kappa\in\mathbb{N}$ such that $\mu_A(\mathcal{H}(\kappa))>0$. For any given $x\in\mathcal{H}(\kappa)$ and $\delta>0$ there exists a hyperbolic periodic orbit $\mathcal{O}(p)$ intersecting $B(x,\delta)$. Because the sizes and angles of the stable and unstable disks of points in $\mathcal{H}(\kappa)$ are uniformly bounded away from zero, we can ensure that we have the following transversal intersections[1]

$$W^u(p)\pitchfork W^s(x)\neq\emptyset\neq W^s(p)\pitchfork W^u(x).$$

This together with the inclination lemma implies that

$$\overline{W^u(p)}=\overline{W^u(x)}\subset A\quad\text{and}\quad\overline{W^s(p)}=\overline{W^s(x)}\subset A.\tag{8.3}$$

Moreover since we can pick any $x\in\mathcal{H}(\kappa)$ we can assume without loss that x has a dense orbit in A (since we took μ_A to be ergodic) and then we can strengthen (8.3) to: there exists a periodic orbit $\mathcal{O}(p)$ inside A such that

$$\overline{W^u(p)}=A\quad\text{and}\quad\overline{W^s(p)}=A.\tag{8.4}$$

[1]Recall the difference between $W^{uu}(p)$ and $W^u(p)$ etc in the flow setting.

Absence of Singularities in A Using property (8.4) we consider, on the one hand, the invariant compact subset of A given by

$$L = \alpha_X(W^{ss}(p)),$$

the closure of the accumulation points of backward orbits of points in the strong-stable manifold of the periodic orbit $\mathcal{O}(p)$. By (8.4) we have $L = A$. On the other hand, considering $N = \omega_X(W^{uu}(p))$ we likewise obtain $N = A$.

Let us assume that σ is a singularity contained in A. By Lemma 8.6, σ is either Lorenz-like for X or Lorenz-like for $-X$.

In the former case, we would get $W^{ss}(\sigma) \subset A$ because any compact part of the strong-stable manifold of σ is accumulated by backward iterates of a small neighborhood γ inside $W^{ss}(x)$. Here we are using the property that the contraction along the strong-stable manifold, which becomes an expansion for negative time, is uniform. In the latter case we would get $W^{uu}(\sigma) \subset A$ by a similar argument reversing the time direction.

We now explain that each one of these possibilities leads to a contradiction to the dominated splitting of the Linear Poincaré Flow on the regular orbits of A, following an argument from Chap. 5. It is enough to deduce a contradiction for a Lorenz-like singularity for X, since the other case reduces to this one through a time inversion.

If $W^{ss}(\sigma) \cap A \setminus \{\sigma\} \supset \{y\}$ for some point $y \in A$ and for some singularity $\sigma \in A$, then we have countably distinct regular orbits of Λ accumulating on $y \in W^{ss}(\sigma)$ (by the definition of A) and on a point $q \in W^u(\sigma)$ (by the dynamics of the flow near σ).

Applying the Connecting Lemma, we obtain a saddle-connection associated to the continuation of σ for a C^1-close vector field Y, which is an "orbit-flip" connection; see Chap. 3.

These connections can be C^1 approximated by "inclination-flip" connections for another C^1 nearby vector field Z, *not necessarily conservative*, as explained in Chap. 3.

However the presence of "inclination-flip" connections is an obstruction to the dominated decomposition of the Linear Poincaré Flow for nearby regular orbits. This contradicts Lemma 2.29 and concludes the proof of Proposition 8.3.

8.1.2 Uniform Hyperbolicity

Here we conclude the proof of Theorem 8.1, showing that proper invariant hyperbolic subsets of a C^{1+} incompressible flow cannot have positive volume.

Proposition 8.9 *Let A be a compact invariant hyperbolic subset for $X \in \mathfrak{X}^{1+}_\mu(M)$, where M is compact manifold with finite dimension. Then either $\mu(A) = 0$ or else X is an Anosov flow and $A = M$.*

The proof of Proposition 8.9 is given as a sequence of intermediate results in the rest of this subsection. Assuming this result we easily have the following.

Proof of Theorem 8.1 From Corollary 8.4 we have the result that a regular invariant subset with positive volume with dominated splitting for the Linear Poincaré Flow admits a positive volume subset which is hyperbolic. Therefore the flow of X is Anosov from Proposition 8.9. □

8.1.2.1 Positive Volume Hyperbolic Sets and Conservative Anosov Flows

For the proof, we recall the notion of partial hyperbolicity from Sect. 5.1.2. We need the following result which will be demonstrated in the next section.

Theorem 8.10 ([8, Theorem 2.2]) *Let $f : M \to M$ be a C^{1+} diffeomorphism and let $\Lambda \subset M$ be a partially hyperbolic set with positive volume, where M is a finite-dimensional compact manifold. Then Λ contains a strong-stable disk.*

Proof of Proposition 8.9 Let A be a hyperbolic subset for $X \in \mathfrak{X}^{1+}_{\mu}(M)$ with $\mu(A) > 0$. Theorem 8.10 applied to $f = X^1$ provides a strong-stable disk γ contained in A. From Lemma 6.20 we see that $L = \alpha(\gamma)$ satisfies $W^{ss}(L) = \{W^{ss}(z) : z \in L\} \subset L$. This implies that $W^s(L) = L$ by invariance.

Consider now $W^u(L) = \{W^u(z) : z \in L = W^s(L)\}$. This collection of unstable leaves crossing the stable leaves of L forms a neighborhood U of L. Hence L is a repeller: for $w \in U$ we have $\operatorname{dist}(X^{-t}(w), L) \xrightarrow[t\to+\infty]{} 0$.

This contradicts the preservation of the volume form μ, unless L is the whole of M. Thus $M = L \subset A$ and X is Anosov. □

8.2 Singular-Hyperbolic Attractors Have Zero Volume

The following result generalizes the results of Bowen-Ruelle [62] which show that a uniformly hyperbolic transitive subset of saddle-type for a C^{1+} flow has zero volume.

Theorem 8.11 *Let $X \in \mathfrak{X}^{1+}(M)$ be a vector field on a 3-dimensional manifold M. Then any proper singular-hyperbolic attractor or repeller for X has zero volume.*

Moreover, we obtain the following dichotomy extending to the continuous time setting a similar result obtained for partially hyperbolic diffeomorphisms in [11].

Theorem 8.12 *Let Λ be a transitive isolated uniformly hyperbolic set of saddle type for $X \in \mathfrak{X}^{1+}(M)$ on a d-dimensional manifold M, for some $d \geq 3$. Then either Λ has zero volume or X is a transitive Anosov vector field.*

Using these results the dichotomy of Theorem 8.12 extends to singular-hyperbolic attractors in three-dimensional manifolds.

Theorem 8.13 *Let Λ be a singular-hyperbolic attractor for $X \in \mathfrak{X}^{1+}(M)$ where M is a 3-manifold. Then either Λ has zero volume or X is a transitive Anosov vector field.*

8.2.1 Partial Hyperbolicity and Zero Volume on C^{1+} Flows

The following result plays a crucial role in the proof of Theorem 8.11. We say that a disk of topological dimension $d \geq 1$ is a *d-disk.*

Theorem 8.14 *Let X be a C^{1+} flow on a d-dimensional manifold with $d \geq 3$ and let Λ be a partially hyperbolic invariant subset such that*

$$\Lambda \cap \gamma \text{ does not contain } d_E\text{-disks for any strong-stable disk } \gamma. \tag{8.5}$$

Then Λ has zero volume.

The proof of the above theorem is a consequence of the following result for compact invariant subsets of C^{1+} diffeomorphisms f with dominated decomposition, whose proof is the content of the rest of this subsection.

Before we state the result let us recall the notion of *dominated splitting for a diffeomorphism f* over a compact f-invariant set Λ, which is very similar to the one given in the definition of (K, λ)-dominated splitting in Chap. 5 since we need only replace the relation (5.1) by

$$\|Df \mid E_x\| \cdot \|Df^{-1} \mid F_x\| < \lambda \tag{8.6}$$

for all $x \in \Lambda$. Analogously *partial hyperbolicity for a diffeomorphism f* is given by a dominated decomposition $E \oplus F$ over a compact invariant subset Λ with uniform contraction along the direction E.

Theorem 8.15 *Let $f : M \to M$ be a C^{1+} diffeomorphism and let $\Lambda \subset M$ be a partially hyperbolic set with positive volume. Then Λ contains a strong-stable disk.*

Let us now prove Theorem 8.14 using Theorem 8.15. Let Λ be a partially hyperbolic compact invariant set for a flow $X \in \mathfrak{X}^{1+}(M)$ where M is a d-manifold with $d \geq 3$. Assume that condition (8.5) is satisfied by Λ.

Arguing by contradiction, if $m(\Lambda) > 0$ then setting $f = X_1$, the time-one diffeomorphism induced by the vector field X, we see that Λ is a partially hyperbolic set for f with positive volume. Hence, by Theorem 8.15, there exists some strong-stable disk γ for f contained in Λ with dimension d_E, which is a strong-stable disk for the flow X^t and contradicts property (8.5). This contradiction shows that $m(\Lambda) = 0$ and proves Theorem 8.14.

8.2.1.1 Pre-balls and Bounded Distortion

Here we give some preparatory results for the proof of Theorem 8.15. We fix continuous extensions of the two bundles E and F to some neighborhood U of Λ, which we denote by $\tilde{E}$ and $\tilde{F}$. We do not require these extensions to be invariant under Df. Given $0 < a < 1$, we define the *center-unstable cone field* $(C_a^F(x))_{x \in U}$ *of width* a by

$$C_a^F(x) = \{v_1 + v_2 \in \tilde{E}_x \oplus \tilde{F}_x \text{ such that } \|v_1\| \le a \cdot \|v_2\|\}. \tag{8.7}$$

We define the *stable cone field* $(C_a^E(x))_{x \in U}$ *of width* a in a similar way, just reversing the roles of the bundles in (8.7). We fix $a > 0$ and U small enough so that, up to slightly increasing $\lambda < 1$, the domination condition (8.6) remains valid for any pair of vectors in the two cone fields:

$$\|Df(x)u\| \cdot \|Df^{-1}(f(x))v\| \le \lambda \cdot \|u\| \cdot \|v\| \tag{8.8}$$

for every $u \in C_a^E(x)$, $v \in C_a^F(f(x))$, and any point $x \in U \cap f^{-1}(U)$. Note that the unstable cone field is positively invariant:

$$Df(x)C_a^F(x) \subset C_a^F(f(x)),$$

whenever $x, f(x) \in U$. Indeed, the domination (8.8) together with the invariance of $F = \tilde{F} \mid \Lambda$ imply that

$$Df(x)C_a^F(x) \subset C_{\lambda a}^F(f(x)) \subset C_a^F(f(x)),$$

for every $x \in \Lambda$. This extends to any $x \in U \cap f^{-1}(U)$ by continuity. Analogously the stable cone field is negatively invariant:

$$Df^{-1}(x)C_a^E(x) \subset C_a^E(f^{-1}(x)),$$

whenever $x, f(x) \in U$.

If $a > 0$ is taken sufficiently small in the definition of the cone fields, and we choose $\delta_1 > 0$ also so small that the δ_1-neighborhood of Λ is contained in U, then by uniform continuity of Df we have $\|Df \mid \tilde{E}_y\| < \lambda^{-1/2} \cdot \|Df \mid E_x\|$ and

$$\|Df(y)u\| \le \lambda^{-1/2} \cdot \left\|Df \mid E_x\right\| \cdot \|u\| \le \lambda^{1/2}\|u\|, \tag{8.9}$$

whenever $x \in \Lambda$, $\operatorname{dist}(x, y) \le \delta_1$, and $u \in C_a^E(y)$.

We say that an embedded C^1 submanifold $N \subset U$ is *tangent to the stable cone field* if the tangent subspace to N at each point $x \in N$ is contained in $C_a^E(x)$. Then, by the domination property (8.6), $f^{-1}(N)$ is also tangent to the stable cone field, if it is contained in U. In particular, if $N, f^{-1}(N), \dots, f^{-k}(N) \subset U$, then $Df^k \mid T_{f^{-k}(x)}N$ is a $\lambda^{k/2}$-contraction according to (8.9), since $\|Df \mid E_x\| < \lambda$ by partial hyperbolicity. Thus, denoting by dist_N the *distance along* N given by the length of the shortest smooth curve connecting two given points inside N, we obtain

Lemma 8.16 *Let $\Delta \subset U$ be a C^1 disk of radius $\delta < \delta_1$ tangent to the stable cone field. Then there exists $n_0 \geq 1$ such that, for $n \geq n_0$ and $x \in \Delta$ with $\mathrm{dist}_\Delta(x, \partial\Delta) \geq \delta/2$, there is a neighborhood V_n of x in Δ such that f^{-n} maps V_n diffeomorphically onto a disk of radius δ_1 around $f^{-n}(x)$. Moreover,*

$$\mathrm{dist}_{f^{-n+k}(V_n)}(f^{-n+k}(y), f^{-n+k}(z)) \leq \lambda^{k/2} \cdot \mathrm{dist}_{f^{-n}(V_n)}(f^{-n}(y), f^{-n}(z))$$

for every $1 \leq k \leq n$ and every $y, z \in V_n$.

We shall sometimes refer to the sets V_n as *pre-balls*. The next corollary is a consequence of the contraction given by the previous lemma, together with some Hölder control of the tangent direction which can be found in [9, Corollary 2.4, Proposition 2.8].

Corollary 8.17 *There exists $C > 1$ such that, given Δ as in Lemma 8.16 and given any pre-ball $V_n \subset \Delta$ with $n \geq n_0$, then for all $y, z \in V_n$*

$$\frac{1}{C} \leq \frac{|\det Df^{-n} \mid T_y\Delta|}{|\det Df^{-n} \mid T_z\Delta|} \leq C.$$

8.2.1.2 A Local Unstable Disk Inside Λ

Assuming that Λ has positive volume and given an embedded disk Δ in M, denote by m_Δ the measure naturally induced by the volume form m on Δ. Choosing an m density point of Λ, we laminate a neighborhood of that point into disks tangent to the stable cone field. Since the relative Lebesgue measure of the intersections of these disks with Λ cannot be all equal to zero, we obtain some disk Δ intersecting Λ in a positive m_Δ subset. Hence, in the setting of Theorem 8.15, we assure that there is a disk Δ tangent to the stable cone field intersecting Λ in a positive m_Δ subset. Let $H = \Delta \cap \Lambda$.

Lemma 8.18 *There exist an infinite sequence of integers $1 \leq k_1 < k_2 < \cdots$ and, for each $n \in \mathbb{N}$, a disk Δ_n of radius $\delta_1/4$ tangent to the stable cone field such that the relative Lebesgue measure of $f^{-k_n}(H)$ in Δ_n converges to 1 as $n \to \infty$.*

Proof Let $\varepsilon > 0$ be some small number. By regularity of m_Δ, let K be a compact subset of H and let A be an open neighborhood of H in Δ such that $m_\Delta(A \setminus K) < \varepsilon \cdot m_\Delta(H)/(1+\varepsilon)$. Then

$$m_\Delta(A \setminus K) < \varepsilon \cdot m_\Delta(K).$$

Choose n sufficiently large so that $V_x \subset A$ for each $x \in K$, where V_x is the pre-ball associated to n. This pre-ball is mapped diffeomorphically by f^{-n} onto a ball $B_{\delta_1}(f^{-n}(x))$ of radius δ_1 around $f^{-n}(x)$ tangent to the stable cone field. Let

$W_x \subset V_x$ be the pre-image of the ball $B_{\delta_1/4}(f^{-n}(x))$ of radius $\delta_1/4$ under this diffeomorphism. By compactness we have

$$K \subset W_{x_1} \cup \cdots \cup W_{x_m},$$

for some $x_1, \ldots, x_m \in K$. Let I be a maximal subset of $\{1, \ldots, m\}$ such that $W_{x_i} \cap W_{x_j} = \emptyset$ for $i, j \in I$ with $i \neq j$. By maximality, each W_{x_j}, $1 \leq j \leq m$, intersects some W_{x_i} with $i \in I$, and thus $f^{-n}(W_{x_j}) = B_{\delta_1/4}(f^{-n}(x_j))$ intersects $B_{\delta_1/4}(f^{-n}(x_i)) = f^{-n}(W_{x_i})$. Hence $f^{-n}(W_{x_j}) \subset B_{\delta_1}(f^{-n}(x_i)) = f^{-n}(V_{x_i})$, that is, $W_{x_j} \subset V_{x_i}$. Hence $\{V_{x_i}\}_{i \in I}$ is a covering of K. Denoting by D the disk $f^{-n}\Delta$ tangent to the stable cone field, by Corollary 8.17 we have

$$\frac{m_D(f^{-n}(W_{x_i}))}{m_D(f^{-n}(V_{x_i}))} = \frac{\int_{W_{x_i}} |\det Df^{-n} \mid T_y\Delta| \, dm_\Delta}{\int_{V_{x_i}} |\det Df^{-n} \mid T_y\Delta| \, dm_\Delta} \leq C \frac{m_\Delta(W_{x_i})}{m_\Delta(V_{x_i})}$$

and, since the area of any $\delta_1/4$-ball is comparable to the area of a δ_1-ball centered at the same point on a given disk (recall that lengths are measured *along* the disk with respect to the induced metric), there is a uniform constant $\theta > 0$ such that

$$\frac{m_\Delta(W_{x_i})}{m_\Delta(V_{x_i})} \geq \theta, \quad \text{for every } i \in I.$$

Hence

$$m_\Delta\big(\cup_{i \in I} W_{x_i}\big) = \sum_{i \in I} m_\Delta(W_{x_i}) \geq \sum_{i \subset I} \theta m_\Delta(V_{x_i}) \geq \theta m_\Delta(K).$$

Setting

$$\rho = \min\left\{\frac{m_\Delta(W_{x_i} \setminus K)}{m_\Delta(W_{x_i})} : i \in I\right\},$$

we have

$$\begin{aligned}
\varepsilon m_\Delta(K) &\geq m_\Delta(A \setminus K) \\
&\geq m_\Delta\big(\cup_{i \in I} W_{x_i} \setminus K\big) \\
&\geq \rho m_\Delta\big(\cup_{i \in I} W_{x_i}\big) \\
&\geq \rho\theta m_\Delta(K).
\end{aligned}$$

This implies that $\rho \leq \varepsilon/\theta$. Since $\varepsilon > 0$ can be taken arbitrarily small, by increasing n we may take W_{x_i} such that the relative Lebesgue measure of K in W_{x_i} is arbitrarily close to 1. Then, by the bounded distortion provided by Corollary 8.17, the relative Lebesgue measure of $f^{-n}(H) \supset f^{-n}(K)$ in $f^{-n}(W_{x_i})$, which is a disk of radius $\delta_1/4$ around $f^{-n}(x_i)$ tangent to unstable cone field, can be made arbitrarily close to 1. □

Let us now prove that there is a strong-stable disk of radius $\delta_1/4$ inside Λ. Let $(\Delta_n)_n$ be the sequence of disks given by Lemma 8.18, and consider $(x_n)_n$ the sequence of points at which these disks are centered. Up to taking subsequences, we may assume that the centers of the disks converge to some point x. By Ascoli-Arzela, these disks converge to some disk Δ_∞ centered at x. By construction, every point in Δ_∞ is accumulated by some iterate of a point in $H \subset \Lambda$, and so $\Delta_\infty \subset \Lambda$.

Note that each Δ_n is contained in the k_n-iterate of Δ, which is a disk tangent to the stable cone field. The domination property implies that the angle between Δ_n and E goes to zero as $n \to \infty$, uniformly on Λ. In particular, Δ_∞ is tangent to E at every point in $\Delta_\infty \subset \Lambda$. By Lemma 8.16, given any $k \geq 1$, then f^k is a $\sigma^{k/2}$-contraction on Δ_n for every large n. Passing to the limit, we find that f^k is a $\sigma^{k/2}$-contraction on Δ_∞ for every $k \geq 1$. In particular, we have shown that the subspace E_x is uniformly contracting for Df for $x \in \Delta_\infty$. The fact that $T_\Lambda M = E \oplus F$ is a dominated splitting implies that any contraction which Df may exhibit along the complementary direction F_x is weaker than the contraction in the E_x direction whenever $x \in \Delta_\infty$. Then, by [196], there exists a unique strong-stable manifold $W^{ss}_{loc}(x)$ tangent to E which is contracted by the positive iterates of f. Since Δ_∞ is contracted by every f^k, and all its positive iterates are tangent to stable cone field, then Δ_∞ is contained in $W^{ss}_{loc}(x)$.

This completes the proof of Theorem 8.15.

8.2.2 *Positive Volume Versus Transitive Anosov Flows*

Now we prove Theorems 8.11, 8.12 and 8.13.

8.2.2.1 Positive Volume Transitive Hyperbolic Sets and Anosov Flows

We start by proving Theorem 8.12.

Lemma 8.19 *Let Λ be a transitive uniformly hyperbolic set for $X \in \mathfrak{X}^{1+}(M)$ such that $m(\Lambda) > 0$, where M is a d-manifold, for some $d \geq 3$. If there exists a point $x \in \Lambda$ in the interior of $W^{ss}_{loc}(x) \cap \Lambda$, then $\Lambda \supset W^{ss}(y)$ for all $y \in \Lambda$. Moreover, the set $W^u(\Lambda)$ formed by the union of all unstable manifolds through points of Λ is an open neighborhood of Λ.*

Here the interior of $W^{ss}_{loc}(x) \cap \Lambda$ is taken with respect to the topology of the disk $W^{ss}_{loc}(x)$. The proof follows [174, Lemma 2.16] and [21, Lemma 2.8] closely; see also the proof of Lemma 6.19.

Proof Let $x \in \Lambda$ be such that x is in the interior of $W^{ss}(x) \cap \Lambda$. Let $\alpha(x) \subset \Lambda$ be its α-limit set. Then Lemma 6.20 ensures that $\alpha(x)$ contains all strong-stable manifolds through its points.

Clearly the invariant set $\alpha(x) \subset \Lambda$ is uniformly hyperbolic.

It also follows from the previous remarks that the union

$$S = \bigcup_{y \in \alpha(x)} W^{ss}(y) \quad \text{or} \quad S = W^{ss}(\alpha(x))$$

of the strong-stable manifolds through the points of $\alpha(x)$ is contained in Λ. By continuity of the strong-stable manifolds and the fact that $\alpha(x)$ is a closed set, we find that S is also closed. Again S is a uniformly hyperbolic set.

We claim that $W^u(S)$, the union of the unstable manifolds of the points of S, is an open set. To prove this, we note that S contains the whole stable manifold $W^s(z)$ of every $z \in S$: this is because S is invariant and contains the strong-stable manifold of z. Now the union of the strong-unstable manifolds through the points of $W^s(z)$ contains a neighborhood of z. This proves that $W^u(S)$ is a neighborhood of S. Thus the backward orbit of any point in $W^u(S)$ must enter the interior of $W^u(S)$. Since the interior is, clearly, an invariant set, this proves that $W^u(S)$ is open, as claimed.

Finally, consider any backward dense orbit in Λ of a point that we call w. On the one hand $\alpha(w) = \Lambda$. On the other hand, $X_{-t}(w)$ must belong to $W^u(S)$ for some $t > 0$, and so $\alpha(w) \subset S$ by invariance. This implies that $\Lambda \subset S$ and, since $S \subset \Lambda$ by construction, we see that $\Lambda = S$. □

Proof of Theorem 8.12 Assume that $m(\Lambda) > 0$. Then Theorem 8.15, applied to the map $f = X_1$ and to the set Λ with dominated decomposition given by the splitting

$$E \oplus \left(E^X \oplus F\right),$$

ensures that there exists a strong-stable disk γ contained in Λ. Analogously, applying Theorem 8.15 to $f = X_{-1}$ and to the set Λ with dominated decomposition given by the splitting

$$\left(E \oplus E^X\right) \oplus F,$$

we get a strong-unstable disk δ contained in Λ.

Now the existence of γ enables us to use Lemma 8.19 and deduce that Λ contains the strong-stable manifolds of each of its points and that $W^u(\Lambda)$ is an open neighborhood of Λ. In the same way, using Lemma 8.19 for the flow generated by $-X$, from the existence of δ we deduce that Λ contains the strong-unstable manifolds of all of its points, that is, $W^u(\Lambda) \subset \Lambda$.

But since $W^u(\Lambda)$ is an open neighborhood of Λ, we conclude that Λ is simultaneously open and closed in M. Hence $\Lambda = M$ by connectedness. This shows that the whole of M is a transitive uniformly hyperbolic set for X and completes the proof of Theorem 8.12. □

8.2.2.2 Positive Volume Singular-Hyperbolic Sets and Anosov Flows

Now we prove Theorems 8.11 and 8.13. To do this we need some preliminary results which show in particular that transitive singular-hyperbolic sets satisfy condition (8.5).

In what follows X is a vector field in $\mathfrak{X}^{1+}(M)$ and M is a 3-manifold.

Lemma 8.20 *Let Λ be a transitive partially hyperbolic invariant set for X with volume expanding central direction. Then*

- *either $W^{ss}(x) \cap \Lambda$ contains no strong stable disks for all $x \in \Lambda$,*
- *or Λ is a uniformly hyperbolic set (and in particular Λ does not contain singularities).*

Proof Let us suppose that there exists $x \in \Lambda$ such that x is in the interior of $W^{ss}(x) \cap \Lambda$. Then Lemma 6.20 ensures that $\alpha(x)$ contains the stable manifold through all its points. It follows that $\alpha(x)$ does not contain any singularity, according to Theorem 6.3, and from Theorem 2.27 the set $\alpha(x)$ is uniformly hyperbolic.

As in the proof of Lemma 8.19 we have

$$S = W^{ss}(\alpha(x)) \subset \Lambda$$

and S is closed. Again we see that S does not contain any singularity σ, for otherwise we would have $W^{ss}(\sigma) \supset W^{ss}(z)$ for some $z \in \alpha(x)$ which would contradict Theorem 6.3 since $W^{ss}(z) \subset \Lambda$. Thus S is a uniformly hyperbolic set.

Then $W^u(S)$ is also an open set as in the proof of Lemma 8.19. Since we are assuming that Λ is transitive, again by the same arguments as in the proof of Lemma 8.19, we have $\Lambda = S$. This shows that Λ is uniformly hyperbolic and in particular it does not contain any singularity of X. □

Proof of Theorem 8.11 Let Λ be a proper transitive singular-hyperbolic set for X. Assume first that Λ does not contain singularities. Then Λ is a proper uniformly hyperbolic subset of saddle type and by Theorem 8.12 Λ has zero volume.

Now if Λ contains singularities, by Lemma 8.20 we know that Λ satisfies property (8.5) in the statement of Theorem 8.14. Hence Λ has zero volume. □

We can obtain a stronger conclusion if we further assume that Λ is an attractor.

Lemma 8.21 *Let Λ be a singular-hyperbolic attractor for X. Then*

- *either $W^{ss}(x) \cap \Lambda$ contains no strong stable disks for all $x \in \Lambda$,*
- *or $\Lambda = M$ is a uniformly hyperbolic set (and consequently X is a transitive Anosov vector field).*

Proof Assume that there exists $x \in \Lambda$ such that x is in the interior of $W^{ss}(x) \cap \Lambda$. From Lemma 8.20 we know that there exists a uniformly hyperbolic set $S \subset \Lambda$ such that $W^u(S)$ is an open neighborhood of S. Moreover we also have $\Lambda = S$.

However, if Λ is an attractor, then $W^u(S) \subset \Lambda$ and so $\Lambda = W^u(S)$. Hence Λ is closed and also open. The connectedness of M implies that $\Lambda = S = M$. In particular X has no singularities and the whole of M admits a uniformly hyperbolic structure with a dense orbit, and thus X is a transitive Anosov vector field. □

Proof of Theorem 8.13 Let Λ be a singular-hyperbolic attractor for a C^{1+} vector field X on a 3-manifold. If $m(\Lambda) > 0$ then according to Theorem 8.15 we find that there exists some strong-stable disk γ contained in Λ.

Hence, since Λ does not satisfy the first alternative of Lemma 8.21, we conclude that $\Lambda = M$ and so X is an Anosov vector field. This concludes the proof of Theorem 8.13. □

8.2.3 Zero-Volume for C^1 Generic Singular-Hyperbolic Attractors

It is possible to extend the previous results on zero volume for singular-hyperbolic attractors for C^{1+} flows to some locally generic subsets in the C^1 topology. Recall that a subset $\mathcal{G}$ of a set $\mathcal{U}$ is *generic* if it may be written as a countable intersection of open and dense subsets of $\mathcal{U}$. Since $\mathfrak{X}^1(M)$ is a Baire space, generic subsets of a given open subset $\mathcal{U}$ of $\mathfrak{X}^1(M)$ are dense in $\mathcal{U}$.

Concrete examples of such open sets on 3-manifolds are given by robust singular-hyperbolic attractors, which comprise the Lorenz attractor, the geometric Lorenz attractors, attractors arising from certain resonant double homoclinic loops [176] or from certain singular cycles [171], and certain models across the boundary of uniform hyperbolicity [173]; see Chap. 10.

Theorem 8.22 *Let Λ be a robust attractor for $X \in \mathfrak{X}^1(M)$ on a 3-manifold M with isolating neighborhood U. Then there is a C^1-neighborhood $\mathcal{U}$ of X and a C^1 generic set $\mathcal{G} \subset \mathcal{U}$ such that $\Lambda_Y(U)$ has volume zero for all $Y \in \mathcal{G}$.*

The proof of this is a consequence of the following.

Theorem 8.23 *Let Λ be an isolated partially hyperbolic set satisfying condition* (8.5) *for $X \in \mathfrak{X}^1(M)$ on a d-dimensional manifold M with $d \geq 3$. Given an isolating neighborhood U of Λ, let $\mathcal{U} \subset \mathfrak{X}^1(M)$ be such that $\Lambda_Y(U)$ is partially hyperbolic and also satisfies condition* (8.5) *for all $Y \in \mathcal{U}$.*

If $\mathcal{U}$ is C^1-open, then there exists a generic set $\mathcal{G} \subset \mathcal{U}$ such that $\Lambda_Y(U)$ has volume zero for all $Y \in \mathcal{G}$.

Proof Let Λ be an isolated partially hyperbolic invariant compact subset for a flow $X \in \mathfrak{X}^r(M)$, for some $r \geq 1$, such that Λ satisfies condition (8.5). We always write U for the isolating neighborhood of Λ.

We consider the sets

- $\mathcal{U} = \{Y \in \mathfrak{X}^1(M) : \Lambda_Y(U)$ is partially hyperbolic satisfying (8.5)$\}$ which we assume is a C^1 open subset of $\mathfrak{X}^1(M)$;
- $\mathcal{V} = \{Y \in \mathfrak{X}^2(M) : \Lambda_Y(U)$ is partially hyperbolic satisfying (8.5)$\}$;
- $\mathcal{U}_\varepsilon = \{Y \in \mathcal{U} : m(\Lambda_Y(U)) < \varepsilon\}$.

Since every C^1 flow X is arbitrarily close to some C^2 flow Y in the C^1 topology (see e.g. [190]) and we are assuming that $\mathcal{U}$ is C^1-open, we conclude that $\mathcal{V}$ is dense in $\mathcal{U}$ in the C^1 topology.

We claim that $\mathcal{U}_\varepsilon$ is open and dense in $\mathcal{U}$ in the C^1 topology. After proving this claim the proof of Theorem 8.23 finishes by setting $\mathcal{G} = \cap_{n\geq 1}\mathcal{U}_{1/n}$. In what follows we prove this claim.

Let $Y \in \mathcal{U}_\varepsilon$ be given. Then for every sufficiently large fixed $T > 0$ we set

$$\Lambda_Y^T = \cap_{t=-T}^T \overline{Y_t(U)} \supseteq \Lambda_Y(U) \quad \text{and} \quad \varepsilon_1 = \varepsilon - m\big(\Lambda_Y^T\big) > 0.$$

There exists $\delta > 0$ such that

$$m\Big(B\big(\Lambda_Y^T, \delta\big) \setminus \Lambda_Y^T\Big) < \frac{\varepsilon_1}{2}.$$

Let $B^{C_1}(Y, \zeta)$ denote the C^1-neighborhood of radius ζ around Y. Using the fact that T is finite, by continuity and compactness we find $\zeta > 0$ such that

$$Z \in B^{C_1}(Y, \zeta) \quad \Rightarrow \quad \Lambda_Z^T \subset B\big(\Lambda_Y^T, \delta\big).$$

We have

$$m\big(\Lambda_Z^T\big) \leq m\Big(B\big(\Lambda_Y^T, \delta\big)\Big) \leq m\big(\Lambda_Y^T\big) + \frac{\varepsilon_1}{2} < \varepsilon,$$

for all $Z \in \mathcal{U} \cap B^{C_1}(Y, \zeta)$. Since $\Lambda_Z(U) \subset \Lambda_Z^T$, we conclude that $m(\Lambda_Z(U)) < \varepsilon$, for all $Z \in \mathcal{U} \cap B^{C_1}(Y, \zeta)$.

This proves that $\mathcal{U}_\varepsilon$ is C^1-open. To prove that $\mathcal{U}_\varepsilon$ is C^1-dense in $\mathcal{U}$, just observe that $\mathcal{V} \cap \mathcal{U} \subset \mathcal{U}_\varepsilon$ by Theorem 8.14. Since $\mathcal{V}$ is C^1-dense in $\mathcal{U}$ this concludes the proof of the claim and ends the proof of Theorem 8.23. □

Proof of Theorem 8.22 This is a consequence of Theorem 8.23 since a C^1 robust attractor for 3-flows is a singular-hyperbolic attractor as shown in Chap. 5. □

8.2.4 Extension to Sectionally Expanding Attractors in Higher Dimensions

The class of sectionally expanding attractors was introduced, by Metzger and Morales in [156], to extend the singular-hyperbolic class of attractors in three-dimensional manifold to higher dimensions; see Sect. 5.2. The same results proved in this subsection are also true for this class of attractors for higher dimensional flows.

Theorem 8.24 *Let Λ be a sectionally expanding attractor for a C^{1+} flow X^t on a d-dimensional manifold, where $d \geq 4$. Then either Λ has zero volume, or $\Lambda = M$ and the flow is Anosov.*

Indeed, a sectionally expanding attractor *without singularities* is a hyperbolic attractor from Theorem 2.27, because of the following.

Lemma 8.25 *The Linear Poincaré Flow over Λ admits a hyperbolic splitting.*

Proof Indeed, if $E^s \oplus E^c$ is the splitting of TM over Λ, then the projections $N^s := \mathcal{O} \cdot E^s$ and $N^u := \mathcal{O} \cdot E^c$ are P^t-invariant, and N^s is uniformly contracted by P^t by the same arguments as in the proof of Theorem 2.27. The expansion along N^u is obtained as follows.

Let v be a unit vector on N^u_x for some fixed $x \in \Lambda$ and let F_x be the subspace spanned by v and $X(x)$. We recall that $X \neq 0$ over Λ by assumption and for some $K > 0$ we have $K^{-1} \le \|X(x)\| \le K$ for all $x \in \Lambda$ by compactness. Let us fix $t > 0$ and consider the basis $\{\frac{X(x)}{\|X(x)\|}, v\}$ of F_x. We note that $DX^t(F_x)$ is a bidimensional subspace F^t_x of $E^c_{X^t(x)}$, where we take the basis $\{\frac{X(X^t(x))}{\|X(X^t(x)x)\|}, w\}$, with

$$w := \frac{\mathcal{O}_{X^t(x)} \cdot DX^t_x(v)}{\|\mathcal{O}_{X^t(x)} \cdot DX^t_x(v)\|} \quad \text{belonging to } N^u_{X^t(x)}.$$

The same argument of Proposition 6.15 applies here: with respect to these othonormal bases we have

$$DX^t \mid F_x = \begin{bmatrix} \frac{\|X(X^t(x))\|}{\|X(x)\|} & \star \\ 0 & \Delta \end{bmatrix},$$

because the flow direction is invariant. Since we can assume without loss of generality that $\|\mathcal{O}\| \le K$ over Λ (taking a bigger value for K if needed), we have

$$\frac{\|X(R(x))\|}{\|X(x)\|}\Delta \ge \frac{1}{K} \cdot \det\left(DX^{t(x)} \mid E^{cu}_x\right)$$

and

$$\begin{aligned}\|P^t_x \cdot v\| &= \|\mathcal{O}_{X^t(x)} \cdot DX^t_x(v)\| = \|\Delta \cdot w\| = |\Delta| \\ &\ge K^{-3}|\det(DX^t \mid F_x)| \ge K^{-3}C\lambda^t.\end{aligned}$$

This proves that N^u is uniformly expanded by the Linear Poincaré Flow. □

Next we argue that sectionally expanding attractors with positive volume for C^{1+} flows cannot have equilibria. Thus from the previous arguments these are hyperbolic attractors with positive volume. Then we deduce Theorem 8.24 using Theorem 8.12.

The possible singularities belonging to a sectionally expanding attractor are generalized Lorenz-like by Remark 5.33.

Assuming that a sectionally expanding attractor Λ has singularities, then they are Lorenz-like and their strong-stable manifolds do not intersect Λ except at the singularities, again by Remark 5.33.

Now we are in exactly the same situation as at the beginning of Sect. 8.2.2 to show that Λ cannot contain strong-stable disks *if Λ contains singularities*. However Λ is a partially hyperbolic set so, if Λ has positive volume, then by Theorem 8.15 Λ must contain a strong-stable disk. Thus Λ cannot contain singularities, as we needed. The proof of Theorem 8.24 is complete.

Chapter 9
Global Dynamics of Generic 3-Flows

Vítor Araújo, Mário Bessa, and Maria José Pacifico

The results in Chap. 5 form the basis of a theory of flows on three-dimensional manifolds and paved the way for a global understanding of the dynamics of C^1 generic flows in dimension 3. Here we present some results from the generic viewpoint, either for C^1 flows on 3-manifolds, or for C^1 *conservative* flows on 3-manifolds. This means that we present some properties satisfied by a generic subset of all vector fields in compact 3-manifolds in the C^1 topology. We recall that a *generic* subset of a topological space is a countable intersection of open and dense subsets. Since $\mathfrak{X}^1(M)$, endowed with the C^1 topology, is a Baire space, then every generic subset in dense. The importance of generic properties stems from the fact that the intersection of any countable number of generic subsets is itself a generic subset, so that we can always add the properties we know to be generic. In this way we obtain topologically large families of vector fields with fairly strong dynamical properties.

The choice of the C^1 topology is a consequence of the present development of deep perturbation tools, such as the Closing and Connecting Lemmas, which are only available in the C^1 topology. However, dealing with the C^1 topology has serious disadvantages. This topology does not have a strong physical meaning when dealing with solutions of differential equations, since perturbations of solutions often arise naturally in higher topologies (C^r with $r > 1$) by dealing with higher order derivatives. Moreover, some results that can be proved using the C^1 topology, such as the dichotomy we present in what follows for C^1 generic conservative vector fields in 3-manifolds, are simply not true in higher topologies, since the "KAM Theory" (see e.g. [26, 248]) directly contradicts this dichotomy; see below.

We first extend a classification result from hyperbolic dynamics to singular-hyperbolic attracting sets. The *Spectral Decomposition Theorem for hyperbolic systems* plays a central role. It ensures that an attracting hyperbolic set having dense periodic orbits must be a finite disjoint union of homoclinic classes. Here we provide a version of this result in the setting of *singular-hyperbolic systems*, presented in Sect. 5, following [170].

Theorem 9.1 *An attracting singular-hyperbolic set with dense periodic orbits and a unique singularity is a finite union of transitive sets. Moreover, either the union is*

V. Araújo, M.J. Pacifico, *Three-Dimensional Flows*, Ergebnisse der Mathematik und ihrer Grenzgebiete. 3. Folge / A Series of Modern Surveys in Mathematics 53,
DOI 10.1007/978-3-642-11414-4_9, © Springer-Verlag Berlin Heidelberg 2010

disjoint or the set contains finitely many distinct homoclinic classes. For C^r-generic flows, $r \geq 1$, the union is in fact disjoint.

We also show that a generic C^1 vector field on a closed 3-manifold either has infinitely many sinks or sources, or else is singular Axiom A without cycles. These results are contained in [168].

Theorem 9.2 *A generic vector field $X \in \mathfrak{X}^1(M)$ satisfies (only) one of the following properties:*

1. *X has infinitely many sinks or sources.*
2. *X is singular Axiom A without cycles.*

A *Singular Axiom A* vector field is such that the non-wandering set of the vector field has a decomposition into finitely many compact invariant sets $\Omega(X) = \Omega_1 \cup \cdots \cup \Omega_k$, each one being either a (uniformly) hyperbolic basic set (i.e. transitive, isolated and with a dense subset of periodic orbits) or a singular-hyperbolic attractor, or a singular-hyperbolic repeller with dense subset of periodic orbits (these are defined in Chap. 5—we note that, in this decomposition, the singular-hyperbolic sets are transitive by definition).

An analogous result was proved by Mañé in [145] for C^1 generic diffeomorphisms on surfaces.

Conservative flows are a traditional object of study from Classical Mechanics, see e.g. [26]. These flows preserve a volume form on the ambient manifold and thus come equipped with a natural invariant measure. On compact manifolds this provides an invariant probability giving positive measure (volume) to all open subsets. Therefore for vector fields in this class we have $\Omega(X) = M$ by the Recurrence Theorem. In particular such flows cannot have Lyapunov stable sets, either for X or for $-X$.

The device of Poincaré sections has been used extensively in the previous chapters to reduce several problems, arising naturally in the setting of flows, to lower dimensional questions about the behavior of a transformation. In the opposite direction, recent breakthroughs in the understanding of generic volume preserving diffeomorphisms on surfaces have non-trivial consequences for the dynamics of generic conservative flows on three-dimensional manifolds.

The Bochi-Mañé Theorem [51] asserts that, for a C^1 residual subset of area preserving diffeomorphisms, either the transformation is Anosov, or the Lyapunov exponents are zero Lebesgue almost everywhere. This was announced by Mañé in [146] but only a sketch of a proof was available [149]. The complete proof, presented by Bochi, admits extensions to higher dimensions, obtained by Bochi and Viana in [53], stating in particular that either the Lyapunov exponents of a C^1 generic conservative diffeomorphism are zero Lebesgue almost everywhere, or the system admits a dominated splitting for the tangent bundle dynamics. A survey of this theory can be found in [52].

Using an important result of Zuppa in [277] proving that C^∞ *conservative* flows are C^1 dense among *conservative flows* (which was recently generalized by Arbieto-

Matheus in [22] providing the C^1 denseness of Hölder-C^1 conservative flows), together with the zero volume results above and several delicate perturbation techniques adapted from the work of Bochi [51], we follow [19, 39] in Sect. 9.3 to prove the *Bochi-Mañé* dichotomy for C^1 conservative flows on 3-manifolds: C^1 *generically either the flow is Anosov or else the Lyapunov exponents are zero Lebesgue almost everywhere*.

The presence of singularities imposes some differences between the discrete and continuous systems. More precisely, let $\mathfrak{X}^r_\mu(M)$ be the space of C^r vector fields defining flows which preserve the volume form μ on M, for any $r \geq 1$, and let $\mathfrak{X}^r_\mu(M)^*$ be the subset of $\mathfrak{X}^r_\mu(M)$ of C^r flows with zero divergence but *without singularities*.

Theorem 9.3 *There exists a residual set $\mathscr{R} \subset \mathfrak{X}^1_\mu(M)^*$ such that, for $X \in \mathscr{R}$, either X is Anosov or else for Lebesgue almost every $p \in M$ all the Lyapunov exponents of X^t are zero.*

Developing the ideas of the proof of this result, the following statement on denseness of dominated splitting, now admitting singularities, was also obtained in the same work by Bessa [41].

Recall the definition of Linear Poincaré Flow in Sect. 2.6. Given an invariant subset Λ for $X \in \mathfrak{X}^1(M)$, an invariant splitting $N^1 \oplus N^2$ of the normal bundle N_Λ for the Linear Poincaré Flow P^t is said to be *n-dominated* if there exists an integer n such that we have the domination relation

$$\frac{\| P^n \mid N^1(p) \|}{\| P^n \mid N^2(p) \|} \leq \frac{1}{2},$$

for *every* $p \in \Lambda$.

Theorem 9.4 *There exists a dense set $\mathfrak{D} \subset \mathfrak{X}^1_\omega(M)$ such that, for $X \in \mathfrak{D}$, there exist invariant subsets D and Z whose union has full measure, such that*

- *for $p \in Z$ the flow has only zero Lyapunov exponents;*
- *D is a countable increasing union Λ_n of compact invariant sets admitting an n-dominated splitting for the Linear Poincaré Flow.*

We prove these two results in Sect. 9.3. Now from Theorem 8.1 already proved in Chap. 8, we observe that C^1 generically the subset D in the second possibility of the statement of Theorem 9.4 has positive volume if, and only if, the flow is Anosov. Hence we can extend Theorem 9.3 to flows with singularities.

Theorem 9.5 *There exists a generic subset $\mathcal{R} \subset \mathfrak{X}^1_\mu(M)$ such that for $X \in \mathcal{R}$*

- *either X is Anosov;*
- *or else for Lebesgue almost every $p \in M$ all the Lyapunov exponents of X^t are zero.*

We remark the Kolmogorov-Arnold-Moser Theorem ensures the persistence of invariant circles with irrational rotations near an *elliptic fixed point* (the eigenvalues of the derivative of the map at the point are both complex with norm one) of a conservative diffeomorphism of a surface. In many cases the family of invariant circles has positive Lebesgue area for all C^∞ nearby maps. Suspending this diffeomorphism we obtain a flow on a compact 3-manifold, which can be made incompressible, with a positive Lebesgue volume set consisting of invariant tori. The flow restricted to these bidimensional tori behaves like a linear irrational flow, without singularities, and with all Lyapunov exponents zero. Hence we obtain a positive Lebesgue measure invariant subset for a conservative flow on a 3-manifold, whose points have only zero Lyapunov exponents, and this is a persistent feature, valid for all nearby flows in the C^r topology, for $r \geq 4$.

This clearly contradicts the statement of Theorem 9.5. So this C^1 generic result cannot be extended to higher topologies.

9.1 Spectral Decomposition

Here we prove Theorem 9.1, stating that an attracting singular-hyperbolic set with dense periodic orbits, and a *unique singularity*, is a finite union of transitive sets.

The straightforward extension of the result on a finite *disjoint* union of homoclinic classes from uniformly hyperbolic to a singular-hyperbolic attracting set with a dense subset of periodic orbits is false, as the next counterexample shows.

Consider a modification of the construction of the geometric Lorenz attractor given in Sect. 3.3, obtained by adding two equilibria to the flow located at $W^u(\sigma)$ as indicated in Fig. 9.1. This modification can be done in such a way that the new flow restricted to the cross-section S has a C^∞ invariant stable foliation and the quotient map in the leaf space is piecewise expanding with a single discontinuity c as in the Lorenz case. The resulting attracting set can be proved to be a homoclinic class just as in the geometrical Lorenz case (see Sect. 3.3.6). In particular, such a set is transitive with *dense periodic orbits* and is also singular-hyperbolic by construction.

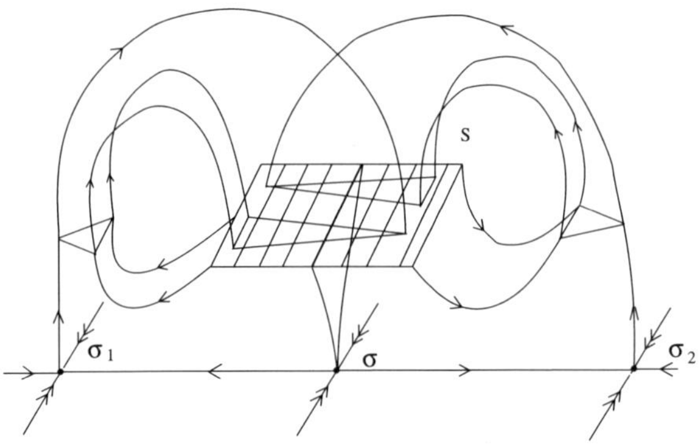

Fig. 9.1 A modified geometric Lorenz attractor

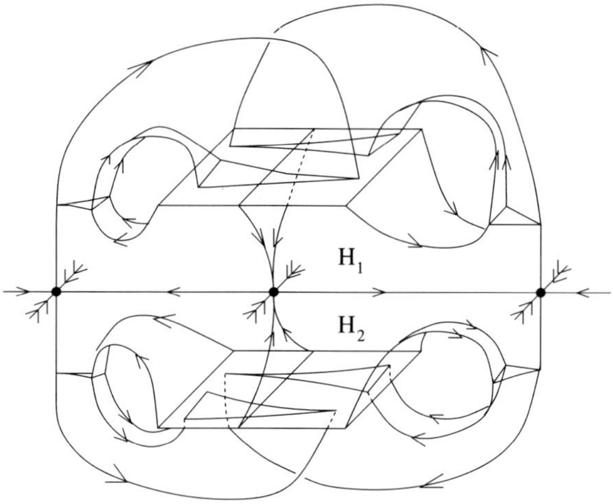

Fig. 9.2 A sketch of the construction of a singular-hyperbolic attractor which is not the disjoint union of homoclinic classes

Now glue two copies of this flow along the unstable manifold of the singularity σ obtaining the flow depicted in Fig. 9.2. The resulting flow can be made C^∞ easily.

In this way we construct an attracting singular-hyperbolic set with dense periodic orbits and three equilibria which is not the *disjoint* union of homoclinic classes (although it is the union of two transitive sets). It is possible to construct a similar counter-example with a *unique* singularity, while this counterexample has *three* equilibria. However the construction in this case is more involved; see [38].

The above counterexample shows that, when dealing with the spectral decomposition for singular-hyperbolic sets, it is possible to obtain a *finite union of transitive sets* rather than a finite disjoint union of homoclinic classes. The next result shows that the former situation always occurs if the attracting set has only one singularity.

Theorem 9.6 *An attracting singular-hyperbolic set with dense periodic orbits and a unique singularity is a finite union of transitive sets.*

Proof Split Λ into finitely many connected components. On the one hand such components are clearly attracting with dense periodic orbits and the non-singular ones are hyperbolic, hence transitive, by the Spectral Theorem for uniformly hyperbolic sets; see e.g. [252]. On the other hand, the singular component satisfies the conditions of Theorem 6.38. Hence this component is either transitive or the union of two homoclinic classes, which are transitive sets. Therefore Λ, which is the union of its connected components, is a finite union of transitive sets. □

Note that by a result of Morales [163] *every transitive set of a flow Y*, close to X, *contained in the isolating neighborhood U of a singular-hyperbolic attractor Λ of X must contain a singularity.* Therefore, since compact invariant subsets in Λ not containing singularities are hyperbolic and admit a spectral decomposition, and the number of singularities in U is finite, the ω-limit set in U for Y has finitely many transitive pieces only, all of which are singular. *Hence, near a singular-hyperbolic attractor, the number of transitive pieces is robustly finite.*

It is natural to ask whether the union in Theorem 9.6 is disjoint. Recall that a vector field is *Kupka-Smale* if all its closed orbits are hyperbolic and their associated invariant manifolds are in general position; see Sect. 2.5.10.

Theorem 9.7 *An attracting singular-hyperbolic set, with dense periodic orbits and a unique singularity, of a Kupka-Smale vector field is a finite* disjoint *union of transitive sets.*

Proof Let X be a Kupka-Smale vector field in a compact 3-manifold and let Λ be an attracting singular-hyperbolic set of X with dense periodic orbits and a unique singularity σ. It suffices to prove that the connected component of Λ containing the singularity σ is transitive. By contradiction, suppose that this is not so.

On the one hand, by Theorem 6.44, we obtain a regular point a in the unstable manifold $W^u(\sigma)$ of σ such that $\omega(a)$ is a periodic orbit $\mathcal{O}(p)$. On the other hand, the unstable manifold $W^u(\sigma)$ is one-dimensional, and so the vector field exhibits a non-transverse intersection between $W^u(\sigma)$ and $W^s(p)$, contradicting the choice of X in the Kupka-Smale class. □

Theorem 9.7 implies that the union in Theorem 9.6 is disjoint for *most* vector fields on closed 3-manifolds. Denote by $\mathfrak{R}^r(M)$ the subset of all vector fields $X \in \mathfrak{X}^r(M)$ for which every attracting singular-hyperbolic set with dense periodic orbits and a unique singularity of X is a finite *disjoint* union of transitive sets. Standard C^1 generic arguments (see e.g. [56]) imply that $\mathfrak{R}^r(M)$ is residual in $\mathfrak{X}^r(M)$ when $r = 1$. The following corollary proves this assertion for all $r \geq 1$. The proof combines Theorem 9.7 with the classical Kupka-Smale Theorem (see e.g. [190]).

Corollary 9.8 *The class $\mathfrak{R}^r(M)$ is residual in $\mathfrak{X}^r(M)$ for every $r \geq 1$.*

Now consider the complement of $\mathfrak{R}^r(M)$. For a compact invariant subset Λ of a vector field X define the family $\mathcal{C}(\Lambda)$ of homoclinic classes contained in Λ. Note that if Λ is hyperbolic then $\mathcal{C}(\Lambda)$ is finite. We are able to give sufficient conditions for the finiteness of $\mathcal{C}(\Lambda)$ when Λ is a singular-hyperbolic set.

Theorem 9.9 *Let Λ be an attracting singular-hyperbolic set with dense periodic orbits and a unique singularity of $X \in \mathfrak{X}^r(M)$. If Λ is* not *a disjoint union of transitive sets, then $\mathcal{C}(\Lambda)$ contains finitely many homoclinic classes only.*

Theorem 9.6 applies to the class of singular-hyperbolic vector fields introduced by Bautista in [36]. A vector field X is *singular-hyperbolic* if its non-wandering set $\Omega(X)$ has dense critical elements and, if $A(X)$ denotes the union of the attracting and repelling closed orbits, there is a *disjoint* union

$$\Omega(X) \setminus A(X) = \Omega_1(X) \cup \Omega_2(X),$$

where $\Omega_1(X)$ is a singular-hyperbolic set for X and $\Omega_2(X)$ is a singular-hyperbolic set for $-X$.

This class of singular-hyperbolic vector fields contains the Axiom A vector fields (uniformly hyperbolic) and the singular Axiom A example resembling the geometric Lorenz attractor, described after Corollary 9.14.

Remark 9.10 An example of a singular-hyperbolic vector field in the 3-sphere $\mathbb{S}^3$ which is not Kupka-Smale can be derived from the example described in Fig. 9.1: we consider the singularity σ_1, which is accumulated by regular orbits only on one side, and weaken its weak-contracting eigenvalue so that $DX(\sigma_1)$ now has three real eigenvalues $-\lambda, 0, \kappa$ with $\lambda, \kappa > 0$. We can perform this perturbation keeping the local stable manifold of σ_1 so that the global picture in Fig. 9.1 is kept. Since σ_1 becomes non-hyperbolic, the vector field is not Kupka-Smale.

The following is a direct consequence of Theorems 9.6 and 9.7, and taken together this completes the proof of Theorem 9.1.

Corollary 9.11 *Let X be a singular-hyperbolic vector field* with a unique singularity *on a compact* 3*-manifold. If $\Omega_1(X)$ is attracting and $\Omega_2(X)$ is repelling, then $\Omega(X)$ is a finite union of transitive sets. If X is Kupka-Smale, then such a union is disjoint. In particular, the union is disjoint for a residual subset of vector fields in $\mathfrak{X}^r(M)$, $r \geq 1$.*

An example of a singular-hyperbolic vector field in $\mathbb{S}^3$ satisfying the conditions of Corollary 9.11, without sinks nor sources, was described just before the statement of Corollary 9.12.

The extension of these results to general singular-hyperbolic attracting sets, with several singularities, is still work in progress.

Proof of Theorem 9.9 Suppose that Λ is *not* a disjoint union of transitive sets. Split Λ into finitely many connected components as before. It suffices to prove that $\mathscr{C}(\Lambda')$ contains finitely many homoclinic classes for all connected components Λ' of Λ. On the one hand, for non-singular Λ' we have nothing to prove, since Λ' is uniformly hyperbolic by Proposition 6.2.

On the other hand, the singular connected component Λ_0 must contain $W^u(\sigma)$ (since it is connected), and $W^u(\sigma)$ has two connected components. Choose points a, a' in each one. Observe that Λ_0 must not be transitive by the assumptions on Λ. Then by Theorem 6.44 there are periodic orbits such that $\omega(a) = \mathscr{O}(a)$ and $\omega(a') = \mathscr{O}(a')$. By contradiction assume that there are infinitely many distinct homoclinic classes in Λ_0.

Then there exists an infinite sequence of pairwise distinct periodic orbits $O_n \subset \Lambda_0$ and an infinite sequence $z_n \in O_n$, and so the set $A = \overline{\cup_n H(z_n)}$ must contain σ. For otherwise $A \subset \Lambda_0 \setminus \{\sigma\}$ is uniformly hyperbolic and the number of homoclinic classes would be finite.

Consider then $x_n \in \mathscr{O}_n$ such that $x_n \xrightarrow[n\to+\infty]{} \sigma$. Since x_n is not σ, the accumulation on σ and the flow-boxes near σ show that the orbit $\mathscr{O}_n$ accumulates also either a or a'. Without loss of generality, assume the former case is true.

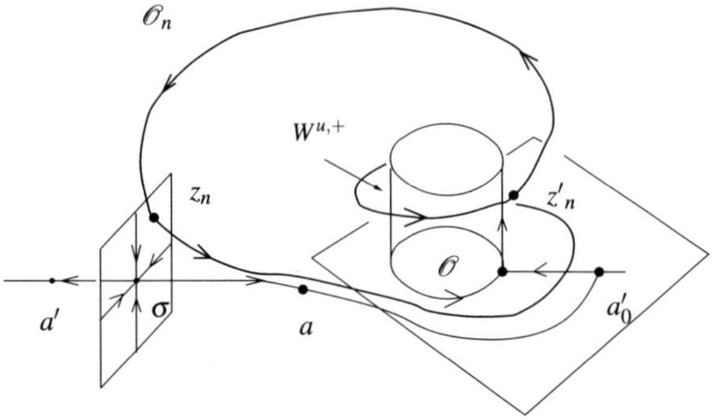

Fig. 9.3 The accumulation on one of the components of $W^s(\sigma)$

Since $\omega_X(a) = \mathcal{O}(p)$ and $\mathcal{O}_n$ accumulates at a, we can find $z'_n \in \mathcal{O}_n$ passing close to O as indicated in Fig. 9.3.

By the inclination lemma we can assume that z'_n converges to a point either in one component $W^{s,+}$ of $W^s(\mathcal{O}) \setminus \mathcal{O}$, or in the other component $W^{s,-}$. Again suppose that we are in the former case. By Lemma 6.43 and the inclination lemma we obtain $z'_n \in \overline{W^{u,+}} = H^+$. But then $H(z'_n) = H(x) = H(z_n) = H^+$ for infinitely many n (since Theorem 2.17 ensures that every homoclinic class contains a dense subset of periodic orbits, all of which are homoclinically related).

This contradicts the choice of z_n. □

9.2 A Dichotomy for C^1 Generic 3-Flows

It is known that a generic *non-singular* vector field $X \in \mathfrak{X}^1(M)$ either has infinitely many sinks or sources, or else is Axiom A without cycles; see Mañé [145] or Liao [133]. The robustness of the geometric Lorenz attractor obtained in Sect. 3.3 shows that this is not true in general if singularities are allowed. Allowing singularities we can improve this as follows. Let $\mathfrak{V}^1(M) \subset \mathfrak{X}^1(M)$ be the set of vector fields that *cannot be* C^1 *approximated by homoclinic loops*. The Connecting Lemma 2.20 implies that any singularity of every $X \in \mathfrak{V}^1(M)$ is separated from the non-wandering set. Using the arguments of Wen [270] and Hayashi [108] we conclude that a generic vector field in $\mathfrak{V}^1(M)$ either has infinitely many sinks or sources or else it is Axiom A without cycles.

Recently Arroyo and Hertz [28] proved that every vector field in $\mathfrak{V}^1(M)$ either can be approximated by one that is Axiom A without cycles, or exhibits a homoclinic tangency associated to a periodic orbit.

9.2.1 Some Consequences of the Generic Dichotomy

Let us describe some consequences of Theorem 9.2. The first one is related to the abundance of three-dimensional vector fields exhibiting either attractors or repellers. As noted by Mañé in [145], a generic C^1 diffeomorphism in the 2-sphere $\mathbb{S}^2$ does exhibit either sinks or sources. It is then natural to ask whether such a result is valid for C^1 vector fields in the 3-sphere $\mathbb{S}^3$ instead of C^1 diffeomorphisms in $\mathbb{S}^2$. The answer is negative as the following example shows.

Write $\mathbb{S}^3 = \mathbb{R}^3 \cup \{\infty\}$ and consider in $\mathbb{R}^3$ an unknotted two-torus $\mathbb{T}^2$. Then the closure in $\mathbb{S}^3$ of each connected component of $\mathbb{S}^3 \setminus \mathbb{T}^2$ is a solid two-torus. Consider a Lorenz attractor in one solid two-torus and a Lorenz repeller in the other. Since a fundamental domain for the Lorenz attractor (respectively repeller) is an unknotted solid two-torus, we can glue the two solid two-torus through the unknotted torus, obtaining a flow in $\mathbb{S}^3$ whose non-wandering set equals the disjoint union of one Lorenz attractor and one Lorenz repeller. Such a flow is singular Axiom A, and it can not be approximated by vector fields with either sinks or sources. However from Theorem 9.2 we deduce

Corollary 9.12 *A generic vector field in $\mathfrak{X}^1(M)$ does exhibit either attractors or repellers.*

The second consequence of Theorem 9.2 is related to a conjecture by Palis in [189], see also Sect. 2.8, asserting the denseness of vector fields exhibiting a finite number of attractors whose basin of attraction forms a full Lebesgue measure subset. Theorem 9.2 gives an approach to this conjecture in the (open) set $\mathfrak{N}^1(M)$ of C^1 vector fields on a closed 3-manifold M which cannot be C^1 approximated by ones exhibiting infinitely many sinks or sources.

Corollary 9.13 *A generic vector field in $\mathfrak{N}^1(M)$ exhibits a finite number of attractors whose basins of attraction form an open and dense subset of M.*

This corollary follows from the no-cycle condition by the classical construction of filtrations adapted to the decomposition of the positive limit set of the flow, as the reader can easily see in [247, Chaps. 2 & 3].

Using the filtration to isolate the dynamics around each basic piece of the singular Axiom A decomposition, since the critical elements are robustly hyperbolic near to each basic piece (recall that singular-hyperbolicity is a robust property of the action of the flow on the tangent bundle), we obtain

Corollary 9.14 *A C^r singular Axiom A flow without cycles is in $\mathcal{G}^r(M)$, the interior of the set of C^r vector fields whose critical elements are hyperbolic, for any $r \geq 1$.*

We note that there exists a classification by Hayashi [107] of the C^1 interior of the set of diffeomorphisms whose periodic points are hyperbolic: they are Axiom A without cycles.

The corresponding result for vector fields is false since the Lorenz attractor is not uniformly hyperbolic. Indeed, we note that we can easily construct a singular Axiom A vector field without cycles and with a singular basic set equivalent to the Lorenz attractor: just take the geometric Lorenz attractor constructed in Sect. 3.3, and embed and extend this flow to $\mathbb{S}^3$ with a repelling singularity at the "north pole" and a sink at the "south pole".

Proof of Theorem 9.2 The argument is based on the following result whose proof we postpone to Sect. 9.2. Denote by $\mathfrak{H}^r(M)$ the interior of the set of vector fields $X \in \mathfrak{X}^r(M)$ such that every periodic orbit and singularity of X is hyperbolic, for any $r \geq 1$.

Theorem 9.15 *Generic vector fields in $\mathfrak{H}^1(M)$ are singular Axiom A without cycles.*

Following the arguments of Mañé in [145], we can obtain Theorem 9.2 from Theorem 9.15. Indeed, let $\mathfrak{G}^1(M) \subset \mathfrak{X}^1(M)$ be the subset of C^1 vector fields such that every singularity of X is hyperbolic. Then $\mathfrak{G}^1(M)$ is open and dense in $\mathfrak{X}^1(M)$ by the local stability of hyperbolic critical elements. For $X \in \mathfrak{G}^1(M)$ define $A(X)$ to be the set of periodic orbits and singularities of X that are sinks or sources.

The set-valued function $\mathfrak{X}^1(M) \ni X \mapsto \overline{A(X)} \in \mathscr{P}(M)$ is lower semicontinuous, again by the local stability of hyperbolic critical elements, where $\mathscr{P}(M)$ denotes the family of compact subsets of M endowed with the Hausdorff distance. Well known topological properties imply that the continuity points $\mathfrak{O}$ of this map form a residual subset of $\mathfrak{G}^1(M)$.

This ensures that *every $X \in \mathfrak{O}$ not satisfying the first item of Theorem 9.2 is in $\mathfrak{H}^1(M)$.*

Indeed, for $X_0 \in \mathfrak{O}$ with finitely many sinks and sources the set $A(X_0)$ is a finite collection of critical elements of X_0. Assume by contradiction that $X_0 \notin \mathfrak{H}^1(M)$. Then we can find a C^1-near vector field Y with a non-hyperbolic critical element ξ. Hence ξ is away from a neighborhood of $A(X_0)$. However $\mathfrak{O} \subset \mathfrak{G}^1(M)$ and $\mathfrak{G}^1(M)$ is open, and thus we can take $Y \in \mathfrak{G}^1(M)$. This guarantees that ξ is *not a singularity* of Y. Then the return map to a Poincaré section of the periodic orbit ξ has two eigenvalues, one of which has modulus 1. Perturbing Y we can find $Z \in \mathfrak{G}^1(M)$ arbitrarily C^1-close to Y (and to X_0) having either an attracting or repelling periodic orbit close to ξ. This contradicts the continuity of the set map $A(X)$ at X_0.

Now from Theorem 9.15 there exists a residual set $\mathfrak{R} \subset \mathfrak{H}^1(M)$ such that every vector field in $\mathfrak{R}$ is singular Axiom A without cycles. The class

$$\mathfrak{Y} = \left(\mathfrak{O} \setminus \overline{\mathfrak{H}^1(M)}\right) \cup \left(\mathfrak{O} \cap \mathfrak{R}\right)$$

is residual in $\mathfrak{X}^1(M)$ by construction (recall that $\mathfrak{G}^1(M)$ is open and dense in $\mathfrak{X}^1(M)$). Note that if $X_0 \in \mathfrak{Y}$ does not satisfy the first item of Theorem 9.2, then $X_0 \in \mathfrak{O} \cap \mathfrak{R}$, since X_0 cannot belong to $\mathfrak{O} \setminus \overline{\mathfrak{H}^1(M)}$ be the previous claim. This means that X_0 satisfies the second item of the statement of Theorem 9.2. □

9.2.2 Generic 3-Flows, Lyapunov Stability and Singular-Hyperbolicity

Here we present a proof of Theorem 9.15. We use the auxiliary Theorems 9.16 and 9.17 below. Recall the definition and properties of Lyapunov stable sets in Sect. 2.5.10.

The first theorem ensures that transitive Lyapunov stable sets containing singularities, if not equal to a critical element, are C^1 generically singular-hyperbolic sets.

Theorem 9.16 *For generic vector fields $X \in \mathfrak{X}^1(M)$, every nontrivial transitive Lyapunov stable set with singularities of X is singular-hyperbolic.*

The second result provides a way to obtain a singular-hyperbolic attractor from a singularity belonging to a Lyapunov stable set of a generic three-dimensional vector field. Together with the previous results, it asserts that the unstable manifold of a singularity accumulates on a singular-hyperbolic set containing the singularity.

Theorem 9.17 *Every Lyapunov stable singular-hyperbolic set with dense singular unstable branches of $X \in \mathfrak{X}^1(M)$ is an attractor of X.*

Here we say that a singular-hyperbolic set Λ has *dense singular unstable branches* if $\Lambda = \omega(x)$ for all $x \in W^u(\sigma) \setminus \{\sigma\}$ and for every singularity $\sigma \in \Lambda$.

Now we explain how Theorem 9.15 is a consequence of Theorems 9.16 and 9.17, but first we need some preliminary results. The first one gives a sufficient condition for a transitive Lyapunov stable set with singularities to have singular unstable branches.

Lemma 9.18 *For generic vector fields $X \in \mathfrak{X}^1(M)$, a transitive Lyapunov stable set with singularities Λ of X, such that the unstable manifold of every singularity in Λ is one-dimensional, has dense singular unstable branches.*

Proof Generically we can assume that $X \in \mathfrak{X}^1(M)$ satisfies the properties presented in Sect. 2.5.10 (in particular X is Kupka-Smale). Let Λ be a transitive Lyapunov stable set of X, σ a singularity of Λ and $q \in W^u(\sigma) \setminus \{\sigma\}$.

On the one hand, since Λ is Lyapunov stable we have $W^u(\sigma) \subset \Lambda$ and in particular $\omega(q) \subset \Lambda$. On the other hand, we have $\dim(W^u(\sigma)) = 1$ by assumption. Then $\omega(q)$ is Lyapunov stable by Property L5 in Sect. 2.5.10.

But Λ is transitive by construction and intersects $\omega(q)$, and so by Lemma 2.25 we have $\Lambda \subset \omega(q)$. Then $\omega(q) = \Lambda$ and Λ has dense singular unstable branches as desired. □

The next one shows that the closure of the unstable manifold of a singularity accumulated by periodic orbits is transitive, provided that the unstable manifold is one-dimensional and its closure is Lyapunov stable.

Lemma 9.19 *Let $X \in \mathfrak{X}^1(M)$ and $\sigma \in S(X) \cap \overline{\mathrm{Per}(X)}$ be such that $W^u(\sigma)$ is one-dimensional and $\omega(q)$ is Lyapunov stable for every q in any of the branches of $W^u_X(\sigma) \setminus \{\sigma\}$. Then $\overline{W^u(\sigma)}$ is transitive.*

Proof We have $W^u_X(\sigma) \setminus \{\sigma\} = \mathcal{O}(q_1) \cup \mathcal{O}(q_2)$ for every q_1, q_2 belonging to different connected components of $W^u(\sigma) \setminus \{\sigma\}$.

On the one hand, since $\sigma \in \overline{\mathrm{Per}(X)}$ we can assume that $q_1 \in \overline{\mathrm{Per}(X)}$ without loss of generality. Then $\omega(q_1) \subset \overline{\mathrm{Per}(X)}$ by invariance. On the other hand, $\omega(q_1)$ is Lyapunov stable for X by assumption. These two properties imply that $\sigma \in \omega(q_1)$, since for $p_n \in \mathrm{Per}(X)$ with $p_n \xrightarrow[n\to+\infty]{} q_1$ we also have $X^{t_n}(p_n) \to \sigma$ for some sequence $t_n > 0$, and we can apply Lemma 2.25.

Therefore $W^u(\sigma) \subset \omega(q_1)$ by the Lyapunov stability of $\omega(q_1)$ once more. But $\overline{W^u(\sigma)} \supset \omega(q_1)$ by construction, and so we conclude that $\overline{W^u(\sigma)} = \omega(q_1)$. This shows that $\overline{W^u(\sigma)}$ is transitive. □

Using this we now show that any hyperbolic singularity accumulated by regular orbits of X is in a singular-hyperbolic attractor or repeller of the flow induced by X.

Theorem 9.20 *For generic $X \in \mathfrak{X}^1(M)$ every $\sigma \in S(X) \cap \overline{\mathrm{Per}(X)}$ belongs to either a singular-hyperbolic attractor or a singular hyperbolic repeller.*

Proof Let $X \in \mathfrak{X}^1(M)$ and let σ be as in the statement. Since X is generic we can assume that σ is hyperbolic. Note that σ must be of saddle-type, for otherwise σ is either a sink or a source, and in each case no periodic orbit would approach σ. Hence either $W^u(\sigma)$ or $W^s(\sigma)$ is one-dimensional.

Suppose the former case is true. The latter case is the same for $-X$. Define $\Lambda = \overline{W^u(\sigma)}$. Property L3 in Sect. 2.5.10 implies that Λ is Lyapunov stable for X because X is generic. Property L5 then guarantees that we are in the setting of Lemma 9.19 and so Λ is transitive.

Therefore Λ is a nontrivial transitive Lyapunov stable set of X. As X is generic, Theorem 9.16 ensures that Λ is singular-hyperbolic. By Theorem 5.10 we know that every singularity in Λ has one-dimensional unstable manifold. We conclude that Λ has dense singular unstable branches by Lemma 9.18, since X is generic. Then Λ is an attractor by Theorem 9.17. □

Now we have the tools to complete the proof of Theorem 9.15 using all the previous results which assume Theorems 9.16 and 9.17.

Proof of Theorem 9.15 For $X \in \mathfrak{X}^1(M)$ denote by $S^*(X) = S(X) \cap \overline{\mathrm{Per}(X)}$ the (finite) set $\{\sigma_1, \ldots, \sigma_k\}$ of singularities accumulated by periodic orbits of X.

Theorem 9.20 ensures that for generic $X \in \mathfrak{X}^1(M)$ and for every $i = 1, \ldots, k$ there is a compact invariant set Λ_i of X such that $\sigma_i \in \Lambda_i$, and Λ_i is either a singular-hyperbolic attractor or a singular-hyperbolic repeller of X.

We claim that $H^* = \Omega(X) \setminus \cup_{i=1}^k \Lambda_i$ is a finite disjoint union of uniformly hyperbolic basic sets. Indeed $H^* \setminus S(X)$ is closed in M, for otherwise we can find a sequence of regular points x_n in H^* converging to some singularity $\sigma \in S(X) \setminus S^*(X)$. But Property L2 gives $\Omega(X) = \overline{\mathrm{Per}(X) \cup S(X)}$, and so σ is accumulated by periodic orbits because $S(X)$ is finite. Hence $H^* \setminus S(X)$ is a closed invariant subset of X without singularities. It is known, after Wen [270], that C^1

generically such sets are uniformly hyperbolic. Property L2 again ensures that $H^* = \overline{\mathrm{Per}(X) \cap H^*} \cup S(X) \setminus S^*(X)$. The Spectral Decomposition Theorem for uniformly hyperbolic sets now guarantees that H^* decomposes into finitely many basic pieces, together with finitely many singularities.

From this we see that $\Omega(X)$ splits into a disjoint union of compact invariant sets each one being either a hyperbolic basic set or a singular-hyperbolic attractor, or a singular-hyperbolic repeller. Hence X is a singular Axiom A vector field. For generic X we can also assume that the vector field is Kupka-Smale, and thus there are no cycles between the transitive pieces in the above decomposition. The proof of Theorem 9.15 is complete depending on Theorems 9.16 and 9.17. □

Proof of Theorem 9.16 Recall that there exists a residual subset $\mathcal{O}$ of the family $\mathfrak{G}^1(M)$ of vector fields whose singularities are hyperbolic, such that the map $X \in \mathfrak{G}^1(M) \mapsto \overline{A(X)}$ restricted to $\mathcal{O}$ is continuous (see the arguments after the statement of Theorem 9.15). Define $\mathscr{R} = \mathcal{O} \cap \mathfrak{H}^1(M)$ which is residual in $\mathfrak{H}^1(M)$.

Given $X \in \mathscr{R}$ and $\sigma \in S(X) \cap \Lambda$ for a non-trivial attractor Λ, observe that every vector field Y sufficiently C^1-close to X has no sources or sinks near to Λ, for otherwise we deduce a contradiction to the choice of X in the continuity set $\mathcal{O}$. All the critical elements of Y are also hyperbolic. Then Y is in the setting of Theorem 2.33, and thus the Linear Poincaré Flow over $\Lambda \setminus S(X)$ is robustly dominated. This means that Λ is in the setting of Lemmas 5.22 and 5.30. Thus we deduce that, *for $X \in \mathscr{R}$, if $\sigma \in S(X)$ belongs to a non-trivial attractor Λ of X, then σ is Lorenz-like for X and $W^{ss}(\sigma) \cap \Lambda = \{\sigma\}$.*

Now let $X \in \mathscr{R}$ have a non-trivial transitive Lyapunov stable set Λ containing a singularity. The previous arguments ensure that Λ is in the setting of Theorem 5.34, and hence Λ is a singular-hyperbolic attractor. □

Proof of Theorem 9.17 We need the following sufficient condition for a Lyapunov stable singular-hyperbolic set, with dense singular unstable branches, to be an attractor.

Lemma 9.21 *Let Λ be a Lyapunov stable singular-hyperbolic set with dense singular unstable branches of $X \in \mathfrak{X}^r(M)$, $r \geq 1$. If Λ admits an adapted cross-section Σ such that every point in the interior of Σ belongs to the stable leaf of some point of $\Lambda \cap \Sigma$, then Λ is an attractor.*

Proof From Lemma 2.26 it is enough to prove that, if x_n is a sequence converging to some point $p \in \Lambda$, then $\omega(x_n)$ is contained in Λ for every big enough n. Now $\omega(p)$ satisfies one of the following alternatives.

1. $\omega(p)$ contains a singularity σ of Λ.

   The orbits of x_n will have σ has an accumulation point. Hence the orbit of x_n also accumulates on some regular point q of the unstable manifold of σ. Since $\omega(q) = \Lambda$ by assumption, we see that for every big enough n the orbit of x_n

crosses the interior of Σ. Then by the assumption on Σ we get $y \in \Lambda$ such that $\mathcal{O}(x_n) \subset W^s(y)$, that is, $\omega(x_n) \subset \Lambda$ for all sufficiently big n.

2. $\omega(p)$ is far from singularities.

   Take S an adapted cross-section to a point q of $\omega(p)$. Then for all big enough n the orbit of x_n crosses the interior of S at some point x'_n very close to q. Since $\omega(p)$ is uniformly hyperbolic by Proposition 6.2, the unstable manifold of q is well defined and $W^u(q) \cap S$ is a line in S crossing all stable manifolds of S in a neighborhood of q. Then x'_n belongs to some of these stable lines. Since $W^u(q)$ is inside Λ by Lyapunov stability, we see that x_n belongs to the stable manifold of some point of Λ. Again $\omega(x_n) \subset \Lambda$ for all sufficiently big n.

□

Now suppose that Λ *is not an attractor*. Then by Lemma 9.21, given any regular point $x \in \Lambda$, we can find an adapted cross-section Σ' such that the intersection $\Lambda \cap \Sigma'$ is contained in the interior of Σ. Indeed, $\Sigma \cap \Lambda$ contains z_0 such that $W^s(z_0, \Sigma)$ does not touch Λ, and then one of the connected components of $\Sigma \setminus W^s(z_0, \Sigma)$, which is also an adapted cross-section containing x, contains z_1 such that $W^s(z_1, \Sigma) \cap \Lambda = \emptyset$. The substrip Σ' between $W^s(z_0, \Sigma)$ and $W^s(z_1, \Sigma)$ only intersects Λ in its interior.

Cover Λ by finitely many flow-boxes near singularities and tubular flow-boxes through adapted cross-sections, around regular pieces of Λ, just as in Chap. 7, but with the family Ξ of adapted cross-sections chosen so that $\Lambda \cap \Xi \subset \mathrm{int}(\Xi)$.

Observe that, since Λ is Lyapunov stable, we can find a neighborhood U of Λ such that $U \cap \Xi \subset \mathrm{int}(\Xi)$ and then another neighborhood $V \subset U$ of Λ satisfying $X^t(V) \subset U$ for all $t > 0$. Then the Poincaré map R defined as in Sect. 6.1 between the sections of Ξ admits only finitely many discontinuity points, at the intersection of Ξ with a compact part of the stable manifolds of the singularities of Λ, since its image cannot touch the boundary of Ξ. We can choose the "waiting time t_2" of R so that the expansion rate on center-unstable cones is at least 4.

Let Ξ^* be the subset of ingoing cross-sections near singularities of Ξ. Fix a point $x_0 \in \Lambda \cap \Xi^* \setminus \cup\{W^s(\sigma) : \sigma \in S(X) \cap \Lambda\}$ and a connected cu-curve γ_0 inside Ξ^* through x_0 not touching the lines of intersection of Ξ^* with the local stable manifold of the singularities. The image curve $R^i(\gamma_1)$, for $i > 0$, is well defined until it returns to Ξ^*, because the image of R does not fall outside of $\mathrm{int}(\Xi)$. Let γ_2 be the next return to Ξ^*. Then its length $\ell(\gamma_2)$ is at least $4 \cdot \ell(\gamma_1)$.

The image of γ_2 is well defined except perhaps at $\gamma_2 \cap W^s_{loc}(\sigma)$ for some singularity σ of Λ. In this case we replace γ_2 by the longest connected component of $\gamma_2 \setminus W^s_{loc}(\sigma)$. Then $\ell(\gamma_2) \geq 2 \cdot \ell(\gamma_1)$.

Inductively we obtain a sequence γ_n, $n \geq 1$, of larger and larger cu-curves in the interior Ξ^*, which is a finite collection of bounded cross-sections. Since the cu-curves cannot be tangent to the stable foliation, and so cannot curl inside Ξ, this is impossible.

This contradiction shows that Λ must be an attractor and concludes the proof of Theorem 9.17. □

9.3 Lyapunov Exponents of C^1 Generic Incompressible 3-Flows

Here we prove Theorems 9.3 and 9.4. Since the proof is rather technical, we present first an outline of the strategy.

Recall the definition of dominated splitting for the Linear Poincaré Flow and that the splitting $N^1 \oplus N^2$ of the normal bundle N is an *m-dominated splitting* for the Linear Poincaré Flow if it is P_X^t-invariant and there is an *uniform* $m \in \mathbb{N}$ such that

$$\Delta(p,m) = \frac{\|P_X^m(p)|N_p^1\|}{\|P_X^m(p)|N_p^2\|} \le \frac{1}{2}, \tag{9.1}$$

for any point $p \in \Lambda$.

We define some useful X^t-invariant sets:

- $\mathcal{O} = \{$Oseledec regular points$\}$, has full Lebesgue measure in M, since a volume form is assumed to be invariant (see Sect. 2.7.2);
- $\mathcal{O}^+ = \{p \in \mathcal{O} :$ the orbit of p has positive Lyapunov exponent$\}$;
- $\mathcal{O}^0 = \{p \in \mathcal{O} :$ the orbit of p has only zero Lyapunov exponents$\}$;
- $\Lambda_m(X) := \{p \in \mathcal{O}^+ : p$ has m-dominated splitting for the Linear Poincaré Flow$\}$;
- $\Gamma_m(X) := M \setminus \Lambda_m(X)$;
- $\Gamma_m^+(X) := \mathcal{O}^+(X) \setminus \Lambda_m(X)$;
- $\Gamma_m^*(X) := \{p \in \Gamma_m^+(X) : p \notin \mathrm{Per}(X)\}$.

The set of Oseledec regular points where (9.1) does not hold will be denoted by $\Delta_m(X)$. Clearly, in $\Delta_m(X)$, the orbits do not have an m-dominated splitting. Nevertheless for some $p \in \Delta_m(X)$ there might still exist some iterate $X^t(p)$ where (9.1) holds. Taking this into account we note that $\Gamma_m^+(X) = \bigcup_t X^t(\Delta_m(X))$.

Let $\lambda^+(X,p)$ be the upper Lyapunov exponent which exists for μ-almost every $p \in M^3$ by the Theorem of Oseledec; see Sect. 2.7.1 and [147]. When there is no ambiguity we denote $\lambda^+(X,p)$ by $\lambda^+(p)$. Given an X^t-invariant set $\Gamma \subseteq M$ we define the "entropy function" by

$$LE : \mathfrak{X}_\mu^1(M) \to [0,+\infty), \quad X \mapsto \int_M \lambda^+(p)d\mu(p).$$

The next lemma gives an equivalent expression for this function.

Lemma 9.22 *Let $\Gamma \subset M$ be an X^t-invariant subset. Then*

$$LE(X,\Gamma) = \int_\Gamma \lambda^+(p)\,d\mu(p) = \inf_{n\ge 1} \frac{1}{n} \int_\Gamma \log \|P_X^n(p)\|\,d\mu(p).$$

In particular, this shows that $LE(X) = LE(X,M)$ is upper semicontinuous as a function of X.

Proof Using the definition of $\lambda^+(p)$ we write $\int_\Gamma \lambda^+(p)d\mu(p)$ as

$$\lim_{t\to+\infty} \frac{1}{t} \int_\Gamma \log \|P_X^t(p)\|\,d\mu(p) = \lim_{n\to+\infty} \frac{1}{n} \int_\Gamma \log \|P_X^n(p)\|\,d\mu(p) =: I_n(X)$$

and the sequence $I_n(X)$ is subadditive: $I_{p+q}(X) \leq I_p(X) + I_q(X)$ for all $p, q \in \mathbb{Z}^+$. Thus it satisfies $\lim_{n\to+\infty}(I_n(X)/n) = \inf_{n\geq 1}(I_n(X)/n)$. □

Denote $LE(X, M)$ by $LE(X)$. The next proposition will be crucial to prove Theorem 9.3.

Proposition 9.23 *Let $X \in \mathfrak{X}^2_\mu(M)^*$, with X^t aperiodic, and suppose that any hyperbolic set has zero Lebesgue measure. Let $\varepsilon, \delta > 0$ be given. Then there exists a C^1 zero divergence vector field Y which is ε-C^1-close to X, such that $LE(Y) < \delta$.*

We assume Proposition 9.23 and prove Theorem 9.3 first. By Corollary 8.2, we have a dense set such that every X is C^2, aperiodic and with hyperbolic sets having full or zero measure. The set of conservative Anosov vector fields, denoted by $\mathscr{A}$, is open. For all $k \in \mathbb{N}$ the set $\mathscr{A}_k = \{X \in \mathfrak{X}^1_\mu(M)^* : LE(X) < k^{-1}\}$ is open because, by Lemma 9.22, LE is upper semicontinuous. By Proposition 9.23, with $\delta = k^{-1}$, we get $\mathscr{A}_k$ dense in $\mathscr{A}^c$, and so the set $\mathfrak{R} = \bigcap_k \mathscr{A} \cup \mathscr{A}_k$ is a C^1-residual set. But $\mathfrak{R} = \mathscr{A} \cup \bigcap_k \mathscr{A}_k = \mathscr{A} \cup \{X \in \mathfrak{X}^1_\mu(M)^* : LE(X) = 0\}$, and therefore for $X \in \mathfrak{R}$ either X is an Anosov vector field or $LE(X) = \int_M \lambda^+(p)d\mu(p) = 0$. This last equality implies that μ-almost every $p \in M$ has zero Lyapunov exponents and Theorem 9.3 is proved.

To prove Proposition 9.23, we consider a large $m \in \mathbb{N}$ (depending on ε) and we use the fact that almost every orbit does not have an m-dominated splitting, for otherwise by Corollary 8.2 X must be Anosov. We start with a local argument and take $p \in \Gamma^+_m(X)$, and also $t \gg m$. By a recurrence result (see Lemma 9.40), we obtain $q \approx X^{t/2}(p)$ such that $\Delta(q, m) \geq 1/2$, say $q \in \Delta_m(X)$. In Sect. 9.3.3 we use the ε-C^1-perturbation Y of X, developed previously in Sects. 9.3.1 and 9.3.2, to map the direction N^u_q into $N^s_{X^m(q)}$. In Sect. 9.3.4 we conclude that this argument allows us to prove that for most points q near p the norm of $P^t_Y(q)$ is smaller than δ. The formula for LE in Lemma 9.22 allows us to compute a bound for $LE(Y)$ in a finite time t. Finally, in Sect. 9.3.5, extend this local procedure to the whole of M through a Kakutani tower argument.

9.3.1 Conservative Tubular Flow Theorem

The following theorem, due to Dacorogna and Moser [75], will be used to obtain a conservative local change of coordinates which trivialize a vector field.

Theorem 9.24 (Dacorogna-Moser) *Let Ω be a bounded open subset of $\mathbb{R}^n$ with C^5 boundary $\partial\Omega$ and $g, f : \overline{\Omega} \to \mathbb{R}$ positive functions of class C^s ($s \geq 2$). Then there exists a diffeomorphism $\varphi : \Omega \to \varphi(\Omega) \subseteq \mathbb{R}^n$ of class C^s which satisfies the partial differential equation*:

$$g(\varphi(q)) \det(D\varphi_q) = \lambda f(q), \tag{9.2}$$

for all $q \in \Omega$ where $\lambda = \int g\, d\mu / \int f\, d\mu$. We also have $\varphi = Id$ at $\partial\Omega$.

Let $T:\mathbb{R}^3\to\mathbb{R}^3$ be the constant vector field defined by $T(x,y,z)=(c,0,0)$ for some $c>0$ and let $\mathfrak{F}$, $\mathfrak{C}$ be the flow-boxes $\mathscr{F}^1_X(p)(B(p,r))$ and $\mathscr{F}^1_T(\overline{p})(B(\overline{p},r))$. We start by giving a brief and informal idea of the proof of Lemma 9.25 below. We would like to find a C^2 volume-preserving diffeomorphism $\hat{\Psi}:\mathfrak{C}\to\mathfrak{F}$ such that $X=\hat{\Psi}_*T$. Given a map $\psi:B(\overline{p},r)\to B(p,r)$ and $w\in\mathfrak{C}$, let t_w be such that $T^{-t_w}\in B(\overline{p},r)$. We define $\hat{\Psi}(w):=X^{t_w}(\psi\circ T^{-t_w}(w))$. It is clear that, if ψ is volume-preserving, then $\hat{\Psi}$ will be a composition of three volume-preserving maps and we are done. However, the choice of ψ to be volume-preserving needs some care and to achieve our purposes we use Theorem 9.24 in an appropriate way.

Lemma 9.25 (Conservative flow-box theorem) *Given a vector field $X\in\mathfrak{X}^s_\mu(M)$ (for $s\geq 2$) and a non-singular point $p\in M$ (eventually periodic with period $\tau>1$), there exists a conservative C^s diffeomorphism $\Psi:\mathfrak{F}\to\mathfrak{C}$ such that $T=\Psi_*X$.*

Proof Assume that $p=\mathbf{0}$ and $X(p)\subseteq\{(x,0,0):x\in\mathbb{R}\}$. Let $X_1(x,y,z)$ be the projection into the first coordinate of $X(x,y,z)$. For a small $r>0$ we define the functions $f:B(\overline{p},r)\to\mathbb{R}$ and $g:B(p,r)\to\mathbb{R}$ such that $f(\overline{y},\overline{z}):=1$ for $(0,\overline{y},\overline{z})\in B(\overline{p},r)$ and $g(y,z):=X_1(0,y,z)$ for $(0,y,z)\in B(p,r)$ (see Fig. 9.4). Since g is of class C^s, we apply Theorem 9.24 to $\Omega=B(p,r)\subseteq\mathbb{R}^2$ and obtain a diffeomorphism $\varphi:\Omega\to\varphi(\Omega)\subseteq\mathbb{R}^2$ of class C^s satisfying the partial differential equation $g(\varphi(\overline{y},\overline{z}))\det(D\varphi_{\overline{y},\overline{z}})=\lambda$, for all $(\overline{y},\overline{z})\in\Omega$, where $\lambda=\int g\,d\mu/\int 1\,d\mu$ and $\varphi|_{\partial\Omega}=Id$.

Now we define the C^s change of coordinates by

$$\hat{\Psi}:\mathbb{R}\times\Omega\to\mathbb{R}^3,\quad (\overline{x},\overline{y},\overline{z})\mapsto X^{\lambda^{-1}\overline{x}}(0,\varphi(\overline{y},\overline{z})).$$

First we claim that

$$\det\left(D\hat{\Psi}_{(0,\overline{y},\overline{z})}\right)=1\quad\text{for all }(0,\overline{y},\overline{z})\in\mathbb{R}\times\Omega. \tag{9.3}$$

By a straightforward computation of the derivative, we obtain

$$\det(D\hat{\Psi}_{(0,\overline{y},\overline{z})})=\begin{vmatrix}\lambda^{-1}X_1(X^0(0,y,z)) & 0 & 0\\ \lambda^{-1}X_2(X^0(0,y,z)) & \frac{\partial\varphi_1}{\partial\overline{y}}|_{(\overline{y},\overline{z})} & \frac{\partial\varphi_1}{\partial\overline{z}}|_{(\overline{y},\overline{z})}\\ \lambda^{-1}X_3(X^0(0,y,z)) & \frac{\partial\varphi_2}{\partial\overline{y}}|_{(\overline{y},\overline{z})} & \frac{\partial\varphi_2}{\partial\overline{z}}|_{(\overline{y},\overline{z})}\end{vmatrix}.$$

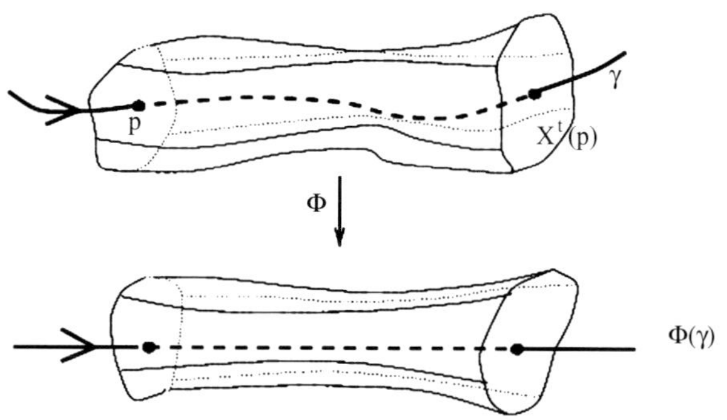

Fig. 9.4 The conservative change of coordinates straightening out all orbits

Now using (9.2) of Theorem 9.24 we conclude that

$$\det\big(D\hat{\Psi}_{(0,\overline{y},\overline{z})}\big) = \lambda^{-1} X_1(0,y,z)\det\big(D\varphi_{(\overline{y},\overline{z})}\big) = g(y,z)\lambda^{-1}\det\big(D\varphi_{(\overline{y},\overline{z})}\big) = 1,$$

and therefore (9.3) is proved. Let us now check that

$$\det\big(D\hat{\Psi}_{(\overline{x}_0,\overline{y}_0,\overline{z}_0)}\big) = 1 \quad \text{for all } (\overline{x}_0,\overline{y}_0,\overline{z}_0)\in\mathfrak{C}.$$

We note that

$$\hat{\Psi}(\overline{x},\overline{y},\overline{z}) = X^{\lambda^{-1}\overline{x}_0}\big[X^{\lambda^{-1}(\overline{x}-\overline{x}_0)}(0,\varphi(\overline{y},\overline{z}))\big] = X^{\lambda^{-1}\overline{x}_0}\big[\hat{\Psi}(\overline{x}-\overline{x}_0,\overline{y},\overline{z})\big],$$

so that $D\hat{\Psi}_{(\overline{x},\overline{y},\overline{z})} = DX^{\lambda^{-1}\overline{x}_0}_{\hat{\Psi}(\overline{x}-\overline{x}_0,\overline{y},\overline{z})} \circ D\hat{\Psi}_{(\overline{x}-\overline{x}_0,\overline{y},\overline{z})}$.

Evaluating $D\hat{\Psi}_{(\overline{x},\overline{y},\overline{z})}$ at $\overline{x}=\overline{x}_0$ we get $D\hat{\Psi}_{(\overline{x}_0,\overline{y},\overline{z})} = DX^{\lambda^{-1}\overline{x}_0}_{\hat{\Psi}(0,\overline{y},\overline{z})} \circ D\hat{\Psi}_{(0,\overline{y},\overline{z})}$.

Now we use (9.3) and the fact that the flow X^t is volume-preserving to conclude that $\det(D\hat{\Psi}_{(\overline{x}_0,\overline{y}_0,\overline{z}_0)}) = 1$. Finally, we take $c := \lambda$ and consider the constant vector field $T := (\lambda, 0, 0)$. Let $(x,y,z) = \hat{\Psi}(\overline{x},\overline{y},\overline{z})$. By a simple computation, we deduce that

$$\begin{aligned}
\hat{\Psi}_* T(x,y,z) &= D\hat{\Psi}_{(\overline{x},\overline{y},\overline{z})}(T(\overline{x},\overline{y},\overline{z}))\\
&= \Big(X_1\big(X^{\lambda^{-1}\overline{x}}(0,y,z)\big), X_2\big(X^{\lambda^{-1}\overline{x}}(0,y,z)\big), X_3\big(X^{\lambda^{-1}\overline{x}}(0,y,z)\big)\Big)\\
&= X\big(\hat{\Psi}(\overline{x},\overline{y},\overline{z})\big).
\end{aligned}$$

Taking $\Psi = \hat{\Psi}^{-1}$ we obtain $T = \Psi_* X$. □

9.3.2 Realizable Linear Flows

The next definition adapts the definition of realizable sequence given by Bochi in [51] and will also be central in the proof of our theorem. In broad terms we consider *modified* area-preserving linear maps acting in the normal bundle at p, $L^t(p) : N_p \to N_{X^t(p)}$ that do exactly what we want. Finally, we ask whether these maps are (γ-almost C^1) realizable as the Linear Poincaré Flow of Y, ε-C^1-close to X, computed on small transversal neighborhoods of one point.

We need to fix some notations before the statement. We recall that the Linear Poincaré Flow is the differential of the standard *Poincaré map*

$$\mathscr{P}^t_X(p) : \mathscr{V}_p \subset \mathscr{N}_p \to \mathscr{N}_{X^t(p)},$$

where $\mathscr{N}_{X^s(p)}$, for $s = 0, t$, is a surface contained in M whose tangent space at $X^s(p)$ is the normal direction $N_{X^s(p)}$ for $s = 0, t$ and $\mathscr{V}_p$ is a small neighborhood of p. We can always guarantee the existence of a continuous time-t arrival function

$\tau(p,t)(\cdot)$ from $\mathscr{V}_p$ into $\mathscr{N}_{X^t(p)}$ by using the implicit function theorem. Due to the presence of singularities, $\mathscr{V}_p$ may be very small.

Given the Poincaré map $\mathscr{P}_X^t(p) : \mathscr{V}_p \subseteq \mathscr{N}_p \to \mathscr{N}_{X^t(p)}$, where $\mathscr{V}_p$ is chosen sufficiently small, and given $B \subseteq \mathscr{V}_p$, the set

$$\mathscr{F}_X^n(p)(B) := \{\mathscr{P}_X^t(p)(q) : q \in B, \quad t \in [0,n]\} = \mathscr{P}_X^{[0,n]}(p)(B)$$

is called the time-n length *flow-box* at p associated to the vector field X. We remark that the sections $(\mathscr{P}_X^t(p)(B))_{t\in[0,n]}$ are pairwise disjoint if $\mathscr{V}_p$ is small enough.

We include here the following useful result that will be crucial later; this enables us to estimate the distortion of the area form pushed-forward between cross-sections by Poincaré maps. Given $n_1, n_2 \in T_q\mathscr{N}_p$ for $q \in \mathscr{N}_p$, we can define a pair of 2-forms induced by the volume form ω according to

$$\hat{\omega}_q(n_1,n_2) := \omega_q(X(q), n_1, n_2) \quad \text{and} \quad \overline{\omega}_q(n_1,n_2) := \omega_q\left(\frac{X(q)}{\|X(q)\|}, n_1, n_2\right).$$

It turns out that $(\mathscr{P}_X^t(p))^*\hat{\omega}_q = \hat{\omega}_{X^{\tau(t,q)}(q)}$ for all $q \in \mathscr{N}_p$. The measure $\overline{\mu}$ induced by the 2-form $\overline{\omega}$ is not necessarily $\mathscr{P}_X^t$-invariant; however both measures $\hat{\mu}$ and $\overline{\mu}$ are equivalent. We call $\overline{\mu}$ the Lebesgue measure at normal sections or *modified area*.

In fact, given $n_1, n_2 \in N_p$ we have

$$(\mathscr{P}_X^t(p))^*\overline{\omega}_p(n_1,n_2) = x(t)^{-1}\overline{\omega}_{X^t(p)}(n_1,n_2),$$

where $x(t) = \|X(X^t(p))\|\|X(p)\|^{-1}$. Since the flow is volume-preserving we have $|\det P_X^t(p)| = x(t)^{-1}$. Therefore we can give an explicit expression for the infinitesimal factor of area distortion by the Linear Poincaré Flow, which in turn implies the following simple lemma.

Lemma 9.26 *Given $\nu > 0$ and $T > 0$, there exists $r > 0$ such that for any measurable set $K \subseteq B(p,r) \subseteq \mathscr{N}_p$ we have $|\overline{\mu}(K) - x(t)\cdot\overline{\mu}(\mathscr{P}_X^t(p)(K))| < \nu$ for all $t \in [0,T]$.*

This provides a way to estimate the area distortion due to the Poincaré map between cross-sections on a very small neighborhood around a piece of orbit of the flow (for a proof see [39, Lemma 3.1.3]).

Definition 9.27 Given $X \in \mathfrak{X}_\mu^1(M)$, $\varepsilon > 0$, $0 < \kappa < 1$ and a non-periodic point p, we say that the modified area-preserving sequence of linear maps $L_j : N_{X^j(p)} \to N_{X^{j+1}(p)}$ for $j = 0, \ldots, n-1$ is an (ε,κ)-*realizable linear flow of length n at p* if, for all $\gamma > 0$, there exists $r > 0$ such that for any open set $\emptyset \neq U \subseteq B(p,r) \subseteq N_p$ we can find

(a) a measurable set $K \subseteq U$ such that $\overline{\mu}(K) > (1-\kappa)\overline{\mu}(U)$, and
(b) a zero divergence vector field Y, ε-C^1-close to X, such that:
   (i) $Y^t = X^t$ outside $\mathscr{F}_X^n(p)(U)$ and $DX_q = DY_q$ for every $q \in U$, $\mathscr{P}_X^n(p)(U)$;
   (ii) if $q \in K$, then $\|P_Y^1(Y^j(q)) - L_j\| < \gamma$ for $j = 0, 1, \ldots, n-1$.

In the previous definition we consider integer iterates, but there is no restriction to consider any intermediate linear maps, such as $L_j : N_{X^{t_j}(p)} \to N_{X^{t_{j+1}}(p)}$ with $t_j < t_{j+1}$ and $\sum_{j=0}^{n-1} t_j = n$. The point p may also be periodic, but with period larger than n. The realizability we deal here is with respect to the C^1 topology.

Next we show how to build some elementary realizable linear flows: the Linear Poincaré Flow itself and also the juxtaposition of two realizable linear flows are realizable linear flows.

Lemma 9.28 *Let $X \in \mathfrak{X}^1_\mu(M)$ and $p \in M$ a non-periodic point.*

(1) *For each $t \in \mathbb{R}$, $P^t_X(p)$ is (ε, κ)-realizable of length t for every ε and κ.*
(2) *Let $\{L_0, \dots, L_{n-1}\}$ be a (ε, κ_1)-realizable sequence of linear maps of length n at p and let $\{L_n, \dots, L_{n+m-1}\}$ be (ε, κ_2)-realizable of length m at $X^n(p)$. Then, for $\kappa = \kappa_1 + \kappa_2 < 1$ the sequence of linear maps $\{L_0, \dots, L_{n+m-1}\}$ is (ε, κ)-realizable.*

Proof Item (1) follows by choosing $Y = X$.

For item (2), given $\gamma > 0$, let r_1, r_2 be the radius according to Definition 9.27 related to the realizable linear flows $\{L_0, \dots, L_{n-1}\}$ and $\{L_n, \dots, L_{n+m-1}\}$ respectively. We take any nonempty open set $U \subseteq B(p, r_1)$. If we have $\mathscr{P}^n_X(p)(B(p, r_1)) \subseteq B(X^n(p), r_2)$ then we have what we need to compose and obtain the realization; otherwise we choose a smaller $r < r_1$.

Given $\nu > 0$, we decrease the radius r if necessary, by using Lemma 9.26, to get $|\overline{\mu}(K) - x(t)\overline{\mu}(\mathscr{P}^n_X(p)(K))| < \nu$ for all $t \in [0, n]$ and any measurable set $K \subseteq B(p, r)$. By definition and choice of the radius $r > 0$, we have the flow-box $\mathscr{F}^{n+m}_X(p)(U)$. Again, by definition, given any $U \subseteq B(p, r)$ we get a measurable $K_1 \subseteq U$ and a vector field Y_1 satisfying (a) and (b) of Definition 9.27. Also for any nonempty open subset of $B(X^n(p), r_2)$, in particular for $\mathscr{P}^n_X(p)(U)$, we get a measurable $\hat{K}_2 \subseteq \mathscr{P}^n_X(p)(U) =: \hat{U}$ and a vector field Y_2 satisfying (a) and (b) of Definition 9.27.

Now we define the vector field $Y = Y_1$ in the flow-box $\mathscr{F}^n_X(p)(U)$, $Y = Y_2$ in the flow-box $\mathscr{F}^m_X(X^n(p))(\hat{U})$ and $Y = X$ elsewhere.

The vector field Y is C^1 since, by definition, $(DY_1)_q = DX_q = (DY_2)_q$ for any $q \in \mathscr{P}^n_X(p)(U)$, and so Y and U satisfies (i). To check (a) we define $K := K_1 \cap K_2$ where K_2 is such that $\mathscr{P}^n_X(p)(K_2) = \hat{K}_2$. By Lemma 9.26, we get $x(n)\overline{\mu}(\hat{U}) < \overline{\mu}(U) + \nu$ and also $\overline{\mu}(U \setminus K_2) < x(n)\overline{\mu}(\hat{U} \setminus \hat{K}_2) + \nu$. So we obtain

$$\begin{aligned}\overline{\mu}(U \setminus K) &= \overline{\mu}(U \setminus (K_1 \cap K_2)) \leq \overline{\mu}(U \setminus K_1) + \overline{\mu}(U \setminus K_2)\\ &< \kappa_1\overline{\mu}(U) + x(n)\overline{\mu}(\hat{U} \setminus \hat{K}_2) + \nu < \kappa_1\overline{\mu}(U) + x(n)\kappa_2\overline{\mu}(\hat{U}) + \nu\\ &< \kappa_1\overline{\mu}(U) + \kappa_2\overline{\mu}(U) + \kappa_2\nu + \nu = \kappa\overline{\mu}(U) + (1 + \kappa_2)\nu.\end{aligned}$$

Therefore the result follows by considering a sufficiently small ν. Finally, (ii) follows by definition. □

Remark 9.29 Using elementary Vitali's covering arguments, we can show that we only have to prove realizability of the linear maps $\{L_0, \dots, L_{n-1}\}$ for $U = B(p', r')$ where $B(p', r') \subseteq B(p, r)$.

9.3.2.1 Small Rotations

Here we construct realizable linear flows of time-1 length at p, which rotate by a small angle ξ the action of the Linear Poincaré Flow, i.e. $L_0 := P^1_X(p) \circ \hat{R}_\xi$ where $\hat{R}_\xi$ is a rotation of angle ξ defined by

$$\hat{R}_\xi := \begin{pmatrix} \cos(\xi) & -\sin(\xi) \\ \sin(\xi) & \cos(\xi) \end{pmatrix}$$

in a canonical base of N_p. Let R_ξ be the 3×3 matrix associated to the linear map $(x, y, z) \mapsto (x, \hat{R}_\xi(y, z))$.

Lemma 9.30 *Given $X \in \mathfrak{X}^2_\mu(M)$, a non-periodic point $p \in M$, $\varepsilon > 0$, $0 < \kappa < 1$ and a fixed time $T = 1$. Then there exists an angle ξ (not depending on p and $\xi = O(\varepsilon)$) such that, for every $\gamma > 0$, there exist $r > 0$ (depending on p) and a zero divergence vector field Y, ε-C^1-close to X, such that*

(a) *$Y - X$ is supported in the flow-box $\mathcal{F}^1_X(p)(B(p, r))$;*
(b) *$\|P^1_Y(q) \quad P^1_X(p)\hat{R}_\xi\| < \gamma$ for all $q \in B(p, r\sqrt{1-\kappa})$.*

Proof We take a nonperiodic point $p \in M$ and define

$$C := \max\{\|DY^1_p\| : p \in M, Y \in \mathcal{U}(X, \varepsilon)\},$$

where $\mathcal{U}(X, \varepsilon)$ is a ε-C^1-neighborhood of X. Using Lemma 9.25 we obtain a C^2 conservative diffeomorphism $\Psi : \mathfrak{F} \to \mathfrak{C}$ and $T = (c, 0, 0)$. Take an uniform bound $\Theta > 0$ for the norm of the first and second derivatives of Ψ computed in time-1 thin flow-boxes and suppose that this constant is also valid for any vector field ε-C^1-close to X. We take the angle ξ such that

$$\xi < \frac{\varepsilon(1 - \sqrt{1 - \frac{\kappa}{2}})}{4\Theta^2}. \tag{9.4}$$

We now fix $\gamma > 0$. To obtain item (a), we note that for any $\alpha < 1$ with $\alpha \approx 1$ there exists $r > 0$ such that $X^{[0,\alpha]}(q) \cap N_{X^1(p)} = \emptyset$ for every $q \in B(p, r)$. We fix such $\alpha > 0$ and let $g : \mathbb{R} \to \mathbb{R}$ be a C^∞ function such that $g(t) = 0$ for $t < 0$, $g(t) = t$ for $t \in [1 - \alpha, 2\alpha - 1]$, $g(t) = \alpha$ for $t \geq \alpha$ and $\dot{g} \leq 2$.

Now for all $q = (0, y, z) \in B(p, r)$ we decrease r so that

$$|y|, |z| < \min\left\{\frac{c\varepsilon}{\Theta^2 \ddot{g}}, \frac{\varepsilon}{2\Theta}\right\}. \tag{9.5}$$

For such $r > 0$, let $G : \mathbb{R} \to [0, 1]$ be a C^∞ function such that $G(\rho) = 1$ for $\rho \leq r\sqrt{1 - \frac{\kappa}{2}}$, $G(\rho) = 0$ for $\rho \geq r$ and $\dot{G} \leq 2[(1 - \sqrt{1 - \frac{\kappa}{2}})r]^{-1}$. We set $\rho = \sqrt{y^2 + z^2}$ and consider the rotation flow $R_{\xi g(t) G(\rho)}(0, y, z)$ acting on N_p, which we denote by $\mathscr{R}_t(q)$, and which is defined by $\mathscr{R}_t(0, y, z) = (0, \hat{R}_{\xi g(t) G(\rho)}(y, z))$. Denote the time derivative by $\dot{\mathscr{R}}_t$. By a simple computation, we obtain

$$\dot{\mathscr{R}}_t \circ \mathscr{R}_t^{-1}(0, y, z) = \xi \dot{g}(t) G(\rho)(0, -z, y). \tag{9.6}$$

We consider the flow $T^t = (ct, 0, 0)$ associated to T and define the map $\Upsilon(t, q) := T^t(\mathscr{R}_t(q))$ for $q = (0, y, z) \in B(p, r)$. Setting $H(t, q) := (t, \mathscr{R}_t(q))$ and

$$F(t, \mathscr{R}_t(q)) := T^t(\mathscr{R}_t(q)) = \Upsilon(t, q)$$

we obtain $\Upsilon(t, q) = F \circ H(t, q)$. Taking time derivatives at $t = s$ we obtain

$$\begin{aligned}
\frac{d}{dt}\Upsilon(t, q)|_{t=s} &= \frac{d}{dt}T^t(\mathscr{R}_t(q))|_{t=s} = DF_{H(s,q)} \cdot DH_s \\
&= \left(\partial_1 F \ \partial_2 F\right)_{H(s,q)} \begin{pmatrix} \partial_1 H \\ \partial_2 H \end{pmatrix}_s = \left(T(T^s(\mathscr{R}_s)) \ DT^s_{\mathscr{R}_s(q)}\right) \begin{pmatrix} 1 \\ \dot{R}_s(q) \end{pmatrix} \\
&= T(T^s \circ \mathscr{R}_s(q)) + DT^s_{\mathscr{R}_s(q)} \dot{\mathscr{R}}_s(q).
\end{aligned}$$

The vector field Z is defined in flow-box coordinates by

$$Z(\cdot) = T(\cdot) + DT^s_{\mathscr{R}_s(q)} \cdot \dot{\mathscr{R}}_s(\mathscr{R}_{-s} \cdot T^{-s})(\cdot).$$

From (9.6) we deduce that the C^1-perturbation is given by $Z = T + P$ with

$$P(x, y, z) = \xi f(x, y, z) \cdot A(x, y, z), \tag{9.7}$$

where $f(x, y, z)$ is the scalar function $\dot{g}(x/c) G(\sqrt{y^2 + z^2})$ and $A(x, y, z)$ is the linear map $(0, -z, y)$. It is straightforward to see that $\operatorname{div}(Z) = 0$. Also, the support of the perturbation P is $B(p, r) \times [0, c\alpha] \subset \mathfrak{C}$.

Now we estimate the C^1-norm of P. We note that $\|P\|_{C^0} \leq \varepsilon/\Theta$ by (9.5). To compute the C^1-norm we take derivatives in (9.7) and, by the product rule,

$$\begin{aligned}
DP &= \xi[\nabla f \cdot A(x, y, z) + f \cdot A] \\
&= \xi\left[\left(\ddot{g}\left(\frac{x}{c}\right) G c^{-1}, \dot{g}\left(\frac{x}{c}\right) \dot{G} \frac{y}{\sqrt{y^2 + z^2}}, \dot{g}\left(\frac{x}{c}\right) \dot{G} \frac{z}{\sqrt{y^2 + z^2}}\right) \cdot A(x, y, z) + f \cdot A\right].
\end{aligned}$$

We use again (9.5) to bound the terms $\ddot{g}(x/c) G(\rho) c^{-1} z$ and $\ddot{g}(x/c) G(\rho) c^{-1} y$.

We remark that the other terms are unaffected by the choice of $r > 0$ small. We take, for example, $\dot{g}(x/c)\dot{G}\frac{y}{\sqrt{y^2+z^2}} z$ and, using polar coordinates given by $(y, z) =$

$(\rho\cos(\beta), \rho\sin(\beta))$, we get

$$\xi\dot{g}\left(\frac{x}{c}\right)\dot{G}\frac{\rho\cos(\beta)}{\rho}\rho\sin(\beta) \le \xi\dot{G}\frac{2\rho^2}{\rho} \le \xi\frac{4\rho}{(1-\sqrt{1-\frac{\kappa}{2}})r}$$
$$\le \xi\frac{4}{(1-\sqrt{1-\frac{\kappa}{2}})} < \frac{\varepsilon}{\Theta^2}.$$

For the other three analogous terms we proceed in the same way to obtain $\|DP\|_{C^0} < \varepsilon/\Theta^2$. We note that we are allowed to take y, z close to zero without interfering with the size of the perturbation. This is a key property of the C^1 topology.

For $q \in B(p, r\sqrt{1-\kappa})$ we have $Z^1(\Psi(q)) = (c, R_{\xi\alpha}(\Psi(q)))$, and so $P^1_Z(\Psi(q)) = \hat{R}_{\xi\alpha} \approx \hat{R}_\xi$ (we just have to choose r suitably small). Hence we obtain the analog of (b) for the trivial vector field.

Now it is straightforward to see that (b) follows by choosing r suitably small (depending on the constants γ, C and Θ) which is possible by what we have seen above.

Finally, we estimate the C^1-norm of P_1. Using (9.5) and the choice of ξ in (9.4) we obtain $\|Y - X\|_{C^1} = \|P_1\|_{C^1} \le \varepsilon$, and Lemma 9.30 is proved. □

Now we commute the composition with the rotation and perturb in the past.

Lemma 9.31 *Let $X \in \mathfrak{X}^2_\mu(M)$, a non-periodic point $p \in M$, $\varepsilon > 0$, $0 < \kappa < 1$, and a fixed time $T = 1$ be given. Then there exists an angle ξ (not depending on p) such that, for every $\gamma > 0$, there exists $r > 0$ (depending on p) and a zero divergence vector field Y, ε-C^1-close to X, such that:*

(a) *$Y - X$ is supported in*

$$\mathscr{F}_X^{-1}(p)(B(p,r)) = \{\mathscr{P}^t_X(p)(q) : q \in B(p,r), t \in [-1, 0]\};$$

(b) *$\|P^1_Y(q) - \hat{R}_\xi P^1_X(X^{-1}(p))\| < \gamma$ for each $q \in \mathscr{P}_X^{-1}(p)(B(p, \sqrt{1-\kappa}r))$.*

Proof We proceed as in Lemma 9.30, this time for X^{-t}, finding a change of coordinates $\hat{\Psi}(x, y, z) = X^{-\lambda^{-1}x}(0, \varphi(y, z))$. Then we consider $R^{-1}_{\xi g(t)G(\rho)}$ for $t > 0$ and we find Z. We define $Z = \Psi_* Y$ and we get

$$P^1_Y(q) = [P^{-1}_Y(Y^1(q))]^{-1} \approx [P^{-1}_X(p)\hat{R}^{-1}_\xi]^{-1} = \hat{R}_\xi[P^{-1}_X(p)]^{-1} = \hat{R}_\xi P^1_X(X^{-1}(p)),$$

by the same arguments used in the proof of Lemma 9.30. □

Now we use the two previous lemmas to build some useful realizable linear flows.

Lemma 9.32 *Given $X \in \mathfrak{X}^2_\mu(M)$, $\varepsilon > 0$, $0 < \kappa < 1$, a non-periodic point p and a fixed time $T = 1$. Then there exists an angle ξ (not depending on p) such that $L_0 = P^1_X(p)\hat{R}_\xi$ and $L_0 = \hat{R}_\xi P^1_X(p)$ are (ε, κ)-realizable linear flows of length 1 at p.*

Proof We prove that $L_0 = P_X^1(p)\hat{R}_\xi$ is (ε,κ)-realizable. Let $\gamma > 0$. By Remark 9.29, we may choose the open set U to be a ball, say $B(p',r') \subseteq B(p,r)$. Now we apply Lemma 9.30 and we get a zero divergence vector field Y, ε-C^1-close to X, such that $Y - X$ is supported inside the flow-box $\mathcal{F}_X^1(p')(B(p',r'))$ and, for every $q \in B(p', r'\sqrt{1-\kappa})$, we have $\| P_Y^1(q) - P_X^1(p')\hat{R}_\xi \| < \gamma$. We note that, since $r > 0$ can be taken arbitrarily small, the arrival time at $N_{X^1(p)}$ for points in $B(p,r)$ is almost 1.

Taking $K = \overline{B}(p', r'\sqrt{1-\kappa}) \subseteq U$, we get

$$\frac{\overline{\mu}(K)}{\overline{\mu}(U)} = \frac{\pi(1-\kappa)r'^2}{\pi.r'^2} = 1-\kappa$$

and so the first statement of the lemma follows.

For the perturbation P, defined in Lemma 9.30, we have $DX_q = DY_q$ for any $q \in B(p',r') \cup \mathcal{P}_X^1(p)(B(p',r'))$. Therefore item (i) on Definition 9.27 is true. Finally, item (ii) follows from item (b) of Lemma 9.30 and the continuity of the Linear Poincaré Flow.

For $L_0 = \hat{R}_\xi P_X^1(p)$ we proceed analogously now using Lemma 9.31. This completes the proof. □

Lemma 9.33 *Given $X \in \mathfrak{X}_\mu^2(M)$, $\varepsilon > 0$, $0 < \kappa < 1$ and a non-periodic point p, there exists an angle ξ such that, for $|\xi_i| < \xi$, $i = 1, 2$, the composition*

$$N_p \xrightarrow{P_X^1(p)\hat{R}_{\xi_1}} N_{X^1(p)} \xrightarrow{P_X^r(p)} N_{X^{1+r}(p)} \xrightarrow{\hat{R}_{\xi_2} P_X^1(X^{1+r}(p))} N_{X^{r+2}(p)}$$

is an (ε,κ)-realizable linear flow of length $r+2$ at p.

Proof Take $\gamma > 0$. By Lemma 9.32, for $\kappa_1 < \kappa$ we get ξ such that $P_X^1(p)\hat{R}_{\xi_1}$ and $\hat{R}_{\xi_2} P_X^1(X^{1+r}(p))$ are (ε,κ_1)-realizable. By item (1) of Lemma 9.28, the trivial map P_X^r is (ε,κ_1)-realizable. Now, if $\kappa_1 = \kappa/3$, then we use item (2) of Lemma 9.28 and obtain the (ε,κ)-realizability. □

9.3.2.2 Large Rotations

Now we find conditions under which we can rotate by large angles. In the previous section we were able to rotate by time-1, but we need to rotate along arbitrarily long times. What then happens if we increase time?

We want to rotate by an angle 2π, and thus we take a time m such that $\xi m = 2\pi$. But ξ is in general very small, and so m must be very large. Note that the choice of m may affect the norm of the perturbation because, for Ψ given by Lemma 9.25, $\|\Psi\|$ depends on m and, in general, increases with m. One of the reasons for this fact is that, for $v \in N_p$, we may have a very small angle $\measuredangle(X^m(v), X(X^m(p)))$.

Furthermore, the dynamics along the orbit may also obstruct the construction of a small norm perturbation. Let us consider a situation in which this last problem is minimized, say when we have simultaneously

(a) no domination, that is $P_X^t(p)$ is "almost conformal" for all $t \in [0, m]$;
(b) almost right angles: $\measuredangle(N^u_{X^t}(p), N^s_{X^t}(p)) \approx \frac{\pi}{2}$, for all $t \in [0, m]$.

Even if we have properties (a) and (b), our perturbations may not have a small C^1 norm, because the normal directions may be sent almost parallel to E^X. If this problem does not occur, then under conditions (a) and (b) we can perform large rotations with just a single perturbation. In fact, this was the strategy for the development of perturbations for linear differential systems in [40, Lemma 3.5].

In general, we concatenate several time-1 small rotations. However, this concatenation implies that κ grows. In [51, Lemma 3.7], Bochi bypassed this problem using a nested rotation lemma. Here we adapt this method to our setting.

Let $\mathscr{E}(p) \subseteq N_p$ be an ellipse centered in p. As in [51], the *eccentricity* E of an ellipse is defined by $E := \sqrt{\frac{\text{major axis}}{\text{minor axis}}}$. We also consider the map $J \in SL(2, \mathbb{R})$ such that $J(\mathscr{E}(p))$ is a disk. For $r > 0$ we define the ellipse $\mathscr{E}(p, r) := J^{-1}(B(p, r))$. Given any ellipse $\mathscr{E}(p, r)$ and $J \in SL(2, \mathbb{R})$ such that $J(\mathscr{E}(p, r))$ is a disk, we denote by $\hat{E}_\xi$ the elliptical rotation defined by $J^{-1}\hat{R}_\xi J$.

Lemma 9.34 *Let there be given $X \in \mathfrak{X}^2_\mu(M)$, a non-periodic point $p \in M$, $\varepsilon > 0$, $0 < \kappa < 1$, a fixed time $T = 1$ and $e \geq 1$. Then there exists $\hat{\varepsilon} > 0$ (not depending on p) such that, for every $\gamma > 0$, there exists $r > 0$ (depending on p) with the following properties.*

If $\mathscr{E}(p, r)$ is an ellipse with eccentricity $E \leq e$ and diam$(\mathscr{E}(p, r)) < \varepsilon$, $\hat{E}_\xi$ is a rotation of the ellipse $\mathscr{E}(p, r)$ satisfying $\|P^1_X(p) - P^1_X(p)\hat{E}_\xi\| < \hat{\varepsilon}$, then there exists a zero divergence vector field Y of class C^1, ε-C^1-close to X such that

(a) *$Y - X$ is supported in the flowbox $\mathscr{F}^1_X(p)(\mathscr{E}(p, r))$;*
(b) *$\|P^1_Y(q) - P^1_X(p)\hat{E}_\xi\| < \gamma$ for all $q \in \mathscr{E}(p, r\sqrt{1-\kappa})$.*

Proof The proof is the same as for Lemma 9.30, but the angle ξ depends also on E. We use Lemma 9.25 and consider the flow-box $\mathscr{F}^1_T(p)(\mathscr{E}(p, r))$. Let $J \in SL(2, \mathbb{R})$ be such that $J(\mathscr{E}(p, r))$ is a disk. Now we take the flow-box $\mathscr{F}^1_T(p)(J(\mathscr{E}(p, r)))$ and define the elliptical rotation by $E_{\xi g(t)G(\rho)} := J^{-1}R_{\xi g(t)G(\rho)}J$. Since $E = \|J\| = \|J^{-1}\|$, ξ should be smaller than the one in Lemma 9.30, which is obtained by taking $\hat{\varepsilon} \approx 0$. □

The next simple lemma says that, if we fix a small ellipse in a small ball $B(p, r) \subseteq N_p$ and consider its arrival into $N_{X^1(p)}$, then this set is almost the image under the Linear Poincaré Flow at p of the same ellipse modulo translations. A similar statement is proved in [51, Lemma 3.6].

Lemma 9.35 *Let $X^t : M \longrightarrow M$ be a C^1-flow, $\zeta \in]0, 1[$ (near 1), and $E \geq 1$. There exists $r > 0$ such that, for all ellipses $\mathscr{E}(q, \tilde{r}) \subseteq B(p, r) \subseteq N_p$ with eccentricity $\leq E$, $P^1_X(p)(\mathscr{E}(q, \zeta\tilde{r}) - q) + \mathscr{P}^1_X(p)(q)$ is contained in*

$$\mathscr{P}^1_X(p)(\mathscr{E}(q, \tilde{r})) \subseteq P^1_X(p)(\mathscr{E}(q, (2-\zeta)\tilde{r}) - q) + \mathscr{P}^1_X(p)(q).$$

Since $P_X^t(p)$ is modified area-preserving, we measure the non-conformality using its norm $\|P_X^t(p)\|$ in the following way: suppose that $d \geq a$ are the major axis and the minor axis (respectively) of the ellipse $\mathscr{E}(p) = P_X^t(p)(B(p,1))$. Then the eccentricity of $\mathscr{E}(p)$ is $E = \sqrt{d/a}$. Since $\|P_X^t(p)\| = d$ and, by volume-preservation, we have $a^{-1} = d \cdot x(t)$, we conclude that $E = \sqrt{d/a} = \sqrt{d^2 x(t)} = d\sqrt{x(t)} = \|P_X^t(p)\|\sqrt{x(t)}$.

The next lemma is a version of [51, Lemma 3.7] and says that bounded eccentricity is crucial to concatenate many elliptical small rotations keeping κ controlled.

Lemma 9.36 *Given $X \in \mathfrak{X}_\mu^2(M)$, $\varepsilon > 0$, $0 < \kappa < 1$ and $E \geq 1$, there exists $\hat{\varepsilon} > 0$ satisfying the following.*

Let $p \in M$ be a non-periodic point and suppose that for some $n \in \mathbb{N}$ we have $\|P_X^j(p)\| \leq E\sqrt{x^{-1}(j)}$ for $j = 1, \dots, n$. If $L_j : N_{X^j(p)} \to N_{X^{j+1}(p)}$ for $j = 0, \dots, n-1$ is a sequence of linear maps satisfying:

(a) $L_{j-1} \dots L_0(B(p,1)) = P_X^j(p)(B(p,1))$ *for* $j = 1, \dots, n$;
(b) $\|P_X^1(X^j(p)) - L_j\| < \hat{\varepsilon}$ *for* $j = 0, 1, \dots, n-1$;

then $\{L_0, L_1, \dots, L_{n-1}\}$ is an (ε, κ)-realizable linear flow at p.

Proof Let us start by fixing some constants. We choose $\kappa_0 < \kappa$ by taking $\lambda \in]0,1[$ near 1 such that $\lambda^{4n}(1-\kappa_0) > 1-\kappa$. We take $\zeta \in]0,1[$ such that $\zeta \in]\lambda, 1[$ and $2 - \zeta \in]1, \lambda^{-1}[$. Let $\hat{\varepsilon}$ be given by Lemma 9.34 depending on ε, E and κ_0.

Using *(a)* and the hypothesis $\|P_X^j(p)\| \leq E\sqrt{x^{-1}(j)}$ (for each j) we consider $\hat{E}_j$ the rotation of the ellipse $P_X^j(p)(B(p,1))$. Then $L_j := P_X^1(X^j(p))\hat{E}_j$ satisfies *(b)* and these are the linear maps which we will (ε, κ)-realize. Let $\gamma > 0$ be given.

By Lemma 9.35 applied n times, and shrinking the radii at each step, there exists $r_1 > 0$ such that, for each j and all ellipses $\mathscr{E}(q, \tilde{r}) \subseteq B(X^j(p), r_1) \subseteq N_{X^j(p)}$ with eccentricity $\leq E$, we have

- $P_X^1(X^j(p))(\mathscr{E}(q, \zeta\tilde{r}) - q) + \mathscr{P}_X^1(X^j(p))(q) \subseteq \mathscr{P}_X^1(X^j(p))(\mathscr{E}(q, \tilde{r}))$, and
- $\mathscr{P}_X^1(X^j(p))(\mathscr{E}(q, \tilde{r})) \subseteq P_X^1(X^j(p))(\mathscr{E}(q, (2-\zeta)\tilde{r}) - q) + \mathscr{P}_X^1(X^j(p))(q)$.

Take $r < r_1$. Now we will define the vector field Y and the measurable set K as in Definition 9.27. By Remark 9.29, we consider $U = B(p', r') \subseteq B(p, r)$. For each j, denoting $\mathscr{P}_X^j(p)(p')$ by p'_j, we define a sequence of ellipses $\mathscr{E}_s^j \subseteq N_{X^j(p)}$ with eccentricity $\leq E$ by

$$\mathscr{E}_s^0 = B(p', sr') \text{ for } s \in]0,1] \quad \text{and} \quad \mathscr{E}_s^j = P_X^j(p)(B(p', sr') - p') + p'_j.$$

Decreasing r, if necessary, these ellipses satisfy the conditions of Lemma 9.35. Thus, for each j we also have

$$\mathscr{E}_{s\lambda^{-1}}^{j+1} \supset \mathscr{E}_{s(2-\zeta)}^{j+1} \supseteq \mathscr{P}_X^1(X^j(p))(\mathscr{E}_s^j) \supseteq \mathscr{E}_{s\zeta}^{j+1} \supseteq \mathscr{E}_{s\lambda}^{j+1}.$$

For each j we apply Lemma 9.34 to p'_j, κ_0, $\mathscr{E}_s^j$ and $\hat{E}_j$, with $s = \lambda^n$. Hence there exists a vector field Y_j such that

(i) $Y_j - X$ is supported in the flow-box $\mathcal{F}^1_X(p'_j)(\mathcal{E}^j_{\lambda^n})$;
(ii) for every $q_j \in \mathcal{E}^j_{\lambda^n\sqrt{1-\kappa_0}}$ we have $\| P^1_Y(q_j) - P^1_X(p'_j)\hat{E}_j\| < \gamma$.

By item (i), the Y_j have disjoint supports, and so we define $Y := \sum_{j=0}^{n-1} Y_j$. We define also $K := \overline{\mathcal{E}}^0_{\lambda^{2n}\sqrt{1-\kappa_0}} = \overline{B}(p', \lambda^{2n}\sqrt{1-\kappa_0}r')$. Hence we obtain

$$\frac{\overline{\mu}(K)}{\overline{\mu}(U)} = \frac{\pi(\lambda^{2n}\sqrt{1-\kappa_0}r')^2}{\pi r'^2} = \lambda^{4n}(1-\kappa_0) > 1-\kappa.$$

Let us see that when we iterate we have a nested sequence, i.e., for all $q \in K$ and each j we have $Y^j(q) \in \mathcal{E}^j_{\lambda^n\sqrt{1-\kappa_0}}$. We have $\mathcal{P}^1_Y(p')(\overline{\mathcal{E}}^0_s) \subseteq \overline{\mathcal{E}}^1_{s(2-\zeta)} \subset \overline{\mathcal{E}}^1_{s\lambda^{-1}}$, and so for every j we obtain $\mathcal{P}^j_Y(p')(\overline{\mathcal{E}}^0_s) \subset \overline{\mathcal{E}}^j_{s\lambda^{-j}} \subseteq \overline{\mathcal{E}}^j_{s\lambda^{-n}}$. Hence for $s = \lambda^{2n}\sqrt{1-\kappa_0}$ we get $\mathcal{P}^1_Y(p')(K) \subset \overline{\mathcal{E}}^j_{\lambda^n\sqrt{1-\kappa_0}}$, and the orbit of q will be inside the domain of each of these rotations.

Finally, to prove that $\| P^1_Y(Y^j(q)) - L_j\| < \gamma$ for all $q \in K$, we use item (ii), and go back and decrease r once more, if necessary. □

9.3.3 Blending Oseledets Directions Along an Orbit Segment

Given $p \in \Gamma^+_m(X)$, we suppose that

$$\Delta(X^t(p), r) = \frac{\| P^r_X(X^t(p))|N^s_{X^t(p)}\|}{\| P^r_X(X^t(p))|N^u_{X^t(p)}\|} \geq c,$$

for $c \gg 1$ and $0 \leq t + r \leq m$. Then the dynamics sends vectors near $N^u_{X^t(p)}$ into vectors near $N^s_{X^{t+r}(p)}$ during that period. The next simple lemma, whose proof may be found in [51, Lemma 3.9], clarifies this behavior. Denote two unit vectors by $n^\sigma_t \in N^\sigma_{X^t(p)}$ for $\sigma = u, s$.

Lemma 9.37 *Given an angle ξ, there exists $c > 1$ such that, if $\Delta(X^t(p), r) > c$, then there exists $v \in N_{X^t(p)} \setminus \{\mathbf{0}\}$ satisfying $\measuredangle(v, n^u_t) < \xi$ and $\measuredangle(P^r_X(X^t(p)) \cdot v, n^s_{t+r}) < \xi$.*

The next lemma gives us the conditions under which we may apply Lemma 9.36. For a proof of the next lemma see [39, Lemma 4.0.11].

Lemma 9.38 *Let $\xi > 0$ and $d > 1$ be given. Let $\measuredangle(N^u_{X^t(p)}, N^s_{X^t(p)}) > \xi$ for each $t > 0$, and let $d^{-1} \leq \frac{\|P^t_X(p)|_{N^s_p}\|}{\|P^t_X(p)|_{N^u_p}\|} \leq d$. Then there exists $E > 1$ such that $\| P^t_X(p)\| \leq E\sqrt{x(t)^{-1}}$ for all $t > 0$.*

Now we are able to mix the Oseledets subspaces by small perturbations along orbits which have no domination.

Lemma 9.39 *Let $X \in \mathfrak{X}^2_\mu(M)$, $\varepsilon > 0$ and $0 < \kappa < 1$. There exists $m \in \mathbb{N}$ such that, for every $p \in \Delta_m(X) = \{p \in \mathcal{O}^+ : \frac{\|P^m_X(p)|N^s_p\|}{\|P^m_X(p)|N^u_p\|} \geq \frac{1}{2}\}$, there exists an (ε, κ)-realizable linear flow such that $L^m(N^u_p) = N^s_{X^m(p)}$.*

Proof First we set up the constants. We take $\xi > 0$ the minimum of the angles satisfying simultaneously Lemma 9.32 and Lemma 9.33 and depending on X, ε and $\kappa/2$. We set $C := \max\{\|DX^{\pm 1}\| : p \in M\}$ and c given by Lemma 9.37 depending on the angle ξ. We take $c > C^2$ also.

Lemma 9.38 gives us $E > 1$ depending on ξ and $d = 2c^2$. Let $\hat{\varepsilon} > 0$, depending on X, ε, κ and E, be given by Lemma 9.36. Let $\beta > 0$ be such that $\|R_{\xi_0} - Id\| \leq C^{-1}E^{-2}\hat{\varepsilon}$ for $\xi_0 < \beta$.

Finally, we take a sufficiently large $m \in \mathbb{N}$ satisfying $m \geq \frac{2\pi}{\beta}$. Now we divide the proof into three steps.

Step I – Small angle between the Oseledets subspaces. We assume that

$$\text{for some } r \in [0, m] \text{ we have } \measuredangle(N^u_{X^r(p)}, N^s_{X^r(p)}) < \xi. \tag{9.8}$$

We take advantage of this fact and we define a realizable linear flow of length 1 in the following way. On the one hand, if $r < m - 1$, the linear map is based at $X^r(p)$ and is defined by $L_0 := P^1_X(X^r(p))R_\xi$. On the other hand, if $r > m - 1$, the linear map is based at $X^{r-1}(p)$ and is defined by $L_0 := R_\xi P^1_X(X^{r-1}(p))$. Now we use Lemma 9.32 and concatenate from the right and left, if necessary, with trivial maps by using item (1) of Lemma 9.28. We obtain $L^m(N^u_p) = N^s_{X^m(p)}$.

Step II – Locally N^s dominates N^u. Now we assume that

$$\text{for some } 0 \leq r + t \leq m \text{ we have } \Delta(X^t(p), r) > c. \tag{9.9}$$

From Lemma 9.37 there exists a vector $v \in N_{X^t(p)}$ such that $\measuredangle(v, n^u_t) < \xi$ and $\measuredangle(P^r_X(X^t(p)) \cdot v, n^s_{t+r}) < \xi$. Since ξ is small, we apply Lemma 9.32 at $X^t(p)$ and at $X^{t+r}(p)$. By the choice of c above, we get $r > 2$ and so we have disjoint perturbations.

Therefore, our first rotation allows us to send $N^u_{X^t(p)}$ onto the subspace $\mathbb{R} \cdot v$. The flow then maps this direction into $P^r_X(X^t(p)) \cdot v$ in time r and, finally, another rotation sends $P^r_X(X^t(p)) \cdot \mathbb{R}v$ onto $N^s_{X^{t+r}(p)}$.

Now we use Lemma 9.28 and concatenate the three realizable linear flows, say rotation-trivial-rotation, by using Lemma 9.33 and we get $L^m(N^u_p) = N^s_{X^m(p)}$.

Step III – Conformal behavior. Finally, we suppose that we do not have either (9.8) or (9.9). We set up the conditions for Lemma 9.38. Since $\Delta(p, m) \geq \frac{1}{2}$ and (9.9) is false we have

$$\Delta(X^r(p), t) = \Delta(X^{t+r}(p), m - t - r)^{-1} \Delta(p, m) \Delta(p, r)^{-1} \geq \frac{1}{2c^2}.$$

Therefore, since $d = 2c^2$,

$$\frac{1}{d} \leq \frac{\| P_X^t(X^r(p))|_{N^s_{X^r(p)}} \|}{\| P_X^t(X^r(p))|_{N^u_{X^r(p)}} \|} \leq d,$$

for every r, t with $0 \leq r + t \leq m$. We observe that, in particular, for $r = 0$, we have $\measuredangle(N^u_{X^t(p)}, N^s_{X^t(p)}) > \xi$ for all $t \in [0, m]$. Now we use Lemma 9.38 and conclude that $\| P_X^t(p)\| \leq E\sqrt{x(t)^{-1}}$ for each $t \in [0, m]$.

Let us take $\xi_0, \xi_1, \ldots, \xi_{m-1}$ with $\xi_j < \beta$ for each j and satisfying also $\sum_{j=0}^{m-1} \xi_j = \measuredangle(N^u_p, N^s_p)$. We define

$$L_j : N_{X^j(p)} \to N_{X^{j+1}(p)}, \qquad v \mapsto P_X^{j+1}(p) R_{\xi_j} [P_X^j(p)]^{-1} \cdot v.$$

Let us check the conditions of Lemma 9.36. Since, by definition, for each j we have $L_{j-1} \ldots L_0 = P_X^j(p) R_{\sum_{i=0}^{j-1} \xi_i}$, we obtain item (a) of Lemma 9.36. Now we have

$$\begin{aligned}
\| P_X^1(X^j(p)) - L_j \| &\leq \| P_X^1(X^j(p)) - P_X^{j+1}(p) R_{\xi_j} [P_X^j(p)]^{-1} \| \\
&= \| P_X^1(X^j(p))[Id - P_X^j(p) R_{\xi_j} [P_X^j(p)]^{-1}] \| \\
&\leq \| P_X^1(X^j(p))\| \| P_X^j(p)[Id - R_{\xi_j}][P_X^j(p)]^{-1}] \| \\
&\leq \| P_X^1(X^j(p))\| \| P_X^j(p)\| \| [P_X^j(p)]^{-1} \| \| Id - R_{\xi_j} \| \\
&\leq C E \sqrt{x^{-1}(j)} E \sqrt{x(j)} \| Id \quad R_{\xi_j} \|.
\end{aligned}$$

(In the last inequality we have used $\| P_X^{-t} \| \leq E\sqrt{x(t)}$.) Therefore we obtain

$$\| P_X^1(X^j(p)) - L_j \| \leq C E^2 \| Id - R_{\xi_j} \| \leq \hat{\varepsilon}$$

and item (b) of Lemma 9.36 is true. From this lemma we have the realizability, and therefore

$$L^m(N^u_p) = L_{m-1} \circ \cdots \circ L_0(N^u_p) = P_X^m(p) R_{\sum_{j=0}^{m-1} \xi_j}(N^u_p) = P_X^m(p) \cdot N^s_p = N^s_{X^m(p)},$$

which proves the lemma. □

9.3.4 Lowering the Norm: Local Procedure

We begin this section by adapting to our setting the results [51, Lemma 3.12] and [53, Lemma 4.5]. The first lemma (for a proof see [39, Lemma 5.0.13]) gives us information about *when* we have a recurrence to a positive measure set. The second lemma is an elementary result which relates the original norm to a new norm.

Lemma 9.40 *Let $X^t : M \to M$ be a measurable μ-invariant flow, $\Delta \subseteq M$ a positive measure set, $\Gamma := \cup_{t\in\mathbb{R}} X^t(\Delta)$ and let $\gamma > 0$.*

Then there exists a measurable function $T : \Gamma \to \mathbb{R}$ such that, for μ-a.e. $p \in \Gamma$, all $t \geq T(p)$ and every $\tau \in [0, 1]$, there exists some $s \in [0, t]$ satisfying $|\frac{s}{t} - \tau| < \gamma$ and $X^s(p) \in \Delta$.

Consider $p, q := X^t(p) \in \Gamma$ and the map $P : N_p \to N_q$ whose matrix written with respect to the *Oseledets basis* (given by $\{n_p^u, n_p^s\}$ and $\{n_q^u, n_q^s\}$) is

$$P = \begin{pmatrix} a^{uu} & a^{us} \\ a^{su} & a^{ss} \end{pmatrix}.$$

Let $\|P\|_{\max} = \max\{|a^{uu}|, |a^{us}|, |a^{su}|, |a^{ss}|\}$.

Lemma 9.41 *We have the bounds*

(a) $\|P\| \leq 4(\sin\angle(N_p^u, N_p^s))^{-1}\|P\|_{\max}$;
(b) $\|P\|_{\max} \leq (\sin\angle(N_q^u, N_q^s))^{-1}\|P\|$.

Now we are able to decrease the norm under a small perturbation.

Lemma 9.42 *Let $X \in \mathfrak{X}_\mu^2(M)$, where X^t is aperiodic and all hyperbolic sets have zero measure. Let $\varepsilon, \delta > 0$, $0 < \kappa < 1$. Then there exists a measurable function $T : M \to \mathbb{R}$ such that for μ-a.e. $p \in M$ and every $t \geq T(p)$, there exists a (ε, κ)-realizable linear flow at p with length t such that $\|L^t(p)\| \leq e^{t\delta}$.*

Proof First we take $m \in \mathbb{R}$ large enough given by Lemma 9.39 and depending on $X, \varepsilon, \kappa/2$. Write $\Gamma := \Gamma_m^+(X)$, $\Delta := \Delta_m(X)$ and note that X^t aperiodic implies that $\mu(\Gamma_m^*(X)) = \mu(\Gamma)$. We have $\mu(\mathcal{O}^0 \cup \Gamma) = 1$, for otherwise there exists a hyperbolic set with positive measure, contradicting our hypothesis. We suppose that $\mu(\Gamma) > 0$, for otherwise, with $\mu(\Gamma) = 0$, μ-a.e. point $p \in M$ would be such that $\lambda^+(p) = 0$ and there is nothing to prove, because a trivial map does the work.

We recall that $\Gamma = \bigcup_{t\in\mathbb{R}} X^t(\Delta)$. For μ-a.e. $p \in \Gamma$, the Oseledets' Theorem in the particular case of a three-dimensional conservative flow gives us $Q(p)$ such that for all $t \geq Q(p)$

(1) $\frac{1}{t}\log\|P_X^t(p)\cdot n^u\| < \lambda^+(p) + \delta$ for all $n^u \in N_p^u \setminus \{\mathbf{0}\}$;
(2) $\frac{1}{t}\log\|P_X^t(p)\cdot n^s\| < -\lambda^+(p) + \delta$ for all $n^s \in N_p^s \setminus \{\mathbf{0}\}$;
(3) $-\log\sin\measuredangle(N_{X^t(p)}^u, N_{X^t(p)}^s) < t\delta$.

By using Lemma 9.40, with $\tau = 1/2$, we get recurrence to Δ approximately in the middle of the orbit segment. However, to get good estimates of the norm of the linear map L^t, points in the orbit after this time must also satisfy items (1) and (2) above.

Let $B_n := \{p \in \Gamma : Q(p) \leq n\}$ for $n \in \mathbb{N}$. We have $B_n \subseteq B_{n+1}$ and $\mu(\Gamma \setminus B_n) \underset{n\to\infty}{\longrightarrow} 0$.

We consider now the following family of sets:

$$C_0 := \emptyset \quad \text{and} \quad C_n := \bigcup_{t\in\mathbb{R}} X^t(\Delta \cap X^{-m}(B_n)).$$

It is easy to see that $C_n \underset{n\to\infty}{\longrightarrow} \Gamma$, and so the measurable function $T : \Gamma \to \mathbb{R}$ will be μ-a.e. defined on each $C_n \setminus C_{n-1}$ for $n \in \mathbb{N}$.

Taking $c > \max\{\log \|DX_p^1\| : p \in M\}$ yields the Lyapunov exponents of any $p \in \mathscr{O}$ less than c (recall the definition of $\mathscr{O}$ at the beginning of Sect. 9.3). For $p \in \Gamma$ we have non-zero Lyapunov exponents, and so we have well-defined Oseledets one-dimensional subspaces N_p^u and N_p^s.

Let $\gamma = \min\{1/6, \delta/c\}$. Now we use Lemma 9.40, replacing Δ by $\Delta \cap X^{-m}(B_n)$ and Γ by $\bigcup_{t\in\mathbb{R}} X^t(\Delta \cap X^{-m}(B_n))$. By this lemma, for each n, there exists a measurable function $T_n : C_n \to \mathbb{R}$ such that, for μ-a.e $p \in C_n$ and for all $t \geq T_n(p)$, there exists some $s \in [0, t]$ satisfying $X^s(p) \in \Delta \cap X^{-m}(B_n)$ and $|\frac{s}{t} - \frac{1}{2}| < \gamma$.

Now we define a sufficiently large $T(p)$ for $p \in C_n \setminus C_{n-1}$ so that

$$T(p) \geq \max\left\{T_n(p), \frac{m}{\gamma}, 6Q(p), \frac{1}{\delta}\log\frac{4}{\sin\measuredangle(N_p^u, N_p^s)}\right\}. \tag{9.10}$$

Let $p \in C_n \setminus C_{n-1}$ and $t \geq T(p)$. Since $t \geq T(p) \geq T_n(p)$, we obtain $X^s(p) \in \Delta$. Hence, by Lemma 9.39, we define a $(\varepsilon, \kappa/2)$-realizable linear flow $L_1 : N_{X^s(p)} \longrightarrow N_{X^{s+m}(p)}$, sending $N^u_{X^s(p)}$ into $N^s_{X^{s+m}(p)}$. Now we concatenate from right to left with trivial maps and, by Lemma 9.28, we obtain a (ε, κ)-realizable linear flow defined by

$$N_p \xrightarrow{L_0} N_{X^s(p)} \xrightarrow{L_1} N_{X^{s+m}(p)} \xrightarrow{L_2} N_{X^t(p)}$$

with $L_0 = P_X^s(p)$ and $L_2 = P_X^{t-m-s}(X^{s+m}(p))$.

To estimate $\|L^t(p)\|$ we consider the linear maps relative to a suitable unitary basis $\{n^u_{X^r(p)}, n^s_{X^r(p)}\}$ for $r \in [0, t]$, which is invariant for the Linear Poincaré Flow, so they have the form

$$L_2 = \begin{pmatrix} c^{uu} & 0 \\ 0 & c^{ss} \end{pmatrix}, \qquad L_1 = \begin{pmatrix} b^{uu} & b^{us} \\ b^{su} & b^{ss} \end{pmatrix}, \qquad L_0 = \begin{pmatrix} a^{uu} & 0 \\ 0 & a^{ss} \end{pmatrix}.$$

The key observation is that $b^{uu} = 0$. Consider the product matrix

$$L^t(p) = L_2 \cdot L_1 \cdot L_0 = \begin{pmatrix} 0 & a^{uu}b^{us}c^{ss} \\ a^{ss}b^{su}c^{uu} & a^{ss}b^{ss}c^{ss} \end{pmatrix}.$$

Claim For $p \in C_n \setminus C_{n-1}$ and $t \geq T(p)$ we have:

(a) $\max\{\log|a^{uu}|, \log|c^{uu}|\} < \frac{1}{2}t(\lambda^+(p) + 4\delta)$;
(b) $\max\{\log|a^{ss}|, \log|c^{ss}|\} < \frac{1}{2}t(-\lambda^+(p) + 4\delta)$.

Proof of the claim To prove that $\log|a^{uu}| < \frac{1}{2}t(\lambda^+(p) + 4\delta)$ we first note that $s > t(1/2 - \gamma) > t/3 \geq T(p)/3 \geq Q(p)$ and so, by Oseledets' Theorem, we have

$\log|a^{uu}| = \log|P^s_X(p)\cdot n^u_p| < s(\lambda^+(p)+\delta)$. Since $\gamma\lambda^+(p) < \gamma c \le \delta$ and $\gamma < 1/2$, we get

$$s\big(\lambda^+(p)+\delta\big) < t(1/2+\gamma)\big(\lambda^+(p)+\delta\big) < t\big(\lambda^+(p)/2+\delta/2+\lambda^+(p)\gamma+\gamma\delta\big)$$
$$< t\big(\lambda^+(p)/2+\delta/2+\delta+\delta/2\big) < \frac{1}{2}t\big(\lambda^+(p)+4\delta\big)$$

and the inequality follows.

To prove that $\log|c^{uu}| < \frac{1}{2}t(\lambda^+(p)+4\delta)$ we consider the fact that $X^s(p)\in X^{-m}(B_n)$, therefore $X^{s+m}(p)\in B_n$ and $Q(X^{s+m}(p))\le n$ by definition of B_n. So we will have the approximation rate given by Oseledets' Theorem if $t-m-s>n$. By (9.10) for $t\ge T(p)$, we have $-m/t\ge -\gamma$. Since $-s/t > -\frac{1}{2}-\gamma$ and $-\gamma \ge -1/6$ we obtain

$$t-m-s = t\left(1-\frac{m}{t}-\frac{s}{t}\right) > t\left(\frac{1}{2}-2\gamma\right) > \frac{t}{6} \ge Q(p)\ge n.$$

Thus $t-m-s$ will be sufficiently large to use item (1) above. Hence

$$\log|c^{uu}| = \log|P_X^{t-m-s}(X^{s+m}(p))\cdot n^u_{X^{s+m}(p)}| < (t-m-s)(\lambda^+(p)+\delta)$$
$$< t(1-m/t-s/t)(\lambda^+(p)+\delta) < t(\gamma+1/2)(\lambda^+(p)+\delta)$$
$$= t(\gamma\lambda^+(p)+\gamma\delta+\lambda^+(p)/2+\delta/2)$$
$$< t(\delta+\delta/2+\lambda^+(p)/2+\delta/2) = \frac{1}{2}t(\lambda^+(p)+4\delta).$$

We note that item (b) is analogous to item (a) and the claim is proved. □

Now we estimate $\|L_1\|_{\max}$. First note that

$$s+m > t(1/2-\gamma+m/t) > t(1/2-\gamma) > t/6 > Q(p)\ge n,$$

and so by (3) we have $(\sin\measuredangle(N^u_{X^{s+m}(p)}, N^s_{X^{s+m}(p)}))^{-1} < e^{(s+m)\delta} < e^{t\delta}$. Since L_1 is (ε,κ)-realizable, we conclude that $\|L_1 - P^m_X(X^s(p))\|$ is small. Therefore, because $t > T(p)\ge m/\gamma$ and $\gamma c\le\delta$, we get $\|L_1\|\le e^{mc}\le e^{t\gamma c}\le e^{t\delta}$. By Lemma 9.41(b) we also have

$$\|L_1\|_{\max}\le \sin^{-1}\measuredangle(N^u_{X^{s+m}(p)}, N^s_{X^{s+m}(p)})\|L_1\|\le e^{2t\delta}.$$

Now we estimate each of the entries of the product matrix:

$$|a^{uu}b^{us}c^{ss}| \le e^{\frac{1}{2}t(\lambda^+(p)+4\delta)+2t\delta+\frac{1}{2}t(-\lambda^+(p)+4\delta)} = e^{6t\delta};$$

$$|a^{ss}b^{su}c^{uu}| \le e^{\frac{1}{2}t(-\lambda^+(p)+4\delta)+2t\delta+\frac{1}{2}t(\lambda^+(p)+4\delta)} = e^{6t\delta};$$

$$|a^{ss}b^{ss}c^{ss}| \le e^{\frac{1}{2}t(-\lambda^+(p)+4\delta)+2t\delta+\frac{1}{2}t(-\lambda^+(p)+4\delta)} \le e^{-t\lambda^+(p)+6t\delta} \le e^{6t\delta}.$$

This implies the inequality $\|L^t(p)\|_{\max} < e^{6t\delta}$. By item (a) of Lemma 9.41 we have

$$\|L^t(p)\| \le 4\frac{1}{\sin\measuredangle(N_p^u, N_p^s)}\|L^t(p)\|_{\max}.$$

But $t \ge T(p) \ge \frac{1}{\delta}\log\frac{4}{\sin\measuredangle(N_p^u,N_p^s)}$ so $\frac{4}{\sin\measuredangle(N_p^u,N_p^s)} \le e^{t\delta}$ and we get $\|L^t(p)\| \le e^{7t\delta}$. Replacing δ by $\delta/7$ we conclude that $\|L^t(p)\| \le e^{t\delta}$ and the lemma is proved. □

9.3.4.1 Realizing Vector Fields

Let $X \in \mathfrak{X}^2_\mu(M)^*$, where X^t is aperiodic and also all hyperbolic sets have zero Lebesgue measure. Given $\varepsilon, \delta > 0$ and $0 < \kappa < 1$, we assume that m is large enough to satisfy Lemma 9.39. By Lemma 9.42, there exists a measurable function $T : M \to \mathbb{R}$ such that, for μ-a.e. $p \in M$ and for every $t \ge T(p)$, there exists a (ε, κ)-realizable linear flow at p with length t such that $\|L^t(p)\| \le e^{t\delta}$.

This means that, for each $\gamma > 0$, we can find $r = r(p,t) > 0$ such that, for every open subset U of $B(p,r)$, there exists a conservative vector field Y, ε-C^1-close to X, satisfying the following. First, $Y = X$ outside the flow-box $\mathcal{F}^t_X(p)(U)$ and, moreover, $\|P^t_Y(q) - L^t\| < \gamma$ for q in a measurable subset $K \subset U$ which is close to U in measure (i.e. $\overline{\mu}(K) > (1-\kappa)\overline{\mu}(U)$ for a small $\kappa > 0$); see Definition 9.27.

Since $\|L^t(p)\| \le e^{t\delta}$ we conclude that $\|P^t_Y(q)\| \le e^{\delta t} + \gamma$ for all $q \in K$ (where $\gamma \approx 0$). The vector field Y *realizes* the abstract map $L^t(p)$, i.e., $P^t_Y(q) \approx L^t(p)$ for a large percentage of points $q \in U \subset N_p$ (see Fig. 9.5).

9.3.5 Lowering the Norm: Global Procedure

Now we use the local construction of realizable linear flows with a small norm in order to decrease the function $LE(\cdot)$ for a zero divergence vector field Y C^1-close

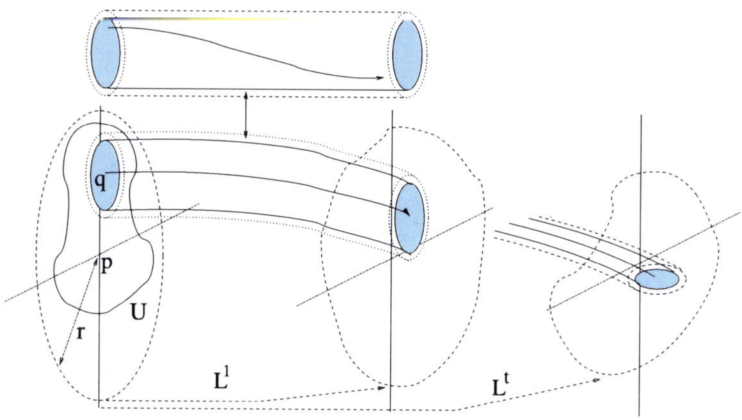

Fig. 9.5 Realizing vector fields given a linear map L^t

to X. We will use the notion of suspension or special flow built under a function, from Sect. 2.3.2.2.

It follows from the Ambrose-Kakutani Theorem [12, 13] that any aperiodic flow is isomorphic to some special flow. This theorem is applicable to the flow of Proposition 9.23. The isomorphism is given by $W : M \to M_h$, where M_h is the flow under the roof function h. The measure $\mu^* = W_* \mu$ is decomposed into the product of Lebesgue measure in $\mathbb{R}$ and an R-invariant measure $\tilde{\mu}$ in the base Σ, that is, $\int_{M_h} f(x,s) d\mu^* = \int_\Sigma \left(\int_0^{h(x)} f(x,s)\, ds \right) d\tilde{\mu}(x)$. Thus we have a simplified representation of our flow $X^s(p)$. Hereafter we assume that our flow has this representation.

9.3.5.1 Sections of Flows and Special Flows

Given a special flow over a section Σ, the set $Q = \cup_{t \in \mathbb{R}} X^t(\Sigma)$ is called the *Kakutani castle*. The *tower of height i*, which is denoted by T_i, is the set below the graph of $h(B_i)$, where $B_i = \{x \in \Sigma : h(x) = i\}$, that is, $T_i = X^{[0,i]}(B_i)$. The next lemma is the flow version of [51, Lemma 4.1].

Lemma 9.43 *Let $X^t : M \to M$ be a μ-preserving aperiodic flow. For every positive measure set $U \subseteq M$ and every $h \in \mathbb{R}$, there exists a $\tilde{\mu}$-positive measure section $B \subseteq U$ such that $X^{[0,h]}(B)$ is a flow-box and B is maximal* (*i.e. no set containing B and with larger measure has the same properties as B*).

Proof See [39, Lemma 6.1.2]. □

For a μ-generic point p, Lemma 9.42 gives us $T(p)$ which, in general, is very large. Hence Lemma 9.43 is crucial to avoid the overlapping of perturbations.

Consider a vector field X under the conditions of Proposition 9.23. For all Y ε-C^1 close to X we define $C := \max\{\|P_Y^1(p)\| : p \in M\}$. We take $\kappa = \delta^2$. Using the function given by Lemma 9.42, we define $Z_h := \{p \in M : T(p) \leq h\}$. Then we get $\mu(M \setminus Z_h) \underset{h \to \infty}{\longrightarrow} 0$, and so for h sufficiently large we have

$$\mu(M \setminus Z_h) < \delta^2. \tag{9.11}$$

We intend to build a special flow with a ceiling function with height not less than h and section inside Z_h. Since, for large h, the set Z_h has almost full measure, by Lemma 9.43, we get a $\tilde{\mu}$-positive measure set $B \subseteq Z_h$. If $x \in B$, then $h(x) \geq h$ and, since $B \subseteq Z_h$, we have $h \geq T(x)$, and so the conditions of Lemma 9.42 are satisfied.

Let $\hat{Q}$ be the castle with base B. We have $\hat{Q} \supseteq Z_h$ in the measure theoretical sense so, by (9.11), we get the inequality $\mu(\hat{Q}^c) \leq \delta^2$. We define the subcastle $Q \subseteq \hat{Q}$ by excluding the towers of $\hat{Q}$ with height larger than $3h$. Adapting [51, Lemma 4.2] (the details are fully presented in [39, Lemma 6.2.1]) we obtain

$$\mu(\hat{Q} \setminus Q) < 3\delta^2. \tag{9.12}$$

9.3.5.2 The Zero Divergence Vector Field Y ε-C^1-Close to X

Now we make use of the realizability of vector fields and the properties of special flows to construct a conservative vector field Y inside the subcastle Q by gluing a finite number of local perturbations supported on self-disjoint flow-boxes. We note that the measures $\tilde{\mu}$ and $\overline{\mu}$ are equivalent. In the next lemma we follow [53, Lemma 4.14].

Lemma 9.44 *Given $\gamma > 0$, there exist Y, ε-C^1-close to X, a castle U for Y^t and a subcastle K for Y^t such that:*

(a) *the castle U is open;*
(b) $\mu(U \setminus Q) < \gamma$ *and* $\mu(Q \setminus U) < \gamma$;
(c) $\mu(U \setminus K) < \kappa(1+\gamma)$;
(d) $Y^t(U) = X^t(U)$ *and* $Y^t = X^t$ *outside the castle* U;
(e) *if q is in the base of K and $h(q)$ is the height of the tower of K that contains q, then*

$$\| P_Y^{h(q)}(q)\| \le e^{\delta h(q)} + \gamma.$$

Proof The castle Q is a measurable set and, since μ is Borel regular, there exists a compact $J \subseteq Q$ such that

$$\mu(Q \setminus J) < \gamma \mu(\hat{Q})/2. \tag{9.13}$$

The compact J is a X^t-castle with the same structure as Q (i.e., preserving the same dynamics of bases and towers as is the case for the castle Q). Now we choose an open castle V such that $J \subseteq V$ with

$$\mu(V \setminus J) < \gamma \mu(\hat{Q}), \tag{9.14}$$

and also with the same structure of Q and J. For every point p_1 in $J \cap B$ we have $h(p_1) \ge h$. Since $(J \cap B) \subset Z_h$ we have $T(p_1) \le h$, and therefore $T(p_1) \le h \le h(p_1)$. So, for all $t_1 \ge T(p_1)$, and for γ fixed, there exists a radius $r_1(p_1, t_1)$ (decrease r_1 if one leaves the open castle V) such that for almost (related with $\kappa = \delta^2$) every point in $U_1 = B(p_1, r_1) \subseteq N_{p_1}$, more precisely for each point $q \in K_1 \subseteq U_1$, we have a vector field Y_1 supported in a small flow-box containing the orbit segment $X^{[0,t_1]}(p_1)$ such that $\| P_{Y_1}^{t_1}(q)\| \le e^{\delta t_1} + \gamma$.

We continue by choosing $p_i' s$ and by Vitali's arguments we fill up J with a union U of self-disjoint open flow-boxes in order to obtain

$$\mu(J \setminus U) \le \gamma \mu(J)/2. \tag{9.15}$$

The set U is a X^t-castle with its section (base of the castle) the union of the U_i. So for each i we get a vector field Y_i supported in a small flow-box containing the orbit segment $X^{[0,t_i]}(p_i)$, ε-C^1-close to X, and such that $\| P_{Y_i}^{t_i}(q)\| \le e^{\delta t_i} + \gamma$ for all $q \in K_i \subseteq U_i$.

We define $Y = Y_i$ inside each flow-box and $Y = X$ outside. Since these flow-boxes are pairwise disjoint, the vector field is well defined and it is ε-C^1-close to X. Note that V is also a castle for Y^t, and U is also a Y^t-subcastle of the Y^t-castle V (having for base the union of all U_i). We take K as the Y^t-subcastle with a section equal to the union of all the K_i. By construction of U, we get items (a), (d) and (e).

Now we prove item (b). Recalling that $V \supseteq U$ and $J \subseteq Q$, by (9.14) we obtain

$$\mu(U \setminus Q) < \mu(V \setminus J) < \gamma\mu(\hat{Q}) \leq \gamma.$$

To prove that $\mu(Q \setminus U) < \gamma$ we use (9.13) and (9.15) and conclude that

$$\mu(Q \setminus U) \leq \mu(Q \setminus J) + \mu(J \setminus U) < \gamma\mu(\hat{Q}) < \gamma.$$

Finally, for item (c) we observe that from item (b) we deduce that $\mu(U) < \mu(Q) + \mu(U \setminus Q) < 1 + \gamma$. The inequality $\mu(U \setminus K) < \kappa\mu(U)$ then leads to $\mu(U \setminus K) < \kappa(1 + \gamma)$. The proof is complete. □

9.3.5.3 Computing $LE(Y)$

We take $t = h\delta^{-1}$ (we may assume that this is an integer). By Lemma 9.22, we obtain $LE(Y) \leq \int_M \frac{1}{t} \log \| P_Y^t(p) \| d\mu(p)$. By the above construction for orbit segments inside the castle K and starting in the base, we guarantee a small upper Lyapunov exponent. Therefore we define the set of points whose orbit stays for a long time in K by $G := \{p \in M : Y^s(p) \in K \ \forall s \in [0, t]\}$ and denote by G^c its complementary set.

Lemma 9.45 *For* $p \in G$ *we have* $\| P_Y^t(p) \| < e^{t(1+6\log C)\delta}$ *for some* $C > 0$.

Proof Let $p \in G$. We split the orbit segment $X^{[0,t]}(p)$ by return-times at B_K (the section of the castle K), say $t = b + r_n + \cdots + r_2 + r_1 + a$, where

$$X^a(p), X^{r_1+a}(p), X^{r_2+r_1+a}(p), \ldots, X^{\sum_{i=1}^n r_i + a}(p)$$

are all in the base B_K. It is clear that $a, b, r_i \in]0, 3h]$ except when $p \in B_K$ $(a = 0)$ and $X^t(p) \in B_K$ $(b = 0)$. Note that

$$\begin{aligned}\| P_Y^t(p) \| = \| P_Y^{b+\sum_{i=1}^n r_i + a}(p) \| &\leq \| P_Y^b(X^{\sum_{i=1}^n r_i + a}(p))) \| \\ &\times \| P_Y^{r_n}(X^{\sum_{i=1}^{n-1} r_i + a}(p))) \| \times \cdots \times \| P_Y^{r_1}(X^a(p)) \| \times \| P_Y^a(p) \|.\end{aligned}$$

But these maps are based at points in B_K. Hence by Lemma 9.44(e) and for $C := \max\{\| DX_p^1 \| : p \in M\}$ (this constant is valid for any vector field ε-C^1-close to X) we get

$$\| P_Y^t(p) \| \leq C^{3h} e^{\sum_{i=1}^n r_i\delta} C^{3h} \leq e^{(b+\sum_{i=1}^n r_i + a)\delta} C^{6h} \leq e^{t\delta} C^{6\delta t} \leq e^{t(1+6\log C)\delta}$$

concluding the proof. □

For points in G^c we use inequality (9.12), Lemma 9.44 and some elementary observations (see [53, Lemma 4.16]) to deduce the following.

Lemma 9.46 *Let $\gamma = \delta^2 h^{-1}$ as in Lemma 9.44. Then $\mu(U \cup \Gamma_m^*(X) \setminus G) < 12\delta$.*

Now using Lemma 9.46 we obtain $\mu(G^c) < 12\delta$. Finally, we finish the proof of Proposition 9.23 and consequently Theorem 9.3:

$$\begin{aligned} LE(Y) &= \inf_{n\geq 1} \int_M \frac{1}{n} \log \| P_Y^n(p) \| d\mu(p) \leq \int_M \frac{1}{t} \log \| P_Y^t(p) \| d\mu(p) \\ &\leq \int_G \frac{1}{t} \log \| P_Y^t(p) \| d\mu(p) + \int_{G^c} \frac{1}{t} \log \| P_Y^t(p) \| d\mu(p) \\ &\leq (1 + 6 \log C)\delta\mu(G) + \log C \mu(G^c) = (1 + 18 \log C)\delta. \end{aligned}$$

Now we replace δ by $\frac{\delta}{(1+18\log C)}$ in the proof, and Proposition 9.23 is complete.

9.3.6 Proof of the Dichotomy with Singularities (Theorem 9.4)

We begin by noting that, since X^t is aperiodic, the measure of all singularities is zero. Moreover, when we estimate the C^1-norm of the perturbation P, defined in (9.7), the choice of $r(p)$ in (9.5) guarantees that, even near singularities (when $c \approx 0$), the perturbation can be done. Furthermore, if in Theorem 9.24 we take Ω with C^∞ boundary and g, f also C^∞, the diffeomorphism φ, provided by Dacorogna Moser, is also C^∞. So our conservative flow-box theorem guarantees a C^∞ conservative change of coordinates Ψ. Note that the perturbation P, defined in (9.7), is also C^∞; moreover we know by [277] that $\mathfrak{X}_\mu^\infty(M)$ is C^1-dense in $\mathfrak{X}_\mu^1(M)$.

9.3.6.1 Adapting the Proof with Singularities, from Theorem 9.3

The following proposition is similar to Proposition 9.23. The main difference is *where* the computation of the entropy function is done.

Proposition 9.47 *Let $X \in \mathfrak{X}_\mu^\infty(M)$ and $\varepsilon, \delta > 0$. There exists $m \in \mathbb{N}$ and a zero divergence C^∞ vector field Y, ε-C^1-close to X, which equals X outside the open set $\Gamma_m(X)$ and is such that $LE(Y, \Gamma_m(X)) < \delta$.*

Proof For a fixed $m \in \mathbb{N}$ we have $p \in \Gamma_m^+(X) \setminus \Gamma_m^*(X)$ if p is periodic, has positive Lyapunov exponent and belongs to $\Gamma_m(X)$. We consider the following simple claim. For a proof see [40, Lemma 3.1].

Lemma 9.48 *Given $\delta > 0$, there exists $m \in \mathbb{N}$ such that $\mu(\Gamma_m^+(X) \setminus \Gamma_m^*(X)) < \delta$.*

We take $m \in \mathbb{N}$ satisfying both Lemma 9.39 and Lemma 9.48 and let $T : \Gamma_m^*(X) \to \mathbb{R}$ be similar to the function of Lemma 9.42. We now define

$$Z_h = \{p \in \Gamma_m^*(X) : T(p) \leq h\}.$$

It is standard that $\mu(\Gamma_m^*(X) \setminus Z_h) \underset{h\to\infty}{\longrightarrow} 0$, and so we may choose $h > 0$ satisfying

$$\mu(\Gamma_m^*(X) \setminus Z_h) < \delta^2 \mu(\Gamma_m^*(X)).$$

Now we increase h, if necessary, and use Oseledets' Theorem to obtain the inequality

$$\|P_X^t(p)\| < e^{t\delta} \quad \text{for all } t \geq h, \tag{9.16}$$

for $p \in \mathcal{O}^0(X)$. Since $X^t : \Gamma_m^*(X) \to \Gamma_m^*(X)$ is an aperiodic flow, we can follow the construction of section 9.3.5.2 and finally compute $LE(Y, \Gamma_m^*(X))$. Analogously, we define

$$G := \{p \in \Gamma_m^*(X) : Y^s(p) \in K, \quad \text{for all } s \in [0, t]\},$$

and denote $\frac{1}{t}\log\|P_{(\cdot)}^t(p)\|$ by $A_p^t(\cdot)$. Then we compute

$$\begin{aligned}
LE(Y, \Gamma_m(X)) &\leq \int_{\Gamma_m(X)} A_p^t(Y)\, d\mu \\
&\leq \int_{\Gamma_m(X)\setminus(U\cup\Gamma_m^+(X))} A_p^t(Y)\, d\mu + \int_{U\cup\Gamma_m^+(X)\setminus G} A_p^t(Y) d\mu \\
&\quad + \int_G A_p^t(Y)\, d\mu.
\end{aligned}$$

Since $Y = X$ outside U, by (9.16) we obtain

$$\int_{\Gamma_m(X)\setminus(U\cup\Gamma_m^+(X))} A_p^t(Y)\, d\mu \leq \int_{\Gamma_m(X)\setminus\Gamma_m^+(X)} A_p^t(X)\, d\mu \leq \delta.$$

Setting $C := \max\{\|P_X^1(p)\| : p \in M\}$ and using Lemma 9.48 and Lemma 9.46, we conclude that $\int_{U\cup\Gamma_m^+(X)\setminus G} A_p^t(Y)\, d\mu \leq 13\delta\log C$. Finally, on G, our construction allows us to obtain $\int_G A_p^t(Y)\, d\mu(p) \leq \delta$ and the proposition is proved. □

9.3.6.2 The Concluding Argument

Let $\tilde{X} \in \mathfrak{X}_\mu^1(M)$ and $\tilde{\varepsilon} > 0$ be given. We will prove that there exists $Y \in \mathfrak{X}_\mu^1(M)$, $\tilde{\varepsilon}$-C^1-close to X satisfying the conclusions of Theorem 9.4. For $\varepsilon = \tilde{\varepsilon}/2$, there exists $X \in \mathfrak{X}_\mu^\infty(M)$ ε-C^1-close to $\tilde{X}$. It suffices to prove Theorem 9.4 for the vector field X and $\varepsilon > 0$.

Proof of Theorem 9.4 Let $X \in \mathfrak{X}^\infty_\mu(M)$ and $\varepsilon > 0$. We will find Y ε-C^1-close to X and a partition $M = D \cup O$ into Y^t-invariant sets such that $\forall p \in O$ we have zero Lyapunov exponents and D is a countable increasing union of compact invariant sets Λ_{m_n} admitting an m_n-dominated splitting for the Linear Poincaré Flow. We define the sequence $\{X_n\}_{n\geq 0} \in \mathfrak{X}^\infty_\mu(M)$, $m_n \in \mathbb{N}$, and eventually $\varepsilon_n > 0$ for $n \geq 0$.

Take $X_0 = X$, $\theta > 1$ (near 1) and $\delta_n \underset{n\to 0}{\to} 0$.

If $\int_{\Gamma_m(X)} \lambda^+(X)\, d\mu = 0$ for some $m \in \mathbb{N}$, then we are finished by taking $Y = X$, $D = \Lambda_m(X)$ and O a full measure subset of $\Gamma_m(X)$. Otherwise, for some $m = m_0$ and $X = X_0$, we have $\int_{\Gamma_{m_0}(X_0)} \lambda^+(X_0)\, d\mu > 0$. Let $\varepsilon_0 \in (0, \varepsilon/2)$ be such that

$$\int_{\Gamma_{m_0}(X_0)} \lambda^+(Z)\, d\mu \leq \theta \int_{\Gamma_{m_0}(X_0)} \lambda^+(X_0)\, d\mu,$$

for all vector fields Z which are $2\varepsilon_0$-C^1-close of X_0 and $Z = X_0$ outside $\Gamma_{m_0}(X_0)$. We observe that such ε_0 exists because $LE(\cdot, \Gamma_{m_0}(X_0))$ is upper semicontinuous and $\Gamma_{m_0}(X_0)$ is invariant, both for X_0^t and Z^t. Recursively, knowing X_{n-1}, m_{n-1} and $\varepsilon_{n-1} \in (0, \varepsilon 2^{-n})$, we define $X_n \in \mathfrak{X}^\infty_\mu(M)$, $m_n \in \mathbb{N}$, and eventually $\varepsilon_n > 0$.

By Proposition 9.47, there exists $m_n \in \mathbb{N}$ and a perturbation of X_{n-1}, $X_n \in \mathfrak{X}^\infty_\mu(M)$, ε_{n-1}-C^1-close to X_{n-1}, with $X_n = X_{n-1}$ outside $\Gamma_{m_n}(X_{n-1})$ and such that

$$\int_{\Gamma_{m_n}(X_{n-1})} \lambda^+(X_n)\, d\mu < \delta_n.$$

We assume that $m_n \geq m_{n-1}$ and note that $\Gamma_{m_n}(X_n) \subseteq \Gamma_{m_n}(X_{n-1}) \subseteq \Gamma_{m_{n-1}}(X_{n-1})$. If $\int_{\Gamma_{m_n}(X_n)} \lambda^+(X_n) = 0$, then we finish the argument by taking $Y = X_n$, $D = \Lambda_{m_n}(Y)$ and O a full measure subset of $\Gamma_{m_n}(Y)$. Otherwise, if $\int_{\Gamma_{m_n}(X_n)} \lambda^+(X_n) > 0$, we choose $\varepsilon_n \in (0, \varepsilon_{n-1}/2)$ such that $B(X_n, 2\varepsilon_n) \subseteq B(X_{n-1}, \varepsilon_{n-1})$ and also

$$\int_{\Gamma_{m_n}(X_n)} \lambda^+(Z) d\mu < \theta \int_{\Gamma_{m_n}(X_n)} \lambda^+(X_n) d\mu,$$

for all vector fields Z which are $2\varepsilon_n$-C^1-close to X_n and $Z = X_n$ outside $\Gamma_{m_n}(X_n)$.

We continue this procedure recursively and if we obtain $\int_{\Gamma_{m_n}(X_n)} \lambda^+(X_n)\, d\mu = 0$ for some $n \in \mathbb{N}$, then we conclude the proof. Otherwise the sequence $\{X_n\}_{n\geq 0}$ converges in the C^1 topology to some $Y \in \mathfrak{X}^1_\mu(M)$. Moreover, since $\varepsilon_n < \varepsilon/2^n$, we conclude that Y is ε-C^1-close to X.

If we set $D = \bigcup_{n\in\mathbb{Z}^+} \Lambda_{m_n}(X_n)$, since $\Lambda_{m_n}(X_n) \supseteq \Lambda_{m_{n-1}}(X_{n-1})$ and $Y = X_n$ at $\Lambda_{m_n}(X_n)$, then Y^t has an m_n-dominated splitting at $\Lambda_{m_n}(X_n)$.

Letting $\Gamma := D^c = \bigcap_{n\in\mathbb{Z}^+} \Gamma_{m_n}(X_n)$, then $\Gamma \subseteq \Gamma_{m_n}(X_n)$. To finish the proof of Theorem 9.4, we now check that $\int_\Gamma \lambda^+(Y)\, d\mu = 0$.

We note that $Y \in B(X_n, 2\varepsilon_n)$ for all $n \in \mathbb{N}$. Hence we have

$$\int_\Gamma \lambda^+(Y)\, d\mu < \int_{\Gamma_{m_n}(X_n)} \lambda^+(Y)\, d\mu \leq \theta \int_{\Gamma_{m_n}(X_n)} \lambda^+(X_n)\, d\mu \leq \theta\delta_n \underset{n\to\infty}{\longrightarrow} 0.$$

We conclude that we have zero Lyapunov exponents in a full measure subset O of Γ. The proof of Theorem 9.4 is complete. □

Now we consider the reason why Theorem 9.4 is stated for a *dense* subset instead of a *residual* subset. In [53], the authors developed a strategy to obtain a residual subset; unfortunately, this strategy does not apply in our case. Let us see why: they start with a C^1 system X which is a continuity point of the entropy function. Then they define the "jump" (see [53, pp 1467]) of this semi-continuous function which is an integral over $\Gamma_\infty(X) := \bigcap_m \Gamma_m(X)$. Being a continuity point implies that the "jump" is zero. So, by definition of "jump", $\mu(\Gamma_\infty(X)) = 0$ or $\lambda^+(p) = 0$ for μ-a.e. point $p \in \Gamma_\infty(X)$ and the statements of Theorem 9.4 are verified. In order to estimate a lower bound for the "jump" we perturb the original vector field X as we did to prove Theorem 9.4. But our conservative flow-box theorem may not be applied to X, unless X is of class C^2, and so this argument only works for $X \in \mathfrak{X}^2_\mu(M)$. *However this set, equipped with C^1 topology, is not a Baire space, and so in general residual sets are meaningless.*

As explained at the beginning of Sect. 9.3, this can be extended to a full dichotomy for C^1 generic incompressible vector fields using the extension of the Bowen result for positive volume invariant sets having a dominated decomposition for the Linear Poincaré Flow.

Chapter 10
Related Results and Recent Developments

Here we briefly present other related results about three-dimensional flows and some recent developments, advancing some conjectures about future developments that we feel are achievable.

10.1 More on Singular-Hyperbolicity

As shown in Chap. 4, Doering in [79] was able to prove, for 3-manifolds, that robustly transitive vector fields are Anosov. Vivier, in [268], extended the results of Doering to higher dimensions, showing that a C^1 robustly transitive vector field on a compact boundaryless n-manifold, with $n \geq 3$, does not have any singularity. Similarly to Doering, Vivier showed that, for n-dimensional manifolds, robustly transitive vector fields admit a global dominated splitting. In Chap. 4 we proved a weaker version of this.

Robust transitivity of a compact invariant set in the C^1 topology for three-dimensional flows is characterized in Chap. 5: such robust sets are singular-hyperbolic attractors or repellers, and are hyperbolic if they have no singularities.

In Sect. 6.3 of Chap. 6 we provided sufficient conditions for a singular-hyperbolic attractor, *having only one singularity*, to be robustly transitive. *The conditions depend on the behavior of the C^1 nearby vector fields* in the trapping region of the original attractor.

Conjecture 10.1 Sufficient conditions can be found which imply the robust transitivity of any singular-hyperbolic attractor (with no restriction on the number of equilibria) depending only on the given vector field.

10.1.1 Topological Dynamics

Some aspects of the topological dynamics of the Lorenz geometric model were studied by Komuro in [124, 125], where it was proved that most geometrical Lorenz

V. Araújo, M.J. Pacifico, *Three-Dimensional Flows,* Ergebnisse der Mathematik und ihrer Grenzgebiete. 3. Folge / A Series of Modern Surveys in Mathematics 53,
DOI 10.1007/978-3-642-11414-4_10, © Springer-Verlag Berlin Heidelberg 2010

attractors do not have the shadowing property, and their expansive properties are investigated. In [123] Klinshpont finds a topological invariant for the Lorenz attractor, allowing him to exhibit an uncountable number of non-homeomorphic Lorenz attractors in the unfolding of a certain homoclinic loop. In [50], Birman and Williams analyze the knot type of the periodic orbits of the geometric model, and in [93] Ghrist and Holmes use the Lorenz attractor to investigate the existence of flows realizing all links and knots as periodic orbits in 3-manifolds, and an explicit ordinary differential equation with such properties is exhibited. The reader is advised to consult the survey [200] by Pesin and Sinai.

Morales in [163] shows that a vector field Y which is C^1 close to a given one X in a 3-manifold exhibiting a singular-hyperbolic attractor Λ must have at least one singularity, and the number of attractors of Y near Λ is bounded above by the number of singularities of X in Λ.

Bautista showed in [35] that the geometric Lorenz model is a homoclinic class and, together with Morales, proved in [37] that every singular-hyperbolic attractor admits a (hyperbolic) periodic orbit.

Arroyo and Hertz, in [28], have advanced a significant step towards an affirmative answer to the Palis Conjecture for 3-flows; see Sect. 2.8. They show that any C^1 vector field on a compact 3-manifold can be approximated by another one showing one of the following phenomena:

- uniform hyperbolicity with the non-cycle condition,
- a homoclinic tangency, or
- a singular cycle.

Arroyo and Pujals, in [27], show that a singular-hyperbolic attractor has a dense set of periodic orbits and is the homoclinic class associated to one of these orbits. These results show that singular-hyperbolic attractors do play the same role as the basic pieces of Smale's Spectral Decomposition. In the same work, Arroyo and Pujals also prove that there exists a relatively open and dense subset of a singular-hyperbolic attractor whose points admit well-defined unstable manifolds which have uniform size. This may prove to be a decisive step towards a criterion for C^1 robustness of singular-hyperbolic attractors which depends only on the attractor. We present a robustness criterium for singular-hyperbolic attractors with only one equilibria in Chap. 6, but depending on the nearby vector fields.

Conjecture 10.2 There is a characterization of robust singular-hyperbolic attractors, and robust 2-sectionally expanding attractors, with any (finite) number of equilibria.

Abdenur, Avila and Bochi [1] prove that nontrivial homoclinic classes of C^r-generic flows are topologically mixing. This implies that, given a nontrivial C^1-robustly transitive set Λ of a vector field X, there is a small C^1-perturbation Y of X such that the continuation Λ_Y of Λ is a topologically mixing set for Y. In particular, robustly transitive flows become topologically mixing after C^1-perturbations. These results generalize a theorem by Bowen [61] on the basic sets of generic Axiom A flows: they are mixing for a C^r generic family of vector fields in compact manifolds.

In this work it is also shown that the set of flows whose nontrivial homoclinic classes are topologically mixing is in general *not* open and dense.

10.1.2 Attractors that Resemble the Lorenz Attractor

In Chap. 5, we present several results showing that every robust attractor of a 3-flow containing equilibria looks like a geometric Lorenz attractor.

Shil'nikov and Turaev, in [263], present an example of a 4-dimensional quasi-attractor and study its perturbations. The quasi-attractor is partially hyperbolic with contracting dominated direction and volume expanding central direction with dimension higher than 2, and contains a singularity with a complex eigenvalue. Such a quasi-attractor can not be destroyed by small perturbations of the system. Since this set has a dense subset of homoclinic tangencies, it is called a *wild strange attractor*, because Newhouse in [182] (see also [193] for a detailed presentation) showed that dynamics with such features for surface diffeomorphism imply the existence of infinitely many attracting periodic orbits for open subsets of arbitrarily close maps.

Conjecture 10.3 Sets of this type exhibit the main chaotic features of the singular-hyperbolic attractors: they support a physical probability measure and are expansive.

Lorenz, in [141], reports a careful numerical study of what seems to be a strange (chaotic) attractor in four dimensions for a system of 2-degree polynomial equations. Rovella in [232] proves existence and persistence of *contracting Lorenz attractors*, that is, with the contracting eigenvalue condition $-\lambda_3 > \lambda_1$; see the following Sect. 10.1.4.

In [187], Pacifico, Rovella and Viana prove that certain parametrized families of one-dimensional maps with infinitely many critical points exhibit global chaotic behavior in a persistent way. Later Araújo and Pacifico, in [20], proved that these maps have a unique physical (absolutely continuous) measure which varies continuously in the space of parameters with very nice statistical properties. An application of the methods developed in these works yields a proof of existence and even persistence of global spiral attractors for smooth flows in three dimensions, to be given in [74].

In [198] Pesin proposed abstract models for attractors with singularities, called generalized hyperbolic attractors, and studied their properties.

Bonatti, Pumariño and Viana, in [58], construct a multidimensional Lorenz-like attractor that is C^1-robust and contains a singularity with at least two positive eigenvalues. Their construction works in dimensions greater than or equal to 5. They also obtain a physical measure for these attractors for an open set of flows in the C^∞ topology.

More recently Metzger and Morales [156] introduced the class of *sectionally hyperbolic vector fields* on n-manifolds containing the singular-hyperbolic systems on 3-manifolds, the multidimensional Lorenz attractors of [58] and the C^1 robustly transitive sets in Li, Gan and Wen [132]. We present these notions in Chap. 5.

Conjecture 10.4 The 2-sectionally expanding attractors satisfy all the properties we have obtained for the singular-hyperbolic attractors in dimension 3. Namely, the periodic orbits are dense in these attractors; they are homoclinic classes associated to a hyperbolic periodic orbit; they are expansive/chaotic; they admit a unique physical measure which is an equilibrium state for the logarithm of the unstable Jacobian of the flow.

10.1.3 Lorenz-Like Attractors Through the Unfolding of Singular Cycles

It is natural to investigate whether an attractor resembling the Lorenz attractor can be obtained as a result of a bifurcation of a singular cycle of a given vector field.

Rychlik and Robinson studied the existence of Lorenz-like attractors in generic unfoldings of resonant double homoclinic loops, for flows in dimension three, in a series of works [227–229, 239]. Rychlik starts with a vector field with a Lorenz-like singularity σ with a connection between both branches of the unstable manifold of σ and the bidimensional stable manifold of σ, such that the singular cycle obtained is of *inclination-flip type*; see Sect. 3.2.2. Robinson considers a resonant connection, that is, the eigenvalues at σ are $\lambda_2 < \lambda_3 < 0 < \lambda_1$ but $\lambda_3 + \lambda_1 = 0$. This means that the singularity neither expands nor contracts volume in the central-unstable direction. In the setting of axially-symmetric vector fields, both cases are co-dimension two bifurcations.

Similarly Ushiki, Oka, Kokubu [264] and Dumortier, Kokubu, Oka [83] show that Lorenz-like attractors appear in the unfolding of local bifurcation of certain degenerate singularities. Analogously Bamón in [31] obtains attractors resembling the Lorenz attractor in higher dimensions unfolding cycles associated to degenerate singularities.

An extension of the results of Robinson, in dimension 3, was obtained in [175] by Morales, Pacifico and San Martin.

Conjecture 10.5 The new 2-sectionally expanding attractors can also be obtained through the unfolding of higher-dimensional cycles associated to (hyperbolic) equilibria.

10.1.4 Contracting Lorenz-Like Attractors

Rovella in [186, 232] presented a parametrized model, similar to the geometric Lorenz model described in Sect. 3.3, which exhibits an attractor for a positive Lebesgue measure subset of the parameter space. This attractor contains a singularity with three real eigenvalues $\lambda_2 < \lambda_3 < 0 < \lambda_1$ but, unlike a Lorenz-like singularity, we have $\lambda_1 < -\lambda_3$, that is, the central-unstable direction at the singularity is *volume contracting*.

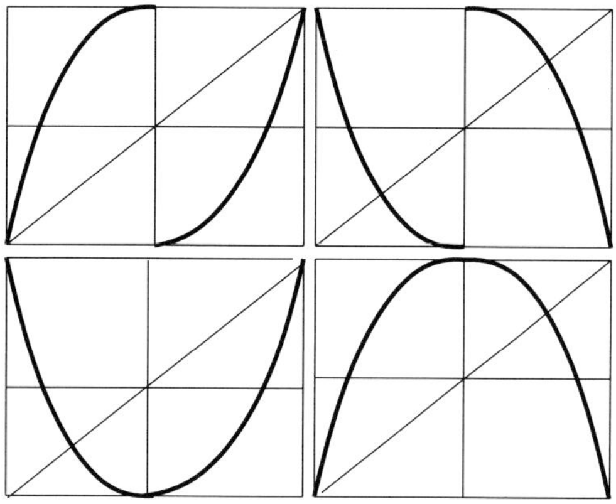

Fig. 10.1 The one-dimensional map for the contracting Lorenz model

This construction is very similar to the geometric Lorenz model, amounting essentially to replacing the one-dimensional map f, whose graph is presented in Fig. 3.24 and obtained through projecting along the contracting foliation, by the map g whose graph can be any of these sketched in Fig. 10.1.

The parameters of these maps describe the vertical coordinates of the critical points of each branch of continuity of the maps in Fig. 10.1.

This example is far from being robustly transitive, but it is 2-persistent. Here 2-*persistent* means that there exists a surface S on the space of C^∞ vector fields such that

$$\lim_{r\to 0} \frac{\operatorname{Leb}\left(\{(u,v)\in B_r(0,0) : \Lambda_{X_{(u,v)}} \text{ is an attractor for } X_{(u,v)}\}\right)}{\operatorname{Leb}\left(B_r(0,0)\right)} = 1$$

for 2-parameter families of C^∞ vector fields $X_{u,v}$ transversal to S at $X_{(0,0)} = X_0$. Moreover, for the parameters above, Lebesgue almost every point in a neighborhood of the attractor has a positive Lyapunov exponent.

Rovella also showed that, in a neighborhood $\mathcal{U}$ of X *outside of the surface* S, there exists an *open and dense subset* $\mathcal{U}_1 \subset \mathcal{U}$ such that for all $Y \in \mathcal{U}_1$ the set $\Lambda_Y = \cap_{t>0} X^t(U)$ *consists of the union of one or at most two attracting periodic orbits, a hyperbolic set of topological dimension one, a singularity, and wandering orbits linking them.*

Metzger in [153, 154] proved the existence of a physical measure and its stability under random perturbations for the contracting Lorenz model. More recently Metzger and Morales, in [155], showed that the contracting Lorenz attractor is also a homoclinic class in a 2-persistent way.

Conjecture 10.6 There exists a higher dimensional example of an attractor for a smooth flow with several positive Lyapunov exponents, which is not 2-sectionally expanding, not robust, but persistent along parameterized families of flows with finitely many parameters. This attractor has a physical measure which is hyperbolic, an equilibrium state with respect to the logarithm of the center-unstable Jacobian of the flow, depends continuously on the attractor (on the parameters where it exists)

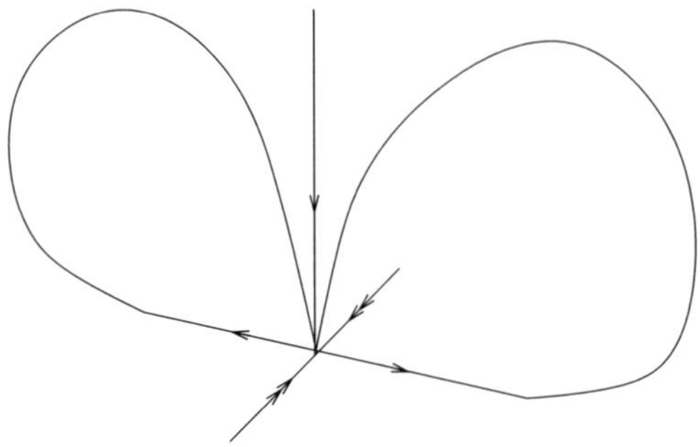

Fig. 10.2 A double homoclinic connection

and has exponentially large deviation estimates and exponential decay of correlations.

10.1.5 *Contracting Lorenz-Like Attractors Through the Unfolding of Singular Cycles*

Recently, in [176], Morales, Pacifico and San Martin proved that, similarly to the (expanding) geometric Lorenz attractors, contracting Lorenz-like attractors can be obtained by unfolding a resonant double homoclinic connection with a *contracting Lorenz-like singularity* σ, i.e. the eigenvalues are $\lambda_2 < \lambda_3 < 0 < \lambda_1$ with $\lambda_3 + \lambda_1 < 0$; see Fig. 10.2.

We note that contracting Lorenz-like attractors persist only in a measure theoretical sense. In this setting, the authors prove the existence of a non-degenerate two-parameter family of vector fields generically unfolding the singular cycle described above, which admits a positive Lebesgue measure subset of parameters such that the corresponding flow exhibits a contracting Lorenz-like attractor.

Conjecture 10.7 There are cycles, for higher-dimensional smooth vector fields, whose unfolding generates examples of non-robust but persistent attractors along finite-dimensional parameterized families, exhibiting several positive Lyapunov exponents.

10.2 Dimension Theory, Ergodic and Statistical Properties

Afraimovich and Pesin in [6] investigate the dimensional properties of "triangular maps" which are a class of maps generalizing the Poincaré first return map P of the geometric Lorenz model.

Concerning fractal dimensions of Lorenz attractors we mention the results of Leonov [130, 131] together with Bouichenko [54]. The first contains explicit formulas for the Lyapunov dimension of the Lorenz attractor and, in the second, a simple upper bound on the Hausdorff dimension of Lorenz attractors is given in terms of the parameters of the Lorenz systems of equations (2.2).

Conjecture 10.8 As in a hyperbolic attractor on a surface, the Hausdorff dimension of any singular-hyperbolic attractor on a 3-manifold satisfies *Bowen's formula*: it is the value $2+\gamma$ where γ satisfies $P_{top}(\gamma \log|\det DX^1 \mid E^s|) = 0$, P_{top} is the topological pressure of the attractor, and E^s is the one-dimensional stable bundle over the attractor.

In [161] Morales shows that every (nontrivial) compact invariant subset of a transitive singular set containing a singularity is one-dimensional, extending a similar result of Bowen in [59] in the setting of uniform hyperbolic flows.

Statistical and ergodic properties of the geometrical model were investigated by, among others, Araújo, Pacifico, Pujals and Viana in [21] and Colmenarez [72], whose results and proofs are contained in Sect. 7.3.

In [276] Young shows that the geometrical Lorenz attractor can be approximated by horseshoes with entropy close to that of the Lorenz attractor.

Conjecture 10.9 It is possible to approximate the topological entropy of a singular-hyperbolic or 2-sectionally expanding attractor by the topological entropy of horseshoes contained in the attractor.

The construction of the geometric Lorenz models forces the divergence of the vector field to be strictly negative in an isolating neighborhood of the attractor. This feature is also present in the Lorenz system of equations (2.2) for the classical parameters. It is then trivial to show that the corresponding attractor has zero volume. Recently it was proved by Araújo, Alves, Pacifico, Pinheiro, in [8], that singular-hyperbolic attractors always have zero volume for flows which are Hölder-C^1, although there is no volume dissipative condition on the definition of singular-hyperbolicity. The precise statement of this result and its proof are contained in Chap. 8.

10.2.1 Large Deviations for the Lorenz Flow

We recall that an invariant probability measure μ for a flow X^t on a compact manifold is *physical* if the points z satisfying

$$\lim_{t\to+\infty}\frac{1}{t}\int_0^t \psi\big(X^s(z)\big)\,ds = \int \psi\,d\mu,$$

for all continuous functions ψ form a subset with positive volume (or positive Lebesgue measure) on the ambient space. These time averages are in principle physically observable if the flow models a real world phenomenon admitting some measurable features. It is then natural to consider the rate of convergence of the time averages to the space average, given by the volume of the subset of points whose time averages stay away from the space average by a prescribed amount up to some evolution time. This rate is closely related to the so-called thermodynamical formalism first developed for (uniformly) hyperbolic diffeomorphisms, borrowed from statistical mechanics by Bowen, Ruelle and Sinai (among others, see e.g. [56, 60, 62, 84, 237, 238]). The main insight behind this is that the family $\{\psi \circ X^t\}_{t>0}$ should behave asymptotically in many respects just like an independent identically distributed sequence of random variables.

Studying suspension semiflows over piecewise expanding base transformations, which naturally appear as representations of singular-hyperbolic flows, as in Chap. 7, Araújo in [18] was able to obtain exponential bounds for large deviations for Lebesgue measure on a neighborhood of the geometric Lorenz attractor (and so for the Lorenz flow in the classical parameters, after the work of Tucker [261]; see Chap. 2). More precisely, if we set $\varepsilon > 0$ as an error margin and consider

$$B_t = \left\{ z : \left| \frac{1}{t} \int_0^t \psi\big(X^t(z)\big) - \int \psi \, d\mu \right| > \varepsilon \right\},$$

then sufficient conditions were found, in terms of the base transformation and the roof function, under which the Lebesgue measure of B_t decays to zero exponentially fast, i.e., whether there are constants $C, \xi > 0$ such that

$$\operatorname{Leb}\big(B_t\big) \le C e^{-\xi t} \quad \text{for all } t > 0.$$

We observe that in this setting Lebesgue measure or volume is *not an invariant measure*.

In [152] Melbourne and Nicol and in [221] Rey-Bellet and Young obtained large deviations principles for *invariant measures* in the same setting, including subexponential or polynomial bounds on large deviations depending on the properties of the base transformation.

Conjecture 10.10 These results are also true for general singular-hyperbolic attractors and should be true, under some mild conditions, for singular-hyperbolic attracting sets as well.

Conjecture 10.11 Exponentially large deviation bounds are true for 2-sectionally expanding attractors with respect to the Lebesgue measure and the physical measure.

10.2.2 Central Limit Theorem for the Lorenz Flow

In [114] Holland and Melbourne, building on the work [151] of Melbourne and Nicol, obtained the Almost Sure Invariance Principle for geometrical Lorenz attractors (and so for the Lorenz flow) which, in turn, implies the Central Limit Theorem.

More precisely, if X^t is the geometric Lorenz flow with physical probability measure μ and ψ is a Hölder continuous function (observable) on the manifold with zero mean $\int \psi \, d\mu = 0$, then there exists a Brownian motion $W(t)$ with variance $\sigma^2 > 0$, and there is $\varepsilon > 0$, such that

$$\int_0^t \psi \circ X^s \, ds = W(t) + O(t^{\frac{1}{2}-\varepsilon}) \quad \text{as } t \to +\infty \quad \text{for } \mu\text{-almost all } x.$$

This result, in turn, implies the Central Limit Theorem: in the same setting as above, for any interval $A \subset \mathbb{R}$

$$\mu\left\{x : \frac{1}{\sqrt{t}}\left(\int_0^t \psi \circ X^s \, ds - \mu(\psi)\right) \in A\right\} \xrightarrow[t\to+\infty]{} \frac{1}{\sigma\sqrt{2\pi}} \int_A e^{-\frac{s^2}{2\sigma}} \, ds;$$

and the Law of the Iterated Logarithm

$$\limsup_{t\to+\infty} \frac{1}{\sqrt{2t \log\log t}} \int_0^t \psi \circ X^s \, ds = \sigma \quad \mu\text{-almost everywhere.}$$

Conjecture 10.12 The Central Limit Theorem holds for every singular-hyperbolic attractor and for 2-sectionally expanding attractors.

10.2.3 Decay of Correlations

After obtaining an interesting invariant probability measure for a dynamical system the next thing to do is to study the properties of this measure. Besides ergodicity there are various degrees of mixing (see e.g. [147, 269]).

Given a flow X and an invariant ergodic probability measure μ, we say that the system (X, μ) is *mixing* if for any two measurable sets A, B

$$\mu\big(A \cap X^{-t} B\big) \xrightarrow[t\to\infty]{} \mu(A) \cdot \mu(B) \tag{10.1}$$

or equivalently

$$\int \varphi \cdot \big(\psi \circ X^t\big) \, d\mu \xrightarrow[t\to\infty]{} \int \varphi \, d\mu \int \psi \, d\mu$$

for any pair $\varphi, \psi : M \to \mathbb{R}$ of continuous functions.

Considering φ and $\psi \circ X^t : M \to \mathbb{R}$ as random variables over the probability space (M, μ), this definition just says that "the random variables φ and $\psi \circ X^t$ are asymptotically independent" since the expected value $\mathbb{E}(\varphi \cdot (\psi \circ X^t))$ tends to the product $\mathbb{E}(\varphi) \cdot \mathbb{E}(\psi)$ when t goes to infinity. The *correlation function*

$$\begin{aligned} C_t(\varphi, \psi) &= \big|\mathbb{E}\big(\varphi \cdot (\psi \circ X^t)\big) - \mathbb{E}(\varphi) \cdot \mathbb{E}(\psi)\big| \\ &= \left|\int \varphi \cdot \big(\psi \circ X^t\big) \, d\mu - \int \varphi \, d\mu \int \psi \, d\mu\right| \end{aligned} \tag{10.2}$$

satisfies $C_t(\varphi, \psi) \xrightarrow[t\to\infty]{} 0$ in this case. The *rate of approach to zero of the correlation function* is called *the rate of decay of correlations* for the observables φ and ψ of the system (X, μ).

The study of decay of correlations for hyperbolic systems goes back to the work of Sinai [250] and Ruelle [235]. Many results were obtained for transformations. For a diffeomorphism f the notion of decay of correlations is the same as above replacing X^t by f^n and letting n go to infinity. Since [60, 235] it is known that the *physical* (SRB) measures for Axiom A *diffeomorphisms* are mixing and have *exponential decay of correlations*, that is, there exists a constant $\alpha \in (0, 1)$ such that given φ and ψ there exists $C = C(\varphi, \psi) > 0$ such that

$$C_n(\varphi, \psi) \le C \cdot e^{-\alpha n} \quad \text{for all } n \ge 1, \tag{10.3}$$

for a suitable class of continuous functions $M \to \mathbb{R}$, in this case the Hölder continuous functions.

In more general cases for smooth endomorphisms (see e.g. [10, 113] and references therein) where the inverse in (10.1) is to be taken as the inverse image of f^n, it is possible to have slower rates of decay.

In contrast to the results available in the case of discrete dynamical systems, obtaining the rate of decay of correlations for flows seems to be much more complex and some results have been established for Anosov flows only recently. Ergodicity and mixing for geodesic flows on manifolds of negative curvature are known since the early half of the XXth century [16, 116, 249].

The proof of exponential decay of correlations for geodesic flows on manifolds of constant negative curvature was first obtained in two [70, 160, 220] and three dimensions [210] through group theoretical arguments.

10.2.4 *Decay of Correlations for the Return Map and Quantitative Recurrence on the Geometric Lorenz Flow*

Quantitative recurrence estimations and logarithm laws can be seen in the following framework: we are interested in a quantitative estimation of the speed of approaching of a certain orbit $X^t(x)$ of the system to a given point x_0. We consider the time

$$\tau_r(x, x_0) = \inf\{t \in \mathbb{R}^+ : X^t(x) \in B_r(x_0)\}$$

needed for the orbit of x to enter the ball with radius r centered at x_0 for the first time, and the asymptotic behavior of $\tau_r(x, x_0)$ as r decreases to 0. Often this is a power law of the type $\tau_r \sim r^{-d}$ and then we may try to extract the exponent d by looking at the behavior of

$$R(x, x_0) = \lim_{r\to 0} \frac{\log \tau_r(x, x_0)}{-\log r}. \tag{10.4}$$

In this way, we have a hitting time indicator for orbits of the system. Another way to look at the same phenomena is by considering the behavior of the ratio of the distance $\frac{-\log d(X^t(x),x_0)}{\log t}$ as $t \to \infty$ (for the equivalence see [92]).

Hitting time results of this kind (sometime replacing balls with other suitable target sets) have been proved in many continuous time dynamical systems of geometrical interest: geodesic flows, unipotent flows, homogeneous spaces, etc. etc. For discrete time systems, results of this kind hold in general if the system has fast enough decay of correlation; see [90]. Mixing is however not sufficient, since this relation does not hold in some slowly mixing system having particular arithmetical properties, as shown in [92].

For the geometric Lorenz flow X^t to the cross-section Σ, with the added regularity assumption that the one-dimensional quotient map f of the return map R to Σ over the stable foliation is such that $1/f'$ is Lipschitz (see the description of the construction of the geometric Lorenz flow on Sect. 3.3), Galatolo and Pacifico in [91] found that R has exponential decay of correlations for Lipschitz observables. Moreover they obtain a logarithmic law for quantitative recurrence: if μ is the physical measure for the geometrical Lorenz attractor, and d_μ is the dimension of μ, then for every regular point x_0 in the attractor, not belonging to the local stable manifold of the equilibrium, and for which the local dimension $d_\mu(x_0)$ is defined,

$$\lim_{r\to 0^+} \frac{\log \tau_r(x,x_0)}{-\log r} = d_\mu(x_0) - 1 \quad \text{for a.e. starting point } x.$$

Here the *local dimension* $d_\mu(x_0)$ of μ at x_0 is defined by

$$d_\mu(x) = \lim_{r\to 0^+} \frac{\log \mu(B_r(x))}{\log r}$$

if this limit exists.

Conjecture 10.13 Relations of this kind hold for every smooth enough singular-hyperbolic attractors, without restrictive assumptions on the smoothness of the one-dimensional quotient maps.

10.2.5 Non-mixing Flows and Slow Decay of Correlations

Let $f : M \to M$ be a diffeomorphism with an invariant probability measure μ and consider the suspension flow X_f over f with constant roof function $r \equiv 1$. Then the probability measure $\nu = \mu \times \text{Leb}$ on $M \times [0,1)$ defines in a straightforward way a X_f-invariant probability measure on X_r *which is NOT mixing*, whatever f may be.

Indeed, consider $A = \pi(M \times [0,1/2))$ and $B = M_r \setminus A$ (recall that $\pi : M \times \mathbb{R} \to X_r$ is the projection defined in Sect. 2.3.2.2). Then the function $t \mapsto \nu(A \cap X^{-t}B)$ for $t > 0$ has the graph as in Fig. 10.3 (here X^{-t} is a shorthand for $(X^t)^{-1}$, the inverse image of the map X^t).

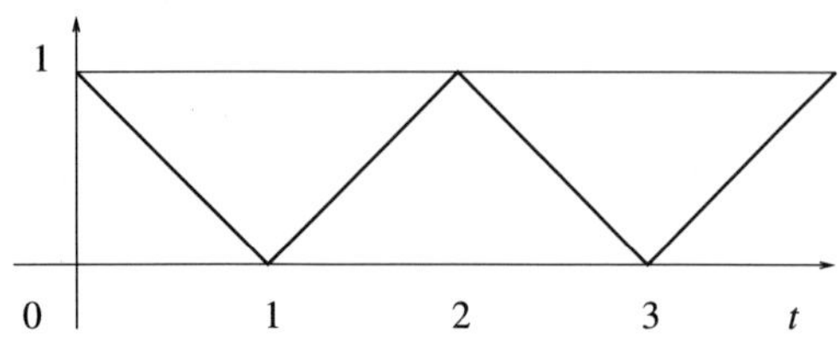

Fig. 10.3 A correlation function for a non-mixing flow

This system is clearly *not* mixing since the sawtooth pattern in Fig. 10.3 goes on for all positive t. Moreover this shows in particular that this suspension flow is not even topologically mixing (see below for the definition).

If however if (X, f, μ) is ergodic, then ν is X_f-ergodic also: indeed, given $A \subset X_r$ such that $(X_f^t)^{-1}(A) = A$ for all $t > 0$ (an X_f-invariant set), then A is saturated, i.e., $p \in A$ if, and only if, $\mathcal{O}_{X_f}(p) \subset A$; thus we may find $\hat{A} \subset X$ such that $A \cap \pi(X \times \{0\}) = \pi(\hat{A})$ is X_f^1-invariant by construction (because $r \equiv 1$), $\hat{A}$ is f-invariant and $\nu(A) = \mu(\hat{A}) \cdot \mathrm{Leb}([0, 1))$. Hence $\mu(\hat{A}) \cdot \mu(X \setminus \hat{A}) = 0$ by the ergodicity of (f, μ) which implies that $\nu(A) \cdot \nu(X_r \setminus A) = 0$.

In addition to the examples of non-mixing suspension flows, which arguably can be characterized as very particular cases, not all Axiom A mixing flows have exponential decay of correlations: Ruelle [236] and Pollicott [209] exhibited suspensions semiflows with piecewise constant ceiling functions over uniformly expanding base dynamics, with arbitrarily slow decay rates of correlations.

Anosov [15] showed that geodesic flows for negatively curved compact Riemannian manifolds are mixing and obtained the *Anosov alternative*: given a transitive volume preserving Anosov flow, either it is mixing (with respect to the volume measure), or it is a suspension of an Anosov diffeomorphism by a constant roof function. We note that Bowen [61] showed that, if a mixing Anosov flow is the suspension of an Anosov diffeomorphism, then it is *stably mixing*, that is, the mixing property remains true for all nearby flows (which are Anosov also by the structural stability of Axiom A flows).

Bowen also showed [61] that the class of C^r Axiom A flows, $r \geq 1$, admits a residual subset $\mathcal{R}$ such that for every $X \in \mathcal{R}$ the spectral decomposition of $\Omega(X)$ is formed by pairwise disjoint pieces $\Omega_1 \cup \cdots \cup \Omega_k$ each of which is *topologically mixing*. That is, given any pair of open sets U, V in Ω_i, there exists $T_0 = T_0(U, V) > 0$ such that $U \cap X^t(V) \neq \emptyset$ for all $t > T_0$.

10.2.6 Decay of Correlations for Flows

Chernov [68] provided a dynamical proof showing sub-exponential decay of correlations for geodesic flows on surfaces of variable negative curvature through a suitable stochastic approximation of the flow (see also [137] for a generalization and previous results [70]).

More recently, a breakthrough was obtained by Dolgopyat [80–82]: smooth (C^r with $r \geq 7$) geodesic flows on manifolds of negative curvature, under a non-integrability condition exhibit exponential decay of correlations. Also Liverani [138] building on the work [80] obtained exponential decay of correlations for C^4 contact Anosov flows.

Using these ideas, applied to the particular case of a suspension over uniformly expanding base dynamics, a conjecture of Ruelle was proved by Pollicott [211]: on a mild (cohomological) condition on the ceiling function, the decay of correlations for this type of suspension flows is exponential for observables not supported on the base. This was extended by Baladi-Vallée [30], clarifying the assumptions on the base and on the ceiling function which suffice to obtain exponential decay of correlations for suspension of one-dimensional expanding maps. All these ideas were used, in a more abstract setting, by Avila-Gouezel-Yoccoz [29] to obtain exponential decay of correlations for the Teichmüller flow on flat surfaces.

Recently Field-Melbourne-Török obtained [87] what they call *stability of rapid mixing* among Axiom A flows, meaning that the correlation function $C_t(\varphi, \psi)$ decays to zero faster than t^{-k} for all $k \in \mathbb{N}$ when $t \to \infty$, for a C^2-open and C^r-dense set of flows among the family of C^r Axiom A flows with $r \geq 2$.

Luzzatto, Melbourne and Paccaut [142] showed that the physical measure for the geometric Lorenz flow is mixing. The speed of mixing for the Lorenz flow is still an open problem.

Conjecture 10.14 Physical measures for singular-hyperbolic attractors are mixing and have exponential decay of correlations. The rates of decay depend continuously on the flow.

10.2.7 Thermodynamical Formalism

The thermodynamical formalism was first developed for (uniformly) hyperbolic diffeomorphisms, borrowed from statistical mechanics by Bowen, Ruelle and Sinai (among others, see e.g. [56, 60, 62, 84, 237, 238]). This was extended to hyperbolic flows by Bowen and Ruelle in [62]. The classical theory relies heavily on the coding of basic pieces of hyperbolic dynamics by subshifts of finite type, for which many tools are available to study in fine detail the relations among its invariant measures. Recently most of this theory was extended to countable shifts by Gurevich [100], Sarig [241, 242] and many others.

The extension of this theory for singular-hyperbolic attractors faces several difficulties: these attractors are modeled by a suspension semiflow whose base transformation is a Hölder-C^1 piecewise expanding but *non-Markov* map, and the roof function is unbounded. In the hyperbolic flow case, the corresponding suspension semiflow has a piecewise expanding Markov map as the base transformation and the roof function is continuous and bounded. In the singular-hyperbolic case, neither the thermodynamical formalism is complete for the base transformation, nor is it clear how to proceed with unbounded roof functions, which imply an extra restriction of integrability on the observables with respect to invariant measure for the base transformation.

Hope of solving this problem in the near future is provided by recent advances in the construction of a thermodynamical formalism for non-uniformly expanding transformations by Oliveira, Viana, Senti, Pesin, Varandas, Bruin, Todd, Pinheiro [64, 65, 184, 199, 202, 265], and ongoing work [17, 188] on the study of equilibrium states for multiples of the logarithm of the derivative, for suspension flows over transformations resembling the Rovella or Lorenz one-dimensional transformations.

Conjecture 10.15 It is possible to build a thermodynamical formalism for Rovella-like, singular-hyperbolic and 2-sectionally expanding attractors.

10.3 Generic Conservative Flows in Dimension 3

Besides the generic dichotomy for conservative flows on 3-manifolds obtained by Bessa [41] and Araújo, Bessa [19] presented in Chap. 9, there are other extensions of generic results first obtained for conservative surface diffeomorphisms.

A result of fundamental importance in the theory of generic conservative diffeomorphisms on surfaces was obtained by Newhouse in [180]. Newhouse's theorem states that C^1 generic area-preserving diffeomorphisms on surfaces either are Anosov, or else the elliptical periodic points are dense. A refined version of this result was presented by Arnaud in [24] in the family of 4-dimensional symplectomorphisms. Even more recently Saghin-Xia [240] generalized the Arnauld result for the multidimensional symplectic case, and in [43] Bessa and Duarte obtained a similar dichotomy for C^1 generic incompressible flows *without equilibria* on 3-manifolds: either the flow is Anosov, or else the elliptic periodic orbits are dense in the manifold. This dichotomy was extended by Araújo and Bessa in [19] to general incompressible 3-flows either with or without equilibria.

Considering C^2 Hamiltonian functions on compact 4-dimensional symplectic manifolds, Bessa and Dias in [44] were able to obtain the Newhouse dichotomy: for any open set U intersecting a far from Anosov regular energy surface, there is a nearby Hamiltonian having an elliptic closed orbit through U. This implies that, for far from Anosov regular energy surfaces of a C^2 generic Hamiltonian, the elliptic closed orbits are generic.

Bessa in [42] extends the results of Bowen [61] and Abdenur, Avila, Bochi [1] (see Sect. 10.1.1) for incompressible flows on compact manifolds showing that there exists a C^1 residual subset $\mathscr{R} \subset \mathfrak{X}^1_\mu(M)$ such that every $X \in \mathscr{R}$ is a topologically mixing vector field.

Bessa and Rocha, in [45], prove that an incompressible and C^1-robustly transitive vector field *among incompressible nearby flows* has no singularities (we stress that this robust transitivity condition is weaker than the assumption of robust transitivity among *all nearby C^1 flows*). Moreover, if the vector field is smooth enough, then the Linear Poincaré Flow associated to it admits a dominated splitting over M.

Conjecture 10.16 The C^1 generic dichotomy for conservative 3-flows admits an extension to higher dimensions: there exists a residual subset $\mathscr{R}$ of $\mathfrak{X}^1_\mu$ such that, for $X \in \mathscr{R}$ and for μ-almost every x, either the Lyapunov exponents are zero at x, or the Oseledets splitting at x is dominated.

Appendix A
Lyapunov Stability on Generic Vector Fields

Here we prove properties L3 through L6 stated in Sect. 2.5.10 following [66].

First we state some general results on continuity points of lower/upper semicontinuous maps. Given a metric space M, consider a set-valued map

$$\Phi : M \to \mathscr{K}(M)$$

with values in the metric space $\mathscr{K}(M)$ of all compact subsets of M, endowed with the Hausdorff distance: for $K_1, K_2 \in \mathscr{K}(M)$

$$d_H(K_1, K_2) = \inf\{\varepsilon > 0 : K_1 \subset B(K_2, \varepsilon) \text{ and } K_2 \subset B(K_1, \varepsilon)\}.$$

We say that Φ is *lower semicontinuous* at $x \in M$ if, for every open set V of M intersecting $\Phi(x)$, we can find a neighborhood U of x in M such that $V \cap \Phi(y) \neq \emptyset$ for every $y \in U$. Similarly, Φ is *upper semicontinuous* at $x \in M$ if, for every compact subset $K \in \mathscr{K}(M)$ satisfying $K \cap \Phi(x) = \emptyset$, there exists a neighborhood U of x such that $K \cap \Phi(y) = \emptyset$ for every $y \in U$. We say that Φ is lower (respectively, upper) semicontinuous if Φ is lower (resp., upper) semicontinuous at every $x \in M$.

A well known result from general topology (see e.g. [127]) asserts that, for Baire spaces M (e.g. complete metric spaces), if Φ is lower (resp., upper) semicontinuous then there is a residual subset $\mathscr{R}$ of M such that Φ is also upper (resp., lower) semicontinuous at every point of $\mathscr{R}$. This result is often used as a device to produce residual subsets, as we do from now on.

Properties L5 and L6 stated in Sect. 2.5.10 are a consequence of the following. We note that property L6 is exactly the same as L5 after reversing time, and so it is enough to prove L5.

Proposition A.1 *There is a residual subset $\mathscr{R}$ of $\mathfrak{X}^1(M)$ such that, if $X \in \mathscr{R}$ and $\sigma \in S(X)$, then the set $\{p \in W^u_X(\sigma) : \overline{\mathscr{O}^+_X(p)}$ is Lyapunov stable for $X\}$ is residual in $W^u_X(\sigma)$.*

Indeed we have the following relation.

V. Araújo, M.J. Pacifico, *Three-Dimensional Flows*, Ergebnisse der Mathematik und ihrer Grenzgebiete. 3. Folge / A Series of Modern Surveys in Mathematics 53,
DOI 10.1007/978-3-642-11414-4, © Springer-Verlag Berlin Heidelberg 2010

Lemma A.2 *If $X \in \mathfrak{X}^1(M)$ and $z \in M$, then $\overline{\mathcal{O}_X^+(z)}$ is Lyapunov stable for X if, and only if, $\omega_X(z)$ is Lyapunov stable for X.*

Proof If $z \in \omega_X(z)$ (i.e., z is recurrent), then $\omega_X(z) = \overline{\mathcal{O}_X^+(z)}$ and we have nothing to prove. So we assume that z is not recurrent. In particular, z is not a singularity.

Arguing by contradiction, assume that $\overline{\mathcal{O}_X^+(z)}$ is Lyapunov stable but that $\omega_X(z)$ is not Lyapunov stable. Then there are a neighborhood U of $\omega_X(z)$ and a sequence $p_n \in M$ converging to $p \in \omega_X(z)$ such that $p_n' = X^{t_n}(p_n) \notin U$ for some sequence $t_n \geq 0$. We can assume without loss of generality that the limit x of p_n' as $n \to +\infty$ exists.

The characterization of Lyapunov stability provided by Lemma 2.25 ensures that x belongs to $\overline{\mathcal{O}_X^+(z)} \setminus U$, since $\omega_X(z) \subset U$. Moreover there exists $t_0 > 0$ such that $t > t_0$ implies that $X^t(z) \in U$ for $t > t_0$. Then $x \in X^{[0,t_0]}(z) \setminus U$.

Using the Tubular Flow Theorem we cover $X^{[0,t_0]}(z)$ by a flow-box B and we note that $\{B, U\}$ is an open cover of $\overline{\mathcal{O}_X^+(z)}$. In addition, because z is not recurrent, we can assume without loss of generality that $z \notin U$, so that $B \cup U$ does not cover $X^{[-s,0]}(z)$ for some $s > 0$ which can be taken as small as desired. Since B is a flow-box, the positive orbits of p_n must leave U and enter B before they approach x. Hence we can find $0 < s_n < t_n$ such that $\tilde{p}_n = X^{s_n}(p_n)$ is not in $B \cup U$. But then every accumulation point of $\tilde{p}_n$ belongs to $\overline{\mathcal{O}_X^+(z)} \setminus (U \cup B)$, a contradiction to the construction of $\{U, B\}$ as a cover of $\overline{\mathcal{O}_X^+(z)} \setminus (U \cup B)$. □

From Lemma A.2 the proof of Proposition A.1 is a consequence of the local result below, see e.g. [190]. Denote by $\mathbf{KS}^1(M)$ the family of all Kupka-Smale vector fields of $\mathfrak{X}^1(M)$.

Lemma A.3 *For every $X \in \mathbf{KS}^1(M)$ there is a neighborhood $\mathcal{U}$ of X in $\mathfrak{X}^1(M)$ and a residual subset $\mathcal{R}$ of $\mathcal{U}$ such that, if $Y \in \mathcal{R}$, $\sigma \in S(Y)$ and $D_Y^u(\sigma)$ is a fundamental domain for the dynamics of Y^t on $W_Y^u(\sigma)$, then the set*

$$\{p \in D_Y^u(\sigma) : \overline{\mathcal{O}_Y^+(p)} \text{ is Lyapunov stable for } Y\}$$

is residual in $D_Y^u(\sigma)$.

Indeed, since $D_Y^u(\sigma)$ crosses every orbit of Y^t inside $W_Y^u(\sigma)$, $W_Y^u(\sigma) \setminus \{\sigma\} = \bigcup_{t \in \mathbb{R}} Y^t(D_Y^u(\sigma))$, and so a residual set R on $D_Y^u(\sigma)$ becomes a residual set $\bigcup_{t \in \mathbb{R}} Y^t(R)$ in $W_Y^u(\sigma)$.

Proof Let $X \in \mathbf{KS}^1(M)$ be given. Then by the hyperbolicity of the singularities $S(X)$ of X, they are finitely many $S(X) = \{\sigma_1(X), \ldots, \sigma_k(X)\}$ and admit "analytic continuations" so that $S(Y) = \{\sigma_1(Y), \ldots, \sigma_k(Y)\}$ for all vector fields Y in a neighborhood $\mathcal{U}$ of X in $\mathfrak{X}^1(M)$.

We denote by u_i the dimension of the unstable manifold of σ_i and assume without loss of generality that $D_X^u(\sigma_i(X))$ is the u_i-sphere of radius $r > 0$ on $W_X^u(\sigma_i)$ around

σ_i, for $i = 1, \ldots, k$, such that $u_i > 0$ (otherwise σ_i is a sink and there is nothing to prove).

Let Σ_i be a cross-section of X such that $D^u_X(\sigma_i) = \Sigma_i \cap W^u_X(\sigma_i)$. Taking $r > 0$ small enough, we can assume that $D^u_Y(\sigma_i) = \Sigma_i \cap W^u_Y(\sigma_i(Y))$ is a fundamental domain for Y^t on $W^u_Y(\sigma_i(Y))$ if we also shrink the neighborhood $\mathscr{U}$ accordingly.

By the Stable Manifold Theory (see e.g. [110]) it follows that there is a C^1 map $\Gamma_i : \mathscr{U} \times D^u_X(\sigma_i) \to \Sigma_i$ such that

$$D^u_Y(\sigma_i) = \mathrm{Graph}(\Gamma_i(Y, \cdot)) = \{\Gamma_i(Y, z) : z \in D^u_X(\sigma_i)\}.$$

Note that the natural projection $\Pi_i : D^u_Y(\sigma_i(Y) \to D^u_X(\sigma_i)$, $\Gamma_i(Y, z) \mapsto z$ is a C^1 diffeomorphism. Now define the set-valued map

$$\Phi_i : \mathscr{U} \times D^u_X(\sigma_i) \to \mathscr{K}(M) \quad \text{by} \quad (Y, y) \mapsto \overline{\mathscr{O}^+_Y(\Gamma_i(Y, y))}.$$

The Tubular Flow Theorem and the continuity of Γ_i imply that Φ_i is lower semi-continuous. Let $\mathscr{R}_i$ be the residual set of $\mathscr{U} \times D^u_X(\sigma_i)$ where Φ_i is continuous, and define the set $\mathscr{R}_i(Y) = \{y \in D^u_X(\sigma_i) : (Y, y) \in \mathscr{R}_i\}$ for any given $Y \in \mathscr{U}$. Then the set

$$\mathscr{V}_i = \{Y \in \mathscr{U} : \mathscr{R}_i(Y) \text{ is residual in } D^u_X(\sigma_i)\}$$

is also residual in $\mathscr{U}$.

Now set $\mathscr{R} = \mathbf{KS}^1(M) \cap \mathscr{V}_1 \cap \cdots \cap \mathscr{V}_k$ which is a residual subset of $\mathscr{U}$, and note that $G_i(Y) = \{\Gamma_i(Y, y) : y \in \mathscr{R}_i(Y)\}$ is residual in $D^u_Y(\sigma_i)$ for all $Y \in \mathscr{R}$ and $i = 1, \ldots, k$. We need to prove that $\overline{\mathscr{O}^+_Y(p)}$ is Lyapunov stable for Y for every triple $(i, Y, p) \in \{1, \ldots, k\} \times \mathscr{R} \times G_i(Y)$ with $i = 1, \ldots, k$.

We argue by contradiction, assuming that there exists $(i, Y, p) \in \{1, \ldots, k\} \times \mathscr{R} \times G_i(Y)$ for some $i \in \{1, \ldots, k\}$ such that $\overline{\mathscr{O}^+_Y(p)}$ is not Lyapunov stable for Y. We can find $y \in \mathscr{R}_i(Y)$ such that $\Phi_i(Y, y) = \overline{\mathscr{O}^+_Y(p)}$ and so $(Y, y) \in \mathscr{R}_i$, that is, Φ_i is continuous at (Y, y).

According to item 1 of Lemma 2.25, we can find sequences $x_n \to x \in \overline{\mathscr{O}^+_Y(p)}$ and $t_n > 0$ such that $X^{t_n}(x_n) \to q \notin \overline{\mathscr{O}^+_Y(p)}$. We note that p is not a critical element of X, that is, it is not a singularity nor a periodic orbit. We observe also that x cannot be a sink, for otherwise $\overline{\mathscr{O}^+_Y(p)}$ would equal $\mathscr{O}^+_Y(p) \cup \{x\}$, and so would be Lyapunov stable. By the definition of q we see that q cannot be a source and, likewise, x cannot be a source either.

We can also assume that x is not a critical element of Y, for otherwise, since $Y \in \mathbf{KS}^1(M)$ the point x would be a hyperbolic critical element of saddle-type for Y and, using the linearization of the flow near x, we see that $\overline{\mathscr{O}^+_Y(p)}$ will contain some point $w \neq x$ in the stable manifold of x and also some point $r \neq x$ in the unstable manifold of x. Both p and w are not singularities nor periodic orbits, and so we can apply the Connecting Lemma to obtain a flow Z which is arbitrarily C^1-close to Y with an orbit connecting the unstable manifold of the continuation of

σ_i to the stable manifold of the continuation $x(Z)$ of x. Then, for the continuation $p(Z)$ of p, the set $\overline{\mathscr{O}_Z(p(Z))}$ is just the orbit segment from $p(Z)$ to $x(Z)$. This contradicts the lower semi-continuity of the map Φ_i at Y, since $\overline{\mathscr{O}_Y(p)}$ contains a point r away from x.

We can also assume without loss of generality that q is not a critical element of Y either, for otherwise, using the linearization of the flow near q, we can replace q by an accumulation point of the positive orbits of x_n in $W^s_Y(q) \setminus \{q\}$ (this is not empty since q is not a source and is hyperbolic).

Let U be a neighborhood of $\overline{\mathscr{O}^+_Y(p)}$ not containing q. Since $K = M \setminus U$ is compact, we have $\Phi_i(Y, y) \cap K = \emptyset$ by upper semi-continuity, but $q \in K$. The definition of q ensures the existence of a sequence $s_n \geq 0$ such that $X^{s_n}(p) \to x$ and we are ready to apply the Connecting Lemma; see Sect. 2.5.7. Recall that the Closing Lemma demands that the accumulation point of the orbits to be connected is *not* a critical element of the flow.

For every $\delta > 0$ there are $x_p = p \in B(p, \delta)$, $x_q = X^{t_n}(x_n) \in B(x, \delta)$, $t_p = s_n \geq 0$ and $t_q = -t_n \leq 0$ such that $X^{t_p}(x_p) \in B(x, \delta)$ and $X^{t_q}(x_q) \in B(x, \delta)$. Hence the Connecting Lemma ensures that we can find $Z \in \mathfrak{X}^1(M)$ arbitrarily C^1 close to Y satisfying $q \in \overline{\mathscr{O}^+_Z(p)}$. This last fact contradicts the upper semicontinuity of Φ_i at (Y, y) since it shows that $\Phi_i(Z, p) \cap K \neq \emptyset$. This contradiction concludes the proof. □

Now we proceed to prove properties L3 and L4 stated in Sect. 2.5.10. Again L4 is just L3 after reversing time, and thus we prove only L3. These properties are consequences of the properties L5 and L6 as in the following argument, very similar to the proof of Lemma A.3.

Let $X \in \mathbf{KS}^1(M)$ be given and consider now the set-valued maps

$$\Phi_i^{\pm} : \mathscr{U} \to \mathscr{K}(M), \quad Y \mapsto \overline{W^{u\pm}_Y(\sigma_i(Y))}, \quad i = 1, \ldots, k,$$

where $\{\sigma_1(Y), \ldots, \sigma_k(Y)\} \subset S(Y)$ are the singularities with one-dimensional unstable manifolds, for every Y in a neighborhood $\mathscr{U}$ of X in $\mathfrak{X}^1(M)$, and $W^{u\pm}_Y(\sigma_i(Y))$ are the connected components of $W^u_Y(\sigma_i) \setminus \{\sigma_i\}$. We observe that $\Phi_i^{\pm}(Y) = \overline{\mathscr{O}_Y(p_i^{\pm})}$ for any chosen points $p_i^{\pm} \in W^{u\pm}_Y(\sigma_i(Y))$. Then the Lyapunov stability of $\Phi_i^{\pm}(Y)$ is true for C^1 generic vector fields Y since L5 has already been proved and we can write

$$\Phi_i^{\pm}(Y) = \overline{\bigcup_{t>0} Y^{-t}(\mathscr{O}^+_Y(p_i^{\pm}))}.$$

Indeed, the continuous dependence of the unstable manifolds on the vector field ensures that each of these maps is lower semi-continuous. Therefore they are continuous on a residual subset $\mathscr{R}_i^{\pm}$ of $\mathscr{U}$. We define $\mathscr{R} = \mathbf{KS}^1(M) \cap \mathscr{R}_1^+ \cap \cdots \cap \mathscr{R}_k^+ \cap \mathscr{R}_1^- \cap \cdots \cap \mathscr{R}_k^-$ which is also a residual subset of $\mathscr{U}$.

We now show that $\mathscr{R}$ is the generic set we are seeking. For $Y \in \mathscr{R}$ and for $\sigma \in S(Y)$ with one-dimensional unstable manifold, we have $\sigma = \sigma_i(Y)$ for some $i \in$

$\{1,\dots,k\}$, and so $\overline{W_Y^{u\pm}(\sigma)} = \Phi_i^{\pm}(Y)$. It is enough to prove that each $\Phi_i^{\pm}(Y)$ is Lyapunov stable for Y since the set $\overline{W_Y^u(\sigma_i)}$ is then the union of two Lyapunov stable sets for Y, and so it is Lyapunov stable for Y also.

Arguing by contradiction, we assume that $\Phi_i^+(Y)$ is not Lyapunov stable for Y (an analogous argument holds for the other connected component). Then we can find sequences $x_n \in M$ such that $x_n \to x \in \Phi_i^+(Y)$ and $t_n \geq 0$ satisfying $Y^{t_n}(x_n) \to q \notin \Phi_i^+(Y)$ as $n \to +\infty$.

As in the proof of Lemma A.3 we can assume that q is not a source. Now x cannot belong to the orbit of a periodic attractor or to a sink, for otherwise $q = x$ would belong to $\Phi_i^+(Y)$. We observe that the only way for the point $x \in \Phi_i^+(Y)$ to be in the orbit of a periodic source (or to be a source singularity) is for $\sigma = x$. In this case the unstable manifold of σ_i is not one-dimensional.

Therefore, if x is a critical element of Y, then x either belongs to a saddle-type hyperbolic periodic orbit or is a hyperbolic saddle singularity. Thus

$$W_Y^u(x) \setminus \{x\} \neq \emptyset \neq W_Y^s(x) \setminus \{x\}.$$

Again this contradicts the continuity of Φ_i^+ at Y, *if the unstable manifold of σ_i is one-dimensional*. Indeed, linearizing the flow near the saddle x, we see that the positive orbit of p_i^+ which accumulates on x must also accumulate on distinct points $w \in W_Y^u(x) \setminus \{x\}$ and $r \in W_Y^s(x) \setminus \{x\}$. Both w and r are not critical elements of Y. The Connecting Lemma ensures the existence of a vector field Z arbitrarily C^1-close to Y having a saddle connection between the continuation of the branch of the unstable manifold of σ_i and the stable manifold of the continuation of x. But this means that the branch of the unstable manifold of the continuation of σ_i for Z does not accumulate on any point of the unstable manifold of $x(Z)$, which contradicts the lower semi-continuity of Φ_i^+ at Y.

Now again x and q are not critical elements of Y, and so the Connecting Lemma enables us to obtain a vector field Z very close to Y in the C^1 topology such that the branch of the unstable manifold of the continuation $\sigma_i(Z)$ of σ_i passes very close to q. This contradicts the upper semi-continuity of Φ_i^+ at Y and completes the proof.

Appendix B
A Perturbation Lemma for Flows

Here we present a proof of Theorem 2.24. This is an unpublished joint work of M. J. Pacifico and E. R. Pujals. Recently another proof by Bonatti-Gourmelon-Vivier appeared in [57]. Let Y be a vector field in the setting of the statement of the theorem.

Given v and $w \in \mathbb{R}^n$, $v \cdot w$ stands for the inner product of v and w. Given $v \in \mathbb{R}^n$ we set $[v]^\perp = \{w \in \mathbb{R}^n, w \cdot v = 0\}$. Given p, let $\Sigma \subset [Y(p)]^\perp$ be a cross-section to Y at p whose size will be fixed later.

Define the parametrized family of maps

$$\hat{A}_t(q) = Y^t(p) + A_t(q)$$

for $q \in \Sigma$. Observe first that $\hat{A}_t$ is C^2 and if Σ is taken small enough then

$$\mathscr{T} = \{\hat{A}_t(q) : q \in \Sigma,\ t \in [a,b]\}$$

gives a neighborhood of $Y^{[a,b]}(p)$.

Lemma B.1 *There exists $r > 0$ such that the following is true whenever* $\operatorname{diam}(\Sigma) < r$. *If $\hat{A}_{t_1}(q_1) = \hat{A}_{t_2}(q_2)$ with $q_i \in \Sigma$ and $t_i \in [a,b]$, then $t_1 = t_2$ and $q_1 = q_2$.*

Proof Assume that $\hat{A}_{t_1}(q_1) = \hat{A}_{t_2}(q_2)$ with $t_1 < t_2$ and $q_1 \neq q_2$. Then

$$Y^{t_1}(p) - Y^{t_2}(p) = A_{t_2}(q_2) - A_{t_1}(q_1). \tag{B.1}$$

On one hand there exists $t_0 \in (t_1, t_2)$ such that

$$\begin{aligned} Y^{t_1}(p) - Y^{t_2}(p) &= (t_1 - t_2)\frac{\partial}{\partial s} Y^{t_0+s}(p)\big|_{s=0} \\ &= \frac{\partial}{\partial s} Y^s\big|_{s=0}(p) = (t_1 - t_2) Y(Y^{t_0}(p)). \end{aligned} \tag{B.2}$$

On the other hand there exists $l \in (t_1, t_2)$ such that

$$\begin{aligned} A_{t_2}(q_2) - A_{t_1}(q_1) &= (A_{t_2} - A_{t_1})(q_2) + A_{t_1}(q_2 - q_1) \\ &= \frac{\partial}{\partial t} A_t\big|_{t=l}(q_2)(t_2 - t_1) + A_{t_1}(q_2 - q_1). \end{aligned} \tag{B.3}$$

V. Araújo, M.J. Pacifico, *Three-Dimensional Flows*, Ergebnisse der Mathematik und ihrer Grenzgebiete. 3. Folge / A Series of Modern Surveys in Mathematics 53,
DOI 10.1007/978-3-642-11414-4, © Springer-Verlag Berlin Heidelberg 2010

Substituting (B.2) and (B.3) into (B.1) we get

$$Y(Y^{t_0}(p)) = -\frac{\partial}{\partial t} A_t\big|_{t=l}(q_2) + A_{t_1}\frac{(q_2-q_1)}{(t_1-t_2)}. \tag{B.4}$$

Observe that $A_{t_0}(Y(p)) = Y(Y^{t_0}(p))$. Since the family A_t depends continuously on t there are $h > 0$ and $\gamma > 0$ such that, if $P_{Y(p)}$ denotes the projection on the direction of $Y(p)$, then

$$\| P_{Y(p)}(A_t^{-1}(Y(Y^{t_0}(p))\| > \gamma \tag{B.5}$$

for all t with $|t - t_0| < h$. Define the numbers

$$K_1 = \sup\left\{\left\|\frac{\partial}{\partial t}A_t\right\|, t \in [a,b]\right\}, \qquad K_2 = \sup\{\|A_t\|, t \in [a,b]\}$$

and $\gamma_0 = \inf\{\|Y(Y^t(p))\|, t \in [a,b]\}$. Observe that γ_0 is positive since p is a regular point. Let $r > 0$ be such that $r < \gamma/K_1$, and $K_1 r + K_2 r/h < \gamma_0$, and take Σ with $\operatorname{diam}(\Sigma) < r$. We split the arguments into a pair of cases.

First case $|t_1 - t_2| \geq h$. Taking norms in (B.4) leads to $\gamma_0 < K_1 r + K_2 r/h < \gamma_0$, which is a contradiction.

Second case $|t_1 - t_2| < h$. Observe that (B.1) and (B.2) imply that

$$(t_1 - t_2)Y(Y^{t_0}(p)) = A_{t_2}(q_2) - A_{t_1}(q_1),$$

which is the same as $(t_1 - t_2)A_{t_1}^{-1}Y(Y^{t_0}(p)) = A_{t_1}^{-1}A_{t_2}(q_2) - q_1$. Thus

$$\begin{aligned} P_{Y(p)}[(t_1-t_2)A_{t_1}^{-1}Y(Y^{t_0}(p))] &= P_{Y(p)}[A_{t_1}^{-1}A_{t_2}(q_2) - q_1] \\ &= P_{Y(p)}[A_{t_1}^{-1}A_{t_2}(q_2) - q_2] \\ &= P_{Y(p)}[(A_{t_1}^{-1}A_{t_2} - Id)(q_2)]. \end{aligned} \tag{B.6}$$

Observe that we have used $q_2 \in \Sigma = [Y(p)]^{\perp}$.
But there exists $l \in (t_1, t_2)$ such that

$$(A_{t_1}^{-1}A_{t_2} - Id)(q_2) = (t_1 - t_2)\frac{\partial}{\partial s}A_{l+s}A_l^{-1}\big|_{s=0}(q_2).$$

Substituting this into (B.6) we get

$$P_{Y(p)}[A_{t_1}^{-1}Y(Y^{t_0}(p))] = P_{Y(p)}\left[\frac{\partial}{\partial s}A_{l+s}A_l^{-1}\big|_{s=0}(q_2)\right]. \tag{B.7}$$

Taking norms in (B.7) and using (B.5) we obtain $\gamma < K_1\|q_2\|$. Since $\operatorname{diam}(\Sigma) < r$ and $r < \gamma(K_1)^{-1}$, this is a contradiction. All together this shows that $t_1 = t_2$ and from (B.4) we see that $q_1 = q_2$. The proof of Lemma B.1 is complete.

□

Now define $\widetilde{A} : \mathscr{T} \subset [a,b] \times \Sigma \to \mathscr{T}$ as follows. For $w \in \mathscr{T}$ there exists, by Lemma B.1, a unique pair $(q_w, t_w) \in \Sigma \times [a,b]$ such that $\hat{A}_{t_w}(q_w) = w$. We define

$$\widetilde{A}(w) = \hat{A}_{t_w+s}(q_w). \tag{B.8}$$

In other words, we have $\widetilde{A}_s(\hat{A}_t(q)) = Y^{t+s}(p) + A_{t+s}A_t^{-1}(A_t(q))$ for $q \in \Sigma$ and $t + s < b$.

Lemma B.2 *The family $\widetilde{A}_s$ defines a C^2 flow in $\mathscr{T}$. Moreover*

$$\frac{\partial}{\partial s} D_w \widetilde{A}_s = D_w \frac{\partial}{\partial s} \widetilde{A}_s. \tag{B.9}$$

Proof Clearly $\widetilde{A}_s$ is C^2. Let us prove that $\widetilde{A}_{s+t} = \widetilde{A}_s \widetilde{A}_t$.

Let $w \in \mathscr{T}$. Then $\hat{A}_{t_w}(q_w) = w$ for a unique $(q_w, t_w) \in \Sigma \times [a,b]$. By definition (B.8)

$$\widetilde{A}_s \widetilde{A}_t(w) = \widetilde{A}_s(\widetilde{A}_t(\hat{A}_{t_w}(q_w)) = \widetilde{A}_s(\hat{A}_{t+t_w}(q_w)). \tag{B.10}$$

Define now $\hat{A}_{t+t_w}(q_w) = \hat{w}$. Note that $\hat{w} = \hat{A}_{t_{\hat{w}}}(q_{\hat{w}})$. By the uniqueness of $t_{\hat{w}}$ and $q_{\hat{w}}$ given by Lemma B.1, we get $t + t_w = t_{\hat{w}}$ and $q_w = q_{\hat{w}}$. Thus

$$\begin{aligned}\widetilde{A}_s(\hat{A}_{t+t_w}(q_w)) &= \widetilde{A}_s(\hat{A}_{t_{\hat{w}}}(q_{\hat{w}})) = \hat{A}_{s+t_{\hat{w}}}(q_{\hat{w}}) \\ &= \hat{A}_{s+t+t_w}(q_w) = \widetilde{A}_{s+t}(A_{t_w}(q_w)) = \widetilde{A}_{s+t}(w).\end{aligned} \tag{B.11}$$

Combining (B.10) and (B.11) we deduce that $\widetilde{A}_{s+t} = \widetilde{A}_s \widetilde{A}_t$.

Now we prove (B.9). Define $\widehat{A}(t,q) = \hat{A}_t(q)$. Then $\widehat{A}$ is C^2 and

(a) $\frac{\partial}{\partial t}\widehat{A} = \frac{\partial}{\partial t} Y^t(p) + \frac{\partial}{\partial t} A_t(q)$ is C^1,
(b) $D_q \widehat{A} = A_t$ is C^1.

Note that (a) and (b) imply that

$$\frac{\partial}{\partial t} A_t = \frac{\partial}{\partial t} D_q \widehat{A}_t = D_q \frac{\partial}{\partial t} \widehat{A}_t = \frac{\partial}{\partial t} A_t. \tag{B.12}$$

Note also that $\frac{\partial}{\partial t} D_q \widehat{A}_t$ and $D_q \frac{\partial}{\partial t} \widehat{A}_t$ are C^1 maps since A_t is a family of invertible linear maps depending C^2 in the parameter .

Now Lemma B.1 shows that $\widehat{A}$ has an inverse map R defined in the image $\widehat{\mathscr{T}} = \widehat{A}(\mathscr{T})$. Moreover R is C^1 since $\widehat{A}$ is C^1.

Again for $s \in [a,b]$ and $w \in \mathscr{T}$ define $\widetilde{A}(s,w) = \widetilde{A}_s(w)$.

Let π_1 and π_2 be the projections on the first and second coordinates:

$$\pi_1 : [a,b] \times \mathscr{T} \to [a,b], \quad (s,w) \mapsto s \qquad \pi_2 : [a,b] \times \mathscr{T} \to \mathscr{T}, \quad (s,w) \mapsto w.$$

Clearly π_i is C^∞, $i = 1, 2$. Since

$$\widetilde{A}(s,w) = \widetilde{A}_s(w) = \hat{A}_{s+t_w}(q_w) = \hat{A}(s + t_w, q_w) = \hat{A}(s + \pi_1 \circ R(w), \pi_2 \circ R(w))$$

we see that $\widetilde{A}$ is C^1, which implies that $\widetilde{A}_s$ induces a C^1 flow in $\mathscr{T}$.

Finally let us verify (B.9). For this, let $\widehat{R}(s,w) = (s + \pi_1 \circ R(w), \pi_2 \circ R(w))$, where R was defined above. Clearly $\widetilde{A} = \widehat{A} \circ \widehat{R}$. Observe that (B.12) and the fact that $\widehat{R}$ and $\widehat{A}$ are C^2 imply that

$$D_w \widetilde{A} = D_{\widehat{R}(s,w)} \widehat{A} \cdot D_w \widehat{R},$$

$$\frac{\partial}{\partial s} \widetilde{A} = D_{\widehat{R}(s,w)} \widehat{A} \cdot \frac{\partial}{\partial s} \widehat{R},$$

$$\frac{\partial}{\partial s} D_w \widetilde{A} = \frac{\partial}{\partial s} D_{\widehat{R}(s,w)} \widehat{A} \cdot \partial_s \widehat{R} \cdot D_w \widehat{R} + D_{\widehat{R}(s,w)} \widehat{A} \cdot \frac{\partial}{\partial s} D_w \widehat{R}$$

and $D_w \frac{\partial}{\partial s} \widetilde{A}$ all exist and are continuous. Thus, by the Schwartz Lemma we obtain (B.9). □

Let Z_A be the vector field induced by $\widetilde{A}_s$, that is, $Z_A(w) = \frac{\partial}{\partial s} \widetilde{A}_s(w)\big|_{s=0}$.

Lemma B.3 *The vector field Z_A is C^1. Moreover*

$$D_w Z_A = \frac{\partial}{\partial s} A_{t_w+s} A_{t_w}^{-1}\big|_{s=0}. \tag{B.13}$$

Proof Since $\widetilde{A}_s$ is a C^2 flow, Z_A is C^1.

Let us calculate $D_w Z_A$. We first calculate $D_w \widetilde{A}_s|_{s=0}$. For this recall that $w = \hat{A}_{t_w}(q_w)$ with $t_w \in [a,b]$ and $q_w \in \Sigma$. To simplify notation we set $t_w = t$ and $q_w = q$. Then, $\widetilde{A}_s(w) = \widetilde{A}_s(\hat{A}_t(q)) = \hat{A}_{t+s}(q)$ and so $D_q \hat{A}_{t+s} = D_{\hat{A}_t(q)} \widetilde{A}_s \,.\, D_q \hat{A}_t$. This implies that

$$D_{\hat{A}_t(q)} \widetilde{A}_s = D_q \hat{A}_{t+s} (D_q \hat{A}_t)^{-1}. \tag{B.14}$$

On the other hand $\hat{A}_{t+s}(q) = Y^{t+s}(p) + A_{t+s}(q)$ implies that $D_q \hat{A}_{t+s} = A_{t+s}$. Substituting this into (B.14) and using the fact that $(D_q \hat{A}_t)^{-1} = A_t^{-1}$ we get

$$D_{\hat{A}_t(q)} \widetilde{A}_s = A_{t+s} A_t^{-1}.$$

Thus

$$D_w Z_A = \frac{\partial}{\partial s} D_w \widetilde{A}_s\big|_{s=0} = \frac{\partial}{\partial s} D_{\hat{A}_t(q)} \widetilde{A}\big|_{s=0} = \frac{\partial}{\partial s} A_{t+s} A_t^{-1}\big|_{s=0}$$

proving (B.13). The proof of Lemma B.3 is completed. □

If $U \subset \mathbb{R}^n$, then U^c stands for the complement of U.

Fix $\varepsilon > 0$ and take $0 < r < \varepsilon$. For each $t \in [a,b]$ let Σ_r be a cross-section to $Y^t(p)$ satisfying $\operatorname{diam}(\Sigma_r) < r$ and $\Sigma_r \subset [Y(Y^t(p)]^\perp$.

Let $\Omega = \bigcup_{t \in [a,b]} \Sigma_r$. Note that Ω is a neighborhood of $Y^{[a,b]}(p)$. Thus there are neighborhoods $U_1 \subset \overline{U_1} \subset U_2 \subset \overline{U_2} \subset \Omega$ of $Y^{[a,b]}(p)$ and a C^1 function $f : \mathscr{T} \to \mathbb{R}$ satisfying:

- $f \mid U_1 = 1$, $f \mid U_2^c = 0$ and $|f| \leq 1$; and
- given $w \in U_2$, for each t_w such that $\text{dist}(w, Y^{[a,b]}(p)) = \text{dist}(w, Y^{t_w}(p))$, we have

$$\|D_w f\| \cdot \|w - Y^{t_w}(p)\| < \varepsilon.$$

Define the C^1 vector field in $\mathbb{R}^n$

$$Z(w) = f(w) \cdot Z_A(w) + (1 - f(w)) \cdot Y(w).$$

Lemma B.4 *Z is C^0-near Y.*

Proof Indeed,

$$Z(w) - Y(w) = f(w) \cdot (Z_A(w) - Y(w)). \tag{B.15}$$

Given w, there are t_w and q_w such that $w = \hat{A}_{t_w}(q_w)$. By (B.8) and the definition of Z_A we get

$$\begin{aligned} Z_A(w) &= \frac{\partial}{\partial s}\widetilde{A}_s(w)\Big|_{s=0} = \frac{\partial}{\partial s}\big(Y^s(Y^{t_w}(p))\big|_{s=0} + \frac{\partial}{\partial s}\big(A_{t_w+s}\big)\big|_{s=0}(q_w) \\ &= Y(Y^{t_w}(p)) + \frac{\partial}{\partial s}(A_{t_w+s})\big|_{s=0}(q_w). \end{aligned}$$

Substituting this last inequality into (B.15) we obtain

$$Z_A(w) - Y(w) = f(w) \cdot \left(Y(Y^{t_w}(p)) - Y(w) + \frac{\partial}{\partial s} A_{t_w+s}\big|_{s=0}(q_w) \right)$$

and then

$$\|Z_A(w) - Y(w)\| \leq \|Y(Y^{t_w}(p)) - Y(w)\| + \left\| \frac{\partial}{\partial s} A_{t_w+s}\big|_{s=0} \right\| \cdot \|(q_w)\|. \tag{B.16}$$

Now we can assume that Σ is sufficiently small so that $\|A_t(q)\| \leq \|A_t\| \cdot \|q\|$ is small for all t and q. We can estimate the first term on the right-hand side of (B.16):

$$\begin{aligned} \|Y(Y^{t_w}(p)) - Y(w)\| &\leq \|Y\| \cdot \|Y^{t_w}(p) - w\| = \|Y\| \cdot \|Y^{t_w}(p) - \hat{A}_{t_w}(q_w)\| \\ &= \|Y\| \cdot \|Y^{t_w}(p) - Y^{t_w}(p) + A_{t_w}(q_w)\| \\ &= \|Y\| \cdot \|A_{t_w}(q_w)\| \leq \|Y\| \cdot \|A_{t_w}\| \cdot \|q_w\| \leq \varepsilon. \end{aligned} \tag{B.17}$$

The second term on the right-hand side of (B.16) can be bounded by

$$\|\partial s A_{t_w+s}\big|_{s=0}\| \cdot \|(q_w)\| \leq \varepsilon, \tag{B.18}$$

if Σ is small. Substituting (B.17) and (B.18) into (B.16) we conclude the proof. □

To finish we need one last lemma.

Lemma B.5 *The vector field Z is C^1-near Y.*

Proof We have

$$D_w Z - D_w Y = D_w f \cdot (Z_A(w) - Y(w)) + f(w) \cdot (D_w Z_A - D_w Y) + D_w Y. \quad \text{(B.19)}$$

The norm of the first term here is bounded by

$$\begin{aligned}
&\|D_w f\| \cdot \|Z_A(w) - Y(w)\| \\
&\quad \leq \|D_w f\| \cdot \|Z_A(w) - Z_A(Y^{t_w}(p)\| + \|Y(Y^{t_w}(p)) - Y(w)\| \\
&\quad \leq \|D_w f\| \cdot \|Z_A\| \cdot \|w - Y^{t_w}(p)\| + \|Y\| \cdot \|w - Y^{t_w}(p)\| \cdot \|D_w f\|
\end{aligned}$$

and the condition on the bump function f implies that both terms in the last expression are small if Σ is small.

To estimate the second term on the right-hand side of (B.19) we recall that Lemma B.3 gives

$$D_w Z_A = \frac{\partial}{\partial s} A_{t+s} A_t^{-1}\big|_{s=0}.$$

On the one hand this is, by hypothesis, near $D_{Y^t(p)} Y$.

On the other hand, since $w = \hat{A}_t(q) = Y^t(p) + A_t(q)$, we also get that w is near $Y^t(p)$ and so $D_w Y$ is near $D_{Y^t(p)} Y$. Combining these last two observations we find that $D_w Z_A$ is near $D_w Y$, concluding the proof of Lemma B.5. □

The proof of Theorem 2.24 is complete.

Appendix C
Robustness of Dominated Decomposition

Here we prove Lemma 2.29. We assume that we are given an invariant subset Λ of a C^1 flow on a boundaryless manifold M such that $\overline{\Lambda}$ is compact, Λ does not contain singularities and the Linear Poincaré Flow on Λ admits a dominated decomposition $N_\Lambda = N_\Lambda^1 \oplus N_\Lambda^2$ where the dimensions of the sub-bundles do not depend on the points of Λ. Alternatively we can assume that Λ is connected.

We already know from Lemma 2.28 that this decomposition is continuous, that is, there are $d_1, d_2 \in \mathbb{N}$ such that $d_1 + d_2 = d := \dim(M)$ and continuous vector fields $(e_i^j)_{i=1,\dots,d_j}$ defined on Λ such that for $j = 1, 2$

- each family $e^j := (e_i^j)_{i=1,\dots,d_j}$ is orthonormal;
- the span of e^j equals N^j at each point $x \in \Lambda$.

We first construct families of vector fields close to these on Λ but defined on a neighborhood of Λ. The compactness of $\overline{\Lambda}$ guarantees the existence of finitely many charts $\varphi_k : V_k \to \mathbb{R}^d$ where each V_k is a non-empty connected open set of M and $(V_k)_{k=1,\dots,l}$ is an open cover of Λ. Let $\varepsilon_0 > 0$ be a Lebesgue number of this open cover.

Given $\eta > 0$ and $x \in \Lambda$, let $0 < \varepsilon_x < \varepsilon_0$ be such that

$$\operatorname{dist}(y, x) < \varepsilon_x \implies \|(e^j)(y) - (e^j)(x)\| \le \eta, \quad j = 1, 2$$

where the norm here denotes

$$\sup_k \left\{ \sum_{i=1}^{d_j} \left\| D\varphi_k(y) \cdot e_i^j(y) - D\varphi_k(x) \cdot e_i^j(x) \right\|_2 \right\}$$

with the supremum taken over all charts $\varphi_k : V_k \to \mathbb{R}^d$ such that $V_k \supset B(x, \varepsilon_x)$, and $\|\cdot\|_2$ is the Euclidean norm on $\mathbb{R}^d$.

Fixing $0 < \bar{\lambda} < \lambda$ and $\xi > 0$ we can also choose $T > 0$ so that

$$C \cdot e^{-\bar{\lambda} T} = \left(\frac{1+\xi}{1-\xi} \right)^{-2}$$

V. Araújo, M.J. Pacifico, *Three-Dimensional Flows*, Ergebnisse der Mathematik und ihrer Grenzgebiete. 3. Folge / A Series of Modern Surveys in Mathematics 53,
DOI 10.1007/978-3-642-11414-4,

and find $\delta > 0$ such that

- for every C^1 vector field Y satisfying $\|Y - X\|_{C^1} < \delta$, and
- for every pair of unit vectors $u_x \in T_x M$, $v_y \in T_y M$ satisfying
  - $x \in \Lambda$, $y \in U$, $Y(y) \neq \mathbf{0}$;
  - $\operatorname{dist}(x, y) < \varepsilon_x$ and
  - $\sup_k \|D\varphi_k(x) \cdot u_x - D\varphi_k(y) \cdot v_y\| < \eta$ (where the supremum is taken over all charts φ_k such that $x, y \in V_k$),

we have for all $-T \leq t \leq T$

$$\operatorname{dist}\big(Y^t(y), X^t(x)\big) < \eta, \qquad \big\|DY^t(y) \cdot v_y - DX^t(x) \cdot u_x\big\| \leq \xi. \tag{C.1}$$

Moreover taking δ small enough we can ensure that the orthogonal projections $O_{Y^t(y)}$ and $O_{X^t(x)}$ are also close, that is,

$$\big\|O_{Y^t(y)} DY^t(y) \cdot v_y - O_{X^t(x)} DX^t(x) \cdot u_x\big\| = \big\|P_Y^t \cdot v_y - P_X^t \cdot u_x\big\| \leq \xi \tag{C.2}$$

for $-T \leq t \leq T$.

Now define the collection of constant functions near $x \in \Lambda$

$$\bar{e}_x^j(y) := e^j(x) \quad \text{for all } y \in B(x, \varepsilon_x/2)$$

for each $x \in \Lambda$ and extend them to the whole manifold M through a bump function so that $\bar{e}_x^j \mid \partial B(x, \varepsilon_x) \equiv 0$.

The union $U := \cup_{x\in\Lambda} B(x, \varepsilon_x/2)$ is a neighborhood of Λ and $(B(x, \varepsilon_x/2))_{x\in\Lambda}$ is an open cover of $\overline{\Lambda}$. Let us take a countable sub-cover $\mathscr{C}$ with the balls centered at some sequence (x_n) and let $\Psi = (\psi_n)_{n\in\mathbb{N}}$ be a partition of unity subordinated to this cover (for these notions, see e.g. [255]). We define

$$\hat{e}^j := \sum_n \psi_n \cdot \bar{e}_{x_n}^j, \quad j = 1, 2. \tag{C.3}$$

Note that Ψ is locally finite: for any given $y \in U$ there exists a neighborhood V of y such that all but finitely many ψ_n are non-zero on V. Since Ψ is subordinate to the cover $\mathscr{C}$ we can be more precise:

$$\psi_n(y) \neq 0 \iff \operatorname{dist}(y, x_n) < \varepsilon_{x_n} \quad \text{for all } n \in \mathbb{N}.$$

Hence the sum (C.3) is well defined on U and for any given $y \in \Lambda$, $\hat{e}^j(y)$ is a finite linear convex combination of some $\bar{e}^j(x_n)$, all of which are close to y. Hence for all such x_n

$$\|\bar{e}^j(x_n) - e^j(y)\| \leq \eta \quad \text{and thus} \quad \|\hat{e}^j(y) - e^j(y)\| \leq \eta. \tag{C.4}$$

In particular we have $\|\hat{e}^j(y)\| \geq 1 - \eta$ which we may assume is bigger than $3/4$.

Now we define cone fields around the subspaces spanned by $\hat{e}^j$ on N_y for all $y \in U \setminus S(X)$, and check that they are invariant under P^t for every flow Y close enough to X in the C^1 topology.

The cone field $C_a^j(x)$ of width $a > 0$ around the subspace $\hat{N}_x^j$ generated by $\hat{e}^j(x)$, for a vector field Y close to X and $x \in U \setminus S(Y)$ and $j = 1, 2$, is given by

$$C_a^j(Y,x) := \big\{ w = w^1 + w^2 + \alpha \cdot Y(x) \in T_x M : w^j \in N_x^j, \alpha \in \mathbb{R}$$
$$a \cdot \|w^j\| \geq \|w^{3-j}\|, j = 1,2 \text{ and } a \cdot \|w^j\| \geq |\alpha| \cdot \|Y(x)\| \big\}.$$

It is impossible that $C_a^1(Y,x) \cap C_a^2(Y,x) \neq \{\mathbf{0}\}$ with $a < 1$, since a non-zero vector v in the intersection could be written as $v^1 + v^2$ with

$$a\|v^1\| \geq \|v^2\| \text{ and } a\|v^2\| \geq \|v^1\| \implies a^2\|v^1\| \geq a\|v^2\| \geq \|v^1\| \implies a^2 \geq 1.$$

We remark that the cone field $(C_a^j(Y,y))_{y\in U}$ is continuous in (Y, y) in the sense that for $\gamma > 1$ there exists $\delta > 0$ satisfying

$$\|Z - Y\|_{C^1} < \delta \ \&\ \mathrm{dist}(z, y) < \delta \implies C^j_{a/\gamma}(Z,z) \subset C_a^j(Y,z) \subset C^j_{\gamma a}(Z,z)$$

because the vector fields $\hat{e}_i^j$ are continuous, $j = 1, 2$, $i = 1, \ldots, d_j$. The domination on the original splitting for X plus the approximation to this splitting of the extension $\hat{e}^j$ of the basis of vector fields ensures that, if we are given $a > 0$ and take T big enough and ξ small enough so that

- $P_X^t \cdot C_a^2(x)$ is a cone of width $Ce^{-\lambda t} \cdot a \ll a$ around the image of $\hat{N}_x^2$ by P_X^t, which is contained in $C^2_{a/8}(X^t(x))$ for all $t \geq T$,

then we obtain

$$P_X^t \cdot C_a^2(x) \subset C^2_{a/4}\big(X^t(x)\big) \quad \text{for all } x \in \Lambda \text{ and } t \geq T.$$

A dual inclusion is also true for the cone around N_x^1. These inclusions together with (C.1), (C.2) and (C.4) ensure that we can also assume that, for Y close enough to X and $y \in U \setminus S(Y)$,

$$P_Y^t \cdot C_a^2(y) \subset C_a^2(Y^t(y)) \quad \text{and} \quad P_Y^{-t} \cdot C_a^1(y) \subset C_a^1(Y^{-t}(y)) \tag{C.5}$$

for all $t \geq T$ if $Y^s(y) \in U$ for all $0 \leq s \leq t$. Indeed, $P_Y^t \cdot C_a^2(y)$ is close to $P_X^t \cdot C_a^2(x)$ for $0 \leq t \leq T$, and so for $t = T$ we have the statement. In addition, if $Y^t(y) \in U$ for $T \leq t < 2T$, then the vectors in $P_Y^{t-T} P_Y^T \cdot C_a^2(y)$ are at a distance of at most η from vectors in $P_X^{t-T} P_X^T \cdot C_a^2(x)$ which is again contained in $C^2_{a/4}(X^t(x))$. The same argument applies for $kT \leq t \leq (k+1)T$, $k \geq 1$.

Let $y \in U$ be given and choose $x \in \Lambda$ close to y, that is, such that $\mathrm{dist}(y, x) < \varepsilon_X$. Then we estimate using (C.2) for $0 \leq t \leq T$ and $i = 1, \ldots, d_1$, $k = 1, \ldots, d_2$:

$$\begin{aligned}\frac{\|P_Y^t \cdot \hat{e}_i^1(y)\|}{\|P_Y^t \cdot \hat{e}_k^2(y)\|} &= \frac{\|P_Y^t \cdot \hat{e}_i^1(y)\|}{\|P_X^t \cdot \hat{e}_i^1(x)\|} \cdot \frac{\|P_X^t \cdot \hat{e}_i^1(x)\|}{\|P_X^t \cdot \hat{e}_k^2(x)\|} \cdot \frac{\|P_X^t \cdot \hat{e}_k^2(x)\|}{\|P_Y^t \cdot \hat{e}_k^2(y)\|} \\ &\leq \frac{1+\xi}{1-\xi} \cdot \frac{\|P_X^t \cdot \hat{e}_i^1(x)\|}{\|P_X^t \cdot \hat{e}_k^2(x)\|}\end{aligned}$$

and

$$\frac{\|P_X^t \cdot \hat{e}_i^1(x)\|}{\|P_X^t \cdot \hat{e}_k^2(x)\|} = \frac{\|P_X^t \cdot \hat{e}_i^1(x)\|}{\|P_X^t \cdot e_i^1(x)\|} \cdot \frac{\|P_X^t \cdot e_i^1(x)\|}{\|P_X^t \cdot e_k^2(x)\|} \cdot \frac{\|P_X^t \cdot e_k^2(x)\|}{\|P_X^t \cdot \hat{e}_k^2(x)\|}$$
$$\leq \frac{1+\xi}{1-\xi} \cdot \frac{\|P_X^t \cdot e_i^1(x)\|}{\|P_X^t \cdot e_k^2(x)\|}.$$

Thus we get

$$\frac{\|P_Y^t \cdot \hat{e}_i^1(y)\|}{\|P_Y^t \cdot \hat{e}_k^2(y)\|} \leq \left(\frac{1+\xi}{1-\xi}\right)^2 \cdot Ce^{-\lambda t}. \tag{C.6}$$

This ensures that $P_Y^t \cdot C_a^2(Y, y)$ is a cone on $N_{Y^t(y)}$ with respect to the splitting given by the subspaces $E^j_{Y^t(x)}$ generated by $(P_Y^t \cdot \hat{e}_i^j)_{i=1,\dots,d_j}$, $j = 1, 2$, around the subspace $E^2_{Y^t(y)}$, whose width is bounded from above by the expression $(1+\xi)^2(1-\xi)^{-2} \cdot Ce^{-\lambda t} \cdot a$. But this is only for $0 \leq t \leq T$.

To obtain a similar exponential bound for $t > T$ we reapply the bound. Let $\ell = [t/T]$ and write $t = \ell T + s$ with $0 \leq s < T$. Then, by induction on the number ℓ of subintervals of length T in $[0, t]$, we get (recall the choice of T and of $\bar{\lambda}$ at the beginning)

$$\frac{\|P_Y^t \cdot \hat{e}_i^1(y)\|}{\|P_Y^t \cdot \hat{e}_k^2(y)\|} \leq \left[\left(\frac{1+\xi}{1-\xi}\right)^2 \cdot Ce^{-\lambda T}\right]^\ell \cdot Ce^{-\lambda s}$$
$$= \left[\left(\frac{1+\xi}{1-\xi}\right)^2 \cdot Ce^{-\bar{\lambda} T}\right]^\ell \cdot Ce^{-\lambda s-(\lambda-\bar{\lambda})\ell T} \leq Ce^{-(\lambda-\bar{\lambda})t}.$$

Moreover (C.5) implies that this cone is contained in $C_a^2(Y^t(y))$ for $t > T$ whenever $Y^s(y) \in U$ for all $0 \leq s \leq t$. A dual statement holds for $P_Y^{-t} \cdot C_a^1(Y, y)$.

Finally we define for all $y \in \Lambda_Y(U) \setminus S(Y)$ the splitting

$$N_y^2(Y) := \bigcap_{t \geq T} P_Y^t \cdot C_a^2\big(Y, Y^{-t}(y)\big) \quad \text{and} \quad N_y^1(Y) := \bigcap_{t\geq T} P_Y^{-t} \cdot C_a^1\big(Y, Y^t(y)\big),$$

which is invariant under P^t for all $t \in \mathbb{R}$ (it is clearly invariant for all $|t| \geq T$ and $P^t P^{-t} = Id$). The domination is given by (C.6) since the subspaces are contained in the cone of width $2a$ around the subspaces F^j generated by $\hat{e}^j$: write $u = u^1 + u^2 \in N_y^1(Y)$ and $v = v^1 + v^2 \in N_y^2(Y)$ with $u^1, v^1 \in F_y^1$ and $u^2, v^2 \in F_y^2$, and compute

$$\frac{\|P_Y^t(u)\|}{\|P_Y^t(v)\|} \leq \frac{\|P_Y^t(u^1)\| + \|P_Y^t(u^2)\|}{\big|\|P_Y^t(v^2)\| - \|P_Y^t(v^1)\|\big|} \leq \frac{\|P_Y^t(u^1)\| + \|P_Y^t(u^1)\|/a}{\big|\|P_Y^t(v^2)\| - \|P_Y^t(v^2)\|/a\big|}$$
$$\leq Ce^{-(\lambda-\bar{\lambda})t} \cdot \frac{1+a^{-1}}{a^{-1}-1}.$$

This goes to zero exponentially fast. We remark that the quotients involving ξ and a can be made as close to 1 as desired by choosing ξ and a sufficiently close to zero, which can be done by shrinking U through setting $\eta > 0$ and $\delta > 0$ small enough from the beginning. In addition, the exponent $\bar{\lambda}$ can be made as close to λ as needed by choosing $0 < \bar{\lambda} < \lambda$ with $\lambda - \bar{\lambda}$ close to zero from the beginning, shrinking the neighborhood U and reducing ξ, δ accordingly.

The proof of Lemma 2.29 is complete.

References

1. Abdenur, F., Avila, A., Bochi, J.: Robust transitivity and topological mixing for C^1-flows. Proc. Am. Math. Soc. **132**(3), 699–705 (2004)
2. Afraimovich, V.S., Bykov, V.V., Shil'nikov, L.P.: On the appearance and structure of the Lorenz attractor. Dokl. Akad. Nauk SSSR **234**, 336–339 (1977)
3. Afraimovich, V.S., Bykov, V.V., Shil'nikov, L.P.: On attracting structurally unstable limit sets of Lorenz attractor type. (Russian). Tr. Mosk. Mat. Obshch. **44**, 150–212 (1982)
4. Afraimovich, V.S., Bykov, V.V., Shil'nikov, L.P.: On attracting structurally unstable limit sets of Lorenz type attractors. Trans. Mosc. Math. Soc. **44**, 153–216 (1983)
5. Afraimovich, V.S., Hsu, S.-B.: Lectures on Chaotic Dynamical Systems. AMS/IP Studies in Advanced Mathematics, vol. 28. Am. Math. Soc., Providence (2003)
6. Afraimovich, V.S., Pesin, Y.: The Dimension of Lorenz type attractors. In Sov. Math. Phys. Rev., volume 6, pages 169–241. Gordon and Breach Harwood Acad. Publ., 1987
7. Alves, J.F., Araújo, V.: Random perturbations of nonuniformly expanding maps. Astérisque **286**, 25–62 (2003)
8. Alves, J.F., Araújo, V., Pacifico, M.J., Pinheiro, V.: On the volume of singular hyperbolic sets. Dyn. Syst. **22**(3), 249–267 (2007)
9. Alves, J.F., Bonatti, C., Viana, M.: SRB measures for partially hyperbolic systems whose central direction is mostly expanding. Invent. Math. **140**(2), 351–398 (2000)
10. Alves, J.F., Luzzatto, S., Pinheiro, V.: Markov structures and decay of correlations for non-uniformly expanding dynamical systems. Ann. Inst. Henri Poincaré, Anal. Non Linéaire **22**(6), 817–839 (2005)
11. Alves, J.F., Pinheiro, V.: Topological structure of (partially) hyperbolic sets with positive volume. Trans. Am. Math. Soc. **360**(10), 5551–5569 (2008)
12. Ambrose, W.: Representation of ergodic flows. Ann. Math. (2) **42**, 723–739 (1941)
13. Ambrose, W., Kakutani, S.: Structure and continuity of measurable flows. Duke Math. J. **9**, 25–42 (1942)
14. Andronov, A., Pontryagin, L.: Systèmes grossiers. Dokl. Akad. Nauk SSSR **14**, 247–251 (1937)
15. Anosov, D.V.: Geodesic flows on closed Riemannian manifolds of negative curvature. Proc. Steklov Inst. Math. **90**, 1–235 (1967)
16. Anosov, D.V., Sinaĭ, Y.: Certain smooth ergodic systems. Usp. Mat. Nauk **22**(5(137)), 107–172 (1967)
17. Apaza, E.H., Tahzibi, A.: Existence of equilibrium states for singular-hyperbolic attractors. In preparation
18. Araújo, V.: Large deviations bound for semiflows over a non-uniformly expanding base. Bull. Braz. Math. Soc. **38**(3), 335–376 (2007)
19. Araújo, V., Bessa, M.: Dominated splitting and zero volume for incompressible three flows. Nonlinearity **21**(7), 1637 (2008)

V. Araújo, M.J. Pacifico, *Three-Dimensional Flows,* Ergebnisse der Mathematik und ihrer Grenzgebiete. 3. Folge / A Series of Modern Surveys in Mathematics 53, DOI 10.1007/978-3-642-11414-4,

20. Araújo, V., Pacifico, M.J.: Physical measures for infinite-modal maps. Fundam. Math. **203**, 211–262 (2009)
21. Araújo, V., Pujals, E., Pacifico, M.J., Viana, M.: Singular-hyperbolic attractors are chaotic. Trans. Am. Math. Soc. **361**, 2431–2485 (2009)
22. Arbieto, A., Matheus, C.: A pasting lemma and some applications for conservative systems. Ergod. Theory Dyn. Syst. **27**(5), 1399–1417 (2007)
23. Arnaud, M.-C.: Création de connexions en topologie C^1. Ergod. Theory Dyn. Syst. **21**(2), 339–381 (2001)
24. Arnaud, M.-C.: The generic symplectic C^1-diffeomorphisms of four-dimensional symplectic manifolds are hyperbolic, partially hyperbolic or have a completely elliptic periodic point. Ergod. Theory Dyn. Syst. **22**(6), 1621–1639 (2002)
25. Arneodo, A., Coullet, P.H., Spiegel, E.A., Tresser, C.: Asymptotic chaos. Physica D **14**(3), 327–347 (1985)
26. Arnold, V.I.: Mathematical Methods of Classical Mechanics. Springer, New York (1978)
27. Arroyo, A., Pujals, E.: Dynamical properties of singular hyperbolic attractors. Discrete Contin. Dyn. Syst. **19**(1), 67–87 (2007)
28. Arroyo, A., Rodriguez Hertz, F.: Homoclinic bifurcations and uniform hyperbolicity for three-dimensional flows. Ann. Inst. Henri Poincaré, Anal. Non Linéaire **20**(5), 805–841 (2003)
29. Avila, A., Gouëzel, S., Yoccoz, J.-C.: Exponential mixing for the Teichmüller flow. Publ. Math. IHÉS **104**, 143–211 (2006)
30. Baladi, V., Vallée, B.: Exponential decay of correlations for surface semi-flows without finite Markov partitions. Proc. Am. Math. Soc. **133**(3), 865–874 (2005)
31. Bamón, R.: A family of n-dimensional differential equations with Lorenz-like attractors. In: Complex Geometry of Groups (Olmué, 1998), Contemp. Math., vol. 240, pp. 13–23. Am. Math. Soc., Providence (1999)
32. Banks, J., Brooks, J., Cairns, G., Davis, G., Stacey, P.: On Devaney's definition of chaos. Am. Math. Mon. **99**(4), 332–334 (1992)
33. Barreira, L., Pesin, Y.: Lectures on Lyapunov exponents and smooth ergodic theory. Proc. Symp. Pure Math. **69**, 3–106 (2001)
34. Barreira, L., Pesin, Y.: Smooth ergodic theory and nonuniformly hyperbolic dynamics. In: Handbook of Dynamical Systems, Vol. 1B, pp. 57–263. Elsevier, Amsterdam (2006). With an appendix by Omri Sarig
35. Bautista, S.: The geometric Lorenz attractor is a homoclinic class. Bol. Mat. (N.S.) (Dpto. de Matemáticas - Facultad de Ciencias - Universidad Nacional de Colombia) **XI**(1), 69–78 (2004)
36. Bautista, S.: Sobre conjuntos hiperbólicos-singulares. PhD thesis, IM-UFRJ, Rio de Janeiro, 2005
37. Bautista, S., Morales, C.: Existence of periodic orbits for singular-hyperbolic sets. Mosc. Math. J. **6**(2), 265–297 (2006)
38. Bautista, S., Morales, C., Pacifico, M.J.: On the intersection of homoclinic classes on singular-hyperbolic sets. Discrete Contin. Dyn. Syst. **19**(4), 761–775 (2007)
39. Bessa, M.: The Lyapunov exponents of generic zero divergence vector fields. PhD thesis, IMPA, Rio de Janeiro, 2005
40. Bessa, M.: Dynamics of generic 2-dimensional linear differential systems. J. Differ. Equ. **228**(2), 685–706 (2006)
41. Bessa, M.: The Lyapunov exponents of generic zero divergence 3-dimensional vector fields. Ergod. Theory Dyn. Syst. **27**(5), 1445–1472 (2007)
42. Bessa, M.: A generic incompressible flow is topological mixing. C. R. Math. Acad. Sci. Paris **346**(21-22), 1169–1174 (2008)
43. Bessa, M., Duarte, P.: Abundance of elliptic dynamics on conservative three-flows. Dyn. Syst. **23**(4), 409–424 (2008)
44. Bessa, M., Lopes Dias, J.: Hamiltonian elliptic dynamics on symplectic 4-manifolds. Proc. Am. Math. Soc. **137**(2), 585–592 (2009)

45. Bessa, M., Rocha, J.: On C^1-robust transitivity of volume-preserving flows. J. Differ. Equ. **245**(11), 3127–3143 (2008)
46. Bhatia, N.P., Szegö, G.P.: Stability Theory of Dynamical Systems. Classics in Mathematics. Springer, Berlin (2002). Reprint of the 1970 original
47. Biragov, V.S., Shil'nikov, L.P.: On the bifurcation of a saddle-focus separatrix loop in a three-dimensional conservative dynamical system. Sel. Math. Sov. **11**(4), 333–340 (1992)
48. Birkhoff, G.D.: Proof of the ergodic theorem. Proc. Natl. Acad. Sci. USA **17**, 656–660 (1931)
49. Birkhoff, G.D.: Nouvelles recherches sur les systèmes dynamiques. Mem. Pont. Acad. Sci. Novi. Lyncaei **1**, 85–216 (1935)
50. Birman, J.S., Williams, R.F.: Knotted periodic orbits in dynamical systems. I. Lorenz's equations. Topology **22**(1), 47–82 (1983)
51. Bochi, J.: Genericity of zero Lyapunov exponents. Ergod. Theory Dyn. Syst. **22**(6), 1667–1696 (2002)
52. Bochi, J., Viana, M.: Lyapunov exponents: How frequently are dynamical systems hyperbolic? In: Advances in Dynamical Systems. Cambridge University Press, Cambridge (2004)
53. Bochi, J., Viana, M.: The Lyapunov exponents of generic volume-preserving and symplectic maps. Ann. Math. (2) **161**(3), 1423–1485 (2005)
54. Boichenko, V.A., Leonov, G.A.: The Hausdorff dimension of attractors of the Lorenz system. Differ. Uravn. (Minsk) **25**(11), 1999–2000 (1989)
55. Bonatti, C., Crovisier, S.: Récurrence et généricité. Invent. Math. **158**(1), 33–104 (2004)
56. Bonatti, C., Díaz, L.J., Viana, M.: Dynamics Beyond Uniform Hyperbolicity. A global geometric and probabilistic perspective, Mathematical Physics, III. Encyclopaedia of Mathematical Sciences, vol. 102. Springer, Berlin (2005).
57. Bonatti, C., Gourmelon, N., Vivier, T.: Perturbations of the derivative along periodic orbits. Ergod. Theory Dyn. Syst. **26**(5), 1307–1337 (2006)
58. Bonatti, C., Pumariño, A., Viana, M.: Lorenz attractors with arbitrary expanding dimension. C. R. Acad. Sci., Sér. 1 Math. **325**(8), 883–888 (1997)
59. Bowen, R.: Symbolic dynamics for hyperbolic flows. Am. J. Math. **95**, 429–460 (1973)
60. Bowen, R.: Equilibrium States and the Ergodic Theory of Anosov Diffeomorphisms. Lect. Notes in Math., vol. 470. Springer, Berlin (1975)
61. Bowen, R.: Mixing Anosov flows. Topology **15**(1), 77–79 (1976)
62. Bowen, R., Ruelle, D.: The ergodic theory of Axiom A flows. Invent. Math. **29**, 181–202 (1975)
63. Bowen, R., Walters, P.: Expansive one-parameter flows. J. Differ. Equ. **12**, 180–193 (1972)
64. Bruin, H., Todd, M.: Markov extensions and lifting measures for complex polynomials. Ergod. Theory Dyn. Syst. **27**(3), 743–768 (2007)
65. Bruin, H., Todd, M.: Equilibrium states for potentials with $\sup\phi - \inf\phi < \mathrm{h}_{\mathrm{top}}(f)$. Commun. Math. Phys. **283**, 579–611 (2008)
66. Carballo, C.M., Morales, C., Pacifico, M.J.: Maximal transitive sets with singularities for generic C^1 vector fields. Bull. Braz. Math. Soc. **31**(3), 287–303 (2000)
67. Carballo, C.M., Morales, C., Pacifico, M.J.: Homoclinic classes for generic C^1 vector fields. Ergod. Theory Dyn. Syst. **23**(2), 403–415 (2003)
68. Chernov, N.I.: Markov approximations and decay of correlations for Anosov flows. Ann. Math. (2) **147**(2), 269–324 (1998)
69. Chow, S.-N., Deng, B., Fiedler, B.: Homoclinic bifurcation at resonant eigenvalues. J. Dyn. Differ. Equ. **2**(2), 177–244 (1990)
70. Collet, P., Epstein, H., Gallavotti, G.: Perturbations of geodesic flows on surfaces of constant negative curvature and their mixing properties. Commun. Math. Phys. **95**(1), 61–112 (1984)
71. Colmenárez, W.: SRB measures for singular hyperbolic attractors. PhD thesis, UFRJ, Rio de Janeiro, 2002
72. Colmenárez, W.: Nonuniform hyperbolicity for singular hyperbolic attractors. Trans. Am. Math. Soc. **357**(10), 4131–4140 (2005)
73. Conley, C.: Isolated Invariant Sets and the Morse Index. CBMS Regional Conference Series in Mathematics, vol. 38. Am. Math. Soc., Providence (1978)

74. Costa, M.J., Pacifico, M.J., Rovella, A., Viana, M.: Persistence of global spiraling attractors. In preparation
75. Dacorogna, B., Moser, J.: On a partial differential equation involving the Jacobian determinant. Ann. Inst. Henri Poincaré, Anal. Non Linéaire **7**(1), 1–26 (1990)
76. de Melo, W.: Structural stability of diffeomorphisms on two-manifolds. Invent. Math. **21**, 233–246 (1973)
77. Deng, B.: The Šil'nikov problem, exponential expansion, strong λ-lemma, C^1-linearization, and homoclinic bifurcation. J. Differ. Equ. **79**(2), 189–231 (1989)
78. Devaney, R.: An Introduction to Chaotic Dynamical Systems, 2nd edn. Addison-Wesley, New York (1989)
79. Doering, C.I.: Persistently transitive vector fields on three-dimensional manifolds. In: Procs. on Dynamical Systems and Bifurcation Theory, vol. 160, pp. 59–89. Pitman, London (1987)
80. Dolgopyat, D.: On decay of correlations in Anosov flows. Ann. Math. (2) **147**(2), 357–390 (1998)
81. Dolgopyat, D.: Prevalence of rapid mixing in hyperbolic flows. Ergod. Theory Dyn. Syst. **18**(5), 1097–1114 (1998)
82. Dolgopyat, D.: Prevalence of rapid mixing. II. Topological prevalence. Ergod. Theory Dyn. Syst. **20**(4), 1045–1059 (2000)
83. Dumortier, F., Kokubu, H., Oka, H.: A degenerate singularity generating geometric Lorenz attractors. Ergod. Theory Dyn. Syst. **15**(5), 833–856 (1995)
84. Ellis, R.S.: Entropy, Large Deviations, and Statistical Mechanics. Reprint of the 1985 original, Classics in Mathematics. Springer, Berlin (2006)
85. Falconer, K.: Fractal Geometry: Mathematical Foundations and Applications. Wiley, New York, USA (1990).
86. Fathi, A., Herman, M.R., Yoccoz, J.-C.: A proof of Pesin's stable manifold theorem. In: Geometric Dynamics (Rio de Janeiro, 1981). Lecture Notes in Math., vol. 1007, pp. 177–215. Springer, Berlin (1983)
87. Field, M., Melbourne, I., Török, A.: Stability of mixing and rapid mixing for hyperbolic flows. Ann. Math. **166**, 269–291 (2007)
88. Fowler, A.C., Sparrow, C.T.: Bifocal homoclinic orbits in four dimensions. Nonlinearity **4**(4), 1159–1182 (1991)
89. Franks, J.: Necessary conditions for stability of diffeomorphisms. Trans. Am. Math. Soc. **158**, 301–308 (1971)
90. Galatolo, S.: Dimension and hitting time in rapidly mixing systems. Math. Res. Lett. **14**(5), 797–805 (2007)
91. Galatolo, S., Pacifico, M.J.: Lorenz like flows: exponential decay of correlations for the Poincaré map, logarithm law, quantitative recurrence. 0901.0574v2. To appear in Erg. Th. & Dyn. Syst. (2010)
92. Galatolo, S., Peterlongo, P.: Long hitting time, slow decay of correlations and arithmetical properties. 0801.3109v4 (Mar. 2009)
93. Ghrist, R.W., Holmes, P.J.: An ODE whose solutions contain all knots and links. Int. J. Bifurc. Chaos Appl. Sci. Eng. **6**(5), 779–800 (1996)
94. Glasner, E., Weiss, B.: Sensitive dependence on initial conditions. Nonlinearity **6**(6), 1067–1075 (1993)
95. Glendinning, P., Sparrow, C.: T-points: A codimension two heteroclinic bifurcation. J. Stat. Phys. **43**, 479–488 (1986)
96. Grobman, D.M.: Homeomorphism of systems of differential equations. Dokl. Akad. Nauk SSSR **128**, 880–881 (1959)
97. Guckenheimer, J., Holmes, P.: Nonlinear Oscillations, Dynamical Systems and Bifurcation of Vector Fields. Springer, New York (1983)
98. Guckenheimer, J., Williams, R.F.: Structural stability of Lorenz attractors. Publ. Math. IHÉS **50**, 59–72 (1979)
99. Guillemin, V., Pollack, A.: Differential Topology. Prentice Hall, New York (1974)
100. Gurevich, B.M., Savchenko, S.V.: Thermodynamic formalism for symbolic Markov chains with a countable number of states. Usp. Mat. Nauk **53**(2(320)), 3–106 (1998)

101. Gutiérrez, C.: Structural stability for flows on the torus with a cross-cap. Trans. Am. Math. Soc. **241**, 311–320 (1978)
102. Gutierrez, C.: On the C^r-closing lemma for flows on the torus T^2. Ergod. Theory Dyn. Syst. **6**(1), 45–56 (1986)
103. Gutierrez, C.: A counter-example to a C^2 closing lemma. Ergod. Theory Dyn. Syst. **7**(4), 509–530 (1987)
104. Hartman, P.: A lemma in the theory of structural stability of differential equations. Proc. Am. Math. Soc. **11**, 610–620 (1960)
105. Hassard, B., Hastings, S., Troy, W., Zhang, J.: A computer proof that the Lorenz equations have "chaotic" solutions. Appl. Math. Lett. **7**, 79–83 (1994)
106. Hastings, S., Troy, W.: A shooting approach to the Lorenz equations. Bull. Am. Math. Soc. **27**, 298–303 (1992)
107. Hayashi, S.: Diffeomorphisms in $\mathfrak{F}^1(M)$ satisfy Axiom A. Ergod. Theory Dyn. Syst. **12**(2), 233–253 (1992)
108. Hayashi, S.: Connecting invariant manifolds and the solution of the C^1 stability and Ω-stability conjectures for flows. Ann. Math. (2) **145**(1), 81–137 (1997). Correction in Ann. Math. **150**(1), 353–356 (1999)
109. Hayashi, S.: A C^1 make or break lemma. Bull. Braz. Math. Soc. **31**(3), 337–350 (2000)
110. Hirsch, M., Pugh, C., Shub, M.: Invariant Manifolds. Lect. Notes in Math., vol. 583. Springer, New York (1977)
111. Hirsch, M., Smale, S.: Differential Equations, Dynamical Systems and Linear Algebra. Academic Press, New York (1974)
112. Hofbauer, F., Keller, G.: Ergodic properties of invariant measures for piecewise monotonic transformations. Math. Z. **180**(1), 119–140 (1982)
113. Holland, M.: Slowly mixing systems and intermittency maps. Ergod. Theory Dyn. Syst. **25**(1), 133–159 (2005)
114. Holland, M., Melbourne, I.: Central limit theorems and invariance principles for Lorenz attractors. J. Lond. Math. Soc. (2) **76**(2), 345–364 (2007)
115. Homburg, A.J., Kokubu, H., Krupa, M.: The cusp horseshoe and its bifurcations in the unfolding of an inclination-flip homoclinic orbit. Ergod. Theory Dyn. Syst. **14**(4), 667–693 (1994)
116. Hopf, E.: Statistik der geodätischen Linien in Mannigfaltigkeiten negativer Krümmung. Ber. Verh. Sächs. Akad. Wiss. Leipz. **91**, 261–304 (1939)
117. Hurewicz, W., Wallman, H.: Dimension Theory. Princeton Mathematical Series, vol. 4. Princeton University Press, Princeton (1941)
118. Kan, I.: Open sets of diffeomorphisms having two attractors, each with an everywhere dense basin. Bull. Am. Math. Soc. **31**, 68–74 (1994)
119. Katok, A.: Lyapunov exponents, entropy and periodic orbits for diffeomorphisms. Publ. Math. IHÉS **51**, 137–173 (1980)
120. Keller, G.: Generalized bounded variation and applications to piecewise monotonic transformations. Z. Wahrscheinlichkeitstheor. Verw. Geb. **69**(3), 461–478 (1985)
121. Keynes, H.B., Sears, M.: F-expansive transformation groups. Gen. Topol. Appl. **10**(1), 67–85 (1979)
122. Kifer, Y.: Random Perturbations of Dynamical Systems. Progress in Probability and Statistics, vol. 16. Birkhäuser Boston Inc., Boston (1988)
123. Klinshpont, N.È.: A topological invariant of the Lorenz attractor. Usp. Mat. Nauk **47**(2(284)), 195–196 (1992)
124. Komuro, M.: Expansive properties of Lorenz attractors. In: Kawakami, H. (ed.) The Theory of Dynamical Systems and Its Applications to Nonlinear Problems, pp. 4–26. World Scientific, Singapure (1984). Papers from the meeting held at the Research Institute for Mathematical Sciences, Kyoto University, Kyoto, July 4–7, 1984
125. Komuro, M.: Lorenz attractors do not have the pseudo-orbit tracing property. J. Math. Soc. Jpn. **37**(3), 489–514 (1985)
126. Kuratowski, K.: Topology. Vol. I. New edition, revised and augmented. Translated from the French by J. Jaworowski. Academic Press, New York (1966)

127. Kuratowski, K.: Topology. Vol. II. New edition, revised and augmented. Translated from the French by A. Kirkor. Academic Press, New York (1968)
128. Labarca, R., Pacifico, M.J.: Stability of singular horseshoes. Topology **25**, 337–352 (1986)
129. Ledrappier, F., Young, L.S.: The metric entropy of diffeomorphisms I. Characterization of measures satisfying Pesin's entropy formula. Ann. Math. **122**, 509–539 (1985)
130. Leonov, G.A.: Asymptotic behavior of the solutions of the Lorenz system. Differ. Uravn. (Minsk) **24**(5), 804–809 (1988)
131. Leonov, G.A.: Formulas for the Lyapunov dimension of Hénon and Lorenz attractors. Algebra Anal. **13**(3), 155–170 (2001)
132. Li, M., Gan, S., Wen, L.: Robustly transitive singular sets via approach of an extended linear Poincaré flow. Discrete Contin. Dyn. Syst. **13**(2), 239–269 (2005)
133. Liao, S.T.: On the stability conjecture. Chin. Ann. Math., Ser. B **1**, 9–30 (1980)
134. Liao, S.T.: Obstruction sets (II). Acta Sci. Natur. Univ. Pekinensis **2**, 1–36 (1981)
135. Liao, S.T.: On hyperbolicity properties of nonwandering sets of certain 3-dimensional differential systems. Acta Math. Sci. **3**(4), 361–368 (1983) (English Ed.)
136. Lima, E.L.: Espaços métricos, volume 4 of Projeto Euclides [Euclid Project]. Instituto de Matemática Pura e Aplicada, Rio de Janeiro, 1977
137. Liverani, C.: Flows, random perturbations and rate of mixing. Ergod. Theory Dyn. Syst. **18**(6), 1421–1446 (1998)
138. Liverani, C.: On contact Anosov flows. Ann. Math. (2) **159**(3), 1275–1312 (2004)
139. Lorenz, E.N.: Deterministic nonperiodic flow. J. Atmos. Sci. **20**, 130–141 (1963)
140. Lorenz, E.N.: On the prevalence of aperiodicity in simple systems. Lect. Notes Math. **755**, 53–75 (1979)
141. Lorenz, E.N.: The local structure of a chaotic attractor in four dimensions. Physica D **13**(1-2), 90–104 (1984)
142. Luzzatto, S., Melbourne, I., Paccaut, F.: The Lorenz attractor is mixing. Commun. Math. Phys. **260**(2), 393–401 (2005)
143. Lyubich, M.: Almost every real quadratic map is either regular or stochastic. Ann. Math. (2) **156**(1), 1–78 (2002)
144. Mañé, R.: Contributions to the stability conjecture. Topology **17**, 383–396 (1978)
145. Mañé, R.: An ergodic closing lemma. Ann. Math. **116**, 503–540 (1982)
146. Mañé, R.: Oseledec's theorem from the generic viewpoint. In: Proceedings of the International Congress of Mathematicians, Vol. 1, 2 (Warsaw, 1983), pp. 1269–1276. PWN, Warsaw (1984)
147. Mañé, R.: Ergodic Theory and Differentiable Dynamics. Springer, New York (1987)
148. Mañé, R.: A proof of the C^1 stability conjecture. Publ. Math. IHÉS **66**, 161–210 (1988)
149. Mañé, R.: The Lyapunov exponents of generic area preserving diffeomorphisms. In: International Conference on Dynamical Systems (Montevideo, 1995), Pitman Res. Notes Math. Ser., vol. 362, pp. 110–119. Longman, Harlow (1996)
150. Markley, N.G.: The Poincaré-Bendixson theorem for the Klein bottle. Trans. Am. Math. Soc. **135**, 159–165 (1969)
151. Melbourne, I., Nicol, M.: Almost sure invariance principle for nonuniformly hyperbolic systems. Commun. Math. Phys. **260**(1), 131–146 (2005)
152. Melbourne, I., Nicol, M.: Large deviations for nonuniformly hyperbolic systems. Trans. Am. Math. Soc. **360**(12), 6661–6676 (2008)
153. Metzger, R.: Sinaĭ-Ruelle-Bowen measures for contracting Lorenz maps and flows. Ann. Inst. Henri Poincaré, Anal. Non Linéaire **17**(2), 247–276 (2000)
154. Metzger, R.: Stochastic stability for contracting Lorenz maps and flows. Commun. Math. Phys. **212**(2), 277–296 (2000)
155. Metzger, R., Morales, C.: The Rovella attractor is a homoclinic class. Bull. Braz. Math. Soc. **37**(1), 89–101 (2006)
156. Metzger, R., Morales, C.: Sectional-hyperbolic systems. Ergod. Theory Dyn. Syst. **28**, 1587–1597 (2008)
157. Milnor, J.: Topology from the Differentiable Viewpoint. Princeton Landmarks in Mathematics. Princeton University Press, New Jersey (1997)

158. Mischaikow, K., Mrozek, M.: Chaos in the Lorenz equations: a computer assisted proof (I). Bull. Am. Math. Soc. **32**, 66–72 (1995)
159. Mischaikow, K., Mrozek, M.: Chaos in the Lorenz equations: a computer assisted proof (II). Math. Comput. **67**, 1023–1046 (1998)
160. Moore, C.C.: Exponential decay of correlation coefficients for geodesic flows. In: Group Representations, Ergodic Theory, Operator Algebras, and Mathematical Physics (Berkeley, Calif., 1984), Math. Sci. Res. Inst. Publ., vol. 6, pp. 163–181. Springer, New York (1987)
161. Morales, C.: Singular-hyperbolic sets and topological dimension. Dyn. Syst. **18**(2), 181–189 (2003)
162. Morales, C.: A note on periodic orbits for singular-hyperbolic flows. Discrete Contin. Dyn. Syst. **11**(2-3), 615–619 (2004)
163. Morales, C.: The explosion of singular-hyperbolic attractors. Ergod. Theory Dyn. Syst. **24**(2), 577–591 (2004)
164. Morales, C.: Examples of singular-hyperbolic attracting sets. Dyn. Syst. **22**(3), 339–349 (2007)
165. Morales, C.: Singular-hyperbolic attractors with handlebody basins. J. Dyn. Control Syst. **13**(1), 15–24 (2007)
166. Morales, C., Pacifico, M.J.: Mixing attractors for 3-flows. Nonlinearity **14**(2), 359–378 (2001)
167. Morales, C., Pacifico, M.J.: Inclination-flip homoclinic orbits arising from orbit-flip. Nonlinearity **14**(2), 379–393 (2001)
168. Morales, C., Pacifico, M.J.: A dichotomy for three-dimensional vector fields. Ergod. Theory Dyn. Syst. **23**(5), 1575–1600 (2003)
169. Morales, C., Pacifico, M.J.: Sufficient conditions for robustness of attractors. Pac. J. Math. **216**(2), 327–342 (2004)
170. Morales, C., Pacifico, M.J.: A spectral decomposition for singular-hyperbolic sets. Pac. J. Math. **229**(1), 223–232 (2007)
171. Morales, C., Pujals, E.: Singular strange attractors on the boundary of Morse-Smale systems. Ann. Sci. Ecole Norm. Super. **30**, 693–717 (1997)
172. Morales, C., Pacifico, M.J., Pujals, E.: On C^1 robust singular transitive sets for three-dimensional flows. C. R. Acad. Sci., Sér. 1 Math. **326**, 81–86 (1998)
173. Morales, C., Pacifico, M.J., Pujals, E.: Strange attractors across the boundary of hyperbolic systems. Commun. Math. Phys. **211**(3), 527–558 (2000)
174. Morales, C., Pacifico, M.J., Pujals, E.: Robust transitive singular sets for 3-flows are partially hyperbolic attractors or repellers. Ann. Math. (2) **160**(2), 375–432 (2004)
175. Morales, C., Pacifico, M.J., San Martin, B.: Expanding Lorenz attractors through resonant double homoclinic loops. SIAM J. Math. Anal. **36**(6), 1836–1861 (2005)
176. Morales, C., Pacifico, M.J., San Martin, B.: Contracting Lorenz attractors through resonant double homoclinic loops. SIAM J. Math. Anal. **38**(1), 309–332 (2006)
177. Munkres, J.: Topology, 2nd edn. Prentice Hall, New Deli (2004).
178. Munroe, M.E.: Introduction to Measure and Integration. Addison-Wesley Publishing Company, Cambridge (1953).
179. Neĭmark, J.I., Shil'nikov, L.P.: A case of generation of periodic motions. Izv. Vysš. Učebn. Zaved., Radiofiz. **8**, 330–340 (1965)
180. Newhouse, S.: Quasi-elliptic periodic points in conservative dynamical systems. Am. J. Math. **99**(5), 1061–1087 (1977)
181. Newhouse, S.: Lectures on dynamical systems. In: Dynamical Systems (CIME Lectures 1978), Progress in Mathematics, pp. 1–114. Birkhäuser, Boston (1978)
182. Newhouse, S.: The abundance of wild hyperbolic sets and nonsmooth stable sets for diffeomorphisms. Publ. Math. IHÉS **50**, 101–151 (1979)
183. Oka, M.: Expansiveness of real flows. Tsukuba J. Math. **14**(1), 1–8 (1990)
184. Oliveira, K., Viana, M.: Existence and uniqueness of maximizing measures for robust classes of local diffeomorphisms. Discrete Contin. Dyn. Syst. **15**(1), 225–236 (2006)
185. Oseledets, V.I.: A multiplicative ergodic theorem: Lyapunov characteristic numbers for dynamical systems. Trans. Mosc. Math. Soc. **19**, 197–231 (1968)

186. Pacífico, M.J., Rovella, A.: Unfolding contracting singular cycles. Ann. Sci. Ecole Norm. Super. (4) **26**(6), 691–700 (1993)
187. Pacifico, M.J., Rovella, A., Viana, M.: Infinite-modal maps with global chaotic behavior. Ann. Math. (2) **148**(2), 441–484 (1998) Corrigendum in Ann. Math. **149**, 705 (1999)
188. Pacifico, M.J., Todd, M.: Thermodynamical formalism for contracting Lorenz flows. J. Stat. Phys. **139**(1), 159–176 (2010)
189. Palis, J.: A global view of Dynamics and a conjecture on the denseness of finitude of attractors. Astérisque **261**, 335–347 (2000)
190. Palis, J., de Melo, W.: Geometric Theory of Dynamical Systems. Springer, New York (1982)
191. Palis, J., Smale, S.: Structural stability theorems. In: Global Analysis, Proc. Sympos. Pure Math. (Berkeley 1968), vol. XIV, pp. 223–232. American Mathematical Society, Providence (1970)
192. Palis, J., Takens, F.: Topological equivalence of normally hyperbolic dynamical systems. Topology **16**(4), 335–345 (1977)
193. Palis, J., Takens, F.: Hyperbolicity and Sensitive-Chaotic Dynamics at Homoclinic Bifurcations. Cambridge University Press, Cambridge (1993)
194. Peixoto, M.M.: On structural stability. Ann. Math. (2) **69**, 199–222 (1959)
195. Peixoto, M.M.: Structural stability on two-dimensional manifolds. Topology **1**, 101–120 (1962)
196. Pesin, Y.: Families of invariant manifolds corresponding to non-zero characteristic exponents. Math. USSR, Izv. **10**, 1261–1302 (1976)
197. Pesin, Y.: Characteristic Lyapunov exponents and smooth ergodic theory. Russ. Math. Surv. **324**, 55–114 (1977)
198. Pesin, Y.: Dynamical systems with generalized hyperbolic attractors: hyperbolic, ergodic and topological properties. Ergod. Theory Dyn. Syst. **12**, 123–151 (1992)
199. Pesin, Y., Senti, S.: Equilibrium measures for maps with inducing schemes. Journal of Modern Dynamics **2**(3), 397–430 (2008)
200. Pesin, Y., Sinaĭ, Y.: Hyperbolicity and stochasticity of dynamical systems. In: Mathematical Physics Reviews, Vol. 2, Soviet Sci. Rev. Sect. C, pp. 53–115. Harwood Academic, Chur (1981)
201. Pesin, Y., Sinaĭ, Y.: Gibbs measures for partially hyperbolic attractors. Ergod. Theory Dyn. Syst. **2**, 417–438 (1982)
202. Pinheiro, V.: Expanding measures. 0811.2545v1 (2008)
203. Pliss, V.A.: The behavior of the solutions of a sequence of second order periodic systems with a small nonlinearity. Differencial'nye Uravnenija **7**, 651–660 (1971)
204. Pliss, V.A.: The structural stability of a sequence of linear systems of second order differential equations with periodic coefficients. Differencial'nye Uravnenija **7**, 261–270 (1971)
205. Poincaré, H.: Sur le problème des trois corps et les équations de la dynamique (Mémoire couronné du prise de S.M. le roi Oscar II de Suéde). Acta Math. **13**, 1–270 (1890)
206. Poincaré, H.: Les Méthodes Nouvelles de la Mécanique Céleste. Tome I. Les Grands Classiques Gauthier-Villars. [Gauthier-Villars Great Classics]. Librairie Scientifique et Technique Albert Blanchard. Gauthier-Villars, Paris (1987). Solutions périodiques. Non-existence des intégrales uniformes. Solutions asymptotiques. [Periodic solutions. Nonexistence of uniform integrals. Asymptotic solutions], Reprint of the 1892 original, With a foreword by J. Kovalevsky, Bibliothèque Scientifique Albert Blanchard. [Albert Blanchard Scientific Library]
207. Poincaré, H.: Les Méthodes Nouvelles de la Mécanique Céleste. Tome II. Les Grands Classiques Gauthier-Villars. [Gauthier-Villars Great Classics]. Librairie Scientifique et Technique Albert Blanchard. Gauthier-Villars, Paris (1987). Méthodes de MM. Newcomb, Gyldén, Lindstedt et Bohlin. [The methods of Newcomb, Gyldén, Lindstedt and Bohlin], Reprint of the 1893 original, Bibliothèque Scientifique Albert Blanchard. [Albert Blanchard Scientific Library]
208. Poincaré, H.: Les Méthodes Nouvelles de la Mécanique Céleste. Tome III. Les Grands Classiques Gauthier-Villars. [Gauthier-Villars Great Classics]. Librairie Scientifique et Technique Albert Blanchard. Gauthier-Villars, Paris (1987). Invariant intégraux. Solutions périodiques

du deuxième genre. Solutions doublement asymptotiques. [Integral invariants. Periodic solutions of the second kind. Doubly asymptotic solutions], Reprint of the 1899 original, Bibliothèque Scientifique Albert Blanchard. [Albert Blanchard Scientific Library]
209. Pollicott, M.: A complex Ruelle-Perron-Frobenius theorem and two counterexamples. Ergod. Theory Dyn. Syst. **4**(1), 135–146 (1984)
210. Pollicott, M.: Exponential mixing for the geodesic flow on hyperbolic three-manifolds. J. Stat. Phys. **67**(3-4), 667–673 (1992)
211. Pollicott, M.: On the mixing of Axiom A attracting flows and a conjecture of Ruelle. Ergod. Theory Dyn. Syst. **19**(2), 535–548 (1999)
212. Pugh, C.: The closing lemma. Am. J. Math. **89**, 956–1009 (1967)
213. Pugh, C.: An improved closing lemma and a general density theorem. Am. J. Math. **89**, 1010–1021 (1967)
214. Pugh, C.: Against the C^2 closing lemma. J. Differ. Equ. **17**, 435–443 (1975)
215. Pugh, C., Robinson, C.: The C^1 closing lemma, including Hamiltonians. Ergod. Theory Dyn. Syst. **3**(2), 261–313 (1983)
216. Pugh, C., Shub, M.: Ergodic attractors. Trans. Am. Math. Soc. **312**, 1–54 (1989)
217. Pujals, E., Sambarino, M.: Homoclinic tangencies and hyperbolicity for surface diffeomorphisms. Ann. Math. (2) **151**(3), 961–1023 (2000)
218. Pujals, E., Sambarino, M.: On the dynamics of dominated splitting. Ann. Math. (2) **169**(3), 675–739 (2009)
219. Pumariño, A., Rodriguez, A.: Persistence and Coexistence of Strange Attractors in Homoclinic Saddle-Focus Connections. Lect. Notes in Math., vol. 1658. Springer, Berlin (1997)
220. Ratner, M.: The rate of mixing for geodesic and horocycle flows. Ergod. Theory Dyn. Syst. **7**(2), 267–288 (1987)
221. Rey-Bellet, L., Young, L.-S.: Large deviations in non-uniformly hyperbolic dynamical systems. Ergod. Theory Dyn. Syst. **28**(2), 587–612 (2008)
222. Robbin, J.: A structural stability theorem. Ann. Math. **94**, 447–493 (1971)
223. Robinson, C.: Generic properties of conservative systems. Am. J. Math. **92**, 562–603 (1970)
224. Robinson, C.: Structural stability of vector fields. Ann. Math. **99**, 154–175 (1974). Errata in Ann. Math. **101**, 368 (1975)
225. Robinson, C.: Structural stability of C^1 vector fields. In: Dynamical Systems - Warwick 1974, Lecture Notes in Mathematics, vol. 468, pp. 262–277. Springer, Berlin (1975)
226. Robinson, C.: Structural stability of C^1 diffeomorphisms. J. Differ. Equ. **22**, 28–73 (1976)
227. Robinson, C.: Homoclinic bifurcation to a transitive attractor of Lorenz type. Nonlinearity **2**(4), 495–518 (1989)
228. Robinson, C.: Homoclinic bifurcation to a transitive attractor of Lorenz type. II. SIAM J. Math. Anal. **23**(5), 1255–1268 (1992)
229. Robinson, C.: Nonsymmetric Lorenz attractors from a homoclinic bifurcation. SIAM J. Math. Anal. **32**(1), 119–141 (2000)
230. Robinson, C.: An Introduction to Dynamical Systems: Continuous and Discrete. Pearson Prentice Hall, Upper Saddle River (2004)
231. Rokhlin, V.A.: On the fundamental ideas of measure theory. Trans. Am. Math. Soc. **10**, 1–52 (1962). Transl. from Mat. Sbornik **25**, 107–150 (1949)
232. Rovella, A.: The dynamics of perturbations of the contracting Lorenz attractor. Bull. Braz. Math. Soc. **24**(2), 233–259 (1993)
233. Rudin, W.: Real and Complex Analysis, 3rd edn. McGraw-Hill, New York (1987)
234. Rudin, W.: Functional Analysis, 2nd edn. International Series in Pure and Applied Mathematics, McGraw-Hill, New York (1991)
235. Ruelle, D.: A measure associated with Axiom A attractors. Am. J. Math. **98**, 619–654 (1976)
236. Ruelle, D.: Flots qui ne mélangent pas exponentiellement. C. R. Acad. Sci., Sér. 1 Math. **296**(4), 191–193 (1983)
237. Ruelle, D.: The thermodynamical formalism for expanding maps. Commun. Math. Phys. **125**, 239–262 (1989)

238. Ruelle, D.: Thermodynamic Formalism. The Mathematical Structures of Equilibrium Statistical Mechanics, 2nd edn. Cambridge Mathematical Library, Cambridge University Press, Cambridge (2004).
239. Rychlik, M.: Lorenz attractors through Shil'nikov-type bifurcation. Part 1. Ergod. Theory Dyn. Syst. **10**, 793–821 (1989)
240. Saghin, R., Xia, Z.: Partial hyperbolicity or dense elliptic periodic points for C^1-generic symplectic diffeomorphisms. Trans. Am. Math. Soc. **358**(11), 5119–5138 (2006)
241. Sarig, O.: Thermodynamic formalism for countable Markov shifts. Ergod. Theory Dyn. Syst. **19**(6), 1565–1593 (1999)
242. Sarig, O.: Continuous phase transitions for dynamical systems. Commun. Math. Phys. **267**(3), 631–667 (2006)
243. Shil'nikov, L.P.: A case of the existence of a denumerable set of periodic motions. Sov. Math. Dokl. **6**, 163–166 (1965)
244. Shil'nikov, L.P.: Existence of a countable set of periodic motions in a four-dimensional space in an extended neighborhood of a saddle-focus. Dokl. Akad. Nauk SSSR **172**, 54–57 (1967)
245. Shil'nikov, L.P.: On the generation of periodic motion from a trajectory doubly asymptotic to an equilibrium state of saddle type. Math. USSR Sb. **6**, 428–438 (1968)
246. Shil'nikov, L.P.: A contribution to the problem of the structure of an extended neighborhood of a rough equilibrium state of saddle-focus type. Math. USSR Sb. **10**, 91–102 (1970)
247. Shub, M.: Global Stability of Dynamical Systems. Springer, New York (1987)
248. Siegel, C.L., Moser, J.K.: Lectures on Celestial Mechanics, Classics in Mathematics. Springer, Berlin (1995). Translated from the German by C. I. Kalme, Reprint of the 1971 translation
249. Sinaĭ, Y.: Geodesic flows on manifolds of negative constant curvature. Sov. Math. Dokl. **1**, 335–339 (1960)
250. Sinaĭ, Y.: Gibbs measures in ergodic theory. Russ. Math. Surv. **27**, 21–69 (1972)
251. Smale, S.: Diffeomorphisms with many periodic points. In: Differential and Combinatorial Topology (A Symposium in Honor of Marston Morse), pp. 63–80. Princeton University Press, Princeton (1965)
252. Smale, S.: Differentiable dynamical systems. Bull. Am. Math. Soc. **73**, 747–817 (1967)
253. Smale, S.: Mathematical problems for the next century. Math. Intell. **20**(2), 7–15 (1998)
254. Sparrow, C.: The Lorenz Equations: Bifurcations, Chaos and Strange Attractors. Applied Mathematical Sciences, vol. 41. Springer, New York (1982)
255. Spivak, M.: Calculus on Manifolds. Addison-Wesley, Reading (1997)
256. Sternberg, E.: On the structure of local homeomorphisms of Euclidean n-space - II. Am. J. Math. **80**, 623–631 (1958)
257. Strogatz, S.: Nonlinear Dynamics and Chaos: With Applications To Physics, Biology, Chemistry, and Engineering, 1st edn. Studies in Nonlinearity. Perseus Books, Reading (1994)
258. Szlenk, W.: An Introduction to the Theory of Smooth Dynamical Systems. PWN—Polish Scientific Publishers, Warsaw (1984). Translated from the Polish by Marcin E. Kuczma
259. Tresser, C.: About some theorems by L. P. Shil'nikov. Ann. Inst. Henri Poincaré **40**, 441–461 (1984)
260. Tucker, W.: The Lorenz attractor exists. PhD thesis, Univ. Uppsala, 1998. Text and program codes available at www.math.uu.se/~warwick
261. Tucker, W.: The Lorenz attractor exists. C. R. Acad. Sci., Sér. 1 Math. **328**, 1197–1202 (1999)
262. Tucker, W.: A rigorous ODE solver and Smale's 14th problem. Found. Comput. Math. **2**(1), 53–117 (2002)
263. Turaev, D.V., Shil'nikov, L.P.: An example of a wild strange attractor. Mat. Sb. **189**(2), 137–160 (1998)
264. Ushiki, S., Oka, H., Kokubu, H.: Existence d'attracteurs étranges dans le déploiement d'une singularité dégénérée d'un champ de vecteurs invariant par translation. C. R. Acad. Sci., Sér. 1 Math. **298**(3), 39–42 (1984)
265. Varandas, P.: Existence and stability of equilibrium states for robust classes of non-uniformly hyperbolic maps. PhD thesis, IMPA, Rio de Janeiro, Brasil, 2007

266. Viana, M.: Stochastic dynamics of deterministic systems. Publicações Matemáticas do IMPA. [IMPA Mathematical Publications]. Instituto de Matemática Pura e Aplicada (IMPA), Rio de Janeiro, 1997. XXI Coloquio Brasileiro de Matematica. [21th Brazilian Mathematics Colloquium]
267. Viana, M.: What's new on Lorenz strange attractor. Math. Intell. **22**(3), 6–19 (2000)
268. Vivier, T.: Flots robustement transitifs sur les variétés compactes. C. R. Math. Acad. Sci. Paris **337**(12), 791–796 (2003)
269. Walters, P.: An Introduction to Ergodic Theory. Springer, New York (1982)
270. Wen, L.: On the C^1-stability conjecture for flows. J. Differ. Equ. **129**, 334–357 (1995)
271. Wen, L., Xia, Z.: C^1 connecting lemmas. Trans. Am. Math. Soc. **352**, 5213–5230 (2000)
272. Wiggins, S.: Global Bifurcations and Chaos. Analytical Methods. Applied Mathematical Sciences, vol. 73. Springer, New York (1988).
273. Williams, R.F.: The structure of Lorenz attractors. In: Turbulence Seminar (Univ. Calif., Berkeley, Calif., 1976/1977), pp. 94–112. Springer, Berlin (1977)
274. Williams, R.F.: The structure of Lorenz attractors. Publ. Math. IHÉS **50**, 73–99 (1979)
275. Wong, S.: Some metric properties of piecewise monotonic mappings of the unit interval. Trans. Am. Math. Soc. **246**, 493–500 (1978)
276. Young, L.S.: On the prevalence of horseshoes. Trans. Am. Math. Soc. **263**(1), 75–88 (1981)
277. Zuppa, C.: Regularisation C^∞ des champs vectoriels qui préservent l'elément de volume. Bull. Braz. Math. Soc. **10**(2), 51–56 (1979)

Index

V. Araújo, M.J. Pacifico, *Three-Dimensional Flows,* Ergebnisse der Mathematik und ihrer Grenzgebiete. 3. Folge / A Series of Modern Surveys in Mathematics 53,
DOI 10.1007/978-3-642-11414-4, © Springer-Verlag Berlin Heidelberg 2010

Printed by Books on Demand, Germany